THE

PUBLICATIONS

OF THE

SURTEES SOCIETY

VOL. 210

THE

PUBLICATIONS

OF THE

SURTEES SOCIETY

ESTABLISHED IN THE YEAR
M.DCCC.XXXIV

VOL. CCX

MATTHEW AND GEORGE CULLEY

FARMING LETTERS 1798–1804

EDITED
BY
ANNE ORDE

THE SURTEES SOCIETY

THE BOYDELL PRESS

CONTENTS

ACKNOWLEDGEMENTS

My chief thanks are due to the Council of the Surtees Society for undertaking the publication of these letters, and especially to Professor Richard Britnell, who encouraged me to embark on the project, commended it to the Society, and has given me a great deal of editorial help and valuable advice.

I am indebted to Sue Wood, Senior Archivist of the Northumberland Record Office, for giving permission for publication of the letters, and to her and her staff for access to these and other materials in their custody, for photocopying services and other assistance. I am grateful to the staff of the Durham Record Office, Durham University Library (Archives and Special Collections) and other libraries and archives where work on this edition has taken me. Lord Barnard and Clare Simpson generously allowed me access to the archives at Raby Castle, and Clifton Sutcliffe searched out and showed me estate account books. The Tankerville papers, deposited in the Northumberland Record Office, were consulted by permission of Charles Bain Smith. I am glad to renew my expressions of thanks to the Norham Local History Society, whose invitation to speak to them about my family occasioned my first encounter with the Culleys, and thus launched me on several years of very pleasurable work. My thanks go to them, and to all those who have given me help and answered questions.

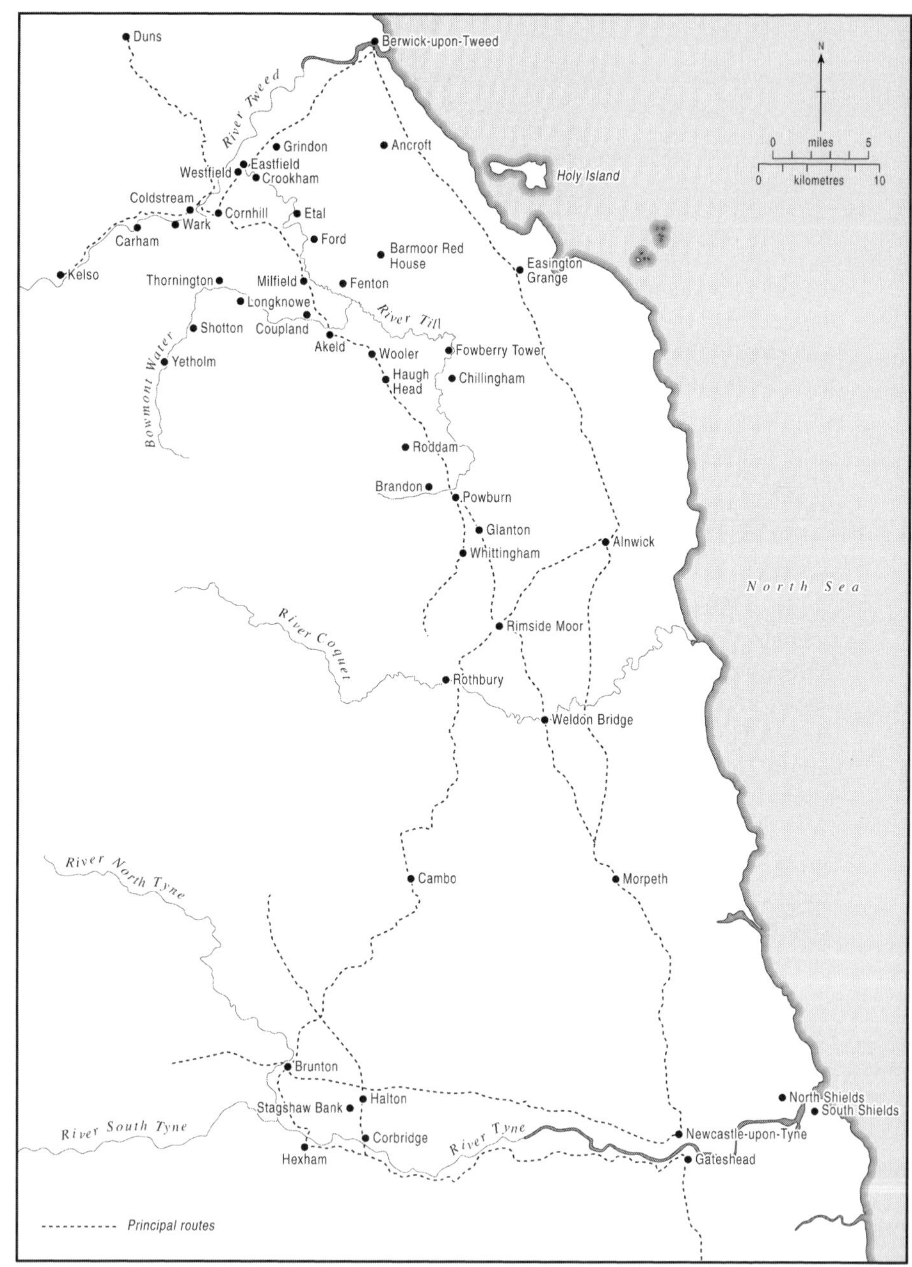

Northumberland

County Durham

ABBREVIATIONS

Bailey, *General View, Durham*	John Bailey, *General View of the Agriculture of the County of Durham*, London 1810
Bailey and Culley, *General View, Northumberland*	John Bailey and George Culley, *General View of the Agriculture of Northumberland, Cumberland and Westmorland*, 3rd edn London 1805, reprinted, with an introduction by D.J. Rowe, Newcastle upon Tyne 1972
Complete Baronetage	The Complete Baronetage, ed. G.E. Cokayne. 6 vols. London 1900–09
Complete Peerage	The Complete Peerage, ed. Vicary Gibbs et al. 14 vols. London 1910–98
Culley, *Travel Journals*	Matthew and George Culley, *Travel Journals and Letters, 1765–1798*, ed. Anne Orde, London 2002
DRO	Durham Record Office
DUL	Durham University Library
DUL, ASC	Durham University Library, Archives and Special Collections
DUL, DDR	DUL, ASC, Durham Diocesan Records
DUL, BT	DUL, ASC, DDR, Bishops Transcripts of parish registers
History of Northumberland	Northumberland County History Committee. *A History of Northumberland*. 15 vols. Newcastle upon Tyne 1893–1926
NRO	Northumberland Record Office
Pigot's Directory	The Commercial Directory of Ireland, Scotland and the four most Northern Counties of England for 1820–1 and 1822. Manchester 1820
ZCU	prefixes references to the Culley Papers, in the NRO

'A Wedder of Mr Culley's Breed'
from Thomas Bewick, *A General History of Quadrupeds*, 1790

INTRODUCTION

The letters printed here were written between October 1798 and July 1804 by two substantial farmers in north Northumberland, Matthew Culley (1731–1804) and his brother George (1735–1813), to John Welch, their steward at Denton, the family property in County Durham. Written severally and jointly, and with some contributions by George's son Matthew Culley junior, the letters give a remarkably full picture of farming in the north of England at the turn of the eighteenth century – crops and livestock, methods, and markets. The letters form part of the collection of Culley family papers now in the Northumberland Record Office, which descended in the family of George Culley. The largest component of the collection relates to the two brothers. It includes journals of their travels in England and Scotland,[1] some farming accounts and leases, and a quantity of correspondence, mostly from between 1783 and 1812. The only substantial quantity of outgoing letters from the two Culleys is the collection printed here. The bulk of the rest of the correspondence is incoming letters: only a few replies and one letter-book have been kept. Among the incoming letters are a few from Arthur Young, the agricultural writer, and Sir John Sinclair, the first President of the Board of Agriculture, and many from customers and friends. There is some correspondence concerning the proposed publication of Matthew Culley's journal of his tour of Scotland in 1773 and 1775. A batch of letters from the livestock breeder Robert Bakewell was extracted from the collection, printed in H. Cecil Pawson's biography of Bakewell,[2] and deposited in the library of the University of Newcastle upon Tyne.

A second smaller component of the Culley Papers is papers of George Culley's son Matthew Culley of Fowberry (1778–1849), with a few of his cousin Matthew Culley of Coupland (1786–1834) mostly from the 1820s and 30s. A third still smaller group relates to George

---

1 Matthew and George Culley, *Travel Journals and Letters, 1765–1798*, ed. Anne Orde, London 2002, henceforth Culley, *Travel Journals*.

2 H. Cecil Pawson, *Robert Bakewell, Pioneer Livestock Breeder*, London 1957.

Culley of Fowberry (1834–83), grandson of the first George Culley's daughter Eleanor Darling, who took the name Culley under his great-uncle's will. There are some family deeds and wills, and some miscellaneous letters. Finally there are some transcripts of letters and journals, and notes. In some cases the original is also present in the Culley Papers; in some it is not. Some of the transcripts and notes seem to have been made for George Culley's great-great-granddaughter Jane Leather-Culley in the 1930s; others were made in 1951 by Dr Angus Macdonald of King's College, Newcastle, now the University of Newcastle upon Tyne. He deposited all except the Bakewell letters in the Northumberland Record Office shortly before he died.

Matthew and George Culley were the eighth and ninth children of Matthew Culley of Denton, near Darlington in County Durham, and his wife Eleanor Surtees. That Matthew, born in 1685, was the second surviving son and youngest child of another Matthew Culley, of Beaumont Hill in the same district. Eleanor Surtees, born in 1696, was the eldest daughter of Robert Surtees of Mainsforth near Sedgefield. Matthew Culley's elder brother John inherited Beaumont Hill. The father of Matthew and George was established, in 1706, on land in Aycliffe held by his father on lease from the Dean and Chapter of Durham; and then in 1722 he bought the small estate of Denton. Three of his eleven children died in infancy, three more as young adults. One daughter and four sons outlived their father. Our Matthew, born in 1731, was the second surviving son; George, the next, was born in 1735. The youngest son, James, was born in 1740.[3]

The Culleys' lives and farming career fall in a significant period of change in British agriculture. The background was a rapid growth in population after a long period of little change. In 1700 the population of England and Scotland was somewhere around 6 million; in 1750 it was about 7 million. At the first census in 1801 it was 10.5 million, in 1821 12.2 million, in 1851 nearly 21 million.[4] That these greatly increased numbers could be fed, until after 1815 largely from sources within the British Isles, and supplied with many of the industrial products of agriculture, in contrast to earlier periods when dearth

---

3 Genealogies in Northumberland County History Committee, *History of Northumberland*, Newcastle upon Tyne 1893–1940, vol. 11, pp. 224–5; H. Conyers Surtees, *Records of the Family of Surtees, its Descents and Alliances*, Newcastle upon Tyne 1925, p. 89. Leases and renewals in Durham Cathedral Muniments, DCD/B/BA, Register 22, f. 48; DCD/F/CA/2, ff. 218, 224.

4 B.R. Mitchell, *British Historical Statistics*, Cambridge 1988, pp. 7–9.

and disease had checked major population growths, attests to greatly increased agricultural production.

The dating, the driving forces, and the relative importance of different elements in the so-called 'agricultural revolution' are all continuing sources of debate among historians. It was not, as the term 'revolution' may suggest, an abrupt process; but the cumulative effect can be described as revolutionary. It is generally accepted that the process began in the seventeenth century; but the weight to be given to the eighteenth century, or the century 1750–1850, is contested. Estimates of output and productivity, of both land and labour, are based on evidence that neither is derived from the whole country nor covers sufficiently long runs of years to be wholly satisfactory. There are no comprehensive statistics of production before 1866, and only the crudest indicators of the composition and activities of the rural population before 1831. Different methods of calculation indicate an increase in output over the eighteenth century of between 60% and 90%, with volume, land productivity and labour productivity all rising together and just about keeping up with population growth.[5]

Whatever the statistical problems, the elements in the growth of output in the eighteenth century are clear. More land was brought into cultivation, improved crops were more widely adopted, more productive methods were taken up. In the second half of the century there was a great increase in the cultivated area. Between 1760 and 1799 at least 2 million acres of waste were brought into cultivation, and more by 1815. Enclosure of commons and open fields had begun many years earlier, but now spread to new areas and gathered speed. In addition to numerous local agreements, the number of parliamentary enclosure acts rose from twenty-six in the five-year period 1750–54 to over 300 in the years 1770–74, and some 2000 between 1793 and 1815. Even without entering into the social implications, it can be said that enclosure did not necessarily bring greater efficiency; but it was necessary for the increase in cultivated area, and it did on the whole accelerate the trend towards more productive methods.

---

5   For recent assessments see Mark Overton, *Agricultural Revolution in England*, Cambridge 1996; id., 'Re-establishing the agricultural revolution', in *Agricultural History Review*, 44(1996), pp. 1–20; id., 'Land and labour productivity in English agriculture, 1650–1850', in *Agriculture and Industrialization*, ed. Peter Mathias and John A. Davis, Oxford 1996. Detailed accounts of developments are in *The Agrarian History of England and Wales*, gen. ed. Joan Thirsk, Vol. 5, *1640–1750*, ed. Joan Thirsk, 2 parts, Cambridge 1984, 1985; Vol. 6, *1750–1850*, ed. G.E. Mingay, Cambridge 1989.

Equally important, possibly more important in the first half of the century, was the spread of improved crops and methods. Roots, legumes and cultivated grasses had become established in the seventeenth century; but they were now adopted more widely and more systematically, and new varieties were pursued. A great deal of publicity was given to turnips. They were an important addition to rotations, allowing wasteful bare fallow to be virtually abolished, Cultivation of the improved crops enriched the soil, and they provided more fodder for stock so that more could be kept on the same acreage. Larger stock numbers in turn gave more manure for the land; heavier manuring raised yields of all crops and made even higher levels of stocking possible. Other factors in improved production were new methods of treating and draining the soil; new methods of cultivation such as drilling, some new machinery, improved implements, and the improvement of livestock quality by selective breeding. Investment was encouraged by a general rise in prices from about 1760 onwards.

Another aspect of the background to these letters is war. Britain was at war with France for most of the years 1793–1815, with a short intermission in 1802–4. For much of the time large British forces were not engaged; and the direct impact on rural life in the north-east was small. Taxation increased. There are some references to men being called up for the militia: George Culley's son joined two successive troops of local volunteers; there was a comical episode of an invasion scare early in 1804. The indirect effect of the war was much greater. Producers of livestock, such as the Culleys, were affected by the state of trade and shipping from the Tyne and Wear, ship-owners buying supplies for their crews or suffering from the activities of press gangs; and all were affected by rises in the prices of foodstuffs, both grain and stock. Corn prices in particular rose rapidly in the period covered by these letters, reaching unprecedented heights in 1800 as a result of a bad harvest combined with restrictions on trade. Corn growers profited; but there was widespread distress and threat of public disorder.[6]

The Culley brothers learned and practised many of the current improvements in agriculture. The letters published here provide comments on their experience, and a day-to-day account of their practice and principles. They had made as boys a pact to farm together.[7] Their father, who was himself evidently a successful and

---

6   See No. 65, n. 4.
7   George Culley to Hon. Henry Grey Bennett, 2 Dec. 1804, Northumberland

enterprising farmer – he made enough money to buy Denton, and sold good quality cattle – did not have land to give them, but was probably able to leave them some capital with which to embark on their chosen career. Almost as important, he arranged for them to visit Robert Bakewell at Dishley in Leicestershire.[8] Matthew went in 1762, the year his father died, George a year later. Robert Bakewell (1725–95) was a leading figure in the developments of the day, noted for good husbandry and above all for his breeding of livestock. Bakewell did not invent selective breeding, which was already well established for thoroughbred horses, but he applied it systematically to the improvement of farm animals, using small closed flocks and herds, and controlling the spread of his blood lines by forming cartels of related breeders and charging high prices for the services of his bulls and tups. Bakewell's methods remain somewhat obscure. He does not seem to have kept detailed records; and whilst a master of self-advertisement, he deliberately shrouded the details in mystery. His results, however, were impressive. His longhorned cattle proved something of a dead end – the future lay with the shorthorn perfected early in the nineteenth century – but his Dishley or New Leicester sheep became very popular with breeders and farmers. The sheep had a high proportion of flesh to bone; they fattened quickly, allowing the farmer a more rapid turnover. They were not suited to all kinds of ground, and their wool was not of the highest quality; but they were an immediate success and, when used as a cross, contributed to other improved breeds.[9]

The visits to Bakewell were very important for the Culleys' farming career. George Culley later became known as Bakewell's favourite disciple. He was an important source of information about Bakewell for Arthur Young and other writers, and helped to spread Bakewell's reputation by his own writing.[10] More immediately, the brothers hired their first tup from Bakewell in 1764 and more within

---

Record Office (hereafter NRO), Culley Papers, ZCU 26. References to the Culley Papers are hereafter indicated simply as ZCU with the relevant bundle number.

8 Memoir of George Culley in *Farmer's Magazine* 14(1813) p. 272; J.C. Loudon, *An Encyclopaedia of Agriculture*, London 1825, p. 166.

9 Biography of Bakewell by H. Cecil Pawson (see n. 2). Discussion of his work is in Robert Trow-Smith, *A History of British Livestock Husbandry 1700–1900*, London 1959; Nicholas Russell, *Like Engend'ring Like: Heredity and Animal Breeding in Early Modern England*, Cambridge 1986; Pat Stanby, *Robert Bakewell and the Longhorn Breed of Cattle*, Ipswich 1995.

10 Arthur Young, *On the Husbandry of Three Celebrated British Farmers*, London

the next few years, claiming by the 1790s that they had not had other than Dishley tups for over twenty years.[11] They brought the Dishley blood with them when they moved to Northumberland in 1767, along with Teeswater ewes from south Durham and good shorthorn cattle.

The brothers' move to Northumberland seems to have been inspired by a desire for greater scope than that offered in south Durham. On their father's death in 1762 the eldest son, Robert, inherited Denton; but he did not farm it himself, becoming instead an attorney in Darlington. Matthew and George might therefore have stayed on as tenants, but Denton demesne farm was under 500 acres and they were more ambitious. In Northumberland large farms were available, and leases for substantial terms of years were usual. In 1766, after first considering the Tyne valley, the brothers took a twenty-one-year lease of a farm at Fenton in Glendale, and settled there with their younger brother James. Fenton had belonged to the Greys of Chillingham, but at the time the Culleys took the farm the property belonged to the granddaughters of Anthony Isaacson of Newcastle.[12] Eight years after entering Fenton the Culleys took a twenty-one-year lease of a second farm, Crookham Westfield, part of the Pallinsburn estate owned by George Adam Askew, and two years after that the adjoining Crookham Eastfield and General's Close. In 1782 they leased Red House for fifteen years,[13] and in 1786 two large and potentially valuable farms, Wark on Tweed and Grindon. The former belonged to the Earl of Tankerville, the latter to John Orde of

---

1811; George Culley to Young, 9 Mar. 1811, British Library, Young Papers, Add. MS 35,131; George Culley, *Observations on Live Stock*, London 1786.

11 George Culley to John Bailey, 27 Apr. 1789; to Mr Calhoun, 7 Jul. 1790, ZCU 31; to Sir John Sinclair, Nov. 1792, ZCU 2; in *Annals of Agriculture*, 19(1793), pp. 306–14.

12 *History of Northumberland*, vol. 11, pp. 336–8.

13 The location of Red House is not certain: there are a number of farms in Northumberland with this name. The most probable is Barmoor Red House near Lowick, the only Red House now in north Northumberland. Against this identification, a note in ZCU 43 summarising the partnership agreement of 1793 (see p. xxii) refers to Red House in the parish of Woodhorn, farther south between Morpeth and the coast. There is now no Red House in the Woodhorn area, and the accuracy of the note may be doubted. ZCU 43 contains typewritten transcripts and notes made in 1951 by Dr Angus Macdonald. Where the transcripts can be checked against the original documents some are found to be not wholly accurate, and Dr Macdonald's typist can be shown to have got some place names wrong, as in the journal of Matthew Culley's tour in Scotland in 1773 and 1775, published in Culley, *Travel Journals*.

Felkington.[14] Thus within twenty years of arriving in Northumberland the Culleys were leasing at least six farms.[15] The total acreage was over 3000 acres, and they were paying nearly £4000 a year in rent.[16] The outgoing tenant at Wark had allowed the farm to run down. Immediately on entering the Culleys set to work on improvement, impressing Lord Tankerville's agent John Bailey with their energy. Within the first six months they had repaired some of the barns and stables and drained eighteen acres, planning more for the next winter. With Bailey's collaboration they were planting hedges and making fences. They were planning to water meadows, and had taken on more labour. Bailey thought that it would take three years to turn the farm about to meet the rent of £1000 a year, but after the fourth year it would be a good farm. In the event, by 1793 Wark was the Culleys' most valuable farm, and a few years later George Culley could describe it as 'a land of Goshen'.[17] Expansion continued in the 1790s, with three more farms. Thornington and Longknowe, taken for twenty-one years in 1793, were part of the Chillingham estate. Shotton, leased for only six years from 1795, belonged to William Cuthbert of Newcastle.[18] The Culleys also began to invest their profits by buying land. In 1795 they bought the estate of Akeld and Humbleton for £24,000 – more than twice the sum it had fetched in 1767. In 1798 and 1799 they bought some additional pieces of land at Denton. Easington Grange was purchased in 1801.[19]

The year 1794 saw the peak of the Culleys' tenancies. In the

---

14 *History of Northumberland*, vol. 11, p. 44; James Raine, *The History and Antiquities of North Durham*, London 1852, p. 321.
15 In addition to the above six, two more, Copfield and Meldon, and Adderstone, are noted on a later list in ZCU 33 as having been leased in the 1780s, but details are lacking.
16 List in ZCU 33. Fenton was over 1100 acres, Grindon was 1400 acres, Crookham Eastfield and Westfield 554, Wark 918, Thornington 819. The rents were: Fenton £750; Crookham Westfield £260, raised in 1796 to £393 15s 0d; Eastfield £339, raised in 1793 to £483; Red House £235; Wark £1000; Grindon £900; Thornington £600.
17 Bailey to Lord Tankerville, 17 Sep. 1786, NRO, Tankerville MSS, Box 4, ZTN/D/1/4/1/2/2; Culley Papers, ZCU 33; see No. 67.
18 *History of Northumberland*, vol. 11, pp. 186–7. Between 1796 and 1804 the Culleys also leased from the Bishop of Durham one oxgang of land at Osmotherley, north Yorkshire, but they did not farm it themselves: Durham University Library, Archives and Special Collections, henceforth DUL, ASC, Bishopric Estates, Lease enrolment books CCB V/1/29, p. 73; CCB V/1/30, p. 559.
19 *Op. cit.*, p. 234; Robert Surtees, *History and Antiquities of the County Palatine of Durham*, vol. 4, London 1840, p. 2.

following year they lost Fenton following a change of ownership, and in 1798 they gave up Red House. But in that year they decided to take in hand the farm at Denton, which Matthew had inherited on Robert Culley's death, unmarried and childless, in 1783. It had been let since their father died in 1762. This decision is the origin of the correspondence printed here.

The three brothers farmed in partnership until February 1793, when James died. Matthew and George then made a new agreement.[20] They lived together at Fenton. None of the brothers married young: they were too busy getting established. George Culley was the first to embark on matrimony. In 1777 he married Jane Atkinson, daughter of another farmer in the area. They had two children, Matthew, born 1778, and Eleanor, born 1779. Jane Culley died in 1780. In 1787 George married again, to Isabella Spours, also from a neighbouring family, but she died within six months. He married for a third time, in 1794, to Hannah Nesbit of Ancroft, who survived him until 1824. Neither the second nor the third marriage produced children. James Culley married Margaret Pickering in 1781. They had one daughter, Eleanor, born in 1784. Matthew Culley found his wife in the Tyne valley: in 1783 he married Elizabeth Bates, daughter of Thomas Bates of Halton near Corbridge. They had five children: Eleanor, born 1785; Matthew, born 1786; Elizabeth, born 1788; Thomas, born 1791, died 1792; and Jane, born 1795. Matthew and his family lived from 1786 at Wark; James lived first at Crookham Eastfield and then at Grindon; George remained at Fenton until 1795, when the lease was given up and he moved to Eastfield.

Some of the character of the two brothers emerges from the correspondence – Matthew careful of and attentive to technical detail, succinct in expression; George fluent in writing (describing himself as 'having the pen of a ready writer') and doing most of the business correspondence, driven to anxiety and ill temper when Welch did not

---

20 Typewritten summary in ZCU 43, dated 21 Mar. 1797. This year is certainly a mistake for 1793. James Culley had died in February (he was buried at Norham on 18 February: DUL, ASC, Durham Diocesan Records, Bishops Transcripts of parish registers, henceforth DUL, BT with parish or chapelry), and a letter from George Culley of 26 Mar. (ZCU 18) to the brothers' lawyer refers to a new deed of partnership. The new agreement was to run until 12 May 1807, when the leases of Wark and Grindon would expire. At that time Matthew Culley would take Thornington, George Eastfield and Westfield. Three arbitrators, Thomas Bates, John Nesbit, and Thomas Vardy of Fenton, were appointed to ensure a fair division of the assets. See below, pp. xxxi–ii, for the division in 1804 and the names of the arbitrators.

write as often as he promised. Both were devoted to their families. George frequently expressed an unforced piety and trust in a benevolent God, who gave the weather and rewarded honest effort. The writer of a memoir of George described Matthew as having a grave, thoughtful disposition, interested in all improvements and good at the technicalities of farming. George was described as specialising in selling and judging stock, often spending four days a week at markets and fairs, even-tempered and pleasant in manner, and often called upon to act as an arbitrator in valuing farm stock. John Welch found Matthew the more independent and stronger in character of the brothers; he thought George more given to self-publicity.[21]

The other main figure in this correspondence is John Welch, whom the Culleys appointed their steward at Denton. Welch was born in 1764, son of John Welch, farmer of Old Penshaw, County Durham, a tenant of the Lambton estate, and his wife Martha. From some of the early letters it appears that the younger Welch worked for the Culleys in the 1790s at Red House, but had returned to Penshaw, presumably when the Red House lease was given up. While in Northumberland Welch married Dorothy Fairbairn of Penshaw. Their first child, a daughter, died in 1798 aged one year: two further children were born at Denton in 1800 and 1801.[22] Between late 1798 and the summer of 1804 the Culleys, especially George, wrote regularly to Welch with instructions, advice, consultation and information about markets. Welch evidently kept the letters, and apparently later returned them to George or his son, so that they have descended in the Culley Papers.

In the 1760s, when the Culleys moved north, the standard of farming in Northumberland was mixed. Arthur Young, on his tour of northern England at the time, found the husbandry of the cultivated lands good and spirited; but he considered such efforts as were being made to improve the extensive wastes as slovenly, and the management of the upland sheep as wretched.[23] Glendale had easy soils and good communications, with a main road running through from Coldstream to Morpeth and Newcastle, and the port of Berwick

---

21 *Farmer's Magazine*, 14(1813), pp. 274–5; Welch note quoted by Stuart Macdonald, 'The role of George Culley of Fenton in the development of Northumberland agriculture', in *Archaeologia Aeliana*, 5th ser. 3(1975), p. 141.
22 DUL, BT, Penshaw, Denton; Durham Record Office (hereafter DRO), parish register, Penshaw.
23 Arthur Young, *A Six Months Tour through the North of England*, 2nd edn London 1771, vol. 3, pp. 91–3.

accessible down the Tweed. To this area the Culleys brought, according to a younger neighbour and disciple, their good livestock from Denton, together with 'superior knowledge and intelligence', and unremitting application and perseverance.[24] They lived frugally, according to George Culley's later recollection, drinking only beer, never wine or spirits; but from the first they spent money on improvements, hedging and hiring tups in a way that frightened their neighbours.[25] In their husbandry they adopted, and helped to establish through the area, a modified version of the Norfolk four-course rotation, a five-course cycle of oats, turnips, wheat or barley, and two years grass. Having first seen turnips drilled by a Mr Pringle near Coldstream in 1766, they made the acquaintance of another practitioner, William Dawson of Frogden, visited him frequently, and sent pupils to study the method.[26] The Culleys were the first farmers in Northumberland to run water over meadows in winter to encourage early growth of grass, and sent one of their men, Harry Rutherford, all the way to Dorset to be taught the technique by the specialist George Boswell.[27] Matthew Culley became particularly keen on watering, as the letters here show; but it was nearly twenty years before his neighbours began to follow the example.

Arable farming was as important for the Culleys' success as the stock for which they are best remembered. George Culley's letters contain as much discussion of the selling of grain and grain prices as they do of stock. They sold considerable quantities of wheat – making more money from wheat than from cattle and sheep at the turn of the century, with grain prices high – substantial quantities of oats, and lesser amounts of barley and peas. They consumed their own turnips, and made a fair business of selling turnip seed. Summary accounts

---

24 John Grey of Dilston, 'A view of the past and present state of agriculture in Northumberland', in *Journal of the Royal Agricultural Society of England*, 2(1841), pp. 153–4. For Grey see below, No. 61, n. 68.

25 See No. 203.

26 George Culley to Arthur Young, n.d., ZCU 3; George Culley, 'A method of drilling turnips', in *Annals of Agriculture*, 20(1793), pp. 159–66. For Dawson see also Andrew Wight, *The Present State of Husbandry in Scotland*, Edinburgh 1778–84, vol. 2, pp. 355–8; Robert Douglas, *General View of the Agriculture in the Counties of Roxburgh and Selkirk*, Edinburgh 1798, pp. 90–1.

27 George Boswell to George Culley, 16 and 25 Mar. 1787, ZCU 12; George Culley to Hon. Henry Grey Bennett, 2 Dec. 1804, ZCU 26; John Bailey and George Culley, *General View of the Agriculture of Northumberland, Cumberland and Westmorland*, 3rd edn London 1805, reprinted, with an introduction by D.J. Rowe, Newcastle upon Tyne 1972, hereafter *General View, Northumberland*, pp. 136–7. Boswell wrote a book, *A Treatise on Watering Meadows*, published in 1779.

are given below. Nevertheless it is as breeders and managers of live-stock that the Culleys are chiefly remembered. When they came to Northumberland the county was home to four breeds of sheep, the Cheviot or hill sheep on the hills of the north, the Blackface or high-land on the high ground in the west and south-west, Mugs (whose faces had a ruff of wool down to the eyes), and a distinctive large long-wooled sheep derived from the Lincolnshire breed on the coast round Bamburgh.[28] As the Culleys expanded into more farms their flocks grew, from 1078 head in 1767 on entering Fenton to 4441 in 1794.[29] Not all of these sheep were of the Culley's own breed. At Shotton at the foot of the Cheviots, and Longknowe with land going up to 340 metres, they kept hill or Cheviot sheep, but used their own-breed tups with them. The brothers did not succeed for some years in persuading neighbouring upland farmers to try one of their tups with Cheviot ewes, even though they offered the service at no cost as an experiment. By the mid 1790s, however, the cross was proving a success in south-east Scotland, although doubts remained about how the sheep would do on higher ground.[30] The Culleys also bought sheep from neighbours for grazing and fattening, and sent some to Denton; but these were not for breeding. They had early on begun to follow Bakewell's practice of letting tups for a season: they had indeed hired one to Lord Darlington at Raby Castle in 1765, even before they moved north.[31] By the 1780s their letting range extended as far as Yorkshire and southern Scotland; and George Culley claimed that it would soon be difficult to find any flock in lowland Northumberland that was not more or less related to the Dishley blood.[32] The Mugs were on the way out. The spread of the New Leicester sheep in the north-east was not the work of the Culleys

---

28 Bailey and Culley, *General View, Northumberland*, pp. 144–54, do not include the Bamburghshire sheep, but their breeders resisted the spread of the New Leicester breed past the turn of the century: *Farmer's Magazine*, 5(1804), pp. 454–6.

29 ZCU 33.

30 George Culley to Arthur Young, n.d., ZCU 3; Thomas Johnston, *General View of the Agriculture of the County of Selkirk*, London 1794, p. 37; Thomas Johnston, *General View of the Agriculture of the County of Tweedale*, London 1794, p. 34; Alexander Lowe, *General View of the Agriculture of the County of Berwick*, London 1794, pp. 17–18; David Ure, *General View of the Agriculture of the County of Roxburgh*, London 1794, pp. 59–60.

31 Raby Archives, account book 1765.

32 Bailey and Culley, *General View, Northumberland*, p. 150; Culley, *Travel Journals*, p. 183.

alone: other breeders bought and hired tups directly from Leicestershire. In the early 1790s the Culleys tried to form a tup association on the model of the Leicestershire association of Bakewell and his friends, in order to control the market. Attempts to reach an agreement with the Leicestershire association on dividing territory failed, however, because the Northumberland men were not able to accept a condition forbidding lettings in Yorkshire to men who dealt at markets. In the north some large successful farmers refused to join the Culleys' association, and a number of mostly smaller farmers denounced the monopolistic tendency and undertook not to deal with its members.[33] There was already enough Dishley blood in the county to make the threat credible, and the Northumberland association had to be dissolved in a rather humiliating retreat. Nevertheless the Culleys continued to do well with their tup-letting as far south as York, and sold some at greater distances.[34]

The Culleys managed their farming to a large extent as a single enterprise. Each farm had a steward or overseer in charge of the day-to-day work. These men met frequently with the brothers in what George described as the senate or privy council; selling was planned jointly; and stock was moved from one farm to another to suit the state of the pastures and the supply of feed. The selling of stock was largely carried out by George, with the help of his son and some senior employees. Grain selling from all the Northumberland farms was almost entirely done by Ralph Brown, steward at Eastfield, who went to markets at Kelso and Berwick. The tups seem to have been kept at Wark and Eastfield. Welch, at Denton, 100 miles to the south, was necessarily more independent. He did his own buying and selling, and hired his own servants; but his accounts were subsumed into the overall accounts of the enterprise. Sheep and cattle were sent regularly from Northumberland to Denton for grazing and fattening, and sale locally and at markets as far south as Skipton and Wakefield. Prices in this area were usually somewhat higher than those at Morpeth, the principal market for Newcastle and Tyneside; and the Culleys were able to take advantage of Denton's more southerly position.

---

33 Correspondence in ZCU 4, 17–18, 31; *Newcastle Courant*, 30 Jun., 21 Jul., 4 Aug. 1792; *Newcastle Chronicle*, 14 Jul.

34 For tup-letting see below No. 64 and *passim*. For sales to a distance see Nos. 18, 207, 211: in 1799 they sent a tup by sea from Blyth to King's Lynn for a customer in Norfolk: George Culley to B. Vipan, 23 Sep. 1799, ZCU 31.

Detailed farm accounts were certainly kept but have not survived. A valuation in 1793 of the stock and implements at Wark, Grindon, Eastfield and Westfield, and Fenton, gives a total of £13,887 15s 0d, with 3087 sheep and 513 cattle, of which 151 were work oxen.[35] Summary accounts for 1799–1802 have been preserved and are given below. They show receipts and payments for rents and the farming operations, but are not very clear about capital transactions and bank borrowing. Nor, unfortunately, are they informative about the numbers of farm servants employed and their wages. A number of names and some individual bargains are mentioned in the letters, but we do not know how many men and women were employed, full-time or day labourers or casual workers seasonally. There are references to servant's cows, and to payments in corn. The figures in letters Nos. 3, 5 and 6 suggest that the wages of a man (hind), hired by the year, ranged from about £12 to £25 per year. In their report to the Board of Agriculture (of which more below) John Bailey and George Culley gave a hind's wages as equivalent to £18 11s a year, paid largely in kind, with a house and garden and other perquisites. Shepherds got pasturage for a number of sheep, and the sale of their lambs and wool. Hinds were obliged to provide a female worker or bondager for seasonal work. Overseers or stewards got £20–30 per year; day labourers got 2s to 2s 6d a day, women 8d to 1s 3d.[36] Some local figures are available for comparison. George Hughes of Middleton Hall paid his hinds a good deal less – between £6 6s and £11 11s per year, almost all in kind; but his situation was different from the Culleys'.[37] Hughes' 1097-acre farm on the edge of the Cheviots had little more than 100 acres of tillage; all the rest was pasture. He would not have needed many men with the higher skills of drilling and cultivation. In County Durham payment in corn was not customary. A man hired by the year, with board and lodging, could earn £21, a woman £8. Day labourers got about the same as those in Northumberland.[38] As for the numbers employed by the Culleys, it does not seem possible to use their total labour costs to produce an estimate: there are too many variables. A couple of local

---

35 ZCU 33.

36 Bailey and Culley, *General View, Northumberland*, pp. 164–6.

37 NRO (Berwick), Simpson Papers, ZSI 57, 58, 59, 317. For Hughes (1747–1834) see *History of Northumberland*, vol. 14, p. 297. He was a major in the Royal Cheviot Legion (see No. 182, n. 35).

38 John Bailey, *General View of the Agriculture of the County of Durham*, London 1810, hereafter *General View, Durham*, pp. 62–3.

cases may be cited. George Hughes seems to have employed two or three hinds and two or three women, hired by the year or half year, but no figure for labourers is recorded. Arthur Young, on his tour in 1768, reported a farm of 1000 acres in the Wooler area, about half and half arable and grass and thus comparable with the Culleys', employing three hinds, three boys, two women, and sixteen labourers, plus presumably seasonal workers in hay-time and harvest.[39] Payment in kind for hinds was standard also in Berwickshire and East Lothian. It continued in north Northumberland at least to the 1840s.[40]

After the initial visits to Dishley, both Matthew and George Culley travelled a good deal about England and Scotland to observe farming practice and enlarge their experience. George claimed in 1794 that he had repeatedly visited most of the stock-breeding districts of England except for Shropshire, Sussex, Devon and Cornwall, and apart also from Wales.[41] Some journals of such travels, kept by both brothers, have survived.[42] There may have been other journals, and there is mention in the correspondence of other visits to the south of England, the Midlands, and even Ireland. The journals show that wherever the Culleys went they looked carefully at crops, stock, and farming methods, and commented on the good and the bad. They took note of canal-building, improvements in communications and trade, and new industry. George Culley also conducted an extensive correspondence, with customers and suppliers, and with farming friends and acquaintances, some of them men whom he had met on his travels and with whom he kept up for years. The Culley papers contain letters from over 100 different correspondents, many of them on business such as selling of turnip seed, purchase of corn and clover seed, and hiring of tups; but a good proportion are from friends such as Benjamin Sayle of Wentbridge, Samuel Deverel of Clifton, and George Boswell of Pidddletown.[43] Not all the letters received were kept: as will be seen, in the letters to Welch there are a

---

39 ZSI 57, 58; Young, *Six Months Tour of the North of England*, vol. 3, pp. 66–70.

40 Lowe, *General View of the Agriculture of the County of Berwick*, pp. 112–15; George Buchan-Hepburn, *General View of the Agriculture and Rural Economy of East Lothian*, Edinburgh 1794, pp. 90–2. Grey in *Journal of the Royal Agricultural Society*, 2(1841), pp. 183–90; W.S. Gilly, *The Peasantry of the Border*, Berwick upon Tweed 1841, pp. 6, 36, appendix 2. For the system of agricultural employment see *Agrarian History of England and Wales*, vol. 6, pp. 672–95.

41 George Culley, *Observations on Live Stock*, 2nd edn London 1794, pp. vii–viii.

42 Published as Culley, *Travel Journals*.

43 ZCU 8–36.

number of references to letters that George had received but which are not preserved in the Culley Papers. For the most part the correspondence, even with friends of long standing, were fairly strictly confined to farming matters, comments on developments, reports on the state of crops and prices, etc. The amount of family news, and comments on public affairs, in the letters to Welch, is unusual. George Culley was an assiduous letter-writer, but he did not keep copies of most of the letters he sent: only one out-letter book is preserved, and a few copies of replies on the back of incoming letters. All this correspondence, as Stuart Macdonald pointed out,[44] was a factor in the spread of ideas and improved farming practice, a factor that can be overlooked but, because conducted between men who trusted one another's experience and judgement, may have been as important as the books being published and the reports on experiments in Arthur Young's *Annals of Agriculture*.[45]

George Culley did, however, also contribute to the literature. In 1786 he published a book, *Observations on Live Stock*. This was a pioneering work, for in the increasing body of works on agriculture there was as yet little reference to livestock breeds or livestock husbandry. There were a few works largely on veterinary matters such as Josiah Ringsted's *The Cattle-Keeper's Assistant*, which reached its ninth edition in 1776.[46] Not long after George Culley's book appeared the Society for the Improvement of British Wool sponsored two reports on sheep, which were published in book form.[47] But there were as yet no comparable works on cattle, and '*Culley on Livestock*' held the field for a good many years. Expanded editions, in 1794, 1801 and 1807, contained material supplied to George Culley by acquaintances and informants such as William Mure, a former farm pupil who became agent to Lord Dacre at Baldoon in Galloway. Sent by his employer in 1792 for a lengthy tour of England, Mure wrote the

---

44 Macdonald in *Archaeologia Aeliana*, 5th ser. 3(1975), pp. 131–41.

45 Young started the *Annals of Agriculture* in 1784. It ran to 46 volumes before ceasing publication in 1815. The *Farmer's Magazine* began publication in Edinburgh in 1800, and contained a good many articles by working farmers. There is a useful bibliography of books on agriculture in W.F. Perkins, *British and Irish Writers on Agriculture*, 3rd edn Lymington 1939. See *Agrarian History of England and Wales*, vol. 6, pp. 361–70.

46 Josiah Ringsted, *The Cattle-Keeper's Assistant*, London n.d., 9th edn 1776.

47 William Redhead, *Observations on the Different Breeds of Sheep, and the State of Sheep Farming in some of the Principal Counties of England*, Edinburgh 1792; John Naismith, *Observations on the Different Breeds of Sheep, and the State of Sheep Farming, in the Southern Districts of Scotland*, Edinburgh 1792.

Culleys long letters that included detailed descriptions of sheep and cattle breeds.[48]

In 1793 the president of the newly established Board of Agriculture, Sir John Sinclair, invited George Culley to do one of the reports that the Board was planning on the state of agriculture in every county in Great Britain. The proposal was for the Cumberland report. George Culley agreed, but asked Sinclair to let him have the assistance of John Bailey.[49] The two were then asked to do the report on Northumberland as well, and both were completed and printed by the end of 1794.[50] The Board of Agriculture sent the reports for comment to selected persons in the counties concerned, and then published them for general distribution. Revised versions were then supposed to be produced, but this work proceeded only slowly. The first reports had been done in a hurry, and some of the authors knew little about the counties they were writing about. Funds were inadequate, and ten years after the start fewer than half the final versions had appeared. Some had to be entirely redone, among them the report on County Durham, originally written by Joseph Granger. A wholly new report by John Bailey appeared in 1810.[51] William Marshall, an agriculturist who had proposed the Board of Agriculture and was aggrieved at not being made its secretary (the post went to Arthur Young), was highly critical of many of the reports. But he made an exception of the Northumberland volume. Marshall does not seem to have known the Culleys (no correspondence with him is preserved in the Culley Papers although there is one possible reference to a letter, in No. 27) and disagreed with them on various matters, such as the use of work oxen; but he acknowledged that George Culley and Bailey were particularly well suited to the task. Their report, he said, was in general so good that 'we shall not see its like again'.[52] George Culley did not think highly of Young as a farmer:

---

48 Letters in ZCU 17–18.

49 George Culley to Sir John Sinclair, 30 Sep. 1793, ZCU 31. For Sinclair (1754–1835) and Bailey (1750–1819) see *Oxford Dictionary of National Biography*. Sinclair did much to improve his estates in Caithness and launched the *Statistical Account of Scotland*. See also below, No. 105, n. 9.

50 Bailey and Culley, *General View, Northumberland*; John Bailey and George Culley, *General View of the Agriculture of the County of Cumberland*, London 1794. The two, with the report on Westmorland, were published as a single volume in 1797. The final version was published in 1805.

51 Joseph Granger, *General View of the Agriculture of the County of Durham*, London 1794; Bailey, *General View, Durham*.

52 William Marshall, *A Review of the Reports to the Board of Agriculture from the*

his comment that Young's 'sheep are scabbed, his cattle ill chosen and worse managed' has often been quoted.[53] But on a visit to Suffolk with Bakewell in 1784 he found Young personally agreeable. He supplied Young with information about Bakewell;[54] and in the 1790s he contributed to *Annals of Agriculture*, often in response to circular letters of enquiry but also volunteering pieces on topics as diverse as transplanting cabbages and drilling turnips, as well as describing his and his brother's breeding and management of sheep.[55]

All this writing made George Culley the better known of the two brothers; the third, James, is hardly ever even noticed. But he never claimed single-handed achievements, and, unlike Young, never gave writing precedence over action. On the contrary it was, in the words of the writer of a memoir published after George's death, the brothers' 'great attention to minutiae, unremitting industry, and superior cultivation' that gained them celebrity as first-rate breeders and agriculturists, bringing them pupils from various parts of the island as well as 'great accumulation of wealth'.[56]

The partnership agreement entered into by the surviving brothers in March 1793 after James Culley's death was to run until May 1807, when the leases of Wark and Grindon would expire. But early in 1804 it was decided to dissolve the partnership and divide the leases and the properties. It may be that Matthew Culley's health was beginning to fail – he died later in the year – and he wished to retire from active farming. The brothers entrusted John Bailey and their respective brothers-in-law Thomas Bates and John Nesbit with the task of valuing the various assets. Of the leases, Wark and Thornington were allotted to Matthew, Grindon, Longknowe, Eastfield and Westfield to George. For the property, the Akeld and Humbleton estate was to go

---

Northern Department of England, York 1808, p. 116. Marshall (1745–1818) wrote a multi-volume *General Survey, from Personal Enquiries, Observation and Enquiry, of the Rural Economy of England*, published between 1787 and 1798. His method was to spend several weeks at a time in one area, observing agricultural practice in detail; but he also did one of the reports to the Board of Agriculture, *General View of the Agriculture of the Central Highlands*, London 1794, which suffered from the same defect of inadequate knowledge of the area as he complained of in other reporters. See *Oxford Dictionary of National Biography*.

53 Culley, *Travel Journals*, p. 222.

54 G. Culley to A. Young, 18 Feb., 9 Mar. 1811, British Library Add. MS 35,131.

55 *Annals of Agriculture*, 14(1790), pp. 180–2, 253–5, 470–5; 15(1791), pp. 625–7; 17(1792), pp. 346–52; 19(1793), pp. 147–50, 306–14; 20(1793), pp. 159–66; 21(1793), pp. 223–9; 23(1795), pp. 519–34; 24(1795), pp. 106–9, 438–48; 34(1800), pp. 289–93; 35(1800), pp. 1–5, 555–7; 37(1801), pp. 280–4.

56 *Farmer's Magazine*, 14(1813), pp. 271–5.

to Matthew, Easington Grange to George. Most of Denton was Matthew's property, inherited from the elder brother Robert; but some additional land there had been bought jointly, and since Akeld was more valuable than Easington Grange, having cost £24,000 in 1795 as against £13,000 in 1803, Denton was allotted to George for a token rent. Even so, there was a discrepancy, estimated by the arbitrators at £13,124, so that whichever brother got Akeld would have to pay the other half the difference.[57]

Matthew Culley was left dissatisfied. He thought he ought to have been allowed the sums of his own money that he had spent on improving the house at Denton, and he insisted that Welch should come up to Northumberland at once to work for him at Akeld and take part in the division of the stock. George Culley had asked Welch to stay on at Denton, and at first objected, saying that at least Welch must stay until after the harvest. Matthew then wrote angrily:

> If you will be obstinate and refuse John Welch his coming to me, that will be an end of all future correspondence, as I have a right to his immediate service in course, as you have detained all or a greater share of our upper servants than comes to your share of them. I believe I am acting in a just way in saying as I do. If you will obstinately persist, contrary to all justice, to keep J.W. I think, and any reasonable person will think the same with me, that you are doing wrong. I have endeavoured to keep from a total outfall, but you seem determined to have your own way in too great a degree, as I have generally submitted to your opinions. What do words mean when not acted up to, or fine speeches. Your saying you had rather loose a great sum of money than differ with me, and now you will not let me have any upper servants but such as you choose to let me have. Are we to divide them as we have done our cattle? Fie the thought! Nothing less than John Welch coming to me, and that in a little time, can keep up a future correspondence between us two brothers, although with regret to me.[58]

For once, George Culley kept a copy of his reply: 'I conceive I have a right to J. Welch's service so long as he is inclined to stay with me, but rather than be at enmity with you, I have by letter advised John to come to you next year. But stay he must with me at Denton until we can procure one in his stead or get it let.' As for dividing the stock, Ralph Brown at Eastfield, Harry Rutherford at Wark, and the shepherds could certainly do it without Welch. 'If not your will be done,

---

57 Award by Bailey, Bates and Nesbit, 11 Apr. 1804, ZCU 26; settlement between Matthew Culley of Akeld and George Culley, 1807, ZCU 46. See Nos. 211, 233–4.
58 Matthew Culley to George Culley, n.d. [? 27 May 1804], ZCU 26.

because I have long been determined not to quarrel with you. If you will quarrel with me I can't help it. I am your affectionate brother George Culley.'[59]

As a result George Culley gave in, stipulating only that Welch must be allowed to visit Denton to advise his successor. That successor was sent from Northumberland. Unfortunately we do not know which of the Culleys' senior employees it was, because the original of the letter to Welch about the arrangement (No. 234) has been lost and the transcriber was apparently unable to read the name. George Culley presumably continued to manage Denton with the Northumberland farms that fell to him, and presumably continued to write long letters to the new steward; but he, if he kept them, did not return them later. The Denton farms were let separately to two tenants in 1806 and 1807.[60]

Matthew Culley remained at Wark until his death on 7 December 1804. His family then moved to Akeld, where his wife died in 1814. Matthew's son went to Cambridge, a clear sign of social advance, and lived at Akeld as a gentleman. In 1831 he married Margaret Anne Tewart, shortly after inheriting Coupland Castle from his uncle Thomas Bates, who had bought the property in 1806. The younger Matthew Culley died in 1834. His son Matthew Tewart Culley (1832–84) was High Sheriff of Northumberland in 1869. He married twice, first a daughter of the Rev. Thomas Knightley, rector of Ford, and secondly his cousin Eleanor Darling, daughter of George Culley's grandson. Matthew Tewart Culley's eldest son Matthew (1860–1920) became a Catholic priest and was succeeded at Coupland by a nephew. The older Matthew Culley's daughters all married after their father's death. Coupland was sold in 1923.[61]

George Culley stayed on at Eastfield. He bought Fowberry Tower near Chatton from Sir Francis Blake for £45,000, and his son Matthew went to live there. It was a grand house, and George Culley worried that his son might be tempted to live in a greater style than his income warranted. After a visit in the summer of 1810 George wrote: 'Whenever I am at Fowberry I am struck with astonishment when I reflect on our beginning in Northumberland 43 years ago. To think of

---

59 George Culley to Matthew Culley, 28 May 1804, ZCU 26.
60 Lease of Denton North Farm to William Cumming, 1806; lease of Denton Demesne farm to George Wetherell, 1807, ZCU 36.
61 *History of Northumberland*, vol. 11, pp. 224–5; *Burke's Landed Gentry*; memorial tablet, Kirknewton; DUL, ASC, Hudleston Clergy Index, under Thomas Knight, Christopher Robinson.

my son inhabiting a *Palace*, although his father in less than 50 years since worked harder than any servant we now have, and even drove a *coal cart!*' George went on: 'I hope and trust that you are possessed of a better understanding and more firmness than to suffer your vanity to get the better of your reason and common sense.' The man who pursued vanity, ostentation, and higher modes of living than his income allowed would always have plenty of followers and false admirers, but would bring ruin on himself in the end, 'besides endangering his future happiness which is the most important of all. ... When a man lives within his income he is enabled to do many *humane, friendly* and *charitable actions*, and it is every man's duty who can afford it to assist the necessitous.' A man must also provide for his dependants, and in general follow the 'common sense' principle: 'To do what is just and right is praiseworthy and approved by both God and man, whilst the doing *unjustly* and *unrighteously* is *wrong and condemnable.*'[62]

George Culley's son never married. He became a magistrate (which his father approved) and added a front and a gatehouse at Fowberry.[63] He died childless in 1849, leaving Fowberry to his nephew George Darling, son of his sister Nelly, whose marriage to James Darling is reported in letter No. 182 and who died in 1806, aged 27. George Darling was supposed to take the name of Culley, but died before he could do so, only a year after his uncle. It was therefore his son George (1834–93) who took the name of Culley and lived at Fowberry, becoming a Commissioner for Woods and Forests and being active in local affairs. He left one daughter, whose only son died childless. Fowberry was sold in 1920.[64]

To sum up the importance of these letters and their writers, there is no doubt that, whilst some of the tributes paid to them at the time should be taken with a pinch of salt, Matthew and George Culley were influential as enterprising farmers in their day, in their own area and further afield. That Northumberland had a reputation in the 1840s for advanced agriculture was in part due to them. More lasting influence is another matter: the type of farming that the Culleys practised so successfully was modified in turn by new methods, machinery, and fertilisers; and the high farming did not survive the

---

62 George Culley to Matthew Culley jr, 3 Aug. 1810, ZCU 30.
63 Nikolaus Pevsner, *The Buildings of England: Northumberland*, 2nd edn London 1992, pp. 286–7.
64 *History of Northumberland*, vol. 11, pp. 224–5; vol. 14, pp. 224–5; *Burke's Landed Gentry*; DUL, BT Ford; memorial tablet, Chatton.

agricultural depression of the 1870s. The 'Culley sheep', the Border Leicester, was not permanently important as a breed on its own; but its descendants, now more often crossed with Swaledales than with scarcer Cheviots, continue. The letters printed here, covering a part of the Culleys' farming careers, provide a vivid picture of, and commentary on, the life of their area at a time of important change.

SUMMARY ACCOUNTS, 1799–1802[1]

Taken Novr 1799 *An Account of our Farming Debts & Credits* Dr.	Nos.	£	s.	d.	Taken Novr 1799 *With each Article set Opposite* Cr.	Nos.	£	s.	d.
Paid Rents and Tythes		4135	17	5	Recd Rents		1049	6	3
Labour		3978	19	11 1/2	Turnip seed	lb. 5286	128	13	3
Sundries		759	8	1½	Sundries		50	5	8 ½
Taxes & Cesses		451	5	4	Hay		2	11	
Wood & Iron		342	17	3 1/2	Quicks	no. 32000	12	17	6
Expences Tolls &c		223	19	7½	Swine	no. 101	135	17	6
Coals		4	1	½					
do.		7	16	3½					
Clover & Rye Grass Seeds		188	7	8	Potatoes	m. 623 ½	40	16	9
Lime		270	4						
Wheat	m. 118	44	2		Wheat	m. 16342	5706	8	9
Barley	m. 50	9	9		Barley	m. 1750 ½	292		1 ½
Oats	m. 796 ½	138	8	11	Oats	m. 11825	2149	7	8 ½
Pease	m. 141	21	11	6	Pease	m. 155 ½	37	4	7
Cattle & Calves	no. 40	112	18	6	Cattle	no. 77	1811	11	7 ½
					Butter		79	19	8 ½

---

1   ZCU 33. 'm' signifies 'measure'. Size not specified, but used consistently.

		£	s	d
Sheep & Wool	no. 117	144	16	¼
	Wool	£780	11	11 ½
	Tups	£309	11	
Horses	no. 12	245	2	11
Principal Money to Funds &c to pay for Estates		8334	3	7 ½
Interests		527	3	
Mother & Sister		50		
Tryal with Col. St. Paul in pts		88	4	10
Cash		151	19	
Hay		1	12	9
Akeld & Humbleton		182	8	8 ½
		20414	14	1 ½

		£	s	d
Sheep	no. 1415	£2879	13	16
Mutton		80	1	7 1/2
	no. 17 ½	3969	16	6
Horses	no. 6	172	15	4
Principal Money		3659		
Interests		575	3	5
Money Borrowed		3705		
Money Lent		1990		
Cash of Mr		33	18	8
		25665	16	

Taken Novr 1800 *An Account of our Farming Debts & Receipts* Dr.

	Nos.	£	s	d
Paid Rents & Tythes		3976	2	1
Labour		4186	16	10 ½
Sundries		809	6	7 ½
Taxes & Cesses		427	1	5
Wood & Iron		154	19	11 ½
Expences Tolls &c		222	11	2 ¼
Lime		135	1	3
Wheat	m. 201	112	13	10
Barley	m. 357 ½	147	2	6
Oats	m. 139 ½	47	2	
Pease	m. 649	322	9	5
Cattle & Calves	no. 57	374		
Sheep & Wool	no./lb. 56	47	13	1 ½
Horses	no. 12	222	6	6
Hay			8	
		£11667	5	11 ¼

Taken Novr 1800 *with both Articles set Opposite* Cr.

	Nos.	£	s	d
Recd. Rents		117	4	1 ½
Turnip seed	lb. 3879	171	12	
Sundries		55	6	3 ½
Willows		3	8	6
Swine	no. 62	122	6	11
Potatoes	m. 1096	235	10	1
Wheat	m. 10302 ½	5973	16	10 ½
Barley	m. 1280 ½	376	10	10
Oats	m. 2581	635	18	2 ¾
Pease	m. 89	23	4	6
Cattle	no. 111	2706	8	½
Sheep	no. 934	1951	12	1 ½
Wool		479	1	9
Tups	no. 54 ½	550	15	8
Horses	no. 6	90	14	
Butter		51	3	6 ½
Recd.		£13544	13	5 ½
Paid		11667	5	11 ¼
Money made by our Farms this Year		£1877	7	6 ¼

Novr 1801 *An Account of our Farming Debts and Credits*			Dr.		Taken 21 Novr 1801 *With each Article set Opposite*		Cr.		
Also of Estates below do. & Money	Nos.	£	s.	d.			£	s	d
Rents & Tythes		4091	14	8	Rents		149	4	10
Labour		5179	16	8	Turnip seed	lb. 9029	525	16	11
Sundries		1018		2 ½	Sundries		77	16	1 ½
Taxes & Cesses		939	1	5	Willows		5	9	2
Wood & Iron		388	5	9 ½	Swine	nos. 32 ½	65	4	2
Expences Tolls &c		271	10	1	Potatoes	m. 683	164	7	4 ½
Clover & Rye Grass seeds		376	8	9	Hay		33	15	2
Lime		69	12		Lime. Denton		143	7	9
Wheat	m. 42	38	9	11	Wheat	m. 14256 ½	10937	3	1 ½
Barley	m. 17	12	6		Barley	m. 1923	727	13	5 ¼
Oats					Oats	m. 7088 ½	1990	19	3
Pease	m. 152	91	15	4	Pease	m. 222	74	16	5
Cattle & Calves	no. 82	553	10	10 ½	Cattle	no. 109	2632	1	6 ½
Sheep & Wool	no. 214	355	3	6	Sheep	no. 1288	3227	4	5 ½
					Wool		773	15	11
					Tups	no. 86 ½	1080		6
Horses	no. 12	243	1		Horses & mares served	no. 45	149	3	6
					Butter		88	5	9
Carried to the other side		13268	16	2 ½		Recd.	£22842	5	4 ¾
						Paid	£13628	16	2 ½
					Money made by the Farms this year		£9213	9	2 ¼
Denton		446	1	3 ½	Rents for Estates		1255	19	6

	£	s.	d			£	s.	d
Mother & Sister	50			Cash		122	10	
Akeld & Humbleton &c (& tryal with Col. St Paul concluded)	211	8	9 ½					
Interest	804	11		Interest		440	1	
Carried to the other side	1512	1	1	Recd.	£11031	19	8 ¼	
				Paid	£1512	1	1	
				Total of Money made by Messrs Culley this year	£9519	18	7 ¼	
				Principal money paid off	£3335	6	8	
				do. do. recd	0800			
				do. do. lent out	5695	3	6	

Taken 22 Novr, 1802

An Account of our Farming Debts & Crdits

Taken 22 Novr. 1802

With each Article set Opposite

Also of Estates & money below	Nos.	£	s.	d		Nos.	£	s.	d
			Dr.					Cr.	
Rents & Tythes		4348	10	1	Rents		164	6	7
Labour		5124	12	2	Turnip seed	lb. 10087	517	9	8
Sundries		7	15		Sundries		78	17	10
		1114	10	11 ¾					
Taxes & Cesses		554	0	7 ¾	Willows		11	9	3
Wood & Iron		345	9	7	Swine	no. 68	183	18	4
Expences Tolls &c		312	6	10	Potatoes		28	18	
Clover & Rye Grass seeds		280	5	6	Hay		12	13	11
Lime		142	15		Lime. Denton		256	7	8

		£	s	d
Wheat	m. 8	2	17	
Barley	m. 26	5	12	2
Oats	m. 62	11	1	
Pease	m. 35	8	8	6
Cattle & Calves	no. 148	1199	2	3
Sheep & Wool	no. 331	662	8	10
	wool	17	13	8 ¼
Horses	no. 13	286	4	
Corn. John & Joshua Todd's waygoing crop on Denton		532	4	4 ½
Carried to the other side		£14955	17	7 ¼
Denton		45	17	2
Mother & Sister		50		
Interest		865	1	5 ¾
Akeld & Humbleton		168	17	7
Carried to the other side		1129	16	2 ¾

		£	s	d
Wheat	m. 13669 ½	35	12	4
		6009	6	3 ½
Barley	m. 2572 ½	596	9	3
Oats	m. 7228	1121	16	7
Pease	m. 685 ½	145	12	2 ½
Cattle & Butter	no. 151	32	2	4
		3340	5	1
Sheep	no. 1431	8	14	9 ¼
		3859	7	3 ¼
Wool		972	7	
Tups	no. 55	698	6	9
Horses (& mares covered)	no. 40	106	12	
Recd.		£18180	13	1 ¾
Paid		£14955	17	7 ¼
Money made by the Farms this year		£3224	15	6 ¼
Rents for Estates		1338		7 ½
Cash		399	18	
Interest		922	19	5
Total Recd.		5855	13	6 ¾
Total Paid		1129	16	2 ¾
Total of Money made by Messrs Culley this year		4755	17	4
Principal Money paid off & for Estates		£13315		
do. do. recd in again		2850		
do. do. Borrowed		140		

'AN ACCOUNT OF THE NAMES OF DOMESTIC ANIMALS AT DIFFERENT AGES' BY GEORGE CULLEY[1]

The general name of the male in Neat Cattle is Bull. During the time he sucks, he is called a bull-calf, until turned of a year old, when he is called a stirk or yearling bull; then a two, three, or four-year-old bull, until six, when he is aged: – but when castrated or gelt, he is called an ox or stot-calf, until a year old, when he is called a stirk, stot, or year-ling; then a two-year-old steer, and in some places a twinter: – at three, he is called a three-year-old steer; and at four, he first takes the name of ox or bullock, though formerly I believe the castrated male was not called an ox or bullock until six years old, when he is looked upon to be the best, tho' some people think an ox improves until seven, eight, or even nine years old. For I apprehend, the taking the name of ox or bullock at four instead of six years old, has taken place since the drawing or working of oxen has been so much disused.

The general name of the female of this kind is Cow: while sucking the dam she is called a cow-calf, quey-calf, or heifer-calf; then year-ling quey or heifer; then a two-year-old quey or heifer, or twinter; the next year a three-year-old quey or heifer; and when four, she is first called a cow, which name is retained until the last. If castrated or spayed, she is called a spayed or cut heifer, or spayed or cut quey in the North parts of the island.

The general name by which the male sheep are known, is Ram or Tup. When lambs, they are called ram or tup-lambs as long as they suck. From weaning, or taking from the ewes, to the shearing or clip-ping for the first time, they are called hogs, hoggerels, or lamb-hogs. Then they take the name of shearing, shearling, sheerhog, or dinmond tups or rams. After that, according to the years they are clipped or shorn, they are called two-shear, three-shear, and so on, which always takes place from the time of shearing. But when gelt or castrated, they are called wedder-lambs while sucking; then

---

1   Culley, *Observations on Live Stock*, 3rd edn London 1801, pp. xvii–xx.

wedder-hogs, until shorn or clipped, when they take the name of shearings; until they are shorn a second time, when they are called young wedders, or two-shear wedders, or more, according to the time they are clipped or shorn.

The general name by which the female sheep are known, is Ewe: while sucking, they are called ewe-lambs, or gimmer-lambs, but when weaned, or taken from the dams, they are called ewe-hogs or gimmer-hogs, until clipped or shorn for the first time, when they take the name of gimmers, which name continues only one year, until they lose their fleeces a second time, when they obtain as long as they live; only, every time they are shorn, they add a year to their age, and are called two-shear, three-shear, or four-shear ewes, according to the times they have been clipped or shorn: And this hold good of all other sheep; for, the age of sheep is not reckoned from the time they are lambed, but from the time of shearing; for, altho' a sheep is generally 15 or 16 months old when first shorn, yet they are not called shearings until once clipped, which is understood to be the same as one year old.

What we call gimmers in the North, in many of the midland parts of England are called theaves; and when twice shorn, double-theaves. There are other varieties of names, in different parts, which I do not recollect. In some places, they call the male lambs heeders, and the female sheeders; and in others, hogs are called tegs, and two-year-old ewes twinters, and three-years-old, thrinters.

EDITORIAL METHOD

The letters are written on sheets of paper, folio size, folded and sealed with a wafer. In editing, contracted words have been expanded, except for standard abbreviations and the Culleys' own signatures. They habitually wrote, for example, B^r or Bro^r for brother, let^r for letter, Jno for John, Thos for Thomas, Ncstle for Newcastle, all of which makes tiresome reading. Capitalisation has been modernised, spelling has not. Words underlined in the manuscript are printed in italics. Some additional punctuation and paragraphing have been supplied: the Culleys filled their sheets right to the edge, only leaving a space for the wafer, and often did not make a break even when one writer took over from another. A good many of the letters were written in part by one of the brothers, in part by the other. George's son, who worked with his father, also contributed, and there are occasional messages from George's daughter and Matthew's eldest daughter. Where the writer changes, the new author's initials are given in square brackets, thus [*M.C.*]. Place names and personal names are as in the original. Where a spelling may make a name not easily identifiable, the modern spelling is given in a footnote. Names are indexed under the modern spelling.

THE LETTERS

1798

1. *George Culley to John Welch*[1]

Eastfield 11th October 1798

John Welch

I am sorry to put you to the expence of postage as I might have sent this by William Brown junior[2] who set of this morning with some more tups for Yorkshire. But at that time I did not know what I am now going to say to you. My brother and I have it in contemplation to take all or a part of Denton into our hands at May next. In that case we would like very well to fix you there, provided you were not over desirous of high wages. This is a matter, however, that we are far from having determined upon. Nevertheless you may as well not engage yourself at present, because we shall be able at Lukesmas fair[3] to explain this matter further. In the mean time you may give us a line per post, or by William Brown who will be through Chester [le Street] again about Tuesday morning, or else there all Monday night. Don't you name this to anyone at present. I am in haste

Geo Culley

1　Welch was at this time living with his parents at Penshaw. He evidently did bits of stock selling for the Culleys. ZCU 6 contains three letters earlier than this one, including one, dated 24 September 1798, asking to be told what Welch had done to get from John Sharp, a butcher from Washington, payment of £15 8s for eight ewes sent to him the previous February.

2　Son of William Brown, the steward or overseer at Grindon.

3　St Luke's fair, Newcastle upon Tyne, 29 October, one of the largest fairs in the north of England for horses and cattle. The British calendar was harmonised with that of Europe in 1752, 11 days being obliterated from the month of September in that year. As a result fairs whose names reflected a traditional association with a particular saint's day now took place 11 days after the day in question, but the customary name continued in use. Thus, for example, Newcastle Lukemas fair was held on 29 October whilst St Luke's day is 18 October; Wooler St Ninian's fair was held on 27 September, St Ninian's day 16 September. The dates of Northumberland fairs are given in Bailey and Culley, *General View, Northumberland*, pp. 171–4.

2.   *Matthew Culley to John Welch*

Wark 10th November 1798

John Welch

After considering matters over in regard to Denton, wish now to take Mr Richardson's[4] part into my own hands, and if you are not already engaged, to have you manage it. We, George Culley and self, will likely breakfast at the Bush in Chester le Street where you met us, and if you are not busy to have you go and see Denton, provided you are clear of your last employer, if you think right. We probably shall be at Newcastle on Sunday the 18th and on Monday at Chester le Street in our way to Denton to get the writings settled. Mr Richardson has as we hear taken [a farm][5] about 600£ a year at St Helen Auckland, at least we heard he was in treaty for it and expected to take. I am with respect   Your obliged

Mattw Culley

3.   *George and Matthew Culley to John Welch*

Eastfield 26th November 1798

John Welch

I think the very first opportunity you should take a ride to Thornton Hall[6] where Mr Peacock[7] lives, and he will show you the farm that you are to have the management of, when you can then form some idea of the system of cultivation that may best suit it, what horses it will require at first, and how many afterwards. Then I would wish you immediately after to come over here and advise with us how to proceed. Remember that we not only mean to bear all your expences in these excursions, but to allow you wages also. When at Denton, please to enquire if Mr Richardson the present tenant would have a house or houses to spare at May Day. My reason for your being particular in that enquiry is that if an house is or will be at liberty I would recommend William Shotton of Redhouse to you if you have no objection, both because I think very well of Willy as a hind or servant, but because he applied to me in case we took any part of Denton into our hands. Besides when you are at fairs or markets I

---

4   Thomas Richardson, tenant of Denton demesne farm.

5   Two words seem to have been omitted.

6   Thornton Hall, near High Coniscliffe, about 1½ miles from Denton. Nikolaus Pevsner, *The Buildings of England: County Durham,* revised edn London 1983, pp. 474–5.

7   Rev. Thomas Peacock, 1756–1851, perpetual curate of Denton 1780–1835. He was a land agent and acted for the Culleys in business matters. DUL, ASC, Hudleston Papers, Clergy Index.

think Willy (though not the properest for an under steward) perfectly confidential and would learn in time to give directions under your orders. I am aware it is a very long way to remove him, but it is very pleasant to have those people under you that are interested in your welfare, which I am persuaded is the case with Willy. If no house can be had this year for Willy Shotton, perhaps you had better take him into a part of the farm house for one year, as it is a very large house.

As we intend to put all or most of the best ground under that course of crops made use of by my brother at Wark, I think it may be right to plow some out from grass in both farms for oats. You must know that the custom of that country is to *hair* or *free* the pasture grounds for the outcoming tenants at Lady Day[8] and eat the meadow ground untill the 14th of May. But Mr Peacock will explain that to you, and also shew you what fields of grass we have permitted him to eat in lieu of a clover field which he supposes he has a right to plow, and I think he had. However there are fields enough which may be plowed for oats and I think ought to be plowed. I only saw the best or low farm, and have formed some guess which fields may with propriety be plowed in that farm, but I have no idea which can be plowed in the upper farm. As no ground can be plowed for oats until Lady Day perhaps it may be right to plow the fallows yourself from the time you can get begun, and get a plowing day from the neighbors for the oats, a thing very common in that quarter as Mr Peacock will inform you, by giving them a piece of boiled beef and an ancker[9] of ale.

Willy Shotton when you were at Redhouse[10] had 8*s* per week by the year and nothing else. After you left Redhouse I think I gave him either 5*s* 6*d* per week or 6*s* and a cow, I forget which. But then I considered him as having the sole charge at Redhouse. If he would take 8*s* as formerly and no cow I would like it best, but if he must have a cow I should think 5*s* per week with a cow plenty, and what he will rather take I should think, as his cow must be kept I fancy with yours. I am only suggesting these ideas and leave you to hire him as well as you can. The distance to move him is the worst part of the present story. I met with a man working at John Todd's lime killns who calls himself Hill or Hills, who went along with the Midletons

---

8   25 March.

9   Anker, a liquid measure of varying size used in northern Europe. A Rotterdam anker, also used in England, was equivalent to 10 old wine gallons or 8½ imperial gallons.

10 See p. xx, n. 13. Welch had worked for the Culleys at Red House until 1798.

when they left the country, and who is now married and lives in Denton I believe, but Mr Peacock will make him out for you. My reason for naming him is for two reasons. First he is a young clever-like fellow, but my principal reason is that you will want someone who understands *drilling*[11] or *ridging* for turnips. Now I do not know that this Hill can ridge, but I think it very probable that he can as he belonged this country and went with or followed the Midletons. At any rate you will want a man of that description, and if Hills be master of it it will save us sending a young fellow from this country provided you can hire him. And if on enquiry you find him competent to the task, I think you should not miss him for a little advance in wages, and you should not neglect to hire him when you are over, provided the man is at liberty and willing to engage with you.

Geo Culley

[**M.C.**] Well John Welch. I think when you buy any draft horses, buy them from 15 to 15½ hands high, usefull strong bodied ones with good legs, especially good forelegs. If they go widish before, let them be inhoughed,[12] not stradling wide behind but in third as the Scotch horses are. We shall want 4 against or at Candlemas.[13] We have a black mare here, may suit to ride a little and may help to harrow. In buying a guinea or 2 is ill spared when a horse or mare pleases.

Mattw Culley

[**G.C.**] John the black mare my brother means of is the one that William Shotton had, and Ralph Brown[14] has rode her lately. We think she may do for a hackney for you and harrow occasionally. We are much at a loss John to know what we are to do with our fat ewes this spring. I fancy we have 600 or upwards, and some of them the best we ever had. Now the misery is that we dare not present the best ones to sale at Morpeth, because the butchers will be selling them again as they are so very tempting. Perhaps there may be between 2 and 300 of these best ones. The rest may be sold at Morpeth at what they will bring, but we have nobody to sell them except myself, who am rather unwilling and unwieldy grown, Ralph Brown is unable, and Matty[15]

---

11 Sowing seed into a groove on a ridge of soil by means of a machine that made holes and covered the seed when sown.
12 Turned in, knock kneed.
13 2 February.
14 Ralph Brown, steward or overseer at Eastfield, did most of the grain selling for all the Culleys' Northumberland farms.
15 George Culley's son Matthew.

is yet too young. Could you not meet the sheep there that may be sold between Christmas and Candlemas? But we can talk this over when you come here. Matty could meet you some times. I have now wrote you a very long letter, yet may have omitted many things that I ought to have wrote about. However if anything occur to you that I may have neglected, be so kind as name them in your answer, which as soon as convenient will be very agreeable to your sincere friend

Geo Culley

PS. I am sorry to tell you that George Stamford has turned out very badly, got into debt to every house in the public way between this and Morpeth, and the people are coming upon us for payment. He also borrowed 8£ from me, and has little wages to take.

1799

4. *George and Matthew Culley to John Welch*

Eastfield 9th January 1799

John Welch

We have sold Mr Lincoln[1] 140 ewes and 6 cattle, and I have told him that if you need money he is to let you have any reasonable sum occasionally. Only do you write to me at the time. My brother also paid 40£ into Surtees's Newcastle bank[2] for you to use if you need. We understand that you got 20£ from that bank. I think you bought John Sharp's horse cheap if he is usefull. We shall not send the ploughs until fresh weather without you write to us for them, because it is of no use to go there to eat hay at a high expence in a storm, and I am rather affraid of the frost continuing too long. Do let us hear from you sometimes, which will oblige your sincere friend

Geo Culley

[*M.C.*] PS. I paid or sent 40£ into Surtees's Bank on Saturday last. M.C.

---

1   William Lincoln, a Sunderland butcher.

2   Surtees and Burdon's Bank, Newcastle upon Tyne, founded 1768: Maberley Phillips, *A History of Banks, Bankers and Banking in Northumberland, Durham and North Yorkshire*, London 1894, pp. 385–400.

5. *George Culley to John Welch*

Eastfield 11th January 1799

John Welch

Harry Turnbull arrived here last night with the three horses, and I will go with him today to Wark to get the ploughs and carts &c got ready to set of with on Monday morning. I will send this letter per post, that John Sharp may not be kept in suspence about his turnips, the offer of which we are much obliged to both you and him for. But we have now sold so many of our best ewes that I apprehend we have no occasion for his or any turnips. I wrote to you by William Lincoln of Sunderland, which letter you may likely receive today if there be anyone from your village[3] at Sunderland. We sold Mr Lincoln 140 ewes and Mr Paul[4] 120, which will now make us very easy for some time. We also sold Mr L. 6 cattle. We will get this young man set away so early on Monday with his two carts that I hope he will be enabled to reach you or rather Chester on Tuesday evening, where I would have you to meet him to direct him the road to Denton, as it would be wrong not to forward the ploughs and horses as fast as possible. I think and hope you have bought the horses with money. I will take care to pay the 30£ to Berwick bank tomorrow, and get Mr Embleton[5] to write to Surtees's people of it, which with the 40£ paid Surtees's people by your master Matthew, which I have mentioned in the letter to you by Mr Lincoln, makes the 70£ up which you have got at Surtees's Bank. You will also see in the letter by Mr Lincoln that you can in future get what money you may want from Mr Lincoln. By unfortunately putting the wafer upon the name of the person you had sent the ring from Mr Langland's[6] we can't make out the name, but I hope there is no fear but it will come safe to hand. However I trust it will in future make you remember to leave a blank for the wafer or seal. You also fold your letters too long. Cannot you see how I fold mine?

Friday afternoon. Harry Turnbull and I have been at Wark and Thornington this forenoon. When we got to Wark my brother and sister were gone to Thornington where we followed them. I left them there and came home. You get the carts and ploughs from Wark. The

---

3   Penshaw.

4   Thomas Paul, a Shields butcher

5   John Embleton, partner in the Berwick upon Tweed branch of Surtees and Burdon's Bank.

6   John Langland, jeweller and silversmith, Newcastle upon Tyne: *The Newcastle and Gateshead Directory for 1795*, Newcastle upon Tyne 1795.

carts are not quite new but the freshest they have. As he will be lightly loaded I have advised my brother to let him take a pair of harrows for you. I advert to what you say respecting William Shotton and his wife. I think very highly of Willy, but he is led by the nose of his wife, and even if she was to consent to go to Denton it is ten to one but she would repent in less than a year. So I give up all thoughts of William Shotton at present. My brother thinks the same, and both of us approve of your hiring this Harry Turnbull if he can be hired in reason. I have been talking to him but he talks wildly 16£ or guineas and board. If he would take some way between 12 and 15£ it might do. You best know and your wife what you could board him for, but we will certainly pay you willingly what ever you charge in reason. He is of an honest stock which is a valuable matter, and I suppose he is or ought to be conversant with the manual parts of husbandry. But you should know him better than us, and I think a pound or two ill saved between a honest and clever man and the contrary. But if you agree, as I hope you will, he should be boarded with some decent people in Denton till you and Dolly[7] go, if such boarding can be met with, because I think clearly that he should go and plow for one at present, as I have no doubt but he understands plowing well, as that is their fort in Cleaveland. If this can be accomplished you still must have a confidential man to live at the High House, sometimes called Lime Kiln Bank house, because that will be a situation of some trust and would have suited William Shotton well and he it. But I have no idea of taking a petted wife there. Consequently you must set your *wits a steep* and consult Mr Peacock &c to find out a faithfull servant for that place. I think I have said all I can to you at present. Be sure you write as often as convenient, which will always be agreeable to your well wisher

Geo Culley

PS. I would not have you get any more money from Surtees's Bank as William Lincoln will now supply you, and after you get to Denton you are to apply to Mr Peacock untill it please God that you can stand upon your own legs. I suppose you will go along with the horses and ploughs &c. and set them a-going. I never heard if you had engaged any man or men to plow &c. Mr Peacock thought, and my brother and I agreed to it as a thing customary in that country, to get a few neighbors' ploughs to help some day to plow your fallows, and give them a piece of beef and an ancker of ale. You speak to Mr Peacock,

---

7 John Welch's wife Dorothy Fairbairn, married 1 Jun. 1795: DRO, Parish Registers, Penshaw.

not only about the propriety of it but the most suitable time for you and neighbors. When you want this black mare perhaps it might be best for you to come here in the coach and take her back. Or we could send her to Morpeth &c, &c. But you will think of that, and about hay, oats &c, &c, &c.

6. *George Culley to John Welch*

Eastfield 27th January 1799

John Welch

I was wishing very much indeed to hear from you, and this moment received yours of 22nd. Five days is a long time, some delay must have taken place before it had got to the post. We all sincerely wish your wife a happy recovery.[8] I am glad that you have hired Harry, though a great wage. A pound or 2 are ill saved in those cases where you know your man. My reason for writing you so early an answer is owing to a letter received yesterday from Mr Thomas Bates,[9] wherein he says that there are 2 powerfull horses to be sold in their neighborhood for which they ask 20 guineas each I think, but my brother has the letter. One 4 rising, the other 5 rising. The first is perhaps too young, but he may do to plow. Now I think, please God Dolly be better, that you should come by Mr Bates's to this place, and you and he buy those horses, one or both, provided you approve of them. Mr Bates will get a person to take them to Denton, and John Todd[10] must be wrote to by you to get a man to go with them and also to procure any grathing[11] wanted until you can get over again. But always remember that I only advise, I don't by any means insist upon your buying those horses in Mr Bates's neighborhood. Only Mr B. has often bought us some very usefull horses in that quarter, and horses are generally better bought amongst neighbors than in the fairs. That is, a man will not sell a neighbor a horse with a glaring fault without he be a very bad man indeed. But at any rate it might be right to see them draw, as a misteatched horse[12] would be a very great disappointment to you. It is very well that John Todd is so friendly to you. Mr Peacock is going to London tomorrow I believe, but how long absent I know not. However you can have money from Mr Lincoln,

---

8   The Welchs' daughter Dorothy was born on 19 January 1799: DUL, BT, Denton.
9   Thomas Bates 1760–1830, brother of Matthew Culley's wife Elizabeth, farmed at Halton and then at Brunton near Chollerford. For the Bates family see *History of Northumberland*, vol. 10, pp. 377–8.
10 Tenant of Denton East farm.
11 Graithing, preparation.
12 One with bad habits.

and if you should buy the horses near Mr Bates he will pay for them. But I had forgot to repeat that you are not to put too much confidence in John Todd untill you know more of him. We neglected to send a few tares[13] by Harry, but you should make or get somebody to make enquiry for a few to sow for giving your horses green, as you will have no clover. It is not very material, but I should rather have been for plowing the grass land for oats and left the fallow for the borrowed ploughs, because bad plowing in fallow is not so bad as bad plowing a lea or grassland. I fancy the man Hills had called here a day or so after Harry was gone, but I was not at home. I hope you bought the knotted chains, as they are the best by a deal in every respect. The ring has never comed from Mr Langlands. He has behaved very comically in this business and has injured our opinion of him. You may call or not as you chuse if you come through Newcastle, but don't you go that way on purpose by any means, because if it never comes to hand it is not deadly, we can get one from Robertson or others. I think your best way and nearest to Mr T. Bates's will be up the south side of the Tyne. There is a turnpike road goes of at Gateshead for Hexham, and you cross at Corbridge for Mr Bates's. Indeed we shall want the pound of mustard, but that is in Gateshead. I cannot say that I am at all inclined to persuade William Shotton to come against his wife's inclination, and the distance would make it exceedingly expensive. But we can talk about that at meeting, and am in interim Yours sincerely

Geo Culley

Mr Lincoln's man was to come today for the rest of the sheep and cattle. If you meet with him, tell him that I shall be very much displeased if he don't send according to promise. I am quite of opinion that the meadow moor should have been the first plowed, because from its toughness and nature of the soil, it will require some time to tender and meliorate.

Agreed with M. Hills as hind for Messrs Culley at Denton from 16th February 1799 to the 12th May, to have 18*d* pence per [?day][14] and coals led. Also from the 12th May 1799 to the 12th May 1800 to have ten shillings a week for one quarter of a year and nine shillings for the other three quarters, house rent paid, coals led, and ground to set one bushell of potatoes.

---

13 Vetch.
14 A word has been omitted. A day labourer's wage was between 1*s* 6*d* and 2*s*.

7. *George Culley to John Welch*

Eastfield 2 March 1799

John Welch

I have been thinking since you left us that you will be too long of applying for and getting your clover seeds to sow, because the season will advance very fast, and as you will not be allowed to harrow I suppose, but only roll, they will have a very bad chance in that dry soil at Denton. You should therefore endeavor by every fair means to prevail upon Mr Richardson to allow you to sow your seeds as soon as possible. And by all means, as soon as you have made your estimate what seeds you will want, get Mr Peacock to write for them without delay, because it may be some weeks I am affraid before they come in at Stockton. Remember it is not like our Berwick port, where ships are coming from London every week. Now as red clover seeds are the cheapest of any other seeds, and will agree well with the generallity of the Denton soil, I would not have you pinch them on any account. I should think not less than 8 or 10 lb. per acre. Remember that that soil will not take so readily with seeds as our lighter soils in this country. Respecting the white clover or hop or yellow, it entirely depends on the time the land is to lie in grass. Either yellow or white will thrive very well upon the limestone soils, but particularly the yellow. But as it is a larger seed than the white, will require more per acre. We generally sow 2 of white and 3 or 4 of yellow per acre along with the red, but I leave it to your own discretion to sow what seemeth best to you. It may also be right to consult any of the farmers in that district as they ought to know best. One measure per acre of clean rye grass is plenty per acre, or even less if the land is not very old plowing. Let me hear from you, which will oblige Yours truly

Geo Culley

8. *George Culley to John Welch*

Eastfield 15th March 1799

Well John

Your large clever letter[15] I received safe per Hopper, and have only to observe that less paper wrote on 3 sides would take up less room and contain as much matter, in the manner I am writing this.

---

15 In this letter of 10 March (ZCU 6) Welch reported on his progress with ploughing, and a neighbour's recommendation of paring and burning (see n. 18) for a boggy field. Welch had been getting clover seed and oats, and asked for a horse.

However have no objection to long letters, and would recommend to you to have one upon *the stocks*, as my poor brother James used to call a letter wrote on different days as things occur to you, and so not sent until filled up. By that means you are able to recollect better, and give an account of things as they come in course, like a journal. This I would recommend to you when you come to live at Denton, except anything urgent is wanted and then wrote by the first post always without paying any regard to expence of postage &c, &c. This is the way I always do when I write to my friends when from home for a few days or weeks. I write day after day anything worth communicating, and when full send away and begin another. Hopper brought me 2 more letters besides yours, one from Mr G. Holmes for 6 sheep, which I sent, the other from J. Sharp for 6 sheep and more oats. I answered both letters but excused myself to Sharp. Indeed it appears strange to me that he could request another favour after his unpleasant behavior last year.[16] You must certainly be content to do as well as you can with your clover seeds sowing this year, only by civil words perhaps you may be able to prevail upon Mr R. sooner. Desire Mr Peacock to talk to him. In future you must get money from Mr Peacock if you please, as I shall draw upon Mr Lincoln for the ballance in a few days, having many heavy payments in rents &c, &c to make in a few days. Did he say to you that he was coming out soon again to buy anything of us? We will take care to send you a horse to Morpeth according to your request, by Wednesday come a week the 27th. I shall be at Morpeth that week myself please God with a very good lot of sheep. If George Nicholson[17] of your town want any, he could take home the horse for you. Tell George will you that I shall have a very nice lot. You may also tell Hopper or any other of your acquaintance. You are certainly right in buying oats for both seed and horses. My brother is not very fond of paring and burning,[18] but on my talking to him of its being so much mossed he assented to pare it.

---

16 See No. 1, n. 1.

17 George Nicholson, a Penshaw butcher. He married Welch's sister Mary in 1800: DUL, BT, Penshaw.

18 A method of converting grass land for ploughing, by slicing off the top layer of turf and burning the sods on the ground. The practice was long established in Co. Durham, less used in Northumberland, where some people said that it ruined the soil. In their reports to the Board of Agriculture John Bailey and George Culley described the method as convenient for breaking up old sward and coarse rushy ground, but said that careful cultivation afterwards was essential: Bailey and Culley, *General View, Northumberland*, pp. 128–30; Bailey, *General View, Durham*, pp. 202–3.

Therefore do which way is most agreeable to your judgement. It will be almost safe to grow you a good crop of turnips provided you can get it burnt in good time. But you are to consider whether you will be able to manage it, as you will have a great deal of fallow next summer, and to secure a good and certain crop of turnips upon burnt land it should be plowed twice over and well harrowed, and plowed rather deeper than the generallity of people plow burnt land, at any rate if only once plowed. If twice, plow deeper the second time after being well harrowed. This is the best way I know of but Mr Christopher Midleton[19] will advise you, who I am disposed to think well of as a good farmer. I am very happy to hear that you have got so well forward with plowing, and that the horses prove pretty well. I should think the clover in the sheep close would grow a good crop of *Dutch* or *Friezeland oats* or Hollands, whichever name they go by in your country. But not Poland oats as they require land in particular good heart. Indeed I should have been for trying 3 or 4 acres of spring wheat in that said clover land, but before you receive this it will be too late. It appears to me a good plan for your plowing day the way you propose, and very right to get the ale from Lamb at the Bait House. Remember to thank the people who assist you in plowing in our name and with our compliments, and be sure you make a return in plowing &c to them when an opportunity offers. We intend to send you a horse or mare that has worked, and it will be at George Fenwick's on Tuesday evening come a week the 26th inst. if all be well. We received the pound of mustard by Robert Dickinson[20] and much in need were we of it. Could take another pound as well by and bye. We had far too many both cattle and sheep in Morpeth last Wednesday. Still I think we sold good sheep above 6*d* per lb. sink,[21] although many midling sheep were left unsold and several cattle. If you have an opportunity of writing by the person who comes to Morpeth for the horse do, as I hope to be up that week, and am in the mean time. Yours in haste

Geo Culley

PS. Mind and don't pinch turnip seed in the burnt land. You can't drill it well but you can and must hoe it as well as you can. Indeed nothing is so unwise as to pinch seed in any land. Should we not find

---

19 Christopher Middleton had farmed in Northumberland (see letter No. 3), now
     at Summerhouse near Denton.
20 Carpenter at Wark.
21 Sink: without the offal, i.e. the skin and the inferior parts of the carcase.

a time to send you some turnip seed some how or other, and how much?

9. *George Culley to John Welch*

Eastfield 3 April 1799

John Welch

We received yours of the 26th in due course, and my wife sent away the things on Monday last the 25th (*directed to you and to the care of Mr West, White Hart Chester le Street*) therefore I hope they will be at Painshaw before you receive this, which I will send by your master Matthew who goes for Denton or rather Thornton on Monday or Tuesday next. I am very glad that you think yourselves able to do your plowing without the help of your neighbors, then there will be no obligation. I hope my brother will be able to prevail on Mr Richardson to allow you to sow the clover and &c seeds directly. If not I think he will use us badly. I never knew clover hurt wheat much, and this strange backward year there can be very little hazard. Corn is advancing here a good deal, and we have all the Yorkshire jobbers out buying all the fat before them, so that I hope the few we have left will be pretty well sold. How do you do for cash? Be sure you ask Mr Peacock when you want, and place it to account. I have no doubt but Mr Peacock is a little delicate in interfering with Mr Richardson about sowing the clover seeds. I don't know who this Bulmer[22] is, I suppose a jobber to Yorkshire. What a strange spring. Such losses in sheep and lambs in all parts of this Island as never was known before I believe, now or in our fathers days. I am in haste

Geo Culley

PS. The stock farmers about Pennistone a high country between Yorkshire and Derbyshire have lost a most all their ewes as well as the lambs, and Mr William Watson of North Midleton said yesterday at the public dinner at Wooler that he would give any man £250 to replace the stock of sheep he had lost this spring. A dog of the greyhound kind has and still worries numbers of sheep. He killed us a wether.[23]

PS. Your mistress desires me to say that she has sent along with half a dozen shirts, two cotton waistcoats which she wishes the boy[24] to wear at nights instead of his shirts.

---

22 A livestock dealer.

23 A castrated male sheep that has been shorn at least twice. For the names of sheep and cattle at different ages, see pp. xlii–xliii above.

24 George Thompson 1795–1804, a young boy, illegitimate son of one Ann

10.   *Matthew Culley and Matthew Culley jr to John Welch*

Eastfield 20th May 1799

John Welch

I got home on Friday night from Mr Bates's of Halton. On thinking over the water courses, a mason's level like an A 10 foot long or 12 foot long with a lead ball or plumb at it will level very well, by knocking down stobs a few inches long and so leveling along the stobs with stakes with inches on them. I think you should level from the hedge below the Horse Pasture to the old mill race where Mr Richardson should have the hay which will give the (inches), but you will hardly do anything now but cut out the old water course over the carrs[25] and so up to the conduit bridge as you go to the High House. But you should cut the old drains out the first thing. As I possibly may come over when you can no longer do without my help, I now think but am not certain that the water across the west carr should not come to the quick hedge but between that and the old drain or mill race until it joins Joseph Hodgson's[26] planting, as it is wrong in the nature of things to force nature. I hope you will cut (stub) the oak in the state for peeling. The bark will pay for peeling if you can get it off without hard labor. There may be people at Gainford &c who understand how to peel oak. We have a cow we can spare, an excellent milker, not very young. If you thought right for her to come with the young stock for your purpose, may calve in a week or 10 days. I hope you got your wife and your furniture safe to Denton, and also that you got the money matters properly done at Sunderland. Give my respects to Mr Peacock, and you must write from Darlington the Great Monday's[27] market how sheep and cattle sell of all sorts. But you must have a letter half wrote before Monday to be ready to finish there and seal and send off here immediately. Your sincere friend

Matthw Culley

[*M.C. jr*] PS. We have sent you ½ a cwt. of turnip seed, if there is not enough you must buy some.

[*M.C.*] If you think ½ cwt. turnip seed too little for your land, as 2 lb. per acre will be little enough on some [of] it, you may say how

---

Thompson, was being fostered by Welch's parents at Mrs Hannah Culley's expense. The reason for the Culleys taking responsibility for the child is not apparent. He died on 7 July 1804. See No. 238.

25 Flat marshy ground.

26 A neighbouring farmer at Denton.

27 The Monday a fortnight after Whit Monday. The dates of fairs in Co. Durham are given in Bailey, *General View, Durham*, pp. 279–80.

mUch you think you may want in your letter from Darlington. John you may let [. . .]28 Dickinson, brother to Robin Dickinson, know that Robin will take his son on such terms as neither to be injured. That is he will let him have meat and board &c not to hurt either. Robin wants no advantage as he is his brother's son, so may if he chuses come north as whenever he pleases. Robin is capable of finishing him in the line of wheels, plows, carts &c in the farmers line of things, nay he also is exceeding clever at roofing houses. We always frame the roofs of our houses, put 2 or generally 3 ribs on a side, especially when the side is rather long or has a long run.

11.   *George Culley to John Welch*

Eastfield 9th June 1799

John

Jemmy Glass the Thornington herd sets of with 20 young cattle for you on Wednesday morning next, and according to our calculations if all things go right with him, I hope he will be with you about Tuesday the 17th. However as he is a stranger it may be right for you to take a ride, or else send a person to meet him by Legs Cross29 and West Auckland &c as he goes the moor road. And I would wish him to return immediately to meet a lot of wether hogs30 which will follow the young cattle directly. And he will then be acquainted with the road and will best know where to get up putting for the hogs. I had a letter from Matty, who gives a very good account of your proceedings, but said never a word about the cattle or sheep going to you. However as we have had some very growing warm weather lately, and heavy stocked here, we have thought right to send them away. Matty would tell you of our sales of fat cattle and sheep. Things are turning very scarce now. No advance lately in the corn, rather the contrary. We have not begun to sow any turnips yet, but shall in a day or 2. They have sown about 20 acres of rape at Wark, about Jock Ridge and the Leamingside, to eat of early and sow with wheat. I am just going to Mr Robertson's tup shew and thence to Berwick to attend the division of Tweedmouth Common, and am yours

Geo Culley

PS. Shall be glad of a line when anything is new.

---

28 There is a blank in the text.
29 A 9th-century cross shaft, almost certainly a boundary cross, standing beside the Roman road 4 miles north of Piercebridge, about ½ mile from Denton cross roads: Pevsner, *Buildings of England: Co. Durham*, p. 111.
30 Castrated male sheep between weaning and first shearing.

12. *Matthew Culley to John Welch*

Wark 28th June 1799

John Welch

Yesterday received your letter by Glass. Am glad the stock got well up to you, and Tom safe back as we are clipping. [You][31] do not say one word how wool was selling at Darlington. I cannot be at Denton before Newcastle fair,[32] at least the assizes must be over as we fully expect to have a tryal for a water course through the south part of Ewart. I hope you got well forward with your turnips. We shall be very late with ours (and cannot supply all our customers with turnip seed), are likely to fall short in our own sowing.[33] We are in want of rain in general, yet our crops look well on our farms at present or mostly so, and our grass in general good within these 3 weeks. At Yetholm[34] yesterday sheep and cattle very flatt, no south country men, from Burroughbridge a very poor account, some sales from the neighbours one with another, many wool buyers. The account from Wakefield is bad, a falling market, Morpeth overstocked on Wednesday last. Do not fail to write here on all material matters especially that of fairs, prices of cattle and of grain &c. Mr Bowser and Matty Culley are going to Scotland may tell Mr Peacock, who you will send a letter by next week. Give our respects to them and let us know how the young man is. Do write once a fortnight of every material occurrence. When you send away one, begin another letter to fill up as things occur. The mill where the wheel scatters the water in going round should be Eden[35] or at least freestone so farr up as the circumference of the water wheel. Do not carry on your work too fast until I be [with][36] you. After the turnips are sown you'l have more time to get home wood or the leading off any other articles you need to the mill &c. Yet the more things of that sort are got the better before short days and wets come on, also before hay and harvest and such

---

31 A word seems to be missing here.

32 12 August.

33 Matthew Culley seems occasionally not to have ordered his thoughts before writing a sentence. This one, presumably, means: 'We shall be very late with sowing our turnips and are likely to fall short of seed; we cannot supply all our customers.'

34 There were two fairs at Yetholm at the end of June or early in July, one at Kirk Yetholm for young and old sheep, the other at Town Yetholm for lambs: Douglas, *General View of the Agriculture of the Counties of Roxburgh and Selkirk*, p. 207.

35 Presumably from the vale of Eden, Cumbria.

36 A word seems to have been missed out here.

busy times. I approve of your only opening the water course to the mill as they can be done after the harvest, when you will see better how to do them more fully and have more labourers. Do not put Joseph Hodgson to the plowing of the grass carrs, a little cut will bring the water forward, lay the earth northward in the grass carrs as much as you can, what is good. Let me know how many work horses you enter at Denton, with 1 riding horse, and how many idle ones under the Act,[37] as they require to know here. Also numbers of dogs. I suppose a curr dog then not taxable. There were many wool buyers at Yetholm especially for short or fine wool. More advance is expected on that than on long. The latter sold (some parcels to the north)[38] at 17s 6d per 24 lb. I have some thoughts of Cumberland slate for a cover, a granary over the mill and if convenient over the dressing floor &c, as much granary as we can have occasion for, the corn to be drawn up by a pulley in bags, then barrowed to any part of the granary. Perhaps a stable may be made on a future day in a convenient place. Nelly Culley[39] in her letter from Ilkley says she heard Mr Place say her father had got the better of him in selling the sheep. The market they say was very full they say at Wakefield. The jobbers had run through and collected sheep and cattle whatever they could get, so may have had too many for market. Besides when any article gets so very dear the money falls short with the lower classes, the consumption is then very small, also people of economy use bacon &c with any one thing they can procure to satisfy hunger. Wishing you and family good health I am &c

Matthw Culley

Respects to Mr Peacock and family, also to Mrs Todd and John.

John Eshelby of High Coniscliff, a person much to be depended upon for honesty &c and knowledge in millwork, recommended to me by his old master Mr John Jackson of Coundon Grange, wrote to me that I was welcome to his advice, says what he knows is in the construction of the water wheel, the first mover, does not pretend to understand the inside work, at least to pretend to much knowledge of the inside work, but is very willing to shew my engineer all that he knows. You may call on him with my very best respects and say I take

---

37 Riding and carriage horses had been taxed since 1784. Work horses were added under an Act of 1796, as were dogs other than single pets: Stephen Dowell, *A History of Taxation and Taxes in England*, 2nd edn London 1888, vol. 3, pp. 226, 263.

38 I.e. some batches of wool sold to Scottish buyers.

39 George Culley's daughter Eleanor.

his letter a very great favour and have ordered you to call on him, and hear and learn of him, also when Robson[40] come he may if he choose call on him if convenient and hear what he has to say. Also says he understands the method of giving water to the wheel. M.C.

By letter from Tommy Hutchison of Bildershaw our cousin (and who if his father had not been an indolent man) might have left his son independent, wants the hay which grows in the rims of your corn land, which I hope will not be much in a future day. May I think let him have any little waste places or hedge banks. I do not much approve of it, however as he is a low situation may let him have any place you judge proper for him to shear or mow. Perhaps he may have the power to let you have dung for the grass for hay. Do write often. You should have sent word how wool sells at Darlington this season.

Dear Sir. With due submission, humbly beg the favor that you will let me have the grass that growes in the rims of your corn fields at Denton. Hay is very scarce and has the appearance to be worse, and if you please to grant me that favor I will pay you your demand for it &c. Thomas Hutchison Bildershaw, 22nd June 1799. Copy of T.H.'s letter

13. *Matthew and George Culley to John Welch*

Wark 11th July 1799

John Welch

I received your letter by Mr Peacock, and understand that Mr Robertson approves of your alteration of his plan. It will roof better and may answer as well in other respects. I think we had better get the Cumberland slate, as according to Mr Peacock's account it will not be any great expence more than tyles. Therefore you with his direction and assistance may apply and get the slates bespoke and ordered by a proper workman. I think he says they are brought to Bowes at 3£ 10s per ton. Perhaps you may hire the leading, or when the turnip season is over get them led with your own carts. Horses are now cheaper than they were, but be carefull when you buy any fresh ones, good ones are scarce. I thought her long legged and shabby, no horse of any sort at all, but our George and you are for great horses. I am for thick usefull ones with a good body and short hams. Long hinder legs must in general tire soon. A little boy at Thornington overturned the cart and killed a big horse for want of having the

---

40 In subsequent letters the name of the builder of the mill is written variously as Robson and Robertson.

roads mended and levelled going down the hill to the burn near the beck where the road from the house runs into the road to Howtil. The mare sent from Shotton is not a good one, was only to work occasionally as a helping one, of our own breed. They that breed will not have them all good, many indifferent ones, and was too young for hard work. I suppose you will not cross Stubs's Carr until I get over as I apprehend we should make a new cut after the season answers, between the hedges and the old race, and plant quicks[41] on the north or south side of it. North perhaps rather may be best. Haws will probably be exceeding plentifull, therefore should gather several sacks of them and sow them very thick on good garden soil, as many of the hedges will be to plant anew, with quicks before we can get our own to plant. If any crabs[42] can be got should have some gathered also and the pips sown, they soon are fit to plant, and outgrow the white thorn generally on good land.

12th Mr Peacock is here. We spoke to R. Dickinson who will come to you in 14 days time in order to roof the mill for you. He may also make you a plow or 2 if you had a maundrill for locks and a few mouldboards from Newcastle got to Denton. You have beams, only may want stilts and shutters &c. R.D. makes an excellent plow, only keep him in tune. His nephew seems seems [*sic*] to be a clever lad. Robson allows him to be clever. I find Mr Peacock and you are on good terms. He is an exceeding clever man in business and is to supply you with money to carry on the farm, says you carry on business well he thinks. As to any stock you may want, must let us know by letter. I remain with regards your obliged friend

Matthw Culley

Respects to your wife and health to the child.

Now on considering things over, in regard to the machine we think it should be got forward, but should not Robson be agreed with by the great,[43] or can you keep his men at their work? Robson has too many irons in the fire to attend to all the work he has. This you must consider well against I come over. In the mean time if any workmen are set to work see they do not idle away their time. It is not likely that Muckle will renew the suit although he desires a reward for what he has done for the machinery. M.C.

[G.C.] John I have given Mr Peacock orders to pay you for George Thompson as I seldom can see you now at so great a distance, and he

---

41 Thorn plants for making hedges.
42 Crab apples.
43 For the task or complete work.

will keep a separate account of it which I can settle with him annually when he comes over here at this season. I much approve of my brother's advice to you about buying short legged firm *little great* horses. I am certain that there is nothing more absurd than buying tall loose long legged horses. Wheat has got a considerable advance these two weeks at London, about 10s per quarter. Ralph sold at 50s our bole[44] the last Saturday. You need not take any notice of it, but we shall most likely take my brother's estate into our own hands, paying my brother a certain rent for it. In regard to Joseph Hodgson's farm, it is an afterthought. At any rate we shall hardly take it. At the same time we think of taking Todd's. Mr Peacock will supply you with cash untill you are able to support yourself. Be sure to set down the sums received from him. I am in haste Yours truly

Geo Culley

14.   *George Culley to John Welch*

Eastfield 13th August 1799

John Welch

On my return home I was sorry to find the enclosed bill returned unpaid. My opinion is that old Fishwick is unwilling to pay his son's debts, otherwise I believe old Fishwick is able enough. My fear is that Salkeld is very poor, and how you are to act is the question. Because if his credit is blown it may set all his creditors upon him. Therefore you must act with caution and delicacy. *Suppose you consult Mr Fenwick,* saying you are accountable to your late masters for the above, and take his advice. I name this for your information and consideration and then leave you to act to the best of your knowledge. I should think if Salkeld have effects it would be best to strike directly. If not you must wait and take your chance. At all events I beg you not to make yourself uneasy, because your health is of more consequence to your family than all the money. Let me hear from you as soon as you can, and believe me your friend and well wisher

Geo Culley

---

44 In *General View, Northumberland,* pp. 180–2, Bailey and George Culley described weights and measures in Northumberland as in a 'sad state of confusion'. At Wooler a boll was equivalent to 6 bushels of all grains; at Alnwick it was 6 bushels of barley and oats but only 2 of wheat. The Culleys seem to use the Wooler measure consistently. From March 1772 farmers and landowners in north Northumberland, especially the Wooler area, undertook to sell only by the boll of 6 Winchester bushels: *Newcastle Chronicle,* 22 Feb., 29 Feb. 1772; *Newcastle Courant,* 22 Feb.

15. *Matthew Culley to John Welch*

Wark 20th August 1799

John Welch

Your mistress and the children[45] met me at Newcastle and on the fair day we took Matty to Houghton le Spring where Mr Fleming brought a young man, son of a Mrs Kirsop of Hexam, she was a most particular acquaintance of my wife's, who was to be Matty's friend. We went next morning to Sunderland or rather Wearmouth Bridge, so to Newcastle. On Wednesday we met Jack Davidson[46] with 2 single horse carts on his way to Yorkshire near Coxwold with 3 tups and 2 ewes for Messrs Taylor[47] and Key, which bargain Matty had made when we were at Newcastle. Fatt cattle 6s per stone. Many lean ones sold, not so high as they have been. Mr William Bates[48] bought 10 of the Browns of Heddon near Stamfordham west of Newcastle at 11£ 15 per, good neat fair oxen. There has been much rain here, by far more than with you. Our corn much laid, especially oats, the Church oats laid flatt and grown up with greens very sore indeed. Wheat sore twisted and laid. Saturday as wet a day as could be for the season, Sunday wet, same on Monday forenoon very wet afternoon. Rivers and brooks all full and deluged with water. Hay sore damaged, we have not one cart of hay in yet. Turnips bad except on dry soils. At Thornington today, good pastures. Bad indeed the turnips, so wet they cannot grow. Your turnips will hardly make up for our deficiencies. George Culley gone for Morpeth with a few ewes and close tups, sold some ewes there last week at 51 to 48. Ralph sold oats about 26s, wheat 51 per boll. I think you should try to get cart and wheels at Cockerton Bridge end. He is, John Dickinson says, a very compleat workman and will make the cart to your liking. We have got ½ a ship load of oats, some oats also from Felton, Mr Nesbit[49] gets the other half, good but very high priced. We will send you the bore rods, they are too small in my augur holes. We shall have to send you some small spayed heifers for feeding, and perhaps if you could sell your shearings[50] now at Denton we might send you some better to feed, of

---

45 Matthew Culley's children Eleanor (Nelly) b. 1785, Matthew (Matty) b. 1786, Elizabeth (Bessy) b. 1788, Jane b. 1795.

46 Hind at Wark.

47 John Taylor of Salton.

48 William Bates of Clarewood, 1738–1812, first cousin of Thomas Bates. He was described by the *Newcastle Advertiser*, 23 May 1812, as a 'man of great agricultural knowledge'.

49 John Nesbit of Ancroft, 1746–1808, brother of George Culley's wife Hannah.

50 Sheep that have been shorn once.

our own full breed. These are 2 or 3 of the queys[51] which were sent to you for incalves,[52] are of our own cows. Perhaps it may be right to get them bulled by a bull of either Robert or Charles Colling[53] provided they were spoken to and could take a reasonable price for their bulling. I cannot describe them, they were sent by mistake. You can ask Charles how much he will take for their bulling. As our old bull is dead we have not one to spare to send you. If we keep any of them for cows we will pay Mr Charles Colling a better price for the bulling of such as we keep, if sold for incalves or butcher hope he will take a common price. We have had such floods as would have washed away all the earth at the head of Mally Ward's[54] garth. Have you cut the water course through the planting and tenement? I think it should cross the north or Houghton [le Side] road not far south of the gate which leads into the glebe, wherabout the lane begins to narrow. Wednesday noon a fine day, now cloudy, but hope this west wind may bear off the clouds. Wishing you and family good health I am with respects to every friend

Matthw Culley

Thursday dull weather after 2 fine days for hay &c.

16.   *Matthew and George Culley to John Welch*

Wark 4th September 1799

John Welch

Well John, I cannot get you a horse, have desired Mr T. Bates to look for one. Perhaps you may find one with you on inquiry. We have as well as some others cut a few oats, some laid and grown up with greens. When the harvest becomes general we cannot say, today have given over or tomorrow, as they are hardly ready. Turnips bad in general, especially all wet lands, ours on dry land and early sown good. Galla Hill so and so only but are now mending but late, as are south side of the town field west of and joining the cottars yards, Downey piece not bad, Pepper Cook growing well but late, Holburn Hill tollerable, the north part of them wet and drowned, Jocks Ridge southeast side bad, the north part, south part rape the dry land good,

---

51 Heifers.

52 Cows or heifers in calf, in this case evidently intended to be sold in this condition.

53 Robert Colling of Barmpton, 1749–1820, and his brother Charles Colling of Ketton, 1750–1836, did much to develop the shorthorn breed of cattle. James Sinclair, ed., *History of Shorthorn Cattle*, London 1907, pp. 24–7, 51–79.

54 In subsequent letters this name is written variously as Ward and Wade.

wet drowned now improves, south side of the middle north Leamingside rape for sheep wet and drowned, part of it good, now it mends again. When are beans and peas to be ripe, broadcast full of weeds, the drilled clean but farr too full a crop for ripening, the season late and they keep growing and blooming, are hurt by the winds as are other corn a little. Several people selling cattle for want of turnips at Dunse fair,[55] many sold at lower prices than last year. Mr Nisbet of Roden, Mr Trumble of Crookham, Mr Robinson of Upsettlington selling for want of turnips. We had Jery Clayton[56] yesterday, offer him 250 shearings at 2£ 2s per, he bid not although inclinable to buy them, to cast 60.[57] They are lively and good, not so good a top as last year but none level.[58] Are much afraid of stock lowering. He complains of turnips being bad and dear with them. Do let us hear often from you how things go. A slow sale at Whitingham[59] and a full market, best cattle were sold to Shields but below 6s. We wish to know how many oxen you can winter on turnips, or small feeding queys. If you could sell the few sheep you have would it be advisable. We can send you some better if we do not sell to Jery or some one. Should you come to Ninian fair[60] or meet me at Newcastle Lukesmas, we have some usefull heifers and small oxen of our own rearing, perhaps you should have some of them for your foggs[61] or aftermaths if they are good according to appearances.

[*G.C.*] Saturday 7th August [*sic*] 99

So far was wrote by my brother. I am going with Mr Askew[62] to look at Nettlesworth near Plawsworth. We set off on Tuesday morning first, to Newcastle that night, and Wednesday I suppose to Nettlesworth as we are to be home again on Thursday night. I don't ask you to meet me because it is so very short a notice, and at any rate we could not be long together. Only if you had to be at your father's before harvest it might be done at the same time. But leave you entirely to your own determination. You will hardly get this letter

---

55 The fair at Duns at the end of August was important for sheep and cattle: Lowe, *General View of the Agriculture of the County of Berwick*, p. 24.

56 A butcher and stock jobber.

57 I.e. entitled to discard 60 from the lot.

58 This refers to the conformation of the sheep.

59 Whittingham fair, 24 August.

60 St Ninian's fair near Wooler, 27 September.

61 Aftermath of grass that springs up after the hay is taken off, left in the field in winter.

62 George Adam Askew, 1771–1838, owner of the Pallinsburn estate to which Eastfield and Westfield belonged. *History of Northumberland*, vol. 11, p. 438.

before Monday night or Tuesday morning. But even then you might do it if convenient to yourself, but as I have said before I must leave that entirely to yourself. I would have comed to you if I had been on horseback, but I am to be in the same chaise as Mr Askew. I was quite against your master buying you a horse here, so pray do the best way you can. I think if you were to go or write to Mr Thomas Bates he would lend you any assistance in his power I am certain. Pray write to us and believe yours truly

Geo Culley

PS. By all means get your carts made in your own country. We have heard nothing of R. Dickinson's return yet. I saw Robson on Thursday who talks of being south soon. This is fine weather now, if it continue will be a great blessing. Everybody are shearing early oats and pieces of barley. William Brown desires remembrance to you, and all your friends here.

PS. I hear of nothing that Jere has bought, a man from near London is buying fresh oxen and steers.

17. *Matthew Culley to John Welch*

Wark 17th September 1799

John Welch

I wrote to you a few days since, and neglected to say that you should before this have said how many feeding cattle can be wintered by you at Denton. The sooner they can be sent from hence the better, as we would be enabled to keep the rest better, and prolong the time of putting them to turnips, which are a very indifferent crop. You are I apprehend under no necessity of keeping your queys for incalving and extra keeping, and as you mowed many acres of hay should have some tollerable foggage and stublings after harvest also for feeding cattle, and the sooner they go to you the better they may drive and have more time to recover the journey before cold weather and long nights come on. This is a very wett morning and the glass low. We have had two exceeding great dashes of wett, Tweed high, our runners here as high or more so than I have ever seen them, the corn distressingly laid, and this heavy rain again threatens a bad harvest, and wheat may sprout and grow in the ear. New oats 25s per bole, delivered 70 bole, have not been able to lead any more, have cut most of our early ones, mowing some laid barley much grown over with greens. Going to Grindon to take our tythes.[63] I see Mr

---

63 Tithe, a church tax levied, for the support of the parochial clergy, mainly on
   agricultural yield, was by the eighteenth century an unpopular tax, highly

Colpitts's[64] death in the paper, am afraid Grainger[65] may be very harsh in letting the corn tythes, so we will lay to grass much of the land (especially the moors and poor clay soils) after good fallowing. Therefore should sow on the poor soils more white clover to suit pasturage, and mow the land that can be watered. If you can sell such or all of the weathers as you can at a reasonable good price, we could send you of our own bred sheep as many as you can winter well, so as to carry them on till your clovers rise which we hope may be early as you have no ewes and lambs to injure your pastures and clovers. Why do you not say to us that you will want money against such and such time, so that we may contrive to send to you?

You must desire John Todd to give my rent to Mr Peacock, otherwise he cannot expect any favor from me. There is an amazing narrowness in Tod. If he does not pay half a year's rent immediately,

---

complex in operation and full of anomalies. Tithes were divided into great (usually corn, hay and wood) and small (usually wool, livestock and garden produce). Rectories were endowed with all the tithes of the parish, vicarages for the most part with the small tithes, the great ones being retained by the impropriators whether clerical (such as bishops or cathedral chapters) or lay, the successors of those who had acquired monastic property under Henry VIII. Such tithes were like any other form of property, that could be bought and sold or leased. Some perpetual curates were given the small tithes by the rector, others had a fixed stipend. Payment in kind was by the middle of the eighteenth century being superseded by money payments, which were fixed either annually or for a term of years. Customary payments in lieu of very small tithes were common, but increasingly sources of dispute as tithe owners sought to realise improved values from rising agricultural production. After years of demand for reform the Tithe Commutation Act of 1836 provided for the conversion of all tithes into a corn rent. See Eric J. Evans, *The Contentious Tithe: The Tithe Problem and English Agriculture 1750–1850*, London 1976; Roger J.P. Kain and Hugh C. Prince, *The Tithe Surveys of England and Wales*, Cambridge 1985. So far as the Culleys' farms were concerned, the great tithes of Denton, a chapelry of Gainford parish, were paid to the Earl of Strathmore, who leased the rectory and tithes of Gainford from Trinity College, Cambridge. The small tithes were paid to the vicar of Gainford by an agreement for his life. In Northumberland the Culleys appear to have leased the tithes of Grindon (Norham parish) and paid those of Eastfield and Westfield (Ford parish). Lord Tankerville bought the tithes of his estate and added the charge to the rents.

64 Thomas Colpitts II, 1732–99, agent to the Earl of Strathmore for the Streatlam and Wemmergill estates. His sons Thomas III, John and George were also agents to the estate. DRO, Strathmore Archives, Introduction, Appendixes 2, 4.

65 Probably Joseph Granger, land surveyor, author of the first *General View of the Agriculture of the County of Durham*. He was not an agent to the Strathmore estate, but was married to Elizabeth, daughter of Thomas Colpitts II, and probably acted as a tithe collector.

may fully expect to loose his farm at May first. I cannot away with such shabby treatment, and will not wait for my rent, you may tell him, any longer. Why am I to pay interest for money or draw on a bank, when rent is due to me? This is *my own particular affair*. Do keep a particular account of the building of the mill, and the money laid out in cuting all the water courses and tail drain, as most probably they may be to be paid by me or mostly so. Also send when an oportunity offers, or have ready, a quarterly account settled and drawn out for my inspection when I come over, which possably may be 20th of next month or sooner. It rains so much that I cannot get to Grindon, are watering the meadows in harvest, our cattle and sheep drove of them totally. I hope you intend to let the water off the coal road into the carrs at the proper places. I must if all be well make use of the water whenever oportunity offers at Denton. Turn your attention that way, to draining and watering in floods especially, the improvements are *obvious*. I approve of your widening the mill race and deepening over the carrs especially, but at first narrow cutts for tryalls. I did not mean the north water to cross the road until south of the glebe 30 or 40 yards or so on where most easy of access, and with a good cover upon the bridge or conduit, without raising the road so as to make it a climb for carts passing over it. I am with respect your obliged master

Matthw Culley

18. *Matthew and George Culley to John Welch*

Wark 24th September 1799

John Welch

Received your letter by Dickinson and this of the 10th instant. The bore rods are so heavy I cannot send them at present, besides it's not important to bore as yet until you have more time on hand. It is well you got the water brought so easy from Birkbeck's[66] field. I wish to have the flood in command down the Hang burn and a stop put into it. It has worn a very dangerous deep course. Why did you not get a cart or 2 of Tod rather than 2 from Mr Peacock? I hope Tod has paid his ½ year's rent to Mr Peacock. I shall take it exceedingly amiss if he has not. He plows his farm well but is very selfish. He hurts my feelings much. You must insist on his paying to Mr Peacock. If he has the

---

66 Joseph Birkbeck, Denton, c.1718–1803. His elder brother John, who died in 1787, was British consul at Nice. Thomas Peacock corresponded on behalf of the family regarding John's widow's will. See Angela Marsden, ed., *A Raine Miscellany*, Surtees Society vol. 200, Durham 1991, pp. 171–221.

money to borrow that is his look out. Am glad that you have got good slates. I wish you may soon get them laid on to save the walls, as the weather is so very wet. We have it as wet here. Our corn is grown before we cut it, have cut 3 fields of winter wheat as sown far from ripe, exceedingly grown through with greens and weeds, are leading today early oats, have not been able to house any this 14 days. The wets have advanced wheat to 3£ per boll at Kelso, good new oats worth 25s, oats are the best crop here. I saw Mr Colpitts's death in the paper, am glad his son succeeds to lett the tythes of Gainford parish, hope your tythe may be worth the money. Shall hardly send so many cattle as you speak off, perhaps 8. It's a good hearing that your turnips prosper. They are a general bad crop I believe. You don't say what Mr Jackson, is it John Jackson of Coundon Grange? He is a decent man, a particular friend of mine. Incalve queys are good to sell when handsome and good buys. We wish not to overload you with stock. Ninian fair is on Friday. You may sell the mare and buy one more usefull to work, are bad together. Shall give you an account of the fair but markets for lean stock are overset. As to money you must write to us in time before you want, and keep your payments good. Send me, or at least have a quarterly account ready to send here by our cart or such a conveyance. We may struggle with our wedders if my brother agrees to sell the ewes at Christmas, we cannot carry all forward, besides our ewes are upon the flooded land (a part of them) with the water flowing over it day by day as much as in winter. Mr Bailey and my brother I think sold their copyright[67] to the printers at London. They have none in hand. This is the 2nd fine day with a very fine wind, are leading oats and threshing for Berwick. I suppose we shall shear on as the season is so very late and backward. I am with esteem &c

Mattw Culley

[*G.C.*] My brother is wrong, we have not sold the copyright of the Report, but it is out of print, and whether it will be worth while to print another edition I know not, we have not yet determined. This is the only fair day, and yesterday, that we have had of a long time, and the harvest is so very late that I dare say we are cutting our wheat rather greener than perhaps we should do, but it grows no whiter this long time. We have cut all our early oats and barley, both good crops,

---

67 Of the *General View of the Agriculture of the County of Northumberland*, first published in 1794. The report was reissued in 1797, bound in one volume with those on Cumberland and Westmorland (by A. Pringle). A third edition came out in 1805. See p. xxx.

and the Angus are ripening very prettily and are a nice crop. But the wheat in general is a very bad dark looking crop, and I am affraid will turn out very badly. We are likely to have a good tup season and are sending one to Blyth today to go into Norfolk.

Ninian fair 27th. This has been the worst fair ever known, and what will be the result I do not know. Hardly any sheep were sold, and few cattle, and all at reduced prices. I think you had better try to sell your few dinmonds[68] even at 5½ per lb. as we shall be so ill of for turnips, having such a large stock in hand of both cattle and sheep. An amazing quantity of both cattle and sheep shewn, and litterally no demand. Now such a prodigious quantity of goods left, will raise the price of turnips exceedingly. New oats however were sold at 26s per bole today, which is some comfort because oats are I think a decent crop, and barley in this country, but wheat and pease are woefully bad. Wheat never offers to grow whiter, we are cutting it in a riggity[69] unhealthy state that ever was. It must be very ungifty[70] at all events. This was a very wet day again from 2 o'clock in the morning until 11 forenoon, and then a brisk west wind broke out. We have had a good few oats at different places, and will thrash and sell directly. This will go by Jack Davison. He will put it into the post at Darlington for you, but will be a night with you on his return. I am in haste

Yours truly Geo Culley

[*M.C.*] Jack will be with you on Sunday, or rather Saturday night and stay over Sunday. He will not fetch the bore rods till he goes empty north for tups.

[*G.C.*] 29th. Ralph sold 100 bolls midling old wheat yesterday at Berwick at 300£ [sic] and 150 bolls new oats at 27s per boll. But O what dismall weather, this is the 3rd days rain again.

19. *Matthew Culley to John Welch*

[Wark n.d.][71]

John Welch

I think when you fill up between the walls at the top next the mill only a little earth should be put in at once, a foot thick, until it dry

---

68 Male sheep between first and second shearing.
69 Rickety, weak.
70 Not prolific, poor yielding.
71 The address and date are lacking. The contents indicate a date in late September or early October 1799. The year 1800 has been written on the letter by a later arranger, but seems to be mistaken.

after pudling, and if there be any fear of the inner wall giving way a prop should be put from the mill gabel to a deal below the wall. On considering over the by water, where it may be best to let it off from the dam, it will only be upper water, so that it may be let off nearly or altogether where we let off the water into the tenement or field Birkbeck possesses, but should be a very broad outlet to let off as much water at once as comes down from the carrs. At least the east waters may be let off in the same way higher up by the tenement, by a board put across with the foot pressing it down. The same way may do where the water goes off from the dam, only there should be a frame swaled[72] and grooved for the boards to stop up and down, with a handle to pull them up by. As to lowering the ram at the carr head from Joseph Hodgson's carrs, I apprehend that if you cut out 3 or 4 feet wide and lay it against the sides at present battened up, that when you are grinding the water will scour away the earth sufficiently in time. As a general rule in many cases it's better to do too little than too much. There will be a necessity to put in a stop somewhere at the carr heads, so as to check the too much water coming to the dam in a flood, lest it should destroy your dam fences altogether. The over much water should be provided against, as your dam is not and will not be sufficiently sadened[73] for some time. Caution is a necessary thing with water. The stop at present may be of the west end of the carrs where a temporary dam is now, made of 2 or 3 boards for the present so as to check the water and force it over the outlets or overflows to the westward. Be sure let there be overflows sufficient westward for floods before you make up again the old water courses at the head of the new race.

Jere Clayton has at last sent the money for 250 of our shearings at 38s per, and wrote that he would draw them yesterday and send them forward, but did not come. He is a teasing fellow. They are much cheaper than them you sold. The weather has been very unpleasant since we got home. Our corn in and covered at Wark, except the beans on the Bar Green, part of those on the south-east town field and some at Leamingside. Some little cut at Shotton, some on the pike at Longknow exceeding green on some undrained land, several pieces cut at Thornington, Eastfield all in I believe. Several pieces cut at Grindon. I can not send Bob Dippie[74] until I get him a few more lessons how to act. I now think we should make the hedge

---

72 Scorched.
73 Saddened, compressed.
74 Hind at Wark, later at Denton.

on the north side of the carr in a straiter line as you were pointing out, and the cross hedge from the thorn where the water broke out down the old water course to the Well field foot. They will be both places to bore in but perhaps it may be as well to cut a drain at the foot of the oat stubble on the stubble end below Lime Kiln Bank. The cross hedge should be first made on the west side of the carr to Well field as I suppose it will be a means of draining some of the springs. Possibly a 5 or 6 foot gutter well sloped (as it may slip and not stand well) with a wide scarsement of 12 or 16 in. raised 3 sods before the quicks are set. As to the main stell,[75] perhaps willows may do best planted on the south side leaning to the north as the earth is cast out on the south side will be to rail or some stake or nice fence to save them being eat by stock. Hope you have got the grey stones[76] home, and geting forward with them. Have not seen Robson yet. Let me know if you saw Matty, and what you said to him, also how he likes his situation. Have you got the things from Darlington and how you like them. Respects to all friends. Your obliged friend

Matthw Culley

PS. Shall wish to know from you or Mr P. what John Tod has done about the farm he viewes, and what he says about the rent of Denton East Farm, also about Birkbeck.

Clayton's sheep are exceeding good and cheap, but then we will be enabled to carry forward our young stock. They have a chance with good luck to pay him very greatly. We have got the wool weighed off by Mr Allan a young man and I believe a partner with Head,[77] has a farm, him and his brother in Wencledale above Massam,[78] seems a decent man. John we will send you a hundred fat ewes in a few days, and you must mind and see every ewe killed yourself as we shall send you all the best nicest youngest ewes we can and such as we think most likely for breeding from. They will be principally gimmers,[79] but perhaps not as fatt as we could wish.

I am &c Mattw Culley

---

75 A large open ditch or drain.

76 The best English millstones were quarried in the Peak District and the Yorkshire Dales: 'grey' stones came from Yorkshire. In the eighteenth century superior 'blue' stones (see No. 22) began to be imported from Germany. They were a form of basalt.

77 Wool merchants.

78 Masham, in Wensleydale.

79 Female sheep between first and second shearing.

20.   *George Culley to John Welch*

Eastfield 6th October 1799

John Welch

As James Glass sets of from Thornington tomorrow morning with 12 cattle for you by the way of Corbridge, Witton le Wear &c, &c as he went before, and we wish you to meet him by the road with somebody if you can, as we are so very busy and he James is an excellent harvest man. We sold oats yesterday at 28s per bol (new) and old wheat at Kelso last Friday at 3£ 15s per bole. Wheat is a sad sample here and very little of it got into the yards yet. Let us hear from you what you have done at Brough Hill.[80] My brother goes from hence on the 18th to York with Nelly Culley of Grindon,[81] and returns by Denton where he will stay until Newcastle fair.

Yours in real haste Geo Culley

21.   *Matthew Culley jr to John Welch*

Eastfield 6 November 1799

John Welch

My father wishes to know how much money you may want before next May Day, as we are paying some money in and would wish to know exactly what our outgoings will be at May Day or as near as may be.

We have sent you 100 ewes and gimmers, all very gay and fine, and very tempting to breed of, therefore would wish you to take particular care who you sell them too, and by all means I must request you will see every ewe killed yourself, as you know very well the butchers are many of them not to be trusted to. We have had so many tricks plaid, that it makes us very frightened about this. There is a part of them very fat, and may be sold as soon as you like as they have been well fed and will prove very well. The rest [. . .][82] as they grow fatter, the sooner you sell them [. . .] better as we can send you plenty more when these [. . .] done. They will do very well to eat turnips along with your feeding oxen and queys. I would not like to have any of them sold to Bulmer or any jobber. You said you thought you could sell some of them to a butcher at Piersbridge. I would rather take a less price for them when you are sure than run any risk

---

80 Brough Hill, Westmorland, fair, 30 September. Bailey and Culley, *General View, Northumberland*, p. 341.

81 James Culley's daughter Eleanor.

82 The edge of the sheet is torn, affecting the beginning of three lines of text.

of their being sold again. Do you think you could kill and sell any of them at home?

We sold Jerry our wedders 250 by letter at 38*s* per, and they went away yesterday. They are very capital and very cheap. They are extremely fat, but some of these ewes are fatter.

I am your obedient servant M. Culley

22. *Matthew Culley to John Welch*

Wark 15th November 1799

John Welch

William Brown sent your letter here this day. We are obliged to you for calling to see Matty. I hope he is in a pleasant situation, but school boys like other folks expect to meet with happier days. I thought him some stouter when I was there. They get plenty of common fare for their money, and have good air. Hope Robson will be with you ere this come to hand, as he was to set out this day for Denton as my brother George sent me word. I hope he will get a pair of second hand blue stones. Mr Wood mill wright at Newcastle said he hoped he could get me a pair. As Muckle has given up the contest for his patent we are subscribing for Wood and Burnit (who supported Rastrick) and Rastrick.[83] As to building an arch or a conduit bridge rather, over the race into Horse Crofts, Robson will be able to assist you where to set it and whether a low arch with a frame and clue,[84] sufficient to let such a quantity of water through at a time (or a little more) as is sufficient to carry the mill wheel properly about to grind corn or threshing machine,[85] may perhaps be as much as may be necessary for your

---

83 Mr Rastrick of Morpeth was a machine maker who made successful threshing machines (see n. 85) in the 1780s and 90s: see *Annals of Agriculture*, 15(1791), pp. 481–93. Thomas Wood and Robert Burnett are listed as millwrights in *The Newcastle and Gateshead Directory for 1795*, and *The Directory for the Year 1801 of the Town and County of Newcastle upon Tyne, Gateshead, and Places adjacent*, Newcastle upon Tyne 1801.

84 Clow, sluice-gate.

85 Threshing machines were in fairly general use in north Northumberland in the latter part of the eighteenth century. Many were worked by hand, some by horses. Water power was coming in, and was recommended by Bailey and Culley (*General View, Northumberland*, pp. 49–58) as the cheapest form where water was available. The first effective threshing machine in Co. Durham was installed in 1795, but they spread rapidly thereafter. In 1810 Bailey recommended (*General View, Durham*, pp. 80–2) steam power as probably the best source where, as in Co. Durham, coal was cheap and readily available and a large farm warranted a powerful machine. Bailey had seen a steam-powered machine at Pallion, and described one recently installed at Chillingham Barns.

bridge to contain at once. Then it may be made only a conduit with long lime stone covers (this is only a hint) because it may last many years if well covered with earth. But I now too late find my error in not imploying the mason Richardson as you advised me. I now learn from Jack Dickinson that he is an exceeding good workman. I am not pleased with Belwoods building the walls of the mill so weak at the bottom, should have been 2 or 2'6 feet 6 inches to the first story and so narrowed with proper offsets to 20 in. and perhaps 18 to 16 at top. Belwood and your slater have been guilty of a great fault of omission in not finishing the pointing or drawing, and the ridge stones and straws. A valuable roof like it should not have been neglected in such a critical situation. I care not how soon you change your mason when it can be done with propriety and you have a job worth Richardson's acceptance. This you will consider off and act as seemeth right. There is so much depends on a good mason, a man of thought and experience in his line of business. The old clue or clour was at high corner of the mill garth adjoining Mally Wade's garth, but I suppose it may be a little east off the old place, as much depends upon the level which you bring over Mally Wade's garth. But as you can cover the bridge with any depth of earth and raise the earth west of the bridge so as to dam the water back to your offlets in a sufficient manner, may on proper consideration make it above the middle or most convenient part of the garth so as to answer the stack garth when one may be wanted in the Goose crofts. Perhaps a winlass or some such may be best to draw up the clue with a long stick or press which may be taken away at pleasure, may be right. Then it may prevent ill disposed people from setting of the water from spite or mischief, as you may have enemies. Probably one or more persons should sleep in the mill at all times or in one of the chambers. Mr Bates had a fire made in his mill by some bad person unknown or persons, but happily had gone out by good luck. The lower down the garth the more water the dam will hold if the clue and dam turn water well. I think a broadish outlet or two may be best made for the tenement, one above the conduit at the glebe gate, and one at the head of mill dam where most convenient, as you think the other a good way above the glebe gate to pass over the road when well gravelled with a syre.[86] Too much water may injure your mill dam most materially if it should get pent anywhere. You should I think carry up a small channel at the head of M. Wade's

---

See also G.E. Fussell, *The Farmer's Tools A.D. 1500–1800*, London 1952, pp. 152–62; *Agrarian History of England and Wales*, vol. 6, pp. 525–8.

86 Gutter or drain.

garth, lay the earth to the side at present so as to bring forward the water at the least expense. It may be afterwards washed away. Then you will know better where to make the bridge. Has Robson made any enquiry about the blue stones on his way by Newcastle? Before you cover your tail race it should be brought up deeper than the level, then the water will go better off and may be longer of filling. In a future day it may be necessary to cover it to the smiddy, and lest the gravel will be to lead out of it after every flood which comes down the Hang burn. These are expensive things but may be necessary. It may be wrong to differ with Matt Staward, but they are proverbyal for doing little work wherever they have been as I hear, and you took them in to meat them. Over full makes saucy men. I know you are anxious to keep them at their work, and I fear there may be too much reason for it with them who have high wages. I could have had the work done by Burnet or Wood who are said to be very clever and knowing in their business, and nearer to you than Robson. Should you not get deals to cover the stable and over the dressing room. You cannot thresh until it be covered. Should the boards be stronger where carts comes on? We laid ours with strong deals where the cart and horses tread, perhaps ash, elm or some hard wood might be best for wearing smooth and last longer than firr where the cart comes only. How will you do for a proper trusty fellow for grinding who will do justice to his master and to the country? Will a young man do best? He may be tryed in the first instance. From here should he be? What do you think of Jack Johnson who lived with us, now of Learmouth Mill? Say how we should do in your next. Yours

Matttw Culley

It was well you had got the oats, as William Brown says people complained that the wind had hurt their uncut oats. John Todd said to me that the farm was as high rented as it would bear and if he keeps it another year at our valuation, he sayth one thing and does another. He may have reasons for so doing. You will let us hear how the ewes get up. We hear Jery Clayton had 2 wethers taken away from him at Rothbury. Perhaps they may be 2 of your ewes. Matty Culley said a part of them were good mutton.

How are you to do for a drying kiln or a drying plate[87] for drying

---

87 Apparently a covered dish or oven for drying grain and pulses over heat, not an integral part of a mill or threshing machine. The objects are not well documented. Grinding mills for crushing or bruising oats for horse feed and beans and peas, worked by hand, were in use (Fussell, *The Farmer's Tools*, p. 185), and one of these may be what Matthew Culley is referring to in No. 25.

oats or beans &c? I did not leave a blank for the seal. The seal falls on these words 'who supported Rastrick', 'also or a conduit bridge'.[88]

Mr Sayle[89] says Leeds fair was exceeding good for fat cattle.

23. *George Culley to John Welch*

Eastfield 23rd November 1799

John

I had forgot to write to you to send us an account of your stock, crops &c, as this is the season in which we always take an account of the stock of cattle and sheep, number of stacks of corn, hay, &c, with the acres of turnips &c, and your farm must necessarily be taken in along with them. I will thank you to have sent me an account of the above as soon after receiving this as may be. People here begin to expect an advance in the prices of fat stock, both cattle and sheep, particularly the former. At Hallow fair Edenburgh the other day they were better sold than was expected, also at Dunse, and by a letter the other day from Mr Sayle he tells me that fat cattle were sold from 6 to 8*s* per stone at Leeds fair. The case is that fat in the first place is not plentifull, and at Edinburgh, Dunse &c the labouring people bought up great numbers of the midling small cattle to kill for Martinmas,[90] finding them cheaper than oat meal or any meal God knows. We are killing after the rate of 5 or 6 ewes every day, which we sell at no more than 4*d* per lb., or in the live carcase (killed by us) at 5¼ or 5½ sink, which though our nicest ewes we prefer to selling such in the market at Morpeth alive. Indeed we must begin next month to sell remainder of our ewes at Morpeth where I am told fat sheep do not sell so very ill, as few good ones come to market but a deal of rubbish. The lad James was not returned the other day home. We are sorry to hear that Jere had 2 or 3 of our wethers died yellow[91] on the road. This shews plainly that sheep should not be drove off red clover, a part of ours were upon wheat stubble clover, very luscious and flowered. I desired Mr Peacock to tell you to write to me about George Thomson what he wants, that my mistress may send him what he may want by Jack as he goes for tups into Yorkshire. As you are now so far of him I am affraid the child may be neglected for necessary clothes. Not that I am affraid of your mother's attentions, but she may depend on your writing about them, which your distance and attention to other

---

88 The words are in fact legible.

89 Benjamin Sayle, Wentbridge, Yorkshire, an old friend and frequent correspondent of the Culleys.

90 11 November.

concerns may make you forget to write. I also desired Mr P. to settle with you about him.

Can't you pull or divide a sheet of paper, and write upon the one half to write to me about the boy, and the other half about stock account or anything alse you have to write about, and then put the half wrote about the child to me in the inside and the other on the outside, same as though it was a whole sheet you know, and folded in a letter same as though the sheet had not been divided. Only remember mine to be on the inside the other. I think you will under-stand me, it is to save postage and is all fair, only put Single Sheet or One Sheet on the outside either above or below the direction. You know this is to prevent anybody seeing the part about the boy but who I chuse. I hope you will write as soon as you have got the stock account right, which will oblige

Yours truly Geo Culley

We have had very thick hazy weather quite calm, with glass very high indeed, which made me suspect a keen frost would succeed, which is the case this morning. It prevents sowing wheat much. I see both mutton and beef advance in London, and wheat 6s per quarter last Monday. What will be the consequence God only knows. But I never was affraid of a scarcity in my time before. They don't seem to get any quantity of foreign corn into the Port of London. I doubt the crops are bad in rest of the north of Europe as well as here. G.C.

24. *Matthew and George Culley to John Welch*

Wark 23rd November 1799

John Welch

I sent you a long letter concerning the bridges &c a week since, hope that you received it some time since. We received both of your letters by Glass this day, sent them both to my brother who has returned them for me to answer. Have had several dry days lately. Hope you have got the harvest over, as you take no notice of leading your oats. I hope that you only once had a little damage on the slates. Belwood had I am afraid neglected to put on the ridge &c as Overend had neglected to point or underdraw the slates. Am glad that you got the ewes safe and well up. Our herd put 11 weathers, mostly them belonging to the herds, yesterday night into the Willow Bush wheat stubbles over eaten clover and seeds, one was dead of Jemmy

---

91 Jaundiced.

Phillips[92] this morning and one of Glass's died in Cornhill also. Both died of a good colour, so none were lost in driving a little and died in the field out of 11 wethers and 21 ewes in the same lott, and you had 100 come up 100 miles safe and sound. My brother says he wrote to you only this morning, so shall send you this after a few days. He mentioned to you the advance of mutton and beef prices from Edinburgh and Dunse, both high for fatt cattle. Indeed there has been little or no grass to keep cattle upon, much less to feed on. Our pastures fall off to nothing, and the late turnips on wett or soft land never mended at all. Am glad that the steer market is so well mended. I have often known the lean oxen mend much in price at or after Martinmas. The price of pork is high, 6s per stone without the hams, the hams 6d per pound. I hope you will get the ewes sold to your neibor butchers so as to see them killed yourself. You must endeavor to get good seed wheat and submit to the prices. I think it wrong to kill any of the ewes, for the reason which you give. Your obliged master

Mattw Culley

PS. I thought best to build a bridge near Mally Wade's garth or lower down, where it be proper for a clue or stop with no more arch than to pass as much water as will carry your mill at a full stroke or power for grinding. Of this Robson and you will be able to judge off. Perhaps the middle or below that of the garth, so as the water does not flow over your bridge top, may be the best place, so as to stop and contain the greatest quantity of water, the wells well filled and puddled. You say Robson came to see you. It was high time. The man had wrote to Robson that the milstones were sent to West Auckland, but Robson neglected to give us notice. I am glad to hear that you get so well forward, hope that the grey stones will be got to work ere you receive this. I think that it may be right to cover only a part of the tail race at present and leave the rest until next summer, as the weather will then suit better and there is often a dead time for labourers between winter work and summer commencing. You have not said what you have done with the Galloway[93] which you bought at Birkbeck's sale. I suppose that you left it with your brother, as William Brown says he did not see it when he was at Denton.

I mentioned a vast number of things in the letter I wrote to you. I do not know but the man may do for a tryal for grinding. If he be honest it is a material matter to us and to our customers. He should

---

92 Herd at Wark.
93 A small strong horse, 13–15 hands.

have a slate, to mark with pencil the moulter[94] which he takes before or after grinding. I know not well which is customary, to take moulter. Should there not be a chest or chests with a partition, so as to put the different moulter into, for your inspection? You will get a large bauk[95] for weighing, and a smaller with proper weights and measures. I got our weighing balks of Tod Hunter, some of them were 2nd hand which he procured for us. You should get the weights of cast metal, at least the large ones, as lead is not so certain. Small ones should be brass, adjusted properly, all of them and numbered would be better I apprehend. Dolly might in your absence give a step to the mill, but you keep her over busy for want of a proper servant girl to assist her, which should be at home with her when you are from home. Harry will give what attention he can but will generally be in the fields as of the greatest importance. These are only hints. There were formerly at Denton many designing people, but we made it a rule to employ honest good people in the principal concerns. John Selby was an honest man we believe, but he can be of little use as he is blind at present, and a blind person is unfit to be about any machinery lest they be catched by any of the wheels. All the machinery, so soon as you can, should be covered for fear of misfortunes, and the people should have no laps to their cloaths, as many misfortunes happen from women's long loose cloaths or men's coat laps. Harry may be proper for the first tryals of threshing. Possibly two feeders in may be necessary if it goes at a full stroke. Mr Smith has got seemingly a clever threshing machine at Shed Law by horses. Most of the wheels are on the side next the first power, all the cogs are metal. It was made by a man of Spittal or Tweedmouth as Robson had so much work on his hands. Take care of not over filling your dams over night with water until they be saddened,[96] so as to prevent the water from leaking through the walls, as water and fire are good servants but bad masters. I believe the level should be first brought over the carr heads by a narrow cut and the earth laid upon the side which when mellowed will wash down or seep with a force of water. I see no great harm in leaving the greatest part of the tail race uncovered until a proper season, but suppose it may be necessary to cover to the bridge, but this will depend upon tryal. I wrote very full of

---

94 Charge for grinding corn.
95 Balk, a squared timber, here the beam of a balance.
96 The next eleven words have been added where indicated by the next note, but probably belong here.

these matters in my last.[97] Put down whatever occurs in your letter at all times, I mean to begin a new letter as soon as you sent away the last one. We will inform ourselves about your cutting machine.[98] Mr Bakewell cut hay, corn and straw all together with his straw cutter. Perhaps hay and straw should be sometimes cut together. Your corn will be generally ruffed[99] by the mill.

28 [November]. Yesterday bought 10 acres of good turnips of Mr Marjoribanks of Lees at 6 guineas per acre, are to be called 7£ per acre. I do not know now but we could have done without them now, only it's good land and a fine situation and a good crop as this year goes. We shall be enabled to do better to our cows and lean stock. We have now a dry season and I yesterday thought late turnips were showing some growing as have grasses, but this morning the frost is increased much, was in general moderate so as not to injure the growth of grass &c. Sown the low haugh[100] on Tuesday, West Broomyknow yesterday, they wrought well. Durst not lime the land this morning, the land has saddened but roads are very slabby and wett. Put all our sheep to turnips even ewe hogs yesterday, most of our feeding cattle also, and mostly tyed up. They would now this seasonable weather have done as well out. It is only some of our wheats that are fit for seed. R. Brown said on Saturday 3½ guineas were offered to Mr Davidson of Akeld for a good sample of seed wheat, such is the scarcity here, so that if we were to send you some by sea to Newcastle it will not answer the expence and trouble. The white or Zeeland is our perfectedest corn. Mr Thomas Bates of Aydon Castle[101] said in a letter to brother George that the market at Newcastle fair was advanced considerably for cattle, 1s per stone for cattle above Newcastle fair prices. I hope the improved plow breasts may answer well. Possably

---

97   The words 'so as to prevent . . . walls' have been interpolated here, but probably belong where indicated above.

98   Cutting machines were invented in the 1760s, in the form of a box with a knife against the end, into which straw was fed by hand: Fussell, *The Farmer's Tools*, pp. 180–2.

99   Roughened.

100   Riverside meadow.

101   Presumably Thomas Bates, 1775–1849, second son of George Bates of Aydon White House (see letter No. 30 and n. 2). This Thomas Bates, later of Kirklevington, became a notable shorthorn breeder. Sinclair, *History of Shorthorn Cattle*, pp. 134–45; Thomas Bell, *The History of Improved Short-horn or Durham Cattle, and of the Kirklevington Herd*, Newcastle upon Tyne 1871; Cadwallader Bates, *Thomas Bates and the Kirklevington Shorthorns*, Newcastle upon Tyne 1897.

your founders may cast by one of them, I am told they were the model only once, cast from the 2nd by plate, but they endeavor to harden the metal by running or milking it often over so as to make it harden to wear better. Also mostly cast thicker than the pattern.

[**G.C.**] John. Your master has left little room for me but I have little to say at present as I wrote to you so lately. I hope to have an account of your stock &c in a day or two, as I can't get forward with the stock annual account for want of it. I congratulate you upon the good weather. It is a pity but we had known of these turnips of Mr Marjoribanks before we parted with our wethers. However it is always a good thing to be strong in meat. I am glad to find by a letter from Mr Peacock that your market advanced on Martinmas day. We have all got it into our heads here that fat stock is to advance fast, but perhaps a severe frost may make an alteration. John Short has been buying 400 sheep below Belford for Glasgow we are told. As we have some tainted ewes by being put upon the watered land designedly, we are going to market those or sell them to Mr William Lincoln who wrote to us yesterday that he wants both cattle and sheep.

I am yours in haste Geo Culley

PS. My Brother's is good advice to you to have a letter upon the stocks, and then you can add a part every day until sent off.

Pay[102] great attention to your ewes as there are some of them rotten, many of those killed at home are, and a part of yours were on the watered land. M.C.

25. *Matthew Culley to John Welch*

Wark 30th November 1799

Well John

Robertson was here yesterday and this forenoon. I have agreed that he shall make you a straw cutting machine directly, as corn is so high and hay so very scarce. It is right to try to save at all ends one can for horse and cattle meat rise. It will not cost much money and it may suit Ralph who is a careful man. But the blue stones cost 30 guineas or so, with the apparatus will cost 50£. Perhaps we should consider well before we lay out so much money, unless in your next you can shew how I am to be reimbursed for such expence. Be so good as send me your calculation soon. Also there may be a fear of there being a want of water to do so much business. We may do with a wet season especially in winter with a proper fellow for grinding and keeping

---

102 This sentence is written at the head of the letter.

accounts, honest and attentive. Mr T. Bates has little trouble, I hear, with his grey stones, as people bring their corn and take the meal home. He grinds for the 16th part[103] and pleases him well. My brother George thinks you should keep leading all the sullage[104] from off the roads you can, as in many places you have only across one field to go, or so on. There can hardly be a better lead, especially on to the barren grass moors and Lime Kiln Bank field which in general is a bare one. The carr will need little I hope except in some parts. We think your 2 more horses should lead all you can at spare times, but as corn and keeping are so very scarce I can hardly consent to keep more horses. Now when the leading to the mill is muddy possably you can lead more sullage. Exersion should be made to lead it rather than lime. It may be tryed mixed with refuse lime. Mr Peacock will let you see what I said about J. Todd. He is a designing man. He is only to have one year from May next. Mr Peacock and you fixed the conditions and rent with me. It keeps me from doing some things on his farm which I wish much to have done to improve it. I want not to check you about the blue stones after all I've said, this is a most advancing time for millers. I mean to have a barley mill in future, as the offal of barley is so good for the horses. I am told that Gateside[105] stones answer best for a barley mill. I own we should have blue stones, which may grind when you have water, and the meal may be kept in sacks or a large bing upstairs. Have you a proper way to pull up the sacks to the loft by folding doors after they are got laid? Perhaps you have Jack Dickinson to lay the floors down. He is an excellent hand I think and works well and closely. When they are done in a future day must I suppose be all tongued with oak laths which are to be had at Stockton or Yarm. I wish you may get the drying plate. Let us know how you approve of it, and whether you approve of the Staindrop mason.

Yours &c Mattw Culley

I have not sent over Bob Dippie as my brother might object to the expence. Perhaps you should throw the metal when you cut the race deeper up the carr to the south side, as a willow tree fence may answer on the south side to fence off the stock in a future day. I think also when you get time that you should open out the bogg as well to the north side of the race in the upper kerr and run a cut from Lime Kiln Bank well to the lane. It will probably raise much water as that

---

103 I.e. takes one sixteenth of the corn ground.
104 Silt.
105 Gateshead.

corner is very wett seemingly. I mean in a slanting direction from the well over the grass inclining to the north as it leads to the lane. The well field should be drained before you plow it, at least the drain above the present hedge which we thought of taking down, should be done and a drain from the hole should be taken down the lower part of the said well field should be opened and filled in again, a covered drain. Water may rise there by boring to drain the carrs also.

I wish much to hear how your drying plate answers when you get it, as we want one here to dry our beans upon. They round in our mill when not dry and do not break, but roll round so as not to answer well, when dry fly in pieces. I never sent the mare indeed have not seen her since I came home, believe she is at Grindon, but of no use to us at present. I ought to send her to you as she would serve to carry you. Get her quiet by gentle usage. I hope we will now be pretty well of for turnips as I have bought 10 acres of good ones at the Lees, which will help to winter our weather hogs, 6 guineas per acre which I am not to mention but say 7£ per. I shall weary you with long letters. Adieu.

26.   *Matthew Culley jr to John Welch*

Eastfield 4 December 1799

John Welch

Yours of the first instant I have just received covering an account of your stock. In respect to the ewes, I would wish you to sell what ewes you have in a short time as they are all tainted, and will not do much good now and may do harm after Christmas. If you can get 6*d* or 6½ *d* it is a very good price. We cannot send you any more as we think they will travel very badly, the roads are so extremely deep pact and the ewes not very fit to travel.

My father and uncle are both from home, my father at Belford on the Tweedmouth court, my uncle gone to Mr Bates to look at some farm for Mr B. It is very likely that you may see him as he has to be in the neighborhood of Durham I believe, and if he wishes you to have any more ewes you may write and I will send you some. We have plenty of meat now, and can sell a good many ewes in the neighborhood, and we can send you none till Mr Lincoln sees them as he is to be out on Sunday week. We have corn very high at Berwick. New wheat I hear sold 4£ 10*s*, oats 28*s*, barley 35*s* and none of them good except oats, and a chance sample of barley. There has not been above 2 or 3 good samples of wheat shown this year. Ratcliff[106] has been out

---

106 A livestock dealer.

and bought everything he could. Green[107] is out buying both fat and lean grass sheep, says the corn not all in about them yet. He has taken a grass farm near Skipton and is buying stock for it now. Clayton has likely got a good bargain of our sheep now as things have turned out. You will remember this is the time of year that we balance our accounts, and you will do yours and send a statement of them. If any of the ewes are marked with a keel on the back or behind the head you should sell them first as they are the worst rotted. Make my compliments to Mr Peacock.

I am your obedient servant Mattw Culley

27.   *George Culley to John Welch*

Eastfield 17th December 1799

Well John

We received both your letters by William Brown as well as the balance in full, for as to the odd shillings I am affraid that you spent much more than that on our account without being repaid. My brother and I are both very pleased and think that you have managed well, and been very fortunate at the last in receiving all the balance of Salkeld except the 10s charged for the cattle staying all night, which he had a right to and I wish him luck of it. Respecting the 16s received from John West for the wether, I never knew butchers give so much as is expected on these occasions. However I think very well of West upon the whole, and perhaps another might have done worse. Fishwick I know nothing of, nor do I think well of him at all. But certainly Salkeld was the man we were to look to if he was able to pay. In fact a chance with a bad man like Fishwick is better than nothing, though often next to nothing. In regard to the piece land you are so kind as to interceed for, I have no doubt of your doing whatever is right. Such men as Mr Fishwick cannot be teized, and God knows what changes may take place at Lambton if the present Mr Lambton[108] die.

Concerning the recommendation of you to Mr Marshall.[109] If I thought you had not deserved it I would not have said what I did for

---

107 David Green, livestock dealer.

108 John George Lambton, 1792–1840, Ist Earl of Durham 1828, Whig politician. He inherited the Lambton estate on his father's death in 1797. *Burke's Peerage; Oxford Dictionary of National Biography.*

109 This correspondence has not been found in the Culley Papers. The reference may perhaps be to William Marshall the agriculturist (see above, p. xxx). No other correspondence between him and the Culleys has been found; Marshall

you. I have little doubt but it would have been the best thing for yourself and family, only it was so very far away for a man and family to go, but a most delightful country and very far back in agricultural knowledge. I did not suppose Dolly would like to go so far, but you may tell her that it was the very place to make her live untill four score years and ten. But without joking, it is a place where many people resort to for the sake of their health, the climate is so fine, and you would have been looked up to much more than ever the Culleys have been for introducing better stock, better cultivation &c, &c into this country. And Mr Marshall is so worthy man that I am persuaded he would not recommend except he thought well of Lord Heathfield.[110] I have no doubt but you would give Mr Marshall a proper answer.

I am to thank you for your account of corn markets, and am glad to hear of such good prices as at Durham. But with us we can sell nothing to pay at all. Wheat 26 to 30*s*, barley 13 to 15*s*, and oats 10 to 20*s* our bole, and all sinking in price except oats. But our corn markets are intirely governed by the London prices. You surprise me with the sudden and low fall in pork with so sudden an advance.

You were perfectly right John in writing two letters, and having answered the first I now proceed to the second. Respecting George Thompson, I am affraid from the tenor of your letter that your mother thinks it a hardship to keep the child, because you say 'I believe if it was not on my account they would rather wish to be without him.' Now John I must beg that the boy may stay no longer than is agreeable to your mother. For how much soever I may think myself obliged to you in this matter, I can never think to lay a greater weight upon your mother than she is able to bear. No doubt about it, we cannot remove him with propriety or prudence in the winter, but as soon as the weather becomes favourable in the spring we will have him brought back if in the least disagreeable to anybody. As soon as convenient I shall be glad to hear from you again, especially upon account of George Thompson. Wishing you and yours many happy returns of the approaching season, I am with sentiments of esteem

Geo Culley

I send this by Thomas Kerr inclosed to William Shotton who I have directed to forward it to you by Mr West &c from Morpeth. When

---

never visited Northumberland, but he could have consulted them if asked by Lord Heathfield to find a skilled man.

110 Francis Augustus Elliott, 2nd Baron Heathfield, 1750–1813. His estate was in Sussex. *Complete Peerage*, vol. 6.

things are not very hurrying you could write by some conveyance or post either.

28.   *Matthew Culley jr to John Welch*

Eastfield 25 December 1799

John Welch

I suppose you will be expecting to hear something from us about the fat ewes. Mr Lincoln came out (as I wrote you) and bought 100 ewes at 38*s* per. They will be above 6*d* per lb. and are not very nice. He also bought 2 queys of us at more than 6/6 per stone and not nice fat, but very fit for cutting up. We have now very few ewes left except the hill ones, and they will not be ready for some time. They are eating Mr Marjoribanks's turnips. We are now I hope pretty strong of meat, and will be able to keep on what stock we have for some time, but I am afraid we will fall short of straw as we have so very little, and that very bad. We have had a good deal of snow on the ground for some time, and a very severe frost. However it has laid very quietly, and could not come on at a better season. I send this by Tommy Wright[111] who is coming into your country to see his friends. I hope you will write by him saying how you come on, what kind of weather you have, and how times are with you. We have had a brother of David Green's out buying, but has done very little. We are expecting a letter every day from Mr Taylor about the tups, when to send for them, and you may expect Jack very soon when we will send you the bore rods &c.

I am your obedient servant Mattw Culley

Thursday December 26th. We have received yours of 23rd inst. this moment, and think it very absurd to send your books so far. You must draw your accounts out into a ledger as you know you used to do, and send us a statement of them. We will think about sending you some stock but will hardly send you any for some time. We will send your letter to my uncle today and he will answer it as he thinks proper. I think you have done right in selling your sheep. You have several queys in calf to sell for incalves. Have got a letter from Lincoln this morning. He will send for his ewes next Sunday.

---

111 Hind at Wark.

29.   *Matthew and George Culley to John Welch*

Wark 28th December 1799

John Welch

You have sent a good long letter. I know from experience that Mr Bates is not exact in writing letters. I expect him here in a few days, but hope he has wrote or sent a man to you ere now, as it's a thing of great moment to you now when your mill seems to answer well. Mr Bates grinds for the 16th part, he takes a sixteenth moulter of their corn whenever they bring it to his mill. This causes no account to be kept. Many people may not be able to bring money with them, then an account should be opened with each one who does not pay, then they will leave you and go to another mill. At some time it may be well enough to grind for money. We are obliged to G. Dent[112] and all your neighbours for their custom, and hope by doing them justice to retain their custom. At same time so long as you get money equivalent to the moulter there is no more trouble. Money is the best change or medium. There is a bare probability of getting Story, who sayth if Mr Atcheson[113] die the mill will not be carried on by trustees but let, so told brother G.C. that in that case he wished to come to us. Of this say nothing at all. Time will shew the event. Mr A. is very poorly in health at present but I hope he may recover. Hot cheeses weaken the stomach and injure the frame, too much a prevalent custom. We are glad to hear of the mill answering so well to your expectations, but as to Joseph Hodgson we mean nothing shabby or mean, but nothing was said as thought about this crop that I remember off, but if he wishes to have his corn threshed, perhaps should pay something reasonable for the threshing of it unless he meant this crop also. May consult with Mr Peacock and ask Joseph how he thought the bargain was, and then write us your opinion. I am hardly a proper person to make a bargain as many particulars do not occur to me which are necessary to be settled before a bargain is concluded. I am sorry for Katy Middleton who had a sorrowful fatiguing time with her parents. Give our respects to her when oportunity offers. Will John Todd expect that you should thresh him a stack for pay when he wants it and you can suit him? Possably you may have opportunity if it will serve you and him to good purpose, for both this is your and his consideration as you have little to thresh this season. In both cases Joseph and he must find hands or allow for them unless Tod pay by

---

112 George Dent, Killerby near Denton.
113 William Atkinson of Yeavering.

the stack or bushell a reasonable price. The mill I suppose will only do one work at a time, not thresh and grind at same time.

As to the offlet of the surplus water, I once thought Robson would make a pipe to carry it off, out of the pipe which serves the wheel. How to do that I know not now. As to the burst of the Waddy I was not without fear, and advised to give great batter with long stones, laid long way in and to lay upon long stones within again, as water is so very powerfull a thing, will if [it][114] get the mastery destroy everything you can do. Is it from the earth you put loose and wet in between the 2 walls next the mill? In such case loose wet earth was very likely to do damage unless sadded and dryed before another portion was added. Did it leak water? I think you were endeavoring to stop some leakage when I was over last. The bridge and clue you built into the Goose Crofts should stop the water at night to the westward, and a clue at the glebe or further north or both, may stop the night water overfilling your dam (absolutely necessary, to prevent danger until your dam be properly saddened). A dangerous matter until a summer be over at least. As to where the offlet should be I know not, a thing necessary to be done, though if the dam could be carried up upon a level or nearly to the north so far as to have the least fall you could get, that the run down would not be so dangerous for wearing the outlet where we let the water into the Hang Burn near the style. That may be too expensive a job, but an offlet may be made more to the south. The weather suits to do this as it's so frosty, but one must be left open to prevent the over filling of the dam. One may be left only sodded on the sides, with a foot or 2 foot high board to put down with your foot and a handle to pull up by. You did right I suppose to order a pair of blue stones by Dudgeon. I hope you get them in good time. The bore rods are ready to be sent by the cart, with binding for the bed stead. I am glad to hear you have got so much water at the head of Well field. You must try to get all you can there as that may help to drain the land below it.[115]

We think you were right in selling the ewes (are afraid the gimmers will not keep long but may try by killing 1 or 2 of them) which were rotted, partly from a preventive from people buying them but more from necessity, a real want of grass when we put them on the watered land were in real want of keeping, grass gone and turnips in a weak state. Our early sown very good. We think you

---

114 A word seems to have been omitted here.
115 An 'x' on the page here indicates that much if not all the paragraph below beginning 'As you have the water' is meant to be interpolated here.

should not have sold the shearings yet, but you had not, I suppose, got Matty's letter. Mr Sayle says good beef at Pomfret[116] fair was 8s per stone, beef very scarce, mutton will be wanted to make up with. Our selling Jere 250 wedders, and the laid turnips will probably enable us to keep our fatt stock further on in the spring, as it's supposed fatt cattle are very scarce. There was little keep at the end of the season, no fog or turnips to give stock, hay uncommonly dear, bad and scarce. Mr Sayle sold old hay at 15d per stone, more in his power 10£ or guineas whichever he chose to take. Jere Clayton wants to engage your cattle or sheep, but we cannot interfere with you and Bulmer, as he buys and pays well. Lincoln bought our ewes, marrows[117] to yours at 38s per, also 2 cows at 32£, were to go this week but not gone, which he is wrong in as the ewes are too far gone, at least some of them. Your white quey well sold. Keep the little one on. She will eat little, and may be well sold, a nice weight, must be 46 stone or above. Oil cake[118] is high, 10 or 12s per ton. Hay can only from necessity be given to cattle, corn out of question except blackened barley, much of that sort Mr B.S. says will not malt. This makes the good exceedingly high.

As[119] you have the water in the bog near the ram in the great carr by putting in an odd pump, hope the springs on the south of the race may be mostly catched by it, especially as the ground saddens and sinks lower. Also think may be a proper time to run furs[120] with a plow in the carr if the plow does not sink the horses. I agree to give the men some reasonable money for fitting the carts although it somehow appears rather clandestine, at least it has not a good appearance, a thing I hate if it proves so. The road is much injured and in bad repair, but they receive much money on it. Some person threatens by the Newcastle paper to indite the road. This will come upon the townships. I hardly like the solicitor upon the road. Pray what is done about Peacock's brother? I fear his family may suffer. You must consult Joseph Hodgson and Tod, and not let the family

116 Pontefract.

117 Companions.

118 Oil-cake, a byproduct of the linseed-oil-extracting industry, was used increasingly from the 1760s as a feed for cattle in winter. By the end of the century demand was widespread and the price was rising. Over 1000 tons were imported in 1792: Trow-Smith, *History of British Livestock Husbandry 1700–1900*, pp. 81–3; *Agrarian History of England and Wales*, vol. 6, p. 1036.

119 An 'x' on the page here indicates that much if not all of this paragraph is meant to be interpolated above where shown by n. 115.

120 Furze.

starve. They are casual poor (I suppose) and you must maintain them if no other supports them. Have they made any complaint to the township? Wishing you and yours and all neighbors the compliments of the season, I am your obliged

Mattw Culley

PS. I wish Mrs Peacock a good recovery and good luck to the child.[121] So Middleton is gone, we stop off one after another. This is not a proper time to send sheep, will say afterwards what stock you are likely to need. I am glad to hear you say from your receits will be able to pay most of the bills. Mr Peacock has we suppose a surplus in his hands, as my brother sent him a bill for 100£ besides the 80£ left him and a good balance also when we settle his account on the 22nd November last.

Robson who I saw on Thursday Jan. 1st at Wooler, will hardly have your straw cutting machine ready for the cart. Desired him not to put on feet as we wished to save carriage of superfluities. If the men cannot put it on, must wait, he may possably be with you some time. I did not come by Newcastle. Mr Bates said [he][122] would order the drying plate and coal varnish, but fear he has not ordered them. Say if he has in your next, if he comes soon shall write to you again. *1800 January 7.* Should have sent this sooner but never got an account of the tups till this moment. Shall send for them on Monday. We received your letters by T. Wright.

[*G.C.*] Matty and Nelly got home Thursday sennight. Mr Robert Selby of Eale[123] died suddenly at Ammerside Law a few days since. My mare fell and cut her knee that day we left Denton when near Halton, so have only a Galloway to ride now.

---

121 The Peacocks' daughter Mary was born on 24 December 1799: DUL, BT Denton.

122 A word seems to have been omitted here.

123 Robert Selby of North Earle, d. 21 Dec. 1799: *History of Northumberland*, vol. 14, p. 175.

1800

30.　*Matthew and George Culley to John Welch*

Wark January 10 1800

Yesterday received Mr P. and your letters, now find my brother has never sent my letter for you. As for the grinder, if he has not cast up suppose that he has found a good birth. No matter, he may hardly be worth having, likely may be uncertain, so would put you out of humour at times as your business in the mill line is great at this time I suppose you mean. We have winds here north easters, sometimes to the south, are plowing but no regular fresh. May thresh for Todd if you have water and can do his work regular.[1] *He will be not good to be your master*, make such bargain as you think eligible by stook or so on, do it well. Mr Bates now for certain has lost his farm. Mr Bates of Aydon,[2] Sir Edward Blacket's steward, his son Tomy has got the farm. I think you had better send away the millwright, till you get the blue stones home. I fear Mr Baker will be hard on your paying for the wood as I now understand they say he is covetous, act properly to him but I dealt with him as a relation. If he is too tight we may go to some other person, are not bound to him, deal with him as any other man under no unreasonable restraint. I have wrote to Mr Peacock in regard to Moncaster. He has little right to the house as her brother is living. I wish to let him have another house for his and her life, if Mr Peacock and you approve, and you employ him and family at such wage as he desires, rather better than worse. But I hope he does not belong you as he has no legal claim I apprehend. Endeavor to get his settlement made good. Had Carlisle kept possession it would have been good to him as possessing 60 years or above, but the sister possesses. Happen had better do things quietly. Johnson of Darlington understands I suppose such a case, but I do not like ungenerous behaviour. Joseph Hodgson should exchange the garth and houses on the south side of the town (that is old John Wright's) for the house and garth on the north side, formerly Dicky Sedgwick's, now and then adjoining Burrell's freehold which was sold off.

1　An 'x' here indicates George Culley's remark below.
2　George Bates of Aydon White House, 1733–1816, first cousin of Thomas Bates, Matthew Culley's brother in law. He was steward of the Matfen estate. Sir Edward Blackett of Matfen, 4th Bt, 1719–1804. See No. 24 and n. 101.

[*G.C.*][3] I am affraid you will not comprehend my brother in the paragraph above, at least I could not understand him well myself. But I think he means that it would not be well for John Todd to be your master. And so I think myself, and hope you will not permit him to be your master. I have no doubt of his being a good farmer, but at the same time I am apprehensive that he is a *narrow selfish* designing man and should be cautiously guarded against. But I have no doubt that your own good sense has found him out by this time.

Geo Culley

[*M.C.*] We send the bore rods, which should perhaps have a close shell as well as the open shell. Also a feather bed, which should be looked at to see if it has taken any wet.

Please send Mr Peacock's letters and parcel to him immediately, that he may have time to send answers for Jack to bring back.

31.   *George and Matthew Culley to John Welch*

Eastfield 11th January 1800

Well John

I received yours safe by Thomas Wright, and though I have not much to say must drop you a line by Jack, or you would have a great right to say I was not kind. The weather seems to have a tendency to frost and snow, which retards farm labour although it is certainly the best time we can have such weather. I am glad to hear that your little stock is looking well. I do assure you that in this country winter meat of all sorts will be very scarce, especially if this weather continue a while. I am sorry to find that turnips are hurt by the late severe frost. If we should happen to get any severe bare frost between [now][4] and Candlemas, I should not be surprised if turnips take much hurt indeed, for that is the trying time for turnips. Morpeth was a pretty good sheep market on Wednesday last, but not so good for cattle although there is no fear but both will be higher that [*sic*: than] any of us ever knew them in the spring. But the difficulty will be to weather the storm. Lincoln's man was storm stayed at H[augh]head near a week, but he deserves to be punished for not sending according to his promise the week before, which was an admirable week for driving. For God's sake be sure you deal with good people for every article. I have got such a fright with Paul[5] that were I able I would go to

---

3   An 'x' here refers to the remark about John Todd above.
4   A word seems to be missing here.
5   Thomas Paul, the Shields butcher, had gone bankrupt (*Newcastle Chronicle*, 19 Oct. 1799), evidently owing the Culleys money on bills.

Morpeth always when we had fat to sell. You should be making enquiry about Bulmer, because jobbers may be good today and very bad in a few weeks. However if he pays as he goes it is everything, and you as a servant have a great advantage because you can say that you have your master's positive orders not to sell without ready money. And mind that I do order you to do that – at least I must beg that you will deviate from that order as seldome as ever you can. I am sure I will never find fault with the prices sold at, without you were to sell them glaringly ill, which there is no fear of. What has reduced pork 1s per stone? We are all wrong, because we have several porkers to sell. I cannot conceive what should reduce pork when all other things are advancing. You may depend upon all good grain advancing through the winter now, and I sincerely wish it may be had in the summer months from abroad to keep away a famine. We send you a new bushell of William Brown's red wheat, but I can't say it is a favorite of mine. But there is no great harm of trying it upon a small scale. I wish all new tryals to be done in that way, without the matter to be tried has been approved by a person we can depend upon. We expect the bore rods and other things from Wark today to go by the carter. But I wish this snowy-like day may not prevent our sending Jack. I think I once recommended to you to have a letter always upon what my brother James, poor man, called the *stocks*. That is, to begin a fresh one as soon as you have finished the last, and write as things occur to you so that it becomes a kind of journal of your proceedings, which I think the best mode of all for your government and our satisfaction. Whenever you write, may as well direct to me as the letters by post always come here first. Except if you have anything particular to say to my brother which I ought not to know, and then I will never open the letter directed to him, I will assure you. There is no fear but a miller will be got, but the difficulty is to get a proper one. The situation and trust is so peculiar and tempting that I think 10£ no object provided you could meet with a man who is both clever and honest. My brother I see has hinted to you about one I think well of, but it is a very delicate matter to touch upon at present. With every good wish to you and yours I am your sincere friend

Geo Culley

[*M.C.*] Be so good as send a copy of the journal of your book or common dayly account, whichever may be such as give you and us least trouble, so as we may understand your accounts properly as to receipts and payments. We send a parcel by Jack for your father, which he will leave at West's, Chesterlee I trust as he goes south, and hope Mr W. will take care to forward it properly. You can make enquiry if your father gets it safe.

[**G.C.**] PS. I hope John you and Mr Peacock have got all settled about George Thompson, as we wish to see the account from Mr Peacock. As you know these matters are all paid for by my wife, and it is therefore necessary that we should have the account sent to us. My brother has said above that you are to send a copy of your day book. This would be much trouble. I think if you send the heads or castings up it will do very well, and your books can be overhauled another time when some of us are over.

G. Culley

32.   *Matthew Culley to John Welch*

Wark 16th January 1800

John Welch

I wrote an answer to your last, which I hope you received ere now. Mr Bates is not come north yet although he told my Nelly at Newcastle that he meant to be here on the holidays. I have now wrote to him about a man in pretty strong terms to grind for you. We have too good a chance for Story as Mr Atkinson keeps very poorly, but better say nothing of the matter. Wish you and family and all neighbours many a happy return of the season.

2nd [*sic* ?20] January a severe blow of snow came on since daylight, happily it is on the forenoon with daylight as it loads on very much. William Brown and the other farms of Longknow and Thornington are very scarce of winter food, so I am afraid we will have to send both fatt and fresh stock to you early in the spring before you may be willing to have the fresh stock. Our folds are very full of young stock and are to take more, also the Longknow ewes were to come to Thornington today and Harry's[6] ewes to Wark. My brother George and we have a surplus or good stock of hay, so have you, and we and the country may have too much occasion for it all if the winter proves, as it now is, a severe one. Lincoln bought 100 ewes and 2 cows, and like his brethren left them 6 or 8 days longer than his engagement. Am afraid he has done wrong, wish they may get well up. It is a long drive for them, only went to Millfield first day, where G.C. sent some hay. We got Matty and Nelly well home, came to Wooler with Mrs Selby's son and daughter, 4 in the chaise, where I and Jack Smith brought them from on Thursday before Christmas.

---

6   Harry Rutherford, overseer or steward at Thornington, the man sent by the Culleys to learn watering methods from George Boswell at Piddletown.

Mr Robert Selby of Eale died suddenly at Amerside Law a few days since, left a widow and 3 daughters and a young son I believe. You must consider your poor people and labourers to get them corn (such as apply to you) on reasonable terms. I cannot tell whether Mr. Bates ordered for you the round drying plate and coal varnish, but he was at Newcastle as he called on Nelly there. I think the carr in oat stubble for fallow this year must be laid down for watering whenever there is a surplus of water, and a proper oflet or more whenever the water can be spared. I hope you attend to the water which comes of the coal road into the carrs. From your obliged friend

Matthw Culley

33.   *Matthew Culley to John Welch*

Wark 24th January 1800

John Welch

Well John my brother sent me up your letter this day of the 19th. I am glad you got the things well by Jack, hope they received no harm by wet weather. The winter is very serious, especially on the hills. We have Thornington ewes, and Longknow ewes are at Harry's as they could not live on the hills. On Wednesday last we had a fine fresh wind from the west (in the evening very mild) blew exceeding strong all or most of the night, yet strong till noon when it fell south, and to east when a very drifty storm succeeded for some hours, got the sheep secured, settled near close of day with a shower of snow and the wind abated to a calm. Today a frost with some sun and always wind, but moderate. Am glad your mill does so well, and threshes well also. Hope Hymers may answer your purpose, but have not seen or heard from Mr Bates since I left him. He was at Newcastle I think on the Saturday, and told Nelly he would be at Wark in the holidays. We now understand Sir Edward Blacket preferred Mr Bates's son Thomas on his father's account for good service, which was likely to be the case as George Bates or son Tomy gave in a proposal for Halton. I suppose the country will blame Mr Bates of Aydon Castle.[7] John Tod uses me very ill. You may tell him that I only promised him the farm conditionally for one year more, provided that he could get no other, now as he has an offer of Morton at the rent he gave in by proposal, that he cannot expect to keep Denton now as Morton is offered to him at 590£ a year. I wish he would stand to his bargain, I

---

7   Presumably John Moore Bates, 1773–1843, elder son of George Bates and brother of the younger Thomas Bates. See No. 30 and n. 2.

am very willing, nay wish that he should leave Denton as an honest man and let me have it. Then I could accommodate Tomy Bates with it. I believe Mr Bates would like well to have it, and where can he, Tod, expect to get so good and cheap a farm as Morton I suppose may be at that rent, and with such things as are belonging to that farm. He behaves very shabbily if he misses Morton and keeps Denton. Greenwich Hospital's farms such as Scremerston, Mr Rome's, Mr Ancrum's and Mr Scott's are let very high as all people say who know them. There is a very great advance on them, but are near coal, lime and market.

I sent the wrong rods by mistake, they belong my brother, perhaps they may want a close shell to draw loose stuff from the bottom when it becomes wet, but may first try if it be wanted. If you got the book on draining[8] there is a spier pointed chissel to push down to try if water will burst up below the boring. As to the mason he wanted genius and cleverness in not springing a stone on each end of the stone that lies over a cover for a door window or hole, so as the weight may be lessened on the weakest part. I leave to you the ½ per foot or yard at present. I agree with you, when you imploy a clever man and one of character you then have a pleasant satisfaction which cannot arise when anything is blundered. Gilkie built a house by the great for Mrs Grey at Kimmerston as you descend the road to Ford. When covered a few days the whole sliped down and fell. He will have to rebuild the whole now. They have good sandstone at Kimmerston but he over sanded the lime with a soft (not a sharp washed sand) earthy sand or loam, which wants the binding quality. A wall to stand forever should have well slaked lime mixed thin with a sharp well washed sand, no earthy particles, a great portion of sand, 4 or 5 parts of sand to 1 of lime. Mr Anderson calls it chunam in the East Indies, and says it will if properly prepared set in water. Wheat was sold at 3£ 5s but very moderate (barley at Kelso 40, oats 32 our bole on Friday last Mr Nisbet said to go north) by Ralph Brown. Our

---

8 Several books on draining land were published in the latter part of the eighteenth century. Thomas B. Bayley, 'On a Cheap and Expeditious Method of Draining Land', in Alexander Hunter, *Georgical Essays*, vol. 4, Edinburgh 1772, was a twenty-page essay. Matthew Culley may well be referring to John Johnston, *An Account of the Most Approved Method of Draining Land; according to the System practised by Mr Joseph Elkington*, Edinburgh 1797, a more substantial and technical volume with illustrations. The system in question received a good deal of attention in the *Annals of Agriculture*.

wheat all in general hen corn. The best is winter or spring corn (what I mean).

Tuesday 28th. Ralph sold 80 boles Wark wheat very indifferent and (20 red 60 white) 30 from Harry very indifferent indeed, and 50 little better than Harry's at 65. Ours stood to 70s he says, but it is expected grain of all kinds may be dearer in seed time. The wheat in the Tweed side is poor skinny corn, little meal in it. I have seen it as unsound, but never see such moderate grain in my time nor so late a harvest. Seed oats and barley, and beans or peas, will be bad to get, such as will grow well, though barley and oats are a much better crop than wheat. My brother says he has wrote to you, believe he has ordered 2 gallons of rum from Mr Moffatt[9] of Newcastle by Mr Howey's waggon,[10] so may enquire for it as soon as this letter by Matty Culley comes to your hand. We have now fine fresh winds and weather, hope to sow soon if the weather stands good. Should you not sow your fallows with wheat without plowing over, if the weather allows you, as they now lye, by having a long swing tree[11] so as each horse goes in the furrow only, which will not tread the land if wet, give it very little harrowing. There is little wheat sown this year for next harvest, therefore should hazard the sowing if it can be done in good time. We have only sown $2/3$ of the Bar or Back Green after pottatoes and beans, low haugh next Duns Mill, and West Broomyknow with wheat, have Dry Tweed, Lambsknow, a part of Leamingside was rape, the west part oat stubble, south parts of town field bean stubble mostly plowed and some parts more of turnip land all to sow with spring wheat. Much of it will harrow nicely if a day fit comes in time for sowing. There has not been so barren a year for corn I suppose for a century or more I suppose (also a part of Jocks Rigg that was rape and turnips ready plowed which I had missed).

I am with respects &c Mattw Culley

I send this by Matty Culley to Newcastle. The weather now [. . .][12] of sowing Lambsknow with white wheat, tomorrow morning if it

---

9 Thomas Moffatt, tea dealer and spirit merchant, Newcastle upon Tyne: *The Directory for the Year 1801, of the Town and County of Newcastle upon Tyne, Gateshead, and Places adjacent.*

10 Thomas Howey & Co were carriers between Edinburgh and Birmingham, London, Manchester, Newcastle, Nottingham, Sheffield, etc. The service ran six days a week. *The Newcastle and Gateshead Directory for 1795.*

11 Swingletree, crossbar on a plough, carriage, etc., pivoted in the middle, to which the traces are fastened.

12 The paper is torn here.

stands good, frosty in the morning. You should get seed and sow when you can. There is little wheat now in this country. Seed of every sort is I hear very bad in quality. Now as to Todd he is a man of little honor I fear. He said to me his farm would bear no advance of rent, then said his wife was uneasy, and he submitted to an advance of rent. I greatly fear the man is without much probity where self interest comes. As he has an offer of Morton, a good or promises to be a good farm he should accept it and give up Denton to me. May tell him I expect he will give back my farm and take the offer he has. The one may be a certainty for a lease. He shall only have mine for one year from May next. Mr Atkinson of Yeavering died on Saturday morning about 8 o'clock, has left a widow and 4 children I believe.

34.　George Culley and Matthew Culley jr

Eastfield 25th January 1800

Well John

We received your packet of long intelligent letters for which we are much obliged, and am glad you propose in future to write a journal-like letter or letters. I sent the long one up to Wark. Your books we have not yet had leisure to examine. Jack was lucky in getting home before a most terrible blast we had on Thursday which has partly filled our lanes in places. By coming in the day time I hope no loss would take place amongst the sheep. It blew the leads off the southmost part of the hip roof of this house. It makes the winter to have the appearance of being very severe now. I am sorry to say that our turnips here are a good deal damaged, very luckily not so at Wark. We think a few of the highland wether hogs in the spring will be better than gimmer hogs, or the latter small ones frequently turn better out than one imagines. Indeed nobody can say what a small weak-looking well-bred hog may turn out to be. The turnips bought by my brother of Mr Marchbanks I find are like to be soon done, and like other bought turnips will prove dear enough. However the 200 ewes must either be marketed in a fortnight or 3 weeks time, or sent to you. Would wish you in an early answer to give us your opinion. Our idea is that as soon as the roads and weather will permit to send you a 100 in the first place if you approve, or you may have them to come by Morpeth where you can meet them, and after trying the market take the remainder home. Grieve Smith,[13] a son of Mr Smith

---

13 Grieve Smith, farmed at Budle, married James Culley's daughter Eleanor in
　　1805: DUL, BT Ford.

of Hay Farm, was here last night and has been at Morpeth these 2 or 3 weeks, and says that good sheep are sold near *7d* sink, from 6 to *9d* per lb. sink according to quality. I am affraid ours are not of the first quality. Or if you think you can parcel them out better at your own home we will send them to you. But of this you can determine and inform us, time enough in your answer. I am very happy to hear that none of your turnips have suffered yet. Your determination to sell in general for ready money I certainly highly approve of. As to C. Charge &c I cannot find fault with or any other good man whose situation you are well acquainted with. But really it is not easy knowing people in these days. It would be very foolish, and very unlike my disposition, to think of confining you to sell to Jere Clayton or any other person. By no means. You are or ought to be a much better judge of the customers in your quarter than I can pretend to be. And though Jere be an excellent payer, as long as Mr Bulmer pays ready money he is certainly as good. And I must agree with you that Jere is a *teaser*. So mind I leave you at full liberty, only to remember the *monies*. Since I was so taken in by Paul I do assure I am become determined to be very cautious. He has been buying some fat cattle of Harry Howey,[14] which he pretended was for his mother or mother-in-law. However he was to send the money, which he did for the first bought, but the next his man got away without payment from the steward at White House. Harry followed and sold the cattle last Wednesday at Morpeth, and Paul threatens a law suit, and if he does sew[15] Harry few will pity him as he had no business to deal with him again before his affairs are settled or his certificate signed. I will have you to keep to that resolution of selling for ready money in general, and that money hard cash or bank notes, or if in bills be sure that the date be short. Bills are a dangerous business at any rate, but it is not easy to refuse them at all times. However as a servant you can say to Bulmer &c that it is your masters' *express commands* that your payments are to be in *gold, Bank of England notes,* or Darlington, Richmond, Newcastle, Durham, Stockton, or in short such country banks as you are acquainted with. I think that fair, because bills I highly disapprove of, and by keeping firm he will continue gold, notes &c, &c. I don't want to give you pain about Cuddy Common &c, &c, but you might as well have made *ducks* and *drakes* of the money. But you

---

14 Harry Howey, Wooler Haugh Head and Brandon White House.
15 Sue.

made us plenty of money other ways, therefore I would do wrong to repeat grievances.[16]

Beans and peas cannot be had here for seed. I have wrote to Norfolk for 20 quarters, but whether I shall succeed I know not. Barley was sold at Kelso last Friday at 50s the Scotch bole or 40 the Berwick, and I have no doubt of the best seed oats being sold at same money in a month's time. We are going to get 8 boles of oats from beside Carlisle again this year, what they call the potatoe oat. Now my brother is very desirous that you should have 3 bushells of them to sow, but I know not how to contrive it. He would have me write to the man to send them to you, which I can easily do, but I think it will be much better for you to write to him, and then you can tell by consulting George Dent, Mr Peacock, &c, &c which way they can be best sent to Barnard Castle directed to you, or perhaps better to George Dent who I am certain will bring them to you to Killerby from Barnard Castle. And I can name this to Mr Faulder our friend, who is to send the oats. If you approve of this, direct your letter to *Mr. Joseph Faulder at Park House near Carlisle*, and point out to him how to direct to George Dent (I think) at Barnard Castle, and I will be bound for it they will come. They are an excellent oat and well worth your while to cultivate, George Humble[17] says they had many more per acre than any other sort, are early and of excellent quality. I will settle with Faulder for them, tell him you are our servant. I think you will continue this from what hints I have given you. Wheat is sold here from 50 to near 70s per bole, barley 30 to near 40 and oats 24 to 32s. No pease, but beans and pease come in by sea kiln dryed and coarse ones, are selling by the merchants at 3£ and 3£ 3s per bole. Certainly if you should have to come to Morpeth, which you are very kind in offering, you must have a good horse, and may be looking out for one if he should happen fall in your way I am yours truly

Geo Culley

[*M.C. jr*] PS. Please let Mr. Peacock know directly that my father received his letter and bill by Jack, and will write to him very soon.

---

16 The words 'Oh goody goody, you only lost £12 here' have been written above the line here in a later hand.

17 Steward or overseer at Wark.

35.   *George Culley to John Welch*

Eastfield 3rd February 1800

Well John

We have received yours of the 31st January this morning and as my brother is here I am desired to answer it directly. In the first place we shall neither send the ewes to you or to Morpeth yet awhile, and if we do, shall give you proper notice and I hope in time, and will also mark them out in lots and give you our opinion of their weight when sent to Morpeth. The hogs we shall not send untill some time after, you may write when you wish to have them. Mutton was sold current at Berwick last Saturday at 6*d* per lb. and best cuts of beef 7*d* per lb., and it is the universal opinion that fat stock never were so scarce, especially beasts in this country. In the Merse or Berwickshire they say there are scarcely any fat cattle at all. It needs no prophetic powers to foretell that fat stock must be higher than ever known, Only the turnips failing by the frost, and changeable weather, will of course send stock to market, and consequently keep prices down, which will make a greater scarcity afterwards. In short fat stock must be scarcer in May than we ever knew them. Mutton seems to be higher with us than you. But I have no doubt but prices must advance with you very soon, as well as here. And although our turnips are going faster from frost &c than we could wish, we may as well eat them as suffer them to waste entirely. Certainly we wish you to keep your cattle as long as you can, except the queys in calf which certainly you should make the best end of you can. And I would have you recollect that you have always a discretionary power to sell or keep as you think best, because before we can advise you an opportunity may be lost. I cannot think of ordering any pease from the south to Stockton. It is so uncertain a business, and very difficult to get without being killn dryed which would prevent their growing.[18] In regard to your buying a horse or mare to ride, certainly a mare will be easyest to come at. But I hope you will not have to come much to Morpeth, perhaps not at all this spring. However we think that perhaps you had better buy a strong horse, or mare, that would draw, and ride the *black mare* which you used to hack formerly, when you have to take such a journey as Morpeth &c. But the best way is to buy none at all if you can do without, so long as corn and hay are so exceedingly dear. I wish you may be able to make out this letter, as I have wrote in a most violent hurry while company were present

---

18 Dried peas would not germinate.

talking to me. However I must remind you to make proper stops and commas which you have neglected sadly in your last letter.[19]

Matthew Culley to John Welch

Wark 11th February 1800

John Welch

Your books will we intend be sent by Matty to Houghton where you may get them when convenient. Mr Bates is obliged to you in regard to Morton farm. I wish I could have accomodated him with Tod's farm this season, but did not consider or know that he would want one. Only I ought to have considered that there was a likelyhood of his wanting one. People here are keeping from selling their fatt stock, which are only in few hands, but likely must sell soon. Seed corn advances rapidly I am told. We shall have to send some ewes soon to Morpeth for you to sell, but not directly. If you cannot get from Morpeth on the Wednesday night, better tarry all night somewhere on the road, than injure yourself or horse. Let not a night's expenses fear you, then go off early next morning for home. I wished much to send you a riding horse from here, but they are too young or somehow not fit to send as yet. Let us know how Northallerton went for horses and cattle of all sorts, fatt and lean, draft as well as road ones. Have sown 3 or more fields with wheat, not in good plight but only tolerable, too frosty for harrowing except a chance day. Our 16 oxen at Wark, heavy and good beef, Laidlaw of Shotton[20] has 6 or 8 outlaid[21] ones, will hardly be able to keep them 14 days longer, can you take them in 3 weeks hence? I have not seen them of some time, can not say what they are like, intend seeing them soon. Do write and say how times go, and what turnips you have. Meat is likely to fall in great scarcity. Should you not plant early pottatoes in your stack orchard behind or west of your house, by digging the land, and then some later ones in the same place. The west side is very good and fresh. Pottatoes will save corn or bread, if such a scarcity prevails as it is feared is too likely to come to pass before harvest. Do not mind the expence of digging any of your garden pieces for carrots, onions, peas and beans &c or for transplanted turnip seed which will be much wanted. Your obliged

Mattw Culley

---

19 The text ends at the bottom of the page with no signature.
20 James Laidlaw or Laidler, overseer or steward at Shotton.
21 Kept out of doors.

Wednesday. M. and sisters set out for Newcastle this morning, hope they may have a good journey. Jemmy Laidler was here yesterday, sayth they can only keep the outlaid feeding oxen 10 days, will be out of meat by 1st April for all their cattle. We will be sore pinched. Would it be right now to sell your little heifers at 8*s* per stone or any of the rest. Turnips are now dearer than ever, oats for seed near 7 guineas per bushell. The oats on the uplands not fitt for seed. Have got a cart load of pottatoe oats, nice seed indeed, from Carlisle by Faulder's purchasing. The weather strong rind, and sharpish frost prevents harrowing. You should not omit to harrow and sow when the weather will allow, even on an afternoon. Little harrowing may do better than much I think. As to your mill and mill race &c cannot now pay proper attention to them. Robson has made a straw cuting machine, nearly finished, if of material service to you will send it by Howey. It's heavy, therefore will be dear carriage, may answer well to cut straw and hay together for your horses, a saving of hay to you. Will hay be so scarce as to be right to sell any, and lead dung and soil in lieu thereof? This is only a hint, for dear as times are likely to be, cattle will hardly pay for hay at 6*d* per stone except in necessity or so on. Neither will oil cake or corn pay I suppose, indeed I think corn should not be given as a general rule any animal but man. It is a heinous crime to waste man's meat when people are afraid of a famine. I know of no substitute so ready as pottatoes, they may and ought to be had long before corn of any kind. We save all ours to sell. When a scarce time comes, none are given to horses, also excellent for them. Man is to be looked to. I wish you to omit some things rather than not plant pottatoes in such garden mould or soil as you have in the old orchard or garden spots. Fill any corner up with them this season of scarcity. Consider your fellow creatures, and your own family will be put up nicely with new pottatoes if good, and you will get the blessing of the poor &c.

When you are at Morpeth pay George Fenwick what money he has paid to Lumsdon's widdow, with 5*s*. over for the ¼ to Candlemas will make 1£ 5*s*.and 6 to 1£ 10. 6, and if need requires as I suppose it may, give her this next ¼ 1.10.6 to 1.15.6. as he sees occasion to give her. She must not be allowed to starve. We give or allow our lower folks bread meal at 20*d* and oat meal at 30*d* per stone. They are at[22] . . . [oatmeal] 5*s* per stone. As to coal varnish and the drying plate I have not spoken for them therefore must write yourself for them.

---

22 There are dashes on the line under 'bread meal', 'do' [ditto] under 'oatmeal', to produce a kind of table.

37. *George and Matthew Culley to John Welch*

Eastfield 17 February 1800

John Welch

We received a letter the other day from Jere Clayton, who is very desirous to buy your fat cattle. You know I wrote to you before that I would never wish to confine you or any servant we have to sell to any particular man. Neither would I wish you to be confined to Bulmer or any other, but to sell to good men and not only for ready money but (observe what I say) that money to be in Darlington, Richmond or other country bank notes as you are acquainted with. But not in bills or drafts *on any account* even if they are in short dates. Because draft or bills are never safe, but gold, silver, Darlington &c bank notes, or Bank of England notes there can be no danger of. Now I must beg and intreat you to attend to what I have said above, because you can say that it your master's express orders not to allow you to take drafts or bills on any man or on any account. Besides I am told that some of the Skipton jobbers are very suspicious. Now although Bulmer has generally paid you ready money I am affraid he has not always done so, at least it appears so by your books, and if you only give him a little credit *once* he will expect more the next time. So [. . .][23] to say that your master's express orders are not to give any credit or take any drafts of anyone at any time. Now although Jere Clayton be a very teazing man to deal with, can't you fix your price and let him teaze on. When a man has his goods at home he can be as firm and as determined as he chuses. But at market it is different, especially when you had to go to Darlington from Red House. Now when you are so near Darlington it will alter the case. Remember when I have said this, that I don't mean to confine you to Jere or anyone. Only I would prefer a teazing good man to a free-buying suspicious one, and all I have said only means this: sell to those that will pay you in good money, not drafts. If you hear nothing to the contrary you must come to Morpeth to meet 50 or 60 ewes on Wednesday the 12th March, or rather the evening before viz. the 11th. But I hope we shall hear from you before then, both how Allerton fair was and how the 1st Monday of March[24] proves. We had thought of sending the 7 cattle to you from Shotton, but my brother is for keeping them a week or 2 longer, but you may expect to have them in 2 or 3 weeks. I am sorry to say that our turnips are in a very bad state indeed, and how we are to put off our fat stock I know not. Indeed this is the most alarming season I ever knew.

---

23 The paper is torn here.
24 A major fair at Darlington.

Wheat now 4£ per bole, oats 36 to 42s, barley about the same, but beans beat all 4 to 4£ 4s per bole. Morpeth was good 7d per lb. sink for mutton last Wednesday, beef I did not hear. But the rot in turnips must sink fat markets soon I think, and what may be prices after God knows, nor who can weather the storm is not easy to say.

[*M.C.*] John I sent a letter by Matty with your books to Houghton, hope you will get them safe. The letter was to be put in the post.

[*G.C.*] If you have not already bought a horse we will try to spare you a hackney.

I am yours truly Geo Culley

PS. Do write and let us know how times are with you.

38.   *Matthew and George Culley to John Welch*

Wark 22nd February 1800

John Welch

Well John we think we are so far from Matty that he may at the Easter vacation be sent for by you to Denton, and you can send him back when the holidays are over. He will see Denton and go into the fields with you, but take care he goes not too near the machinery as several misfortunes have happened by this. He may also assist in planting some willow trees as an amuzement to him. We can allow you for his board while he stays at Denton. He must take some spare cloathes and shirts with him. Have you done anything about a drying plate for oats &c, and about the blue stones? You will now have a better guess about the water as the weather has now been dry for some time. We have got most of our seed wheat done. As it had been long plowed it harrowed very fine, the frost pulverised it to a nicety. We hope that you will have got much of your wheat sown where the land is dry. Have you ordered seeds from London by Mr Peacock? If you could get perennial rye grass in the country or near Durham, may be better than London seeds. The full large big seed is often of the annual sort. It grows very strong and vigorous but goes off after mowing &c. The first season is very pleasing to look at. John Dickinson will stay now with his unkle by way of improving himself, is a nice young man I think,

[*G.C.*] So far was wrote by my brother. Now I have to tell you that our turnips are wasting so fast that we are determined to begin next week with our ewes to Morpeth, where we must request you to come on Tuesday evening the 4th of March. George Johnson has the key of the folds at Mr Whitfield's, and Matty I hope will meet you there, and Ralph has promised me that he will draw the ewes and mark them in lots and give Matty Culley his opinion of their weights. Last

Wednesday was a most particular good market, full 8*d* per lb. sink for good sheep. Ralph Fenwick sold his shot or cast wethers, not very good I am told, 22 lb. per quarter at 58*s* each, which is as near 8*d* sink as can be. You can either come from Darlington fair to your father's all night on the Monday, which will make it easy for you the next day, or come all the way from home on the Tuesday. This I must leave to your own determination. I rather wonder that we have not heard from you since I wrote a very long letter and sent you. However you must not neglect to come to Morpeth. I am in haste

Geo Culley

PS. It is more than probable that this waste in turnips will throw more sheep than common into the market, but you must do as well as you can and be thankfull. We shall only have two weeks sheep for market at present.

39.   *Matthew Culley to John Welch*

Wark 25th February 1800

John Welch

We sent you a letter yesterday desiring to meet Matty at Morpeth with ewes on Wednesday sennight, the 1st Wednesday in March. Now you must desire Mr Peacock from me to get Tod's rent immediately as it is time he should pay, besides he is not the man he should be. If you have no occasion, or Mr Peacock, may rest it in Weatherel's bank[25] until occasion requires. I will want it no longer. You have neglected to write and give us an account of Candlemas at Northallerton. My brother and William Brown are almost out of turnips, so necessity makes us sell, and they have sent 220 wethers to us and William Brown sent us 100 hogs, are to send us some oxen, and we must try to sell Byrne our oxen. Perhaps if you can spare 2 or 3 days should come on to Wark with Matty from Morpeth. We shall send you 7 outlaid oxen directly, they are pretty good, not so fatt as yours are I hope, were wintered at Shotton until a few days ago. The dry frosts have rotted all the turnips or nearly so, but oxen no, 15 mostly very good above 80 stone. Whether you should sell hay or keep oxen I know not. Selling hay will make more money I suppose. We cannot well send you sheep yet until your clover &c will keep them. Of this you must give us timely notice so as the cattle or sheep for the clover may have time to travel to Denton, against the meat

---

25 Wetherell, Mowbray & Co., bank at Darlington. Originally Richardson & Mowbray, founded before 1778, name changed to Wetherell Mowbray August 1799: Phillips, *History of Banks*, pp. 333–60; *Newcastle Chronicle*, 3 Aug. 1799.

may be good and fit for them to go on to. Copland Mill is now advertised, therefore Story who was (and is now at the mill) servant to the late Mr Atkinson will be at liberty against May or Whitsuntide, who offered to come to us, and the market for hiring begins in March at Cornhill. Therefore should let us know whether the man you now have suits, will stay on or better hire Story. We should wish neither of them ill by keeping them in suspence. This you must turn in your mind and determine which way may be likelyest to be done. Can the man you now have be depended upon? We believe Story may. Jack Story and Robin Story are persons to be depended on as we think, and are his brothers. If Story is hired (or whoever you hire) should you not hire or get a house for him as near the mill as possable? The reasons are obvious and plain, more convenient, more in the way to prevent accidents or damages, and to let on or off the water, more security to the mill. Mr Bates said the man you have could put the mill right when need required, as a millwright. Have you tried him in picking the stones &c? These are only hints for you, and which way will take the attention most from your shoulders and Harry's. With respects to Mr Peacock, family and all friends I am your obliged friend

Mattw Culley

Mr Peacock should also speak to Joseph Hodgson to take his rent at a reasonable time when it suits Joseph. He may have had indulgence from Mr Bowes,[26] yet on 2nd thoughts Mr Boweses party could give much time. How have you done about Tod's threshing his corn with you, and also Joseph's? Pray how goes on now Birkbeck? He should pay now about, used to pay nobly in great time. If money be wanted by you should get of them, as ready prompt payments are necessary and keep a person firm and satisfactory in his own mind. Always keep your payments clear with rich and poor. Sometimes a tenant may be able to pay part without being put about, but Tod must pay directly on demand. If Matty wants a little money may give him 2 or 3s to pay for necessaries. We wish him not to waste his money, or to be in want. Mr Fleming gives him to pay for necessaries I believe, charge it to my account. You should measure the land taken from Tod in the Weems bottom. There was a hole fell into the drain across the low well spring, is it filled up? That part of the field should not be

---

26 Probably George Bowes of Streatlam, who died in 1760. His daughter and heiress Mary Eleanor married, in 1767, John Lyon, 9th Earl of Strathmore: *Burke's Peerage; Oxford Dictionary of National Biography*.

plowed again when laid off with seeds properly, rye grass and white clover principally.

26th A hard frost. Robson never got the straw cutter finished, a heavy business will take up much room. How can it be got sent, is some cwts. and cumbersome. Pray how does the springs in the carrs do? Does the water rise and land full? What hedges have you in? Does much water rise in the wet place below the limestone bank quarry? Have you bored it? Do not leave it short. Sink it well down on both east and west side, so as to give the water all the vent you can. Do not fill it up this season.

40.   *George Culley to John Welch*

Eastfield 28th February 1800

John

Yours of 23rd came yesterday morning, but I only came home from the H[augh] Head last night, where and at other places I have been close confined along with Mr Bailey on the division of Tweedmouth common which I hope will soon be done now. I am rejoiced that you have got so much of your wheat sown. Nothing like striking while the iron is hot. The weather is now very frosty and cold again here, however we have got a vast wheat sown at the different farms lately, pretty nearly up with the turnips. Everybody begin to think that horses, especially work horses, will be very high in the spring. You are right to put of with buying a horse. Depend upon it I don't want to confine you to sell to any one person whatever, either Clayton or Bulmer &c, only get your money in cash or such notes as you are acquainted with, but no bills by any means. It does not signify whether Bulmer or any other draw the bills, but they indorse them last, consequently are responsible. But if you get gold, Bank of England or country bank notes you have no further trouble. And *once for all* you have only to say that it is your masters' express orders that you are to take no bills on any account. *Surely where no credit is given there can be no loss.* Although you have sold your gimmers for less price than is now given in this country, we find no fault. It would be cruel to do so. You got as good a price as was going, and your money well paid, which is everything. It is not high prices so much as certain payments that makes people rich, and what is more, *much more*, it makes people *easy, comfortable,* and *as happy as this life can make us,* while bad debts and uncertain payments gives *uneasiness, distress* and *sleepless nights.*

I am extremely happy to hear that your turnips keep pretty safe. Ours are in a sad condition indeed and all this light country. Wark are much better. And still happier that you can take Shotton's 7 beasts. I own they are a small lot, but we have no more outlaid ones, and I think it would be wrong to send any inlaid ones yet, without you could house them when they get up to you. We have some in our shades,[27] pretty full of hair, that we shall not know what to do with presently, as they will not be fat enough to sell before our meat be quite done. Indeed I wish much, when you are so near as Morpeth, that you could have slipped over for a day or 2, and you could then judge better than by a hundred letters. Perhaps you could not come the first week, if you have left no orders at home, except you were to write a line home per post. Give this a thought and come if you can, because the sooner you come the better. Indeed the wisest way would have been for you to have staid, and gone back to Morpeth again, because we shall only have two weeks' sheep at present. The money you make of your sheep I wish to have left with Mr Orde,[28] but that need not stop you because you can leave it with Mr Bennett who I have often paid Mr Orde's rent to. Mr Whitfield will shew you where he lives, just on the other side the street opposite to Mr Delson's. He will give you a kind of memorandum for it. His rent is 450£ for the ½ year. He has got 232£ and I will send as much by Matty the next week as will make up the whole rent with what you leave with Mr Bennett, and then Matty the following week can go and settle with Mr Orde overnight if he be at home, if not with Mr Bennett which will be the same thing. I am sadly affraid corn of all kinds will get much over high. I hope you have got your oats from Mr Faulder by this. Ours are come and very good they are, but the first that came were the best. I shall send Mr Faulder a bill for the whole amount in a day or 2. The beans comed from Mr F. are also very good, and they are not to be had here for money. I have no doubt but you would receive my last letter, and be at Morpeth next Tuesday evening. If the market will not afford 8*d* sink take a ½ penny less, or even a penny rather than set them up. If old Galley be so good a worker don't sell her by any means. I would caution you against buying any queys for Mr. F. without he or son come over, and then lend them any assistance in your power. But to buy for him unseen I advise you against. It is a pity but you had blue stones, I am sure you have taken three pairs. I am glad your miller pleases so well, but I am always affraid of drinking men. I made it a

---

27 Shed, lightly constructed wooden building.
28 William Orde of East Orde, Morpeth and Nunnykirk, 1735–1814, owner of
    Grindon: *Burke's Landed Gentry*.

rule all my life never to keep a drunken servant. I am with every good wish to you and yours

Geo Culley

I now think we have 3 smallish cows which are tied up, and although not fat are in a way of mending and would get fat with a little grass, and would travel cleverly along with the 7 oxen, and if you can tye them up it would be well, and turn them out by degrees to turnips &c. If not they might do in a fold and untill they hardened by degrees at any rate. I think them fitter to go than any of our tied up oxen, and would be some little relief to us who will be fast for keeping for both fat and lean stock. I never remember dreading a spring so much as this.

41. *Matthew and George Culley to John Welch*

Wark 29th March 1800

John Welch

Well John I think that you sold the sheep very well at Morpeth, considering what poor stuff they were. At same time it helps to shew what it is to buy bad or midling stock of any kind. Mark Thompson's sheep were bought at less money, and were smaller when bought. Only he had had inferior sheep, tups, from us for some years, and his land and management were in general bad, his good land most valuable but always over stocked. 8s is a very high price for beef, but ship beef must be had. George Hutchison intended to go forward to South Shields (if he reached Morpeth) as his son is working somewhere there at masonry and his sister &c lives there. Now I wish you could enquire of John Forster[29] quietly and let us know particulars about Molly Chapman. We are told by her daughter in law a widow Chapman, that Molly had 3£ per month allowed by her husband George Chapman when away in the east country, who is said to be a drinking man. But if he has a wife who drinks that may make him drink more. The widow Chapman says she wastes all she gets in drinking, and when my sister sent her some cloaths she pawned them. All goes for drink, she and one of her daughters drink, and one of her lasses is very sober. The widow, by Mr Wallis's[30] help, expects to get (I think she said 150 or 250£) by his means her husband's money, who died in a ship of war in the East Indies, and the money was due to him. She seemed a decent woman, but one story is good

---

29 A Shields butcher.
30 Rev. Richard Wallis, 1753–1827, curate of Bishopwearmouth 1779–83, vicar of Seaham 1783–1827: DUL, ASC, Hudleston Clergy Index.

till another is told, but I wish much to know the truth of the matter. It is hard on George Chapman to bear all the blame when much may be owing to his wife. Mr Wallis, clergyman and magistrate, is she says a very good friend, and when he heard her deceased husband was a distant relative of ours, has endeavored to get her the money due to her. So far a fine seed time, this a soft morning.

Saturday 29th March. A *good* lambing season lately, as the weather was mild and dry. Spring wheat breers[31] finely. We will I hope get the money matter with Mr Winship put right soon. I hope the hogs will get well up and safe. Turnips now put a top, and grasses grow. We put the most of tup hogs to clover a few days ago (Tuesday). Will put more of the wedders to grass as [. . .][32] are now at grass. I hope we will get very little rain, dry weather generally suits the north best at this season. We have had so much wet as has enabled us to run the water over the land. Friday 4th April. This is a fine dry day, has dried the ground finely. My brother was here this morning, has sold Binney[33] the oxen at East and Westfield 11 @ 30£ per 335£

14 @ 26£ per   354£
1 @   21.10

Possably we may be able to struggle with the 8 fatt outlaid oxen now, as the weather is so good that the clovers and pastures take our sheep pretty well now, and hope they will improve if the weather keep moderately fine. Our turnips put a top which makes them go further although very small. We give our lambed ewes some turnips with the grass, helps to keep them in milk. The lambs seem to thrive well. We are glad to hear the children are well. With respects to all friends

Your obliged friend Mattw Culley

PS. Your mistress is now much better, you may tell Matty when you see him. I have some thoughts of being with you before Easter Monday, but am not certain of it yet. Perhaps you may be able to keep your cattle on a little longer, or a part of them, if we can keep on the 8 outlaid oxen. My brother will finish this letter. Robson was saying he would be with you about Easter I think, or send you a man to put up the blue stones and bolting machine.

---

31 Sprouts.
32 The paper is torn here.
33 In other letters written variously as Binney, Beaney, Benny and Benney, a livestock dealer and butcher.

Eastfield Saturday 5th April 1800

Well John

I think you and Mr Faulder bought your queys well. I shall not be surprised John if you get more than 9*s* perhaps 10*s* for your fat cattle by and bye. My reasons are that literally speaking there are no fat cattle left in this country, and don't you think that they will begin by and bye to be scarce in yours. It is a strange price 10*s* per stone, but Mr Sayle sold at that price last year in Smithfield and I shall be much surprised if they are not sold at that price or more at Wakefield and *Darntonfield*[34] too this year before long. You have no occasion to hurry now as we shall not send you the outlaid cattle at Wark at all now. We can do with them, as we have sold all ours now. At any rate I would see Easter Monday over before I made an offer if I can advise you. Pray what must be the result in the months of May, June and even July? I don't know how you are in your country, but we have no following cattle in this country nor fresh ones. You can tell what were at Durham fair[35] where many fresh ones used to be shewn. Depend upon it the like never happened before in our days, nor perhaps may not again of a while, and why not avail ourselves of the times. Besides it is not like a critical time, when stock is likely to come in or is a good grass year in June &c when there are many fresh cattle to follow. There is at least 2 months before any grass beef can come in. Now after this short sermon wrote in a great hurry for Ralph to take to Berwick, do what seemeth good to you. I am sure of one thing, that you will do the best of your judgement and knowledge. Ralph bought a nice hay or dry cow yesterday at 7.12.6, will be fat in the end of July or beginning of August. By the bye, should we not buy you half a score such to send you in a month or 6 weeks time? Tell me in your next. But don't let me forget to tell you a strange but true tale. Mr. Nisbet told our Matthew yesterday that the other day he sold 100 wethers for 300£, only 20 lb. per quarter. This is just 9*d* per lb. What are things to come to! This is no joke. Mr Scott sold some sheep the week before to Edinburgh very high. Both he and Mr Nisbet had their sheep at turnips near Dunbar. I am in real haste

Geo Culley

---

34 Darlington.
35 31 March, a great fair for horses and cattle.

42. *Matthew Culley to John Welch*

Halton 23rd April 1800

John Welch

The man which Mr Bates sent to look at the stones you got from B[arnard] Castle and was at Denton says your nuts below the milstone is by far too small or carries the stone, should be larger. Therefore have agreed with him him [*sic*] to put up your milstones. Robson neglected to send men in time to do the work. Therefore may let them know how that I have agreed with one Phips to be at Denton soon to do the job. May tell them it is not out of disregard altogether to them, but there was an unnecessary delay in the master promising to send men and disapointing the work, besides I apprehend you will find Phips both millwright and practical miller, has improved Mr Bates's mill so as to grind faster and better, with less water, by puting on a larger nutt, says our wheel should have been between the mill and threshing part which he will describe to you better than I can. Besides you will find him a clever hardworking fellow, willing and active, and not so sausy about their meat. Pay Robson's men for their day if they be come, I mean for the few days they may now have been, but although unpleasant to set them off I believe the work will not be so well done as by the man I send, and I have hardly been so well used by Robson, indeed he undertakes more business than he can look after and at so wide a distance. If they are come, behave to the men civilly and tell them I have agreed with another person as their master has not sent men to his word and has trifled more than he should do in geting the blue stones set forward. I hope this man will please you and will be nearer to do anything which may be wanted. Time will shew which is most deserving of being employed. Let us hear how Matty got to school and how Mr and Mrs Fleming &c are.

The weather is now good and grass grows. Hope you will write to give an account of what stock may be of use to you. Shall not be home for a few days yet as Mr Bates has an offer of a farm adjoining Brunton, only the river between him and it. Here are 3 good mares to sell, belonging a widow Bell, joins Mr Bates. Have desired him to buy one of them at the sale on [. . .][36] I have told Mr Jopling who says Mr Colling wants some heifers in calf or calved. I think they call the gentleman Lyon of Binchester near Bishop Auckland, that he may have your best quey that is calved at 20 guineas. Perhaps he may buy

---

36 The date is left blank.

more of them. He is nephew I believe to the late Mr Wren and succeeded to his friend and relation. Do write soon, let us hear how things go and what you are in need off. Your obliged master

Mattw Culley

43.   *George and Matthew Culley to John Welch*

Eastfield 2nd May 1800

John Welch

Yours of 29th ultimo came this morning, and we think you sell your fat cattle at wonderfull prices, and higher than perhaps we can do here at present. Nevertheless as I promised Mr Binney the refusal of our few outlaid oxen I would wish to be as good as my word. If he should not buy them, as I think he will not, I will then write to you and we will send them to Denton if you wish it. I would have been better pleased if you had told us the weights of your cattle as near as you could. You tell the prices and the price per stone in Darlington last Monday, but neglect to say the weight per stone of your own queys. My Matty and I went up with your letter to Wark this morning and we thought it best to send you the fat cows 7 that we had tied up but have been at grass now a few days and are pretty forward, and I fancy my brother will send you other 7 or 8 from his cows, 3 bought of James Darling[37] this day at 14£ each, but we think 2 of them capital and the 3rd not a bad one. They have had a few turnips and have been clear of milk many weeks. I am not certain but he will also send you Miss Hill and a yellow cow of their own, both fresh but some milk which will leave them by driving to Denton, also other 3 or 4 clever grazing cows will be bought, we think, at from 7£ to 9£ per. If a red one with a white face come, she was bulled today, and if a small bodied black polled quey come I would wish you to sell her as I am affraid she is a common buller. We have 4 or 5 more not unclever cows which we once thought to have sent you but were unwilling to overload you. However you might have sold them as they are, as I think them likely cows to graze, and not sold save one. But we could buy more if you chuse and send, but would gladly have you write oftener and not spare pen and paper, and we will most willingly pay postage. I was much disappointed at not hearing from you sooner. We durst not send the little ox as he would have killed himself with bulling the cows and would always be bulling. But you will get him and the others after if you like. They will not stay long in hand I

---

37 James Darling, Hetherslaw, b. 1775, married George Culley's daughter Eleanor
   on 9 June 1803: see No. 184.

would hope, if they were with you. They will also send about 40 or 50 hogs only, and would wish you to send to meet the man with them if you can do it. I thought if we had sent you a few more feeding cows, as we can buy them much lower now than after, that you might have run some of the worst in the Lime Kiln Banks land, until some of the early ones had gone of and then they would have filled up the vacancy. But we durst not do it untill we hear again from you, which I hope will be very soon and often at this time of the year. We would have you buy a cow for your hind about Barnard Castle or some-where, as we have not a suitable one for you. Why could not you have wrote before the market and again after it? What signifies a little trouble and paper, you that can write so well? We are going to Morpeth with 50 clipped sheep against next Wednesday, Matty and I, as we are very bare of grass, good as the weather is. Morpeth is very good I believe but not so high as you speak of. You certainly have bought the 4 little oxen low for these times. I doubt they are over poor to work. We had also a very wet day on Friday but charming mild dripping weather at present. Your corn markets beat ours far. We have had none above 4£ per boll, but wheat is the only grain we have, no pease, next to no oats, but still not so high as yours, and very little barley. What will be done I know not without some large quantity of corn come in. The people are murmuring in places. Thank you for your account of Nelly. So it is likely that Trenham's house will fall into our hands? I shall be glad to hear of George Dent's recovery. Tell Mr Wright civilly that for a small acknowledgment we will allow them a road through leave at the Colling orchard, but not otherwise without they can prove a right, which I am pretty certain they cannot do. But I hope they will submit to ask leave, which is much better than going to law! So if they pull down you must build again. Do write often which will oblige yours

Geo Culley

It would have been lucky if my brother had got John Todd's farm into his own hands this year. The 2 Fittes would have paid a deal of money in grazing as we have bought and could still for a while buy cows here that would make a deal of money. Don't you think that there are very few fresh or following cattle this spring, which will keep fat stock longer high. Being a good crop of lambs may affect the price of mutton a little, but I think very little. I have often known us put fat cattle into a meadow field for 2 or 3 weeks after being laid in, and not do much harm to it, not so much as sheep would do, because they don't eat so near as sheep. Have you not a little few oats to sell? If so they will be strang[e]ly sold. There is no doubt but oats are the dearest of all other grain, and likely so to be. There are none literally speaking in this country

[*M.C.*] I would wish you to get the cows bulled, the sooner the better, as it is almost certain they will be sold before they are too forward with calf.

[*G.C.*] I would advise selling these cows as they get decently fat, because we cannot be sure of these unknown times continuing. Not that I think times will be low in a few months without a great grass year which is very likely now. And it is always right to be content with reasonable profits.

44.   *George Culley and Matthew Culley jr to John Welch*

Eastfield 13th May 1800

Well John

I keep scolding and requesting you to write often, and I hope in time to be able to convince you of the importance of communication by letter. How would the merchants dealing in corn and various articles be able to plan and speculate if they had not the earliest information and intelligence of what is going forward in different parts of the world? Yours and ours, you may answer, is very confined compared to what I am speaking off. True, but in our confined circle it is of as much importance compared as the other. I well remember profiting 1s per stone for our quantity of wool by means of a letter from Darlington only, and 1s per stone in beef, or 1d per lb. in mutton, is well worth being acquainted with. I am certain your good sense will easily see the propriety of these remarks. But I never saw it of such consequence as at this moment. We had such a market for fat yesterday at Alnwick[38] as was never before known in the north. 11s to 13s per stone sink, one cow sold by weight at 14s. If you can beat these prices we will most willingly send you over the few bullocks. If not we have engaged Benney to handle them at the Whitsun Bank eve[39] 3 weeks to come yesterday. If we should not bargain then we will send them to you. But my accounts from Wakefield are 9 to 11s sink for beasts, and 8d to 9d for sheep per lb.sink only. Now if Mr Sayle's account is true we are very much beyond you indeed. We got 9d and better for our sheep on Wednesday and was called a bad market. But there are several fews of sheep but literally speaking no fat. Bartholomew Kilvington, a decent little butcher from Sunderland, asked me yesterday if your fat cattle were sold at Denton. I told him

38 Alnwick fair, 12 May.
39 Whitsun Bank fair, Wooler, Tuesday in Whitsun week.

they were, but that we had sent you a few very nice cows. He bought some very green beasts yesterday, and I am of opinion that when you are disposed to sell your cows it might be right to write a line to Mr Lincoln, who I think very highly off. And I again say that without Wakefield advances that Sunderland and particularly Newcastle people can and will give the best prices. The Morpeth butchers for Newcastle market gave as much as 13s per stone sink for a pair of oxen yesterday. 4 oxen bought of Mr Hope at 35£ per 3 weeks ago by Mr Spours, were sold yesterday at 50£ per, 5£ returned, consequently they gave Mr S. 53£ for 3 weeks. *Marrow*[40] *me that if you can*. Mr Clunie offered Mr Taylor 12 steers at 24£ per [. . .][41] weeks ago and more, sold 6 of them at 33£ yesterday. A fortnight since Ratcliff or another bid Mr Buston 40£ per for 4 oxen, which a few days before were offered at 35. Mr B. like a wise man said he would not sell until he heard how markets were. He sold them yesterday for 50 guineas apiece. These are *strange and wonderfull times*, and what they are to be we know not. But can't think they will be any higher. And lower they can't be without people tie up their mouths, and sailors must be fed, want that will. Mr Buston sold his 4 to the Morpeth men. They would [have][42] give more for Mr Spours than he got, they declared. But I suppose all the best pieces were sold at Newcastle at 1s per lb. Wheat prices seem at stand 4£ our boll at Berwick, as much corn is expected in by shipping. But at Kelso 5£ per boll for wheat has been given for 3 weeks. They sold some at Wark at 5£. You know very well we don't like to restrict you by any means, but I must recommend our friend Lincoln if suitable to him and you. Ralph sold 4 old tups yesterday, clipt, at 4.10s per a very fair price for old venison. As I wish to give you the earliest information will send this by post today and hope for a letter from you soon. You know I advise you to keep a letter in hand always, or what we call upon the stocks. What does 5£ postage signify between you and us compared to advantages gained by early intelligence, and the postage if you write every week can't amount to the half of 5£. Single servants have comed much down in price in this part this May, for obvious reasons. I am with best wishes yours

Geo Culley

We have had a few cold fining days lately, but certainly a fine spring on the whole. You may acquaint the Mr Collings when you see them, with the prices of Alnwick market. It will perhaps please them,

---

40 Match.
41 There is a blot on the page here.
42 A word seems to have been omitted.

and it is always right to make as many *friends* and as few *enemies* as possible in this world. It is an old and very good maxim. Be sure you make my compliments to the Mr Collings, also make my respects to Mr Wright and tell him we will send him half a cwt. turnip seed which we beg his acceptance off. We shall also send half a cwt. for Pattison. We can't spare more this year, and a cwt. for yourself to sow and sell the remainder. But by no means don't sow it too thin this year, as I am affraid of many defective seeds. If you see our Nelly soon tell her we begin to wish her home for various reasons that I have not time to mention now. May read her this paragraph, and tell her we are all well and expect to hear from her soon. Very good men, 20 or 23 years old, were hired at Wooler and Alnwick fairs at 8£, such as we hired 2 years since at from 12 to 16£.

[*M.C. jr*] As we are very much pinched for meat we would wish you to take some more cows. You could put them into your moors or sell those fat ones you have, as we are quite out of hay and nearly out of straw and oats. Do write soon and tell us what you can take &c.

45. *Matthew and George Culley to John Welch*

[Wark, *c.*13 May 1800][43]

John Welch

I hope you got the cows and sheep safe. I wish that the miller and millwright may answer your purpose. Robson and his men are chagrined. Matt Staward wrote an excuse to Robson, he says he does not see where any improvement can be made. Mr Bates's man says the mill should have been on the east end of the mill or water wheel and the threshing part where the mill is now, says the long laying axel adds much friction, R. and his men say not so. Endeavor to make yourself master of what the present millwright says about the mill and machine being placed on each side the water wheel, as one story stands good until another is told, and everyone is fond of his own way. Things cannot be totally altered now, but you may when both men are tried, form some opinion of your own which man is cleverest in his business. H. Rutherford says their water wheel goes with less water and quicker by putting on a larger pinion or nutt, but the water wheel by going oftener round in 10 minutes consumes more water (agreeable to reason) but does more work upon the whole with a

---

43 The address and date are missing. The date 15 May has been added by a later arranger, but 13 May is more likely.

given quantity of water. R. says she is constructed on Smeaton's[44] principles. It may be so, but no man is wise in all things. Mr Bates and his grinder say there is much improvement in the mill by this man. Mr B. says it would have saved him 30 or 40£ if he had had this man at first. R. has so many places to go to that he cannot see after them and the work is carried slowly forward. It has seemingly made him more attentive to send his brother to attend on G. Culley's machine which he is making at Eastfield. You had kept Matty too long from school. I understood that he was to go on Monday at the latest.

14th received your letter by the herd this day. Am glad the cattle and hogs got well up. If you could not attend Mr Bates's sale and Mrs. Bell's, no matter if you can do without a horse. I hope Phips will be with you soon. I suppose you will be scarce of water now so that the blue stones will not be of much use except when it is plentifull. Grass grows slow, the easterly winds hurt it much. I am glad you have bought Emerson's house. I think it worth the money as it will give you more work people in the town. How will Natt do for a cow's grass? If Mr Peacock gets James Trenham's house you will be clear of all sales in Darlington I apprehend. If Birkbeck drops his lease is at an end, on the farm. Do write often, mind not the postage. I am with regard

Your obliged Mattw Culley

PS. We are happy to hear that George Dent is out of danger. Give our best regards to him.

Eastfield 15th May 1800

[*G.C.*] John the above came here this morning with a letter from you which I fancy had comed by Glass on his return from you. As I

---

44 John Smeaton, 1724–92, civil engineer, responsible for, among other works, the Forth–Clyde canal, the bridge over the Tweed at Coldstream, and the Eddystone lighthouse: *Oxford Dictionary of National Biography*. In 1759 Smeaton presented to the Royal Society *An Experimental Enquiry concerning the natural powers of water and wind to turn Mills and other machines depending upon circular motion* (London 1760; another edition was published in 1794), in which he demonstrated the superiority of the overshot wheel, fed from above, but showed how the efficiency of an undershot wheel, fed from below, could be improved. A number of works on mill-wrighting were published in the next few years. Thomas Fenwick, a colliery viewer, published *Four Essays on Practical Mechanics*, Newcastle upon Tyne 1801, in which he calculated the power of water wheels of different diameters and with different flows of water. Andrew Gray published *The Experienced Millwright*, Edinburgh 1806. Robertson Buchanan, an engineer, published *Practical Essays on Mill-Work and Other Machinery*, Glasgow 1814. Buchanan agreed with Smeaton, and criticised the calculations of John Banks, *A Treatise on Mills*, London and Kendal 1795, as not borne out in practice.

only wrote to you on Monday I do not have much to say, but I should have been sorry if you had sold the best cows before you received my Monday letter. Because you will there see that beef is much *higher* and *scarcer* than mutton, that both are much dearer at *Morpeth* than *Wakefield*, and that the above dearness in mine and other people's opinion arises from the *coal trade entirely*. Consequently that meat is sold higher at Newcastle and Sunderland than anywhere else in the Island, London excepted. And that made me advise and recomend to you (I don't like to command) to give Mr Lincoln a chance, because I think Mr Lincoln has always behaved to us with more honor than any other butcher I recollect. I would have wrote to Lincoln that you have some cows fat, only that I think it will come best from yourself, and besides I know not but you may be under some promise to *Bulmer*. However if you are, you can be under no obligation to sell to him at 11s when I am persuaded Lincoln can and will give you 13s per stone. I am certain that except they have a very sudden advance in beef &c at Skipton and Wakefield, that they will have no chance with Newcastle and Sunderland.. And as to your talk of money falling scarce is a joke and nonsense as long as the coal trade remain good. Besides we have plenty of money in England.

[*M.C.*] I bought 10 ewe hogs[45] at Beaumont Hill Mr Alder's[46] sale. White of Norham bought some ewes and lambs for your neibour Middleton there, but I did not think of sending those I bought until the Monday and Mr White had sent his away for Summerhouse on the Sunday. May perhaps send them on a future day. Cost 29£ per, discount to be taken off. We are very much put about for meat, and have laid in no land for hay yet. If you are overstocked sell some cows, as we have some fresh cows to send to you. Grass grows not so fast as our stock eats it hardly.

[*G.C.*] Not that I want to prevent you selling as many cows as you can and as early, for a times price, because we are really distressed for meat and must send you over some more coarse cows in a very little time if this frosty cold weather continue. But I think a few of the midling cows might be put of in the carrs, moors &c, &c. Don't suppose that we think you sold your queys badly. By no means. You have as much and a better right to say that I sold our last cattle badly. But you know there is a strange difference between 11s and 13s per stone. 2s per stone is *wonderfull*. I never imagined beef would get

---

45 Female sheep between weaning and first shearing.
46 William Alder, Bowmont Hill, Paston: *History of Northumberland*, vol. 11, p. 187.

above [. . .]⁴⁷ per stone, which was equally as foolish as Mr Wastell⁴⁸ saying 30 years ago that 'mutton could never rise above 4*d* per lb.' But we know nothing what is to be. But we know that this is a wonderfull time, and that there is a real scarcity of fat cattle, but not sheep. And although it will and must go far, yet our sailors are to be fed, and depend upon it there is money plenty. We can create more but we can't create cattle nor corn without a certain time allowed us. Not that I would have you go over stiff by any means. At the same time I am very much disposed to think that fat cattle will be *high sold* (although not so high as at present) all along, and it is certain that there fewer fresh or following cattle in this country than I ever knew. But don't let that prevent you selling at 13*s* per stone or even less. But every man has a right to avail himself honestly of times and seasons, and such as these we or our forefathers never saw before. What may come after we know nothing. But I must say that in all human probabillity there will be no drop in fat cattle these 2 months. There is no drop in our corn markets, only the factors are not fond of buying barley, and oats we have none. There is no fear of your few oats selling, as there are few or no oats to be got from abroad, and I am affraid this importation will be soon over and then the American &c corn will be coming in the month of harvest when it may do more harm than good. No doubt but the cows would suffer with driving so far, but I assure you that James Darling's 2 cows would have sold for fat at Alnwick at 11/6 or 12*s* per stone. Yours truly

Geo Culley

After this I shall be quite against sending trifles of turnip seed to George Holmes or any other person. We will rather send you a large quantity to retail untill you can get some grown yourself, which I hope you will contrive to do next year. It is a losing game this year, because the turnips could have cut to make more of them. But that is nothing so long as a man is a profitter upon the whole, and it has been a very lucrative trade to us on the whole. But retailing to distance is quite out of our way and tires me sadly who have most of letters to write &c, &c. John I must request you to give Nelly 25£ as she has

---

47 There is a blot on the paper here.

48 William Wastell of Great Burdon near Darlington, 1708–88. He was a successful grazier who believed in the importance of breeding, and supplied bulls to farmers from whom he bought stock for finishing. He was a first cousin of Matthew and George Culley as his mother, Jane Culley, b. 1676, m. 1703, was an older sister of their father. Bell, *History of Improved Short-horn or Durham Cattle*, pp. 14–16; Bates, *Thomas Bates and the Kirklevington Shorthorns*, p. 9; DRO, Haughton le Skerne Register.

some things to buy at Newcastle as she comes home. If you can't spare it, tell Mr Peacock to take up a little at the bank untill you sell your cows &c, and give her. However if that will not do, I will remit you a bill for the amount as soon as I hear from you again. Only I thought this the simplest and easyest way, and I will account to my brother for it as it is Nelly's own account unconnected with my brother's and us. And tell Nelly that I will write to her in a day or 2 but expect a letter from her about her coming home. And this makes me wish you to furnish her with the 25£ because if I was to inclose her a bill she may be come away before the letter reach her. We wish her home on account of the colours being given soon to the troop her brother is in,[49] and he is going south with Tommy Charge the beginning of June.

46.   *Matthew and George Culley to John Welch*

Wark 18th May 1800

John Welch

Yesterday a very wet day, nearly fair this morning. The land rather soft enough for plowing. So the cart will go to Eastfield tomorrow and reach you on Thursday if all be well. We are (what has not happened to us for 40 years) out of hay and mostly out of corn except wheat. That is our oats will hardly carry us through to harvest, barley the same, and beans scarcest of all. The beans was a real failing crop. Sowed one field of barley early which promises to be very early and looks well the 2nd short side. There will be seemingly a great loss in many oats and wheat fields by the grubs, wire worm &c. Our Dry Tweed much injured seemingly by them. William Brown has some quite hurt. Thomas Dods has plowed some down, Mr Bailly also, or sown over again with the drills. We think we will sow some rape amongst Grindon wheat, its on clover lay which he plowed for wheat as he could get no wheat sown on his fallows in the autumn. Our cattle in general so poor can hardly go a day's work out. But we should not complain, as many people are sadly off for fodder. John Younger who always stocked hard has lost most of his cows, those alive give no milk. Our pastures may mend now but I am afraid we have no clover land laid in enough to cut for our horses &c, and I wish we have not overstocked you, if any way too hard stocked sell any cows, queys or such things as you judge best to quit. I hope you have got Phips ere now. My brother will give you an account of what

---

49 Matthew Culley jr was in a troop of volunteers raised by George Askew of
   Pallinsburn.

things come by the cart. We also have 10 ewe hogs, might go in your moors. They were bought of Mr Alder at Beaumont Hill alias Stand Alone, and some cows which were bought in for you to graze or sell. You should write once a week or 10 days in general, do not mind postage. Hope you have agreed with Emerson about the house and pieces of ground. I am with esteem your obliged friend

Mattw Culley

Eastfield Wednesday 21st

[*G.C.*] When my brother began this letter he expected the cart to set off on Monday last and be with you on this evening. But after considering, thought it best to go on the Thursday, reach you on Saturday, rest Sunday and then return. I was at Wark yesterday when I got this letter wrote as above, George[50] said he was not sure that he could put your 2 cwt. turnip seed in one bag but would if he could. However 3 cwt. is to be sent to you, out of which ½ cwt. is for Pattinson and ½ cwt. for Mr Wright Cleasby as a present with our compliments. The remainder 2 cwt. for your own use and to sell the residue. And I would recommend you to save a couple of acres next year if possible to sell in your own country, as it is sold higher in general with you and is upon the whole a very lucrative production. It should if possible be sowed in a sheltered place, as wind hurts it much when in flower. Also upon good land, and left very thick or rank in hoeing. We had fine warm showers again yesterday, which suits this country well. Nevertheless we are foolishly overstocked, and bought several cows for you, which with those we have dried of our own are not going well. I have several times said in my letters to you that you might put them of in your poorer pastures untill you are relieved by selling the fattest of those sent, but have received no answer from you on that account, while we are so ill of for grass that we shall be obliged to send them to you at all events if we do not hear from you soon. But surely these rains must suit your limestone lands as well as other dry countries, and at any rate you would sell them again at Darlington unfed if no better can be done. They are well bought in my opinion at least, they would pay plenty here provided we could have kept them, but it is not in our powers. You may say why did we buy them? I answer. Because my brother when over at Denton wrote for us to do so. Matty Culley is at Morpeth today. I expect him home tonight when I will tell you in this letter what he has done. I fancy there a good many sheep up this week. Have you wrote

---

50 George Humble.

to Mr Lincoln about your cows, or do you approve of what I wrote to you upon that head? John if you will be so kind as forward the letter for Mr. Peacock which is sent by the bearer of this as soon as it comes to your hand, he may have it in his power to return me an answer by the bearer. John if you have not proper weights of all sizes to weigh turnip seed &c &c, by all means get a set directly. There is no doing without necessaries. I am going to meet your master at Thornington this morning and will finish this at night when please God Matty will be home.

John I have luckily received yours of 18th inst. and will send you a cow for T. Wright along with those for feeding. Will likely be sent of some day next week. I am glad you approve of writing to Lincoln. Hewitson must have bought those oxen well of Mr Colling, and the rascal Paul still better. What a wonderfull difference between your beef markets and ours. *Not much difference* in mutton, and may alter if your country can supply Newcastle and Shields. But most certainly they are not in this country. To be sure the consumption is small compared, but the shipping must be supplyed, and in my opinion that is the *primum mobile*. The coal trade has been so good, is the cause of this scarcity in Northumberland. However I will be glad to take 12*s* for our oxen as they are only plain beef except the little ox at Wark which is now got to be very nice. You are certainly right in putting the best cows into Mount Holy, for as you say it is much if such a time cropps again even in your days let alone mine. My brother says the quey you have left is much better than the one you sold, and I think you sold her well. We will send you no more sheep untill you want, nor cattle till you ask us, as these will relieve us. I thank you for giving Nelly 25£s and still more for telling us she is well again. Without much more corn come in, the markets will rise again because this is not equal to the consumption. Yours truly

Geo Culley

1. A machine for cutting straw and hay. 2 cwt. of turnip seed for yourself in 1 bag. ½ cwt. for Mr Wright to be weighed off for him. ½ cwt. for Pattison to be weighed off for Pattison.

John be so kind as forward Mr Peacock's letter to him as soon as it comes to your hand, as it contains some important business. I am sorry my brother's new millwright has not comed to you yet. Perhaps he had better have kept to Robson still.

47. *George Culley to John Welch*

Eastfield Wednesday evening 10 o'clock 21st May 1800

John Welch

Matty is comed, when we had given over expecting him and I had sealed your other letter. He has had a very bad market, a drop of a halfpenny or penny per lb., and cattle much dropped. When things get to such extravagant prices it's no wonder that sudden turns or bad markets take place in proportion. The ship owners say that before they give such high prices they will lie by their ships awhile. I would not have you too stiff with Mr. Lincoln. Be so kind as forward the letter to Mr Peacock as soon as it come to hand, as it's of importance. I am in haste

Geo Culley

48. *George Culley to John Welch*

Eastfield 26th May 1800

John Welch

As Matty Culley I trust will be with you before you receive this letter, and will I hope give you any information necessary, I would not have wrote at all but to describe as near and as well as I can the cows and queys sent per bearer, which are as follows. 9 farrow or dry cows for feeding, one ditto which is calved but in milk for your hind, and two queys only 2 years old this grass, which being too young to feed, you may either get them bulled for incalves or keep them unbulled for feeding, and run them a year in your moors or any way that appears most eligible to you. They are got by our own bulls but out of servants cows. We have sent you all we have except one oldish cow which has not been very well but is better. However we have kept her. We could have bought more and very well as we think, but did not wish to overload you. And we are still plentifully stocked in the north. However how it is with you I know not, but we have plenty of rain, hail and thunder, and thank God warm weather now. Nevertheless there are great complaints of both wheat, barley and oats being eat entirely by grubs, slugs, snails &c, &c, mostly oats, and those after clover. It is certain that many people have plowed down large quantities and sown it with barley, and some keep it for turnips. These damages are all or mostly upon damp or wettish lands. We have a little in a wett boggy part of our Sandhole field, which we have rolled in the right but don't know the effect yet whether beneficial or not. Nelly Culley is thank God arrived this moment per coach well. Yours in haste

Geo Culley

49. *Matthew and George Culley to John Welch*

Wark 27th May 1800

John Welch

I am sorry that your miller is such a babler as you find nothing of dishonesty about him you say. What a pity he should not be master of himself. I never intended in the least to make any alterations in the threshing machine. Whether we should alter the nut on the grey stones will depend on you as I am afraid it can not be done without taking off the cog wheel when a larger nutt is put on, which may be a considerable expence. I wish you to make no alterations without due and serious consideration, as I now suppose you are a good judge of mill and mill work from the great practice you have had, and I leave this and other things in the mill to you. Is the breach in the dam mended in part? The dressing part will be necessary for the blue stones, the bolting machine. Robson said he had ordered them, and I supposed that they would have been with you before now. Story comes here on Friday and his coming to you in a future day may be considered about. I am better pleased with buying Emerson's house than Trenham's, if a good person does not offer for it and this is an unpleasant time to begin a publick house when corn, hay and provisions of all sorts are so very high and not likely to be plentiful next year. You have never said that you have had a meeting with he who lets the vicarial tythes. It's very well corn has declined in value. I hope your incalve queys will bring a good price when fitt for market.

[*G.C.*] So far brother Mattw Culley. Now John I am so convinced that a drinking or rather a drunken man is incurable that if I have any influence with your master I would most certainly part with him. And I am very glad that James Story may now be come at fairly, and although he may not be *so knowing* in the mill way as your man yet, I believe he is far from unclever, and honesty and sobriety goes a long way with me. I have made it a rule all my life never to keep a drunken servant. They are not only bad themselves but are very apt to corrupt others and lead them astray. I am also sorry that the cutting machine is so complex. I abominate all complicated machines. James Story made us more of Newton Mill than we ever made before by much, and he conducted Mr Atkinson's so well that Mr A. had no notion of letting his mill or parting with Jemmy so long as he would stay, although they differed to a great degree about some parts of his hiring. Mr A. supplyed most of Holy Island with oatmeal and wheat flour, to please full as well as Mr Graham of Berwick. I hope Matty will be with you before you receive this, although I mean to send it away tomorrow and write again after Whitsun Bank. I have just received a letter from Morpeth by Mr Joe Hughes from Matty and his money, and I am glad that he has had a rather better market, although I fancy we had better have sold our oxen sooner. However we did it for the best and they are not over fat, I would have you be selling. I understand by Tomy Batters, who is just come from York, that things are coming down both fat and lean. The fat can only fall from the very unheard prices because they are certainly scarcer than ever.

Yours Geo Culley

PS. I am sorry that you are hurt with grubs and worms as well as this country. Wheat is up and oats also at London again. My fear is that there may still be a want before harvest. I am much obliged for your attention to my daughter. I paid the 25£ into our stock of cash here which you were so kind as give her by my wish. G.C.

50. *George Culley to John Welch*

Eastfield 31st May 1800

John Welch

Since I wrote to you in my brother's letter which I sent away yesterday we received a very unpleasant letter from young Benney of Shields, that he was sorry that he could not fulfill his promise in coming to see our fat cattle as he had bought a lot of old Mason.[51] I confess that it has chagrined me a great deal. But now it can't be helped, and the next thing is what is best to be done. We have determined to shew 4 of them at Morpeth on Wednesday first, as I am to be there with 50 sheep. But if I can't sell them for 10s per stone or more. I will send them forward to you. I will certainly not send them if I can only sell them decently, but things are in such a way now that it amounts almost to a stagnation with fat cattle. The ship owners will not give these high prices, say 'they had better lye up their ships than lose by them'. Indeed things were at such an extreme as not only never were known before but such as could not stand, although fat never was so scarce and in all probabillity will continue so still for some time. But there is a point beyond which things cannot go. I am most mortifyed at keeping these cattle so long after Alnwick fair, purposely to oblige this ungrateful animal, and I believe have vexed myself more than becomes me. Because if we lose 5£ per head by them it can't be helped, and is perhaps plenty after all. However I thought it right to apprize you of this, as they may possibly come to you. I hope this will reach you at Darlington, and if Matty be there

---

51 Jack Mason, a livestock dealer.

shew it to him, and tell him all his friends at Wark and here are well but poor old aunt,[52] who I think is going down the hill. Be so kind as write directly, and if you could find time on Monday to write 3 lines saying how your market is, and direct to me at Mr Whitfield's Morpeth, I would get it there on Wednesday evening. You can afterwards write more fully, and I would wish you to write again when at home and say whether we may send you the remainder of the fat cattle, as we shall still have 5 behind, reckoning the little Wark ox that has done wonders while the others have not done near so well as I expected. Besides these my brother has 2 or 3 nice cows, Miss Hill and a yellow cow &c. Now it is possible that although our ship folk are cooled and will not give the prices, that the south country people may give better or at the least keep it longer up. But I am sure you will let me know your mind and advise with us to the best of your knowledge. I know you are already overstocked, but you will speak out and help to ease my vexation. I am in haste Yours

Geo Culley

51. *George Culley to John Welch*

Morpeth 3rd June 1800

John Welch

I have this moment received yours for which I thank you kindly. I do assure you I expect a bad market for cattle tomorrow, and will willingly take 10s per stone sink before I will send them to you, or even a little less. We had not a shilling bid, nor scarcely the price asked for the 4 we shewed at Whitsun Bank today. Indeed it was a most shocking market for all kinds of stock, both cattle and sheep. I should not wonder at Mr Lincoln being a little offended, and I never liked that coxcomb Benney, who has behaved abominably. You are certainly right to encourage Chrisp, he is a petted fellow but very safe. I would really advise you to be selling if possible, because although I should be so lucky as sell tomorrow we shall likely send the others to you. I have received no letter from Matty since the one, and money from him at Wooler last Thursday after this market, but I hope I shall see it upon my return, and if he be only well I shall be thankfull. We have had wonderfull fine weather untill Saturday the wind got north, and it has been bitter cold ever since. Corn on all damp soils looking very ill indeed. You may depend upon it that all

---

52 Jane Culley, 1720–1816, Matthew and George Culley's sister, who lived with George at Eastfield.

kinds of grain will advance again very soon and fast, perhaps too fast as the corn has advanced 20 per cent up the Bal[t]ick very lately.

Wednesday morning 4th June. Well John it is most extraordinary that here is not one beef butcher to be seen. I don't believe we shall any of us be asked the price. Consequently must send them forward, which I am extremely sorry for. But whatever price you sell them at I will be satisfied. This is such a turn as never was. I would willingly have taken 10s per stone or less, but we can't have an offer. I suppose the 2 white ones to weigh about 78 stone each, the quite white one 80 and the other 77, the other 2 74 stone each, black one the finest. I suppose they would have been readily sold at Alnwick at 12 or 12/6 per stone, but the game is up. We will not send you the remainder and 3 cows, that is 4 oxen and 3 cows, 7 in all, untill we hear from you again, because we can keep them a while. But I beg you will be selling as soon as possible, and if you can't get 10s take 9, for I am persuaded they will come to 8s per stone soon. Such a stupid time never came before, still if they are sold even badly it is better than losing by Paul. Send Tom Aynsley[53] back as soon as you can, and write us your sentiments. I am in haste Yours

Geo Culley

PS. Here are a vast of sheep and lambs in, but yet I have sold my sheep at near 9s per lb.

PS. If you get the letter before Tom reaches Denton I will direct him to go by Merrington, Heighington and across Walworth Moors.

52.　*Matthew Culley to John Welch*

Wark 6th June 1800

John Welch

I met your master at Wooler yesterday. He got home on Wednesday, too long a ride for him and his mare, but had left the market as soon as he got money for sheep and sent the oxen for you. I hope you would receive my brother's letter before you see Bulmer. I am afraid you are too hard a salesman. Do not sell him things too dear, and make him a fair return when he is too hard put to it, by a dear bargain from you so long as he pay you well. I am afraid that we hardly used Lincoln well, but must endeavor to use him better, as I believe he is a good fellow and pays well, but must have his profits. I hope you have dealt with Bulmer on Saturday, which may be done before you receive this if it be tomorrow, but be advised, do not deal

---

53　Hind at Eastfield.

too hardly, always sell when a good chap offers in falling markets. But it is wrong to stay too much by one person, for a plain reason. No one can buy but when he has occasion for stock, and his brother chapman may want when he does not. We ought not to blame such a light-headed thing as Benny, a weather cock. We may blame ourselves, we kept them too long, prices were really too high to stand, we see it now too late. To shoot at the highest mark often disappoints us. A good maxim to take good prices when they are going. I am afraid that we have over stocked you. In such case sell any of your stock that you can get quit off and submit to the times, the sooner the better. I am sorry your miller is so very uncertain. Mr T. Bates said he kept his places badly. There is no doing with babling men. As to Story, he has taken some house, but where I am not able to say as he is not come to Wark. I offered him a cart, but the wife said our cart and Nell Barry's would not carry all their goods. You had not a house at the time you wrote (he has a fine lad who works at plow &c). Now if the man you have be quitted, Story would have to get his bed and board somewhere, but if your man will not do you must consider what way for the best should be done. A man who loves drink is not [sic] to be dispensed with in general, as you are seldom safe with him. I think Phips has put you off very much. I wish he may do your work as well as he should, but has put you off very much. M. Culley in a letter to his father seems pleased to see Denton look so well, plenty of grass, winter rain'd land carrys much stock in the spring, the grasses fresh with vigor beyond expectation. I know Denton grounds will not bear bare eating, especially when drought comes. To overstock is very bad management, which has been our loss this this [sic] spring. Harry, Jamieson,[54] Laidlaw out of meat, had to buy hay for them, and want corn for all our servants also in general on all our farms. Wark run out of hay although we had such an overplus at the end of last summer, much of our stock too poor by such means, very very bad management. Have been until now sore run for meat, might have sold 100 or possably 200 fat hogs if we had had clover for them. No meat equal to clover for feeding stock. But yet we ought to be thankfull as we were better off than most off our neighbours. Matty says you have raised much water in your carrs by boring, that the land is much firmer than last year. Let me know where you bored, and what water you have

---

54 Steward or overseer at Longknowe.

raised by it. I am with respects to all neighbours and friends, and wishing you well &c

Mattw Culley

PS. I hear Trotter has failed, I hear sold his house to Gilbert Taylors unkle who wants 10£ of him. He is said to be much in debt, possibly 200£ to the hinds &c (house &c sold at 30£ or 40£ or more) a good house and houses but only 1 piece of ground and a bean butt. The grub, wire worm and slugs have injured corn much on strong and wet land. Yesterday Dunse fair, only a slow market. Have sown some rape and Sweedish turnip seed.[55] This is a good season when you can spare men to putt up that breach in your dam with dry earth well wet when put in and made firm, no matter for a wall on the outside, also for cutting the stell a foot or more wider and deeper on the lowest, or where easiest thrown out on your carrs, so as to hold more water in the night. We should divide our fatt cattle to more customers, or rather take a chap when he comes here with a good inclination to buy. Nelly is gone to Holy Island. Matty wrote from Harrogate. A cold wind, mercury falls, some little wet.

53. *George Culley to John Welch*

Phenix Morpeth Tuesday evening 17th June 1800

John Welch

I am much obliged to you for the letter I have received from Darlington and still more so that you sold these oxen. I hope you will sell the cows still for more, but even 25£ is not a bad price. I did not bring the rest of our oxen this week, as I thought the next being Horse Week would be a better chance. They are not so heavy as those you had. I will take 35£ per for them rather than send them forward, but if I can't get that I will send them. I am affraid the markets are worse to the south, as Bulmer did not come. Yet London keeps good, and I am convinced that neither cattle nor sheep (fat) are plentifull in this country. Yet here is a strange quantity of lambs and sheep here tonight. People hurry them in from the great prices, and very right. I have some nice clipt hogs and very good ewes, but I fancy we shall drop tomorrow 2*s* per head or more. I thank you for your information

---

55 Swedish turnip or rutabaga, first introduced into England in 1767. After a slow start they spread rapidly towards the end of the century, but were still not much grown in Northumberland in the 1790s. They were extremely hardy, could be lifted and chopped, or eaten off the ground after the common turnips were finished. *Agrarian History of England and Wales*, vol. 6, pp. 299–300; Bailey and Culley, *General View, Northumberland*, pp. 105–06.

about corn, it will be dear I am affraid, yet I would not have you miss your oat market. However you must attend to turnip sowing, never mind other people, do you sow on. Dry weather is the best for sowing always. I am only affraid of the weather being too dry, I mean droughty, at least it has that appearance to me. I think your wool prices are not bad. I do not understand why they have not got all the wool. We can prove the delivery to Mr Howey, and Mr Howey can prove the delivery to Newcastle. I think they must be negligent. I don't well understand them. I thought the buyers always wrote to the sellers. However I shall consult my brother about writing. If you hear more about wool let me know, and do write oftener. I will be here again next Tuesday, please God, so may either write here or to home, I will be glad to receive it. What are we to do with 3 or 4 famous cows? I once named them to you before, and said it would be all one trouble to send them and oxen to Denton together. It is Miss Hill and 2 better, very fair beef.[56]

54.   *George Culley to John Welch*

Eastfield 21st June 1800

John Welch

I received your letter at Morpeth last Wednesday with the agreeable account of your having sold the 4 oxen to Tomy Blacket, and I can't say but I should have been still better pleased if you had sold your cows also. Matty Culley returned safely yesterday thank God, and tells me that he had recommended to you to buy us 4 horses. I hope this will reach you before you buy any, because it is now far too late to buy them, because before we could have them our turnips will be nearly done, and the keeping of horses is so very heavy at this time that we certainly had better let this alone untill next spring, or winter at any rate. However if you have bought any we must certainly take them. Besides the interest of the money should be taken into account. I wrote to you from Morpeth last Wednesday, and intend to send this from there again on Wednesday next. We had a very heavy sheep and lamb market last Wednesday, owing I think to having too many lambs in. People hurry them to market on account of the high prices, and very wise it is so to do. I am going to shew 5 oxen and 3 cows at Morpeth on Wednesday, which I will certainly sell if I possibly can. But if not shall send all or such as I have left to Denton. But I am determined to sell if I can, and I hope to sell because from these west-

---

56 The signature is missing.

erly winds we would expect the ships down. But speaking of winds brings to my mind the very tempestuous wind we had blow most of yesterday. I wish it has not done a great deal of harm amongst the growing corn, especially that which is got to any length, such as rye, good wheat or very forward early oats, and beans and pease in particular must have suffered very much where they are comed to any size. I would advise you by all means to sell everything you can, fat as well as feeding cows, because I do believe we are going to have very dry weather.

Should we hire you James Story to come at Lammas[57] if we can, and the other man can be quitted? I mean James Story as a single servant untill he sees how he likes the situation. Because I am affraid if we miss him we may not have an opportunity to hire another so good. What should we give him as a single servant? If you have not laid out your money, perhaps you had better give it to Mr Peacock to lodge in the bank at Darlington untill wanted, because he is more conversant with these things than you are. You will not neglect to write us word how wool markets go with you, prices of stock and corn, also concerning your pastures and meadows. We think you had better sell several of your feeding cows which were sent from here. As to Head saying there are two packs of our wool never come to their hand, it was sent by Howey, who if occasion requires will shew that it was delivered to order on the road. Are not some of your shearings fit to sell? Some of them may be tollerable, and pray do not sell too hard, be content with good prices now rather than less afterwards. Fear not but that we can send you plenty of stock whenever you want any. Robson and us cannot settle he says until you send us an account of your bill for their meat &c. Do send your account by Mr Peacock when he comes here. You will want some colts and horses if we get Birkbeck's land next May, and Tod's we are to have. I would have recommended Tod to a farm in Cleveland, but the gentleman wants a Tweedside farmer who understands this country's method of cultivation.

Sunday morning 22nd June. As I am desirous that you should receive this letter in time to prevent you buying any horses if possible, I will send it per post today and you will I hope receive it at or from Darlington tomorrow, and if I have time will write you another line from Morpeth on Wednesday. I am yours

Geo Culley

---

57 1 August.

PS. If you have bought any horses, perhaps it may be as well for to keep them in your carrs or somewhere untill afterwards. Certainly there can never be any harm in buying a 4 or 5 year old horse if of a good kind at any time, or even a 3 years old one or more. To be sure my brother observes you will need horses next May for John Todd's and Birkbeck's farms. The long good corn has suffered a good deal I am affraid by the wind on Friday.

55.   *George Culley to John Welch*

Morpeth Wednesday 25th June 1800

John Welch

I hope you would receive the letter I wrote to stop you from buying horses. I sold the 5 oxen to Crisp at 34£10 per. I could not make better of it about 9/6*d* per stone. But I could never be bid more than 22£ per for the 3 cows. If I could have made 24£ per I would not have sent them to you. There were a vast of fat cattle in about 70, very green most of them, but some better than ours and several left unsold. I was told that Wakefield is very good for fat cattle, consequently I hope you will be enabled to sell them better than I can do here. Nevertheless let me advise you to sell by all means, they can't be better and they may be worse, although they are very scarce in this country, and the misfortune is that people shew them very green which hurts the better ones. These are 3 very good cows, but having had so little time makes them appear hollow and swamp and light in the market, and to be sure green ones will weigh badly. We have had a very good sheep market, nearly as good as before last week, which was a bad one. I do really think that both sheep and cattle are very scarce. I think to send this of per post, but you may meet Jemmy or not, he will be at Denton on Friday evening. Yours

Geo Culley

Mr Welch.[58] This is to inform you that I have bought the 3 cows after they came out of Morpeth. I am yours &c John Forster

56.   *George and Matthew Culley and Matthew Culley jr to John Welch*

Wark 30th June 1800

Well John

I received yours of the 27th inst. in due course, and particularly thank you for it because I think it was very industrious of you indeed

---

58 Written on the back of the letter in another hand.

to write so long and usefull a letter after your fatigue at Borroughbridge &c. You are a man of deeds John and so far am I from blaming you for buying those horses that I have much pleasure in commending you. The horses are most certainly cheap compared, if they have not too much black in them. I ought never to say much about horses because I always profess to know little about them. But this much I think I know, that blacks are in general the softest kind of horse we have in this Island. I am fond of the country horses in Yorkshire or Durham, that is the breed with some blood in them, and I will hope these to be of that sort. Your master here, our Matty and myself have just been reading and considering over your letter here this morning, and in the first place we all agree that you should take the tythes of Denton (vicarial I mean) at all the money they ask, or as much less as you can take them for, and for the vicar's life most certainly, because if the vicar drops no lease will stand good. Then we send over Jemmy Glass from Morpeth for 2 of the horses for William Brown, as William is very weak in force poor man and far back I am told. We leave it to you to send any 2 you chuse, but think you should not send the young horse for one, as you will likely manage him better than we may do. I hope you will receive my letter on the Saturday, as I begged Forster to put it in the post. You must know that in my hurry I had perhaps luckily given Jemmy Glass the letter open when I set him away with the cows. So when Forster bought them I requested he would write below that he had bought them and then put a wafer on it and send it per post. I got no more than 24£ per for the 3 cows, and I am now highly pleased that I sold them, because we have now a confirmed drought which I have been long looking for. But let us be thankfull that we have sold so much. Things must and will come down now every day although still scarce. It would have been very lucky if you had sold your 4 cows. However I don't blame you because it was done for the best. But the greatest fear of you is that you will be too hard a seller. It is better *to me sell than me keep.* But it is impossible to do always right. We have now sold all the fat cattle and sheep we well can do now, because our gelt ewes are about done and our hogs are not fat, that is we have sold all or nearly all that are marketable. I think differen[t]ly from you about meat not coming down, because corn is so dear. They certainly go together to a certain length, but then the poor man must have bread corn at all events, and although he would like also to have a piece of mutton &c with it, he must be content to take so much as the remainder of his money will buy after buying bread. But bread must first be had. But I have no doubt but your natural good sense will direct you to sell at what you can get in these critical times. This has been the most remarkable year I ever knew, or perhaps ever shall know again. However we have done very well upon the whole, and let us be truly thankfull.

I also would have you sell those of your dinmonds that are marketable, not but our kind of sheep will pay if anybody's will, but yours are only ½ bred ones. There is nobody here has sold any wool yet scarce that I know off. The wool men say it is worth no more than last year's price. But nobody are willing as yet to take less than 2 to 3s above last year's price, and what will be the result I know not. As our last year's customers have not wrote, nor thought it worth their whiles to take any notice about it to us, which is always the case in this country, we consider ourselves at liberty to sell to whomsoever we can. I would really advise you to sell your oats when you have got your turnips done, because the prices are undeniable, and early oats will drop the others whenever they can be brought to market. They are just beginning to shoot here, as well as some pieces of wheat, and you may depend on it that people will take them to market at an early time as the prices will be very high at the first. Turnips come very well in the fine light soils that have worked small. We begin to hoe our Sweedish turnips this morning, very promising indeed. I am glad that you have got your blue stones up at last. Water will come again some time. I have not spoken to Story, nor will not until I receive your instructions, and I trust you will also send in Robson's account which I want much to have settled. Pray take your tithes if you can.

I am yours Geo Culley

PS. John we use a great deal of things for the tups, caps and jackets. Now there were people about Cockerton &c who used to work and sell what was then called beggars inkle[59] and is the very thing. If such a thing can now be got pray send us a large quantity by the tup carts if any come your way. Any old wives in Denton will tell you what it is. Mally Wade or Ward if living, or Nanny Dawson or old Birkbeck. I think it was then ½ d per yard but will be more now. G.C.

[*M.C.*] Well John I have little to say at present, but fatt stock will every day decline in price, especially after having things so very high. My brother luckily sold the oxen 5 and 3 cows 8 on Wednesday, and I wished very much for you to sell your fatt cattle, also some of your feeding ones. To be overstocked is not a good way, and I am afraid we have overstocked you with both fat and lean or fresh cattle. There was a full market of sheep and cattle at Yetholm, and were mostly

---

59 A kind of linen tape.

sold but at low prices. Some decent hogs at 18*s* 6*d*, better ones at 22 to 23*s* per head. The drought now begins to be serious here. Hay most likely will be a very light crop and straw also will be short and thin, and there is no stock in hand of either hay or straw. My advice is to sell stock by submitting to the present prices for fear they will be worse. Your obliged friend

Mattw Culley

[*M.C. jr*] Wednesday July 2nd. We have had a very bad market, a drop of full 1*d* per lb. perhaps 1 ½ per. A vast number of lambs in and very badly sold, which made the sheep market much worse, and there were a great many sheep and all very bad ones. Lamb sold at 6*d* per lb. and less, cattle several in and some sold at near 8*s* per stone but more at 7/6*d*. I send this by James Glass who comes for the two horses. I think you may as well send the mares, as they will be tractabler than the young horse. Very few butchers, and those very nice to buy as they are afraid of things coming down. You will mind and write home to my uncle by James. I am yours in haste

Mattw Culley

57. *George and Matthew Culley to John Welch*

Eastfield 30th July 1800

Well John

Our Jack Davison is at Raby Castle this week with 2 tups for Lord Darlington.[60] Had we known that you had bought Nelly Culley a Galloway, he might have brought it home with him nicely, but we only received yours of the 27th last night, therefore did not suspect that anything was to come back from Denton. I should think the best way now will be to send it to Newcastle Lammas fair, and some of us can lead it home in our hands. Indeed it will be very expensive coming by Newcastle. Perhaps it might have been as near and less expence to have sent it all the way by Mr Thomas Bates's all night at Brunton, or we to have sent all the way for it, to have comed that way. But I will see my brother before we determine, and before this letter is sent to you, when we will fix and I will then tell you our determina-

---

60 William Harry Vane, 1766–1842, 4th Earl of Darlington, succeeded to the title 1782. Marquis of Cleveland 1827, Duke 1833: *Complete Peerage*, vol. 4. Both his father and he did much to improve the Raby estate. The Culleys had had dealings with the estate over a number of years: Lord Barnard bought cattle from Matthew Culley senior in 1741; the 3rd earl hired a tup from Matthew and George in 1765: Raby Archives. Arthur Young visited Raby in 1771. He admired the castle, and commended the earl's experiments and farm yard: *Six Months Tour through the North of England*, vol. 2, pp. 428–42.

tion. I am also glad to find that you have been selling some more of your cows. You may depend upon it that they will be every day worse to sell, particularly so long as this drought continues. The want of meat obliges people to go to market with their stock before they be half fat in this country, and it will be the same in others. Matty Culley was at Alnwick fair on Monday last,[61] where was a great shew of very midling fat cattle, and sold for little more than 6s per stone. Mr John Forster butcher of South Shields, to whom you sold the 4 cows to, did not put up his appearance there, which I do not like so well. However I shall draw upon him on Saturday at Berwick for the amount, as we are rather a little in want of money as Mr Askew's rent will be to pay at Lammas and some other things. I have wrote to Mr Forster on that account. Do you ever see him since he bought the cows of you? I suspect he is pinched, but yet I would gladly hope he would pay. I drew for the 72£ he ought for our cows, but have not heard a word from him on the subject yet. I understand sheep are not sold for much more than 6d.sink now at Morpeth, but lambs are said to be in great demand. Things will and must be low so long as this drought stands, and it has every appearance of standing. We had a small shower about 12 o'clock last Monday, which did us service for our young turnips. Indeed I am happy to say that I never saw the young turnips grow faster any year in my life than they have done on the turnip soils since the weather became so hot, and if we do but get moderate showers we have a very great chance of a very good crop upon all the light soils. Indeed the older sown ones are much more in want of rain. Our Sweedish ones are wanting rain very much, quite stopped in the dry parts, but otherwise are very promising indeed. I believe about Wooler on their dry gravels they are much worse than we are here, and their oats are irrecoverable burnt about Wooler, Akeld, Ewart and all Milfield plain. Hay is a scarce article I do assure you, selling about Berwick for 1s to 1/4d the Scotch stone of 22 lb. But I remember in the serious drought in the year 1762 hay got in your country to 1s per stone at this season, and yet never rose higher although it was most shocking crop that year. Indeed we did not mow any at Denton that year until October, and poor stuff it was. You know the old proverb that 'Foreseen droughts never happen,' and it is to be accounted for thus. What we see before our eyes we guard against with the most *rigid economy*, and certainly very wise and right. My old friend Mr Bakewell used to say 'that we learned a great deal of economy in dry

---

61 Alnwick fair, the last Monday in July, in 1800 28 July.

years,' and it must be the case with people of any consideration. Still much depends upon the winter being favorable or severe. However if our turnips hold their *feet* I shall not be much affraid, although we never had so little hay by *half* at least I am sure, and straw is and must be exceedingly short, will go in little room as we say. Well let us be thankfull that it is no worse, and use every reasonable means to guard against famine. I have only to advise and recommend you to sell all you can and buy nothing. I do advise you, I will prevent all I can here the buying of any stock. As to your little few sheep I would not mind, because they will live in your carrs or anywhere until winter, and even after untill snow come, and they will pay for keeping if any will. I am glad you are selling your few oats at last, prices are very high still. It is said some of our Berwick merchants have bid 2£ our boll for all the oats they can get delivered until Christmas. I do not know this for certain but believe there is truth in it. The great want of hay, and no old oats in hand, must have an effect. Yet the wheat markets are strangely dropped here. But drought is always favorable to wheat and rye, it always fills well and is gifty, and does not burn like oats and barley. Pease and beans also load well in dry weather.

Who has been in *Holderness*, Mr Abbs or Mr Peacock? We will be glad to see George Dents son or any friend of his along with Mr Peacock. You rather surprise me John about the carrs being so dry, but it may be much owing to the uncommon dry season. It was a lucky job that you agreed for the vicarial tithes of Denton during the life of the present vicar. I hope you have it upon stamps. You should consult Mr Peacock on these things, as he knows more about these matters than either you or I. And let us never be ashamed to ask advice. We can't know everything in this life, therefore it is always *wise* and *prudent* to ask advice of those *that do know*. My brother is charged this year 6s per acre at Wark for all his hay tithe, and you never saw half so bad a crop at Wark of hay. Some pay 10–11s per acre for clover tithe, nay I was told one man was charged 14s per acre. There is no knowing what tithes may come to. The more industrious you are, the more you are *punished* in the tithe way, and the more you improve your farm or stock &c &c the more you are taxed in the tithe way. *There is no other tax as grievious as the tax of tithes.* Yet as they cannot be avoided it is wise 'of a bad bargain always to make the best'. I will go to Wark today some time to let your master see your letter and this, and then will write about the Galloway getting over, and send you this from Wooler tomorrow. I think my brother named to you that writing to either him or me was the same, as we always see each other's letters, and as the letters come quickest here, it is perhaps best to write to me. Have you ever sold your piece of wool? If not by all means sell it. As it is all hog wool it should be worth at least 1s per stone more than common. May try without making a firm bargain to Neddy Pease[62] &c, as they are good people. Make my brother's and my respects to Messrs Pease, and desire them to look at it. We sold ours the other day to Nathaniel Handasyde at 20s per stone of 24 lb., but we are to return 10£ again which will be 3d or 4d per stone abatement, so that we will be less by that than 10d per lb. But yours should be a 1d per lb. more by right I think or nearly so. Mr Nisbet sold his to Natt at 20s per stone, and Mr Gregson got 21s but I believe few else have got so much. Mr Gregson is all pasture wool, which is always accounted the best. I am yours truly

Geo Culley

[**M.C.**] Wark 30th July. Well John I think you should dam the water on the south side of your main stell where it turns off to the south, and force it into the water course to the mill at this scarce season for water. Our springs in general are all now dry, and many almost dry. Hope you would get some rain on Monday last as the wind was south then. I agree with what my brother says, especially at this time to sell anything you can in the cattle way or sheep, especially fatt ones, at this very dry time. I know not what to say for this country, the drought is so very great. The corn is burning up on all hot soils, especially near Milfield and Ewart. Our pastures are as dry as sand and quite parched and burnt. We had no rain here on Monday. Our corn in general stands good at Wark, but the straw will be short and scanty here and everywhere.

[**G.C.**] We will send a man all the way for Nelly Culley's Galloway but cannot say when, but it will be soon. This is really a hotter wind than common today, it must burn everything up. It is an ill wind that blows nobody any good. We have a famous harvest for our turnip seed which is now nearly all stacked. I still think it would be right of you to grow a couple of acres of turnip seed, because you can sell it always so much dearer than we can do. If Mr Bates can make so much of it, why may not you, and we can't send it to you without consider-

---

62 Edward Pease, 1767–1858, elder son of Joseph Pease, Darlington Quaker woollen manufacturer. Edward Pease was later instrumental in founding the Stockton and Darlington Railway. See Anne Orde, *Religion, Business and Society in North East England: The Pease Family of Darlington in the Nineteenth Century*, Stamford 2000; M.W. Kirby, *Men of Business and Politics: The Rise and Fall of the Quaker Pease Dynasty of North-East England, 1700–1943*, London 1984.

able expence. You will see that your master has wrote a part of the above, and we are going to send this to the turnpike house to go by the post from Berwick tomorrow.

G. Culley

58.   *Matthew and George Culley to John Welch*

Wark 13th August 1800

John Welch

Newcastle fair. Horses of value sold at high prices, fatt oxen and cows very scarce, butchers allowed of a great difference in value from Alnwick fair. All sold before we got to the hill at 10 o'clock. We came from Brunton, Mr Thomas Bates's that morning. Tomy Blacket does not complain of your cattle, says Crisp employed White to bid you a bad price. Mr Blacket steped forward and bought of you, Crisp was then obliged to set out for Morpeth, where he was glad to buy of my brother. We got here last night, have sent 1 cart load of new barley to Thornington, one to Grindon for servants besides what our own people have got. Cut a few Church oats, and are going to shear such as are ripe in the park joining Margaret Bolton's, a very nice crop indeed. Wherever there is straw the crop promises well. I never saw corn fill better than this year, especially barley which is the most promising crop of grain in this, and also said to be so in the south. So far was wrote by Mattw Culley Wark.

[*G.C.*] 16th August 1800. Well John. This very severe drought alarms me much for the turnips, and if they fail I know not what will be done, as we have so very poor a crop of hay. But God's will be done. We have behaved *most absurdly* in respect to wheat in hand, but nobody could have supposed so sudden and great a drop. However it cannot now be helped. We were so engaged with our turnip season that we could not well spare a day to carry corn to Berwick &c. But we certainly might have hired carts to have done it, and there we made the mistake. We have not less than 20 acres of oats completely burnt upon this farm, and I dare say Harry is bad enough at Thornington. Wark oats are very good, but if it please God only to send us rain to save our turnips in part so as to enable us to winter our stock, I hope it will make truly thankfull.

Geo Culley

PS. I had James Story here the other day, and he was to give me an answer tonight whether he would come to Denton or not. He and son seem very willing, but not the wife. I advise him to go to Denton and try how he likes for a few weeks, and then hire.

[*M.C.*] John. Phips says he is afraid of the mill axel. Robson's men, he says, have weakened it much by making it 8 sides (when perhaps it was unnecessary) that it seems over weak, and springs water or oozes a little when it goes. And therefore says you should or ought to have an axel by you. Perhaps if you had Mr Thomas Eshellby of Coniscliff he might try looking at it (or some milwright) who going, may give you his opinion of the matter. Phips is afraid when it turns to work both pairs of stones at once that the stress will be too much for its small end to bear. It will be serious for you to want the mill altogether when the axel fails, till another is got, and to put in will loose much time. This is worth attending to, but if it can be done without why get one to lay by you. This you'l consider about at your leisure, and turn in your mind. M.C

We think Northallerton fair, which is near hand as to time, old Bartle day,[63] a good place to sell incalves at (if you have a number worth going with) as there are many capital men come there who want good ones, and if you suit them they may find the way to Denton in a future day. Besides it will bring you acquainted with that market, and some people which you ought to see. Harry Barugh or Barf (and his brother have[64] the management of a farm for their sister at Romonby close by Northallerton) are particular friends of ours, and remarkably honest people, will assist you with their advice and tell you what they will bring. They are dairy men, also Harry buys sometimes for a particular man. Harry's house is a home to us, also a tup customer. Mr.Taylor of Salton, Mr Cleever junior, Mr Key, Mr Dorkin all neighbors from Maltonside.

59.   *George Culley to John Welch*

[Eastfield][65] Saturday evening 16th August 1800

Well John

Young Story is this moment come to tell me that his father being often poorly, they think it best for the young man William Story to go to you. So he is to set of on Wednesday first, and if all be well hope he will be with you on Thursday evening as we lend him a horse, which must stay till he come back again. And if you find that he can manage your mill we will try to hire the whole family. But if I have them to hire, you must be so kind as tell me to the best of your knowledge at what rate I should hire them. The father is a failing man, but could manage the mill well enough I suppose, with the assistance of the son

---

63 St Bartholomew's day, 24 August. Northallerton fair, 4 September.
64 The parentheses ought probably to start before 'have'.
65 The address is omitted.

when needed, and at other times the young man can work for you. Now if you could contrive them a house, and the young man approve of the country and the situation, they could all come to you before the winter puts in. So you must either find one or build one, for I would gladly hope these people will please, as they are honest folks and the lad is of an amiable disposition. Berwick market rather better today, but wheat was only sold for 2£ 2s to 55s per bol. New oats 1£ 1s. But Mr Clunie told Ralph Brown that wheat will advance again, though not near so high as it had been. Young William Story must board with you or somebody else, and we to be answerable for the board. Let me hear from you soon after the young man arrives. I am yours in haste

Geo Culley

60.   Matthew Culley to John Welch

Wark 27th August 1800

John Welch

Yesterday Dunse fair, a great many cattle sold to go south. Armstrong and brother with Tomy Scott bought, Scott said about 200 here and in the country, lower than other years but prices were not very low. Mr Fenwick Compton sold 2 heifers and 4 oxen to Joseph Mills for 100£. T. Mason bought 3 cast poll cows 17£ per and 4 queys all of G. Logan at 11 guineas per, about 40 to 42 stone per, the cows are old. Some oxen at Whittingham 6.6 per, cows at 6.9 and 1 at 6s very fatt ones. A man called Fish, a big heavy fellow, bought a many at low prices I was told, he and brother compounded but never paid. The low priced were many of them sold, those in fresh order. Ned Smith sold 48 pretty good oxen at 16 guineas per, very low we thought, to Adam Boag nephew to William Brodie above 60 stone per, Mason bought 6 oxen of John Mole very good fatt at 19£ per, not 60 hardly (2 cows at 23£ cheap). We thought the Morpeth man bought some good Kyloes at 9£ 9s per of Mr Bell of Dunse. Ralph Brown sold 100 bolls wheat to Mr Graham of Berwick at 3£ 5 per boll, to Lowry at 3£ 5 and 5s referred of Etal Mill, seems to advance a little. Our crop will be very good I believe, especially pottatoe oats in the close west of Dry Tweed. Andrew Bolton[66] sayth he never saw so good a crop in it. These rains retard leading but has saved the turnips. We hope now ours will be good. Our rape is very good considering the land, are beginning to eat it with ewes. A few tup hogs are on it. The rains and north wind have done little good for pasture, only the grass is some

---

66 Hind at Wark.

fresher. We got a good quantity of oats in on Monday, 6 or 7 small stacks, threshed 60 bolls barley of 8 acres 2nd short side, and 3 acres peas of Broomyknow about 24 measures per acre, very nice corn both of them Are cuting wheat the hot land ripe, then the soily greenish, not level ripe in the twist Broomyknow, cut some of the middle low haugh, have mown much of our barley and some oats. Shearers[67] very plentifull, cannot get employment in many places, especially in the Lowthians where its said the crop and all the way to Edinburgh is very poor, as is the Carse of Gowry they say, perhaps much owing to a bad seed time and severe drought. So corn has the appearance of being scarce this year, as the crops are thin and weak on most of the grounds. Lord Tankerville[68] has bought the great or corn tythe of Wark Sunnylaws and Wark Common, for which I am to pay for my lease 5£ per cent for 4700£, not cheap but will save me some trouble in regard to my own tythes. Indeed my tythes would be worth much this year, but I have been very well used in my tythes. I hope your lad the miller got safe to you and will please. I am with respects to you and Dolly &c

Mattw Culley

PS. I hope Mr R. Dent got well home. Make our respects to them when you see any of them. He made too short a stay, hope he may make a longer soon if he can spare time to come and tarry a while, we shall be glad to have him. There is a necessity of shearing clean this year when there is a probability that the fodder will be much wanted. Be carefull to shear low and clean. I have had some work to get ours to cut oats low, and wheat lower than usual, as it will cover further when cut 3 or 4 inches lower, or in other words straw 3 feet 6 in. will go further by much than straw 3 feet long, as the 6 in. at bottom is considerably thicker.

61. *George and Matthew Culley to John Welch*

Eastfield 7th September 1800

Well John

We were glad to hear from you by John Davison. For fear I should forget, I must tell you that by a letter received today from Mr Sayle from Norfolk he gives a most shocking account of the corn, 'says that it is all grown or growing whether cut or uncut, and the wheat all *rusted* or *mildewed*, as black as one's hat. Very little wheat got into the

---

67 Seasonal reapers of corn.

68 Charles Bennett, 1743–1822, 4th Earl of Tankerville, succeeded to the title 1767: *Complete Peerage*, vol. 12.

stackyards, some barley so much grown as to be fit for nothing but pigs'. Thank God we have nothing of that sort here yet. But after a good deal of rain last Thursday, and 2 hazy days since, we have a very wet day this good Sunday and I am sadly affraid our rains are only coming, because our weather glass has fallen for several days although the wind has been all the time east or north east. Notwithstanding this, we are only selling our old wheat at 11s per bushell. Our merchants will buy none, and they have sold most of what they have in hand to the bakers, which has filled them full. But surely it must rise this unfortunate weather. However 'it is a bad wind that blows nobody good'. This will help to drive the mill round John. You have sold the black polled quey exceedingly well. Mr Darling's cows have not done well, I believe they were starved a while before we got them, though they had done well before. But a check takes a good deal of making up. I have a notion that the potatoe oats will require deepish land. Ours has not much, and although they have a surprising crop at Wark of them in the East Middle Close, yet G. Humble says they are much smaller than they had them the last year, and several places were a little burnt. I should incline to think that Angus oats will be the best for you on your oldish land, both from what you say and what I have observed of these oats for a long time. I dare say the potatoe oats will be valuable but we don't know enough of it yet to judge properly of its merits, or what soils suit it best. I am very happy to hear that your turnips are so much mended. Ours are strongly recovered.

I rather think it not right to sell your little sheep yet, or at any rate cowardly, because I am morrally certain that all fat sheep will be well and better sold by and bye. If you think these little sheep are not good enough for keeping, and your rape wants eating, I would advise you to put them upon it and not eat it too much and it will bud again. Or if on the contrary you think your little sheep have good enough keeping we will send you 100 ewes more or less to eat your rape, and then finish them on turnips and sell them about Christmas to some of your Piercebridge or Heighington butchers in the way you did last year, or any other way so long as you are certain of them being butchered, provided you think you can keep that number of ewes and your little wethers, because I am averse to your selling your little wethers at present. They are sure to swell and grow and *weigh* more than you are aware of, and we will keep the ewes here longer, or send you fewer, as best suits you, as we have very good keeping for our ewes yet. Only it is wrong to be too late in the autumn before we send you such sheep as are proper to send you. Respecting your piece wool you need be in no hurry if you have convenient lying for it, because the Darlington people will buy it afterwards. Mr Sayle says, and Mr Deverel[69] and others, that long wool is advancing. Deverel had very near 11*d* per lb. bid for his lately, and yours being all hog and rather fine should be worth as much as any long wool. However it is no object at any rate so little a quantity as you have. Now before I am done with sheep I would suggest to you whether it might be wise for you to keep a few ewes either for holding stock or fat lambs now you have got your wool and lamb tithes taken. You have a good deal of soundish land about the High House which ought to do very well for even holding stock sheep, that is scarcely good enough for feeding. And recollect how well Mr Peacock has frequently been paid for fat lamb upon such bad unsound stuff as his. We could either buy you 50 or more ewes for a tryal in the fat lamb season, or send you so many of our cast ewes from Longknow or Shotton as you chuse, which will bring a nice lamb and then make fat. I would wish you to turn this over in your mind and to let us know. We could send a mettled young tup all the way well enough with the ewes.

And this brings to my mind the advantage it would be to see you over at some certain periods. So much can be done and said when together in consultation, which can hardly be expressed or conveyed by letter. To be sure it takes 4 days coming and returning beside the expence, which is all nothing to the loss of your time from home. This I also wish you to turn over in your mind, and write me your sentiments. You can't be *supposed to come until after the harvest*. I think John Pattison should not pay more than 1/6*d* per lb. or as much less as you think right for his turnip seed. He is to be ranked as a retailer, but then he can hardly be put on the footing of Ned Potts &c who take 8, 10 or 12 cwt. of us annually. Mr Brodie had 3 ½ cwt. and yet paid 1/6 or 8£ 8*s* per cwt., and the money paid to us long ago. Pattison had only ½ a cwt. I believe, but he may have as many cwts. as he likes next season, as others have it and I am disposed to think as well as you that turnip seed will be high again next season, and as a confirmation the prices are getting up every week at London. As to yourself you may have as many cwts. as you chuse, even to a cart load, either single horse cart or 2 horses if you think right. We could if you chuse send it by the cart or carts that have to come for the tups. This also is to be considered of and wrote in time, as we shall have to break into

---

69 Samuel Deverel, Clifton, Notts.

the stack. Your mistress thanks you for the trouble you have had about walnuts, and can do very well without them.

I think you had better ask William Story what he expects per week, with or without his meat. Perhaps 2s a day without his meat during the summer time, and 1s/4d or 1s/6d during the winter months. If he has his meat that must be deducted. Robson allowed you for his men 8s per week. Or it may be best to hire him by the lump untill May next, when I hope his father and mother will be to send. But when I have said all this, it must be left to your own discretion, and very proper because you are certainly better able to judge not only of the merits of the young man, but of the price of labor in your district compared with this, and you are not to lose a hog *'for a halfpenny worth of tar'*, although it is your duty to hire him as low as is reasonable. We will see his father and tell him what you say. But I am surprised that William did not write a line to his father or mother. If he was accustomed to write as much as you or I be would have done it. I am glad your bare fallows work well now, but you should not neglect sowing your moors or wet lands when opportunity offers. We have had most singular seasons of late years, not but we have had such like before, but we are apt always to think the worst of the present time. One thing I am and have long been clear in, viz. to be content with such seasons as it pleases the Almighty to send, and leave all the rest to Him. Or in other words to have a *full trust* and *confidence* in the Almighty Being at all times, and we will be sure to have peace and comfort in this world and happiness hereafter. This is the best advice in my humble opinion for either you or your real well wisher

Geo Culley

PS. Mr Peacock has wrote to me to buy James Grey's[70] ewes of Milfield again for him (100) if not more than the last year's prices. I have no idea how cast or draft ewes will rate, but I think myself that they will be low. Not but cast ewes in fair condition may be better sold than we are aware of, because fat lambs have paid so well this year. Besides ewes in condition may be either turnipped or put to tup for fat lamb. Not that I expect Mr Grey's to be very forward, only this

---

70 James Grey 1743–1813, uncle of John Grey 1785–1868, who inherited the Milfield estate at the age of 8. His mother ran the estate during his childhood. As a boy he was befriended by George Culley. Later he was active in public life. From 1833 to 1863 he was administrator of the northern estates of Greenwich Hospital, and lived at Dilston. Josephine E. Butler, *Memoir of John Grey of Dilston*, Edinburgh 1869; *History of Northumberland*, vol. 11, p. 247.

weather will raise a task and freshness which will mend ewes and all sheep much. I am told that Meggy Mare is a famous miller's mare for carrying pokes &c, and she is litterally as hard as the *nether millstone*. G.C.

[*M.C.*] Well John you must consider what cattle you think may be wanted for your winter feeding. M. Culley thinks the 2 years old steers[71] you had from us should have a few turnips, with straw, to carry them on for spring feeding or summer feeding if you think that will be right on consideration.

[*G.C.*] This is the 8th and we have had constant hazy weather and rains ever since ½ past 4 o'clock on Thursday last in the morning. But the last night and now 12 o'clock of this day it has rained most heavily, so that we fear that we shall be as bad as the account Mr Sayle gives. Indeed it is a very bad look out.

[*M.C.*] My brother speaks of your coming over, but then your farm will be neglected, as 3 days coming or going and 2 or 3 days here makes 6 days absence. In harvest this will not do, although we would be glad to see you at a proper time when you can find a vacancy. I cannot say I am fond of your breeding lambs, because if we are over-stocked you can relieve us as you now are, but ewes and lambs are an eating stock. Grasing may pay better in my opinion, but you may consider this over, and if found right may try 20 or 25 ewes on your outland. If cows pay well on your good land, there are sometimes ewes and lambs bought at moderate prices, or decent hogs in the spring. We have got half a field of barley east side of Pauper Crick Know, some 5 little stacks of oats and barley at Tod Acres, the south or mid Gallus Hill oats and barley, a few peas on the Broomyknow early ones exceedingly well loaded. When you sow peas get some of the seed from London or at a distance at least from home. All the seed peas bought from the south so far as I see and hear are well for straw and greatly loaden. Laidler has some nice ones in the flat before John Hood's house at Shotton, but they are mostly out yet so likely will be lost or nearly so. Swines meat peas are most difficult to win in bad weather. Pottatoes are I am afraid a very bad crop, are so scanty from the drought, very small produce, and very small like nuts, have it's said since the rains came put a top afresh, but the pottatoes do not thrive under ground or grow bigger. This will likely be a time to rot sheep on all wet or clayey soils, as it is generally allowed that the harvest months are the most dangerous, especially when they are

---

71 Young oxen between two and four years old.

scarce of meat, not so bad when there is plenty of old grass for the new to come through, which shews the danger of overstocking, besides old grass is of great value for winter store stock turned out in the day and straw at night in the fold yard. This breeds more dung by taking in the young store stock early, keeps them in condition also, and takes little or no more meat. Tuesday morning.

Mattw Cully

Tuesday. It now rains, and I can hardly see over the park hill it is so close and misty. I do not remember such a bad harvest as it now is and seems to be, although we have no advance of the water so low no fish can come up the Tweed. The Bowmont (and Till have had freshers) was back on Thornington haugh, the corn cut and uncut must grow, as the eastan go full of water the same as if in a steep nearly for some days now. I forgot the 2nd short side barley in and threshed, 7 ½ bolls per acre at 2£ per, 14£ per acre. Very early sown so was early reaped, most extra fine corn, white as this paper, makes extra good bread. Servants here &c farms have almost eat the 60 boles. It was an unpleasant sight to get them good bread so very early. We usually sowed barley as soon or sooner than oats at Denton on the dry land, indeed on all land, generally the best corn, not always the most bushells, does not do well except the land be in fine tilth. We got some rape from West, he should be settled with for the turnip seed at 18*d* per lb. I am afraid the people take an advantage in seling, they take more than market price (10*s* per bushell is 4£ per quarter, 10 quarters the last, 32£ but I believe it may run 40£ or 50£ per. Tuesday afternoon. Still wet and close. Possably your tythes may do at the money, perhaps may let them in a future day. He is so young a steward, he can not do anything decisive. Give my respects to him. Thomas Bates's sayle is on the 3rd October. Will you be there? I think I shall. Sheep and cattle in condition likely to get fatt, probably may be well sold. We are glad to hear Matty was well. It depends on time and his mother whether he goes or not to Denton next holiday time. It's long to Christmas. Our respects to Robert Dent, Mr Peacock &c and all friends M.C.

Corn must grow as grass &c grows wonderfully fast, mercury rises, only for cold I fear a north wind we must submit.

62. *Matthew Culley to John Welch*

Wark 7th September 1800

John Welch

Davidson got here on Friday with your letters and Mr Peacock's &c. This is an unpleasant day for harvest, wet and damp. We have only got in some oats, no wheat led at all. We have cut a good deal, and lately the weather has been so hazy we could do little at shearing, although we have much corn ripe. Turnips are growing wonderfully ever since the rains came, were generally at a stand as well as the grasses. Now every pasture grows wonderfully. Sheep are doing well as are the small cattle, and the feeding ones are upon the waterd land. They are mended partly, but are much backwarder than last year. Your account of the price of beef is higher than we were told, 50 we were told of. You will be able soon to form a guess what cattle and sheep you can be able to winter on turnips. As to hay you must by no means use any at all of it, but when you cannot absolutely avoid it. Turnips now may be a decent or good crop, but all depends on them. Hay is scarce but good, straw is scarce but will likely be bad also from the wett and roky weather has damaged it very sore. Our pease straw was as good as hay the first got, the latter tollerable, but now what's out will be much injured. The grain wasted much if this wet continues longer. A north wind, with the mercury falling more and more every day. If the corn takes harm, the grass grows, everything green grows wonderfully. But we (till this day) had not so much rain as at Eastfield &c on Thursday, but I hope our corn has not taken much harm yet. I had wrote thus far when I found my brother had wrote, so deferred sending this.then.

Ninian fair 27 September. A very wet morning and rained so much that we only left Eastfield about 9 o'clock, so that the market was much over before we got there. Many Yorkshire men buying sheep and fresh cattle, also some neighbours from near Newcastle buying cows for fogg. Things were well sold. Joseph Moo's shearings sold at 34s 6d they (Moo) thought 17 lb. per quarter, very good, none near so good in the market. Ewes at 22s to 27 or 28 per head. Green had bought many, and part went some time since. Mr Wood's ewes at Tresson 24s bought some say 2 or 3000. The demand seems great. Jedburgh on Thursday was good for fresh cattle they say. Turnips are wonderfully improved, so are pastures and stubbles, but the weather has been wet and warm, a very growing time. Barf bought 40 score hill wethers of Mr Robson of Belford, also 20 score its said of somebody else. Green bought Oliver's highland wethers, John Davison of Newcastle got no shearings this day, wanted much to buy good ones. I believe we can winter our cattle and sheep as our turnips are so good. You should let us know how many oxen you can winter on turnips, with your weathers you now have. I fear we have tainted many of our feeding ewes. We find this year that when we have eat a break of rape not too close, and then put away the sheep to another break, the old one has recovered so as to be as good or better than at first. By this means it takes much eating. If you want any stock to eat your turnips they cannot now be got hardly to pay well for keeping. You should always let us know these things in time. We may send you some oxen and queys. Matty bought Jemmy Grey's ewes at 22s per ewe, we suppose well worth the money, on this day went (Mr. Peacock thought them too dear) for a young Scotchman who was some time with my brother at Eastfield.[72] We have had good harvest and very wet also, then some very good, so as to get in most of our corn, and lately wet but open weather, and now a very wet morning. We have got mostly in at last, only a few loads of wheat, oats and barley, but not a day's leading. My brother has got mostly in, Grindon all cut and much out, half led only I suppose. Harry &c half led in, crops go in little room. Our beans are out only half cut, rather pulled as they are so short, but are well loaded. Stackyard not near filled, but the corn is gifty. Ralph Brown sold some new white wheat for seed yesterday (Friday) for 3. 12. 6 to 4£ out of Dry Tweed, fine but soft. I mean to be at Mr T. Bates's sale of ewes at Brunton on Friday first the 3rd of October, but suppose you will hardly get there to meet me if you should get this letter in time. We shall want to buy some good strong colts at Newcastle for working in the spring, if to be met with there. You must if all be well meet us there a day or 2 before the fair or sooner, to assist Matty and me to buy. Will you want any stout good colts yourself, or working country horses or mares, as likely you will want some for Tod's and Birkbeck's farms. I hope Mr Peacock will give you Joseph Birkbecks rent for me at Newcastle. With respects to your family and all enquiring friends I am &c

Mattw Culley

PS. We have not heard from Matty I think since Mr Peacock sent us word after he returned from the north. What cows and incalve queys have you sold since you last wrote? Fat cows are likely to be well sold. A Quaker from Leeds called himself Chapman wanted wool. If he calls have ready money, is I believe now better sold for money.

Mr Bailey says he has a 7 row drill[73] for you. Am afraid that we can

---

72 The reference is to the system of apprenticing a young man, often a farmer's son, as a pupil to a farmer of good reputation. The pupil might get his keep and/or some pocket money, but no wages. For the practice in the 19th century see *Agrarian History of England and Wales*, vol. 7, pt 1, ed. E.J.T. Collins, Cambridge 2000, pp. 623–4. The young Scotsman in question may have been William Mure: see p. xxix.

73 An implement for drilling seeds.

not get it sent to Denton before Christmas, when we send for the tups. Was much obliged to Harry and you for calling on Matty. Your mistress will want him home at the vacancy to get him new cloathed &c.

63. George and Matthew Culley and Matthew Culley jr to John Welch

Eastfield 20th October 1800

John Welch

We received yours by Jack, and am desired by my brother to tell you that he wishes you to keep the horses you are for sending here, until old Martinmas when we are to have Mr Bates's two, and yours can come that way that time, which is much the best road in every respect for them in my opinion at all times. It is a pity that you did not keep the milked cow &c, as they would have paid well for Auckland fogg &c, and that is the best place for you to get fog at any year almost as there are many graziers in the vicinity of Darlington to take grass at any time almost. But I am clearly of opinion that it is wise to give fat cattle turnips at an early period, let them be as valuable as they will. Cattle should never be allowed to sink or lose condition if possible.

[*M.C.*] Wooler fair[74] was so extra good for sheep that I believe all were sold except George Hughes's[75] ewes, for which he was I suppose offered 32s per for or nearly so much. Consequently turnips will and now are offered for lower prices, very few sheep besides them were unsold. Possibly a few turnips at a moderate price may fall in your way now or in the winter. Ralph Brown sold some new wheat very high on Saturday to Brown, 4£ and od (5 bolls for seed at 4£ 10s). Some old along with the new at 4£. The old worth 50s only from Thornington. We think these prices may decline after seed time. We send the horse and mare away on Friday to be early shewn on Monday morning. The fair is on Wednesday for cattle. I mean we may probably go in a chaise (or coach) if we can get a partner. Wool is or was lately (Nat. Handysyde says) worth 12d per lb., but now the stocks fall as we likely may have to make a separate peace with France if the emperor give up the contest.[76]

---

74 17 October.

75 George Hughes, 1747–1834, Middleton Hall farm. See above, p. xxvii.

76 After the French victory at Marengo in June 1800 an armistice was in force between Austria and France. Peace negotiations in the autumn broke down and hostilities were resumed. Renewed French victories, at Hohenlinden and

[*G.C.*] Bill Snowdon asked 40*s* for his dinmonds in the morning, was bid 38*s,* then asked 45*s* was bid 42*s* and took them home unsold that I know of. 10 gimmers bought by my brother at Bowmont Hill sale were ordered to Wooler fair unknown to me. When we got there Andrew Young was asking 33*s* per for them. My brother immediately asked 40*s* per. He had to go to hire a maid and I sold them to Ned Henderson at 38*s* per. In short it was in every respect the best fair I ever saw at Wooler. George Logan of Fishwick sold a large lot of ewes at 40*s* per. I am in haste

George Culley

PS. What was the matter with Mr Peacock's child that died so very suddenly?[77] It is my *decisive advice* to sell wheat at these prices, let the prices be what they may after. I consider 4£ as good as 5£ per boll in summer.

[*M.C. jr*] We hope to see you before the cow hill.[78]

64.   *Matthew Culley jr, George and Matthew Culley to John Welch*

Eastfield 11 December 1800

John Welch

We have this moment received a letter saying that the three tups we have in Yorkshire will be at Mr Henry Barugh's of Carlton near Thirsk tomorrow the 12th inst. Therefore you must be so good as send a cart of directly on receiving this to Mr Barugh's and fetch the three tups, and also write immediately to let us know when you receive Lord Darlington's 2 tups that we may send for them, as the season is now getting very late and his Lordship certainly must have had plenty of use of them now.[79]

[*G.C.*] It is very extraordinary John that notwithstanding all we say to you, that you still neglect to write to us. We fully expected to have heard from you how Great Monday market was at Darlington.[80] Were there no Wakefield jobbers there? I rather hope that you had a decent market, because old Mason was at Morpeth on Wednesday last. There has been a strange quantity of fat cows and queys shewn at Morpeth, it is said these weeks since Martinmas. But sheep,

---

in Italy, led to the conclusion of the peace treaty of Luneville on 9 February 1801.

77 Jane Peacock died on 6 October, aged four. The cause of death was croup. DUL, BT Denton.

78 Newcastle fair, 29 Oct., held on the Town Moor or Cow Hill.

79 There is no signature.

80 The first Monday in December.

especially good ones I am told, begin to be wanted. We think of selling a 100 ewes to Lincoln &c soon. Turnips are all bought up in this quarter already, and we begin to think that there must be more stock laid on this year than the last, especially cattle. Corn still advances. Ralph Brown sold at 95s.our boll wheat last Saturday, and there is no doubt but wheat will be sold at 5£ or upwards next Saturday, and all other grain in proportion. This embargo laid on by the Emperor of Russia[81] will advance corn markets. I am in haste

Geo Culley

[*M.C.*] John Welch. I think of being with you after Christmas, as I may come over when the children return to school. I would not have you do so much on the west side of the carrs joining the lane, as I mean to bring the water over it if I can, in grass as it now is. And I am determined where to run the hedge over the north hall field to Carlisle's house. In haste &c

Mattw Culley

[*G.C.*] Wark December 12th. I hope you will write immediately on receiving this, and spare me the trouble of scolding you so often. Nelly Dickinson died this day. Do see or write to Christopher Eales[82] of Raby, as we wish the sheep home directly. Yours truly

Geo Culley

---

81 On 7 November Tsar Paul of Russia laid an embargo on British vessels in Russian ports. In the middle of December treaties were signed between Russia, Prussia, Denmark and Sweden reconstituting an Armed Neutrality.
82 Steward to the Earl of Darlington at Raby Castle home farm.

1801

65. *George and Matthew Culley and Matthew Culley jr to John Welch*

Eastfield 10th January 1801

John Welch

By a letter from Mr Stubbins[1] of the 29th December last 'he says that he has sent a turnip cutter[2] for us to Mr R. Colling by his cart which was sent for his tup'. It is a pity that Mr C. had not told us of it because it would have come free of expence by Jack. However you had best send or go for it yourself, and direct it for us by the waggon, as we are in immediate want of it. Perhaps this will be best done by yourself on a Monday by going early and bringing it to Darlington, can have a proper direction put upon it and given to the proper carrier that day. I never saw one, therefore can give no account of it, but suppose it is not very large. I should hope it is in a box. If not perhaps it may be prudent to get something of that sort made for it to keep it safe, and better to direct upon. But your own good sense will suggest the best mode to you. Upon asking Matty I find it is about 2 stones weight, consequently you must desire Mr Colling with our compliments to get it to Darlington for you by one of his carts. Could you not employ *Harry* or some other *confidential* person to sell the corn for you at the markets, which would keep you more at liberty? Now when your farm is becoming larger, and consequently your business more extensive, it is highly requisite that you should have a *confidential* person under you, not only to look after the work people but market corn &c &c. If you can't think upon one in your own neighborhood, we had better try to send you one. But I think you will find one out by a little attention. Selling stock is a very different thing, which requires not only experience but superior abilities. But many a man may be found capable of selling corn after a little practice, provided he is to be *trusted*, that is incapable of selling stock. I never

---

1   Nathaniel Stubbins, Holme Pierrepont, Notts.
2   In the early 1760s the Society for the Encouragement of Arts offered a premium of £20 for a machine to slice turnips. By the end of the century a variety of turnip cutters was on the market, the principle the same as that of the straw cutter, i.e. a box into which the turnips were fed, with a knife or knives to slice them. Few were yet cheap enough for general use: Fussell, *Farmer's Tools*, pp. 89–90, 180–2.

thought Mr Colling's Robert anything remarkable, and yet he sold all Mr C.'s corn, and as to selling stock I thought him very incompetent. I once stood by him when selling a few sheep at Darlington, and he appeared to me to do it far from well. You will find your presence at home more necessary every day when your business becomes more extensive. Your two farms lie rather detatched from each other, and may require in hoeing times, harvests &c sometimes two different sorts of workers, which will require two different *overseers* or what we here call *stewards* improperly. I only offer these hints for your consideration, sensible that your own good sense will see the propriety of it and furnish the means necessary. For sure I am that there is no trusting a parcel of workers, especially women, without a guide or head. Even in plowing, where more than 3 or 4 plows are employed together, I am convinced one of them should have the *charge* of the raft as we call it here,[3] and we have long followed that practise. I have long been convinced of the propriety, and even necessity, of captains over *tens* and captains over *hundreds* &c.

We are quite against your selling your shearing sheep, for 2 most obvious reasons. First they will improve very much, and more towards spring if you can spare them a few seeds or other grass, but seeds always grow them best. 2ndly mutton has every appearance of advancing at this time, and will advance in all probabillity more than beef. You say 'that you want cattle to tread or make manure', a very important matter indeed, especially for raising turnips. But you must observe that sheep improve ground by having turnips led off or by eating them on either, much more in our opinion than by cattle either led to them in a field or to shed &c. But then you very justly say that this does not give you muck for your turnips, I grant it, and yet think the consumption by sheep most profitable in any point of view. Besides ours are outlaid cattle which are to go to you, consequently very improper to be housed. I would also suggest to you that to eat one half of the turnips upon where they grow, where the land is weak or not over rich, is an excellent method, by leading one ridge more or less of, and eating the rest on, or leading of 5 or 6 drills and leaving as many more or less according to the condition that the ground is in at the time. However as soon as we have your answer we will either send you the Shotton cattle directly, or keep them a month more or less according to your instructions. But John don't be surprised if before we have your answer we send you a 100 fat ewes to sell as

---

3   An 'x' here indicates the note by George Culley at the end of the letter.

soon as you can meet with a customer. You need not mind who buys them, either home butchers or Holdgate[4] or Bulmer or any other good man, as we don't mind their being killed by anybody. But remember to take no bills. Your language to old *Head* was not only proper but gave me very great pleasure. William Lincoln is to be here on Monday first to handle a 100 very good ewes, 21 lb. per quarter, for which I mean to ask him 2 ½ guineas each, and take no less than 50*s* at any rate. He says he can't afford more than *7d* per lb. sink. If so we shall not bargain, but I think he will not leave them. I am much mistaken if they are not half as fat again as J. Hodgson's. I wish they were with you in same condition as they are now, but that is impossible. However I believe if well drove they will suffer less than most sheep. Should he not buy them they shall immediately set of for Denton over the moors in the mind I am in. Should he buy them we will send you another 100, but not more than 18 lb. per quarter. Our decided reasons for sending you these sheep are that your prices are better than ours. The price that Holdgate gives J. Hodgson is above *7d* ¾ rather, and there are none sold at Morpeth I am told for more than *7d* sink. Very full markets, and some unsold every week almost. Why should we not avail ourselves of your situation and assistance for quitting both cattle and sheep when it answers or when your prices exceed ours by so much as to pay for driving. Besides it saves me going to Morpeth, and although not unwilling and perhaps abler than I have been some other winters, am willing to be excused on fair terms. You very properly say that sheep can be drove or travel at less expence than cattle, and that will *operate* very much against the Wakefield people (jobbers) coming this far north for cattle at any rate during the winter months, and even sheep I think, and although the prices of beef in particular were higher here last spring than with you, it is more than probable it may not happen again of a while, particularly as the Russia trade is done away for a time.

I hope we shall have an immediate answer from you on receiving this. It is impossible John that you and us can conduct business at the distance we are from each other on the same concern or *interest* as it were, so well as if we were nearer together. But the best way certainly is to communicate to each other by letter as often as possible. Indeed there is no other way that I know of. But I have said *so much* on this head already that I have no doubt but you will attend to what I have said, and not neglect writing in future. For what is 5£ paid for postage

---

4   A livestock dealer.

to be compared to the advantage gained by writing letters, and I am sure neither you nor I will consider writing as any trouble, as we both carry *'the pen of ready writers'*.[5] It is a little odd that oats should be dearer with us than with you, but we always come the nearest to your price in oats of any grain. You exceed us much in barley. But I should imagine the liberty given to bakers or millers to mix grain or flour[6] will advance barley &c in the same proportion that it sinks the price of wheat. Corn is certainly lower, and many people think it will continue to lower. But I am of your opinion, that without a good importation it will not lower much. However when corn gets to these strange or unknown prices, or any other commodity, surprising and unexpected changes often take place. Consequently it is always wise to sell in these critical times, and at any rate to keep as little in hand as possible, and on no account to set any corn up at the markets in these *strange, perilous* and *critical* times. Head can't have any right to expect the refusal of our wool. You were certainly right in not telling Tommy Hutchinson what I had wrote. Tommy Charge is in this country, and our Matty returns with him and will likely be at Northallerton fair. It is now likely that both my brother and Matty will be in your country together, but don't let prevent your writing, then or afterwards. I rejoice at my friend George Dent's recovery but lament his son. However youth if not too much tainted will recover again. Certainly get your tithe agreement signed and done.

Geo Culley

[*M.C.*] John Mr Bates's horses are a thing I cannot judge off now, but I expect to see Mr Bates soon, here or on my return from you. Why have you not sold your 2 cows? Never keep cows after Christmas if you can possably avoid it. They never pay until after Candlemas.

---

5   Psalm 45.2: 'My tongue is the pen of a ready writer.'
6   Under the corn law of 1791 imported wheat paid a duty of 24s 6d when the price was under 50s per quarter, 2s 6d when the price was between 51s and 54s, and 6d when the price was above 54s. Export of wheat was allowed when the price was less than 46s per quarter, and a bounty of 5s was payable when the price was below 44s. The Act did not work any better than its predecessor of 1773, owing to a combination of factors. The war disrupted trade, and a succession of bad harvests sent prices of grain much higher than the low duty levels, reaching unprecedented heights in 1800. In almost every year in the 1790s temporary measures were needed to relieve scarcity, including free importation. Government attempts to manage importation were a failure and had to be abandoned. The use of mixed grain for bread was encouraged, without success, in 1795–6 and again in the winter of 1800–1. The harvest of 1799 was bad; that of 1800, although better, did not bring relief. D.G. Barnes, *A History of the English Corn Laws from 1660–1846*, London 1930, pp. 58–9, 71–85.

Never let your wishes to make of them let you keep them on. I hope
to set out with my Matty on the 18th or thereabout, shall likely want a
hack to return north on.

Mattw Culley

[*G.C.*] John you will see that my brother has wrote a little part
down, but some company coming interrupted him, and to make it
more intelligible I have signed his name and interlined mine above.
However as he and our Matty are both coming to you in a little time, I
hope they will be able to explain what appears now deficient. Yours

Geo Culley

PS. Do write as soon as possible after receiving this, and at any rate
say what you think about sending you fat ewes. I have no doubt of its
being right, except the expence and loss of condition by driving. We
have delightful weather at present, which is an inducement to us to
send the ewes. G.C.

PS. I hope you would take care to dry the turnip seed which Jack
Davison brought to you if it got any wet. A little damp spoils it. In
that case you had best pour it thin upon a *good* floor. I say *good*
because it will pass through a very small hole or crevice, and if your
floors are not good you should put cloths below it. At any rate it is an
excellent way to pass it through a corn or winnowing machine with a
very gentle wind, towards spring or before selling it, because it
prevents it *miling* or *moulding* &c.

[*M.C. jr*] I see John Pattison has not paid you in full for his turnip
seed, that is he has only paid you 2£ 16s in the room 4£ 4s. I hope you
will take care to receive of him in full

M. Culley

[*G.C.*][7] The principal or leading plowman should set out the ridges
a proper size &c &c, also have the management or direction of the
drilling &c[8] and although I say that dinmonds or wether shearings
will improve much after this season of the year, I don't think the same
of old ewes. An old ewe seldom improves much after Candlemas. But
gimmers will, and therefore we have taken out all our gimmers to
keep last.

---

7   An 'x' indicates that this note refers to the passage noted at n. 3.
8   A further 'x' here presumably indicates the end of the added note.

66. *George Culley to John Welch*

Eastfield 14th January 1801

Well John

If you won't write, that is no reason why I should not. We thought it right to acquaint you per post that we have this day set of 100 ewes for you by James Glass, but when they may reach you is another matter. We don't tye him to his stages but leave that to his own discretion. The weather is so uncommonly fine that we thought much to await your answer to the long letter which we sent you last. Mr Lincoln came and bought 120 ewes at 50s per, very cheap I hope, as he allows them to be 21 lb. per quarter when got to Sunderland, so that they are under 7 ½ per lb. and very fat indeed. He also bought 5 cattle, 4 midling ones valued at 6s per stone sink, and one good cow at 7s. The ewes don't go till Monday come a week, the 26th inst., and the cattle also. The ewes we have sent you are very much worse than those sold to Mr Lincoln, but on that account will travel the better. We think they may weigh 17 lb. per quarter, and I would recomend to you to sell them as soon as you can after their arrival, and to whomsoever you can sell them best. We are not uneasy about their being kept to breed from. However although I advise you to sell them immediately I certainly would leave you a discretionary power to act as appears best to you, that is either to sell or keep them a while. Only in the mind I am in, and if your markets keep better than ours for mutton, I would advise more fat sheep coming to you. Not only the remainder of the ewes, which by the bye will now be mostly gimmers, and most off them very nice gimmers, but the dinmond wethers also, because either you must have them or we must sell them rough at Morpeth as we can't pretend to clip them here. However as you have part seeds on your own farm and some on John Todd's besides some good grassland, it may be right for you to consider whether they may not pay you as much for keeping to clip as any other stock you can either buy here or in your own neighborhood. This I beg you to turn over in your mind and give us your opinion freely upon it. If you determine to take them, they should come to you before the weather be well warm and in that case we can either keep the outlaid cattle longer or you may have them whenever you think right. Tell us your mind on this also in your answer. Your little wethers I am still not fond of your selling cowardly, because although they will not pay so much as our full bred wethers by the ware,[9] yet they will I fancy pay you well and be good to sell. But all

---

9 By the piece.

these things you must consider, and then write your opinion freely to us. The corn markets are going to advance again we hear by a letter from London the other day. I know not what this day may be at Morpeth, but I don't hear that any mutton here fat [has][10] been sold there above 9*d* per lb. sink. Yours

Geo Culley

Thursday 15th. Yours of 13th this moment come to hand, and I have to thank you for forwarding the turnip cutter. You will perceive that I have said you may sell your ewes to who you like, jobbers or others, without seeing them killed. They won't be too heavy. William Lincoln has got a good bargain, I now see clearly, and I am better pleased that he has than another. By a letter received from Mr Sayle this morning, sheep were sold for near 9*d* per lb. sink at Rotherham last Monday. But if you get 8*d* or even 7 ½ I shall be content. Be not over hard, reasonable things are best. I hope the 100 ewes will weigh 17 lb. per quarter too at least, provided they get well up. James will give them time I am certain, at least he promised so, and I hope you will take some more soon, either gimmers or wethers. Mr Sayle says that cattle will advance also, but I don't think so. Only one thing brings on another. But there are numbers of cattle feeding in many parts. Corn is up 8*s* per quarter wheat last Monday at London, and barley in proportion. I am only affraid of too high prices till a great import arrive. But do sell as you thrash I beg. I am glad to hear you speak with such glee about your little sheep, depend upon it they will be and are nice ones. Your clovers will swell them and ours and the gimmers, and will pay you more in my opinion than anything you can put on. I am rejoiced to hear that your work is so properly arranged. Nothing like method and system. No man could ever conduct any business without a proper arrangement. I am convinced that our success has (under the Almighty) been entirely owing to that. There is no doubt of the Emperor making a separate peace, and that will cause a drop in all corn by and bye probably, especially if the former let us also make peace, which must be the case soon I think.[11] At any rate it is right to be selling. I am glad you have got the mare so quiet. I thank you for your *long, full* and *intelligent* letter. Let me hear from you again as soon as convenient, and everything will go as well can be expected. I have no fear, you know all sublunary things are

---

10 A word seems to have been omitted.
11 After the signature of the Austrian-French peace treaty of Luneville on 9 February 1801, Anglo-French negotiations were also expected. A treaty was not concluded until 25 March 1802.

given to change, and you and I are mortals, and as such liable to err. But as long as we act straight forward 'and do to others as we would wish them to do to us', we will be rewarded I trust as good and faithful servants. That we may always do as we ought to do is the earnest prayer of your friend and master

Geo Culley

67. *George and Matthew Culley and Matthew Culley jr to John Welch*

Eastfield 29th January 1801

Well John

I am happy to join this family in congratulating you on the birth of your child,[12] and we all wish your wife a good recovery, you and your son health and the blessing of the Almighty. I was up at Thornington yesterday with an anxious wish to hear from you, but no account then of J. Glass. However last night a letter arrived from you with a postmark, which must have come by Glass. However Ralph Brown is gone there after drawing what ewes and gimmers we have here, to make out a lot and to prevail upon James to set off with another drove as soon as possible. Before I send this letter I will name the day that is fixed for their departure. You deserve our thanks for not only selling the ewes so well, but so soon. I am certain that we could not have sold them here so high by at least 4s per. Nay I am morally certain that if the 120 ewes sold and now gone to Mr Lincoln could have been safe drove to Denton (and I think Glass would have done it) that you would have sold them for 56s per. They were one of the nicest lots we ever sold. But I wish the man luck of them. However I would have blamed you very much if you had not shewed them to Bulmer when you did, because I am of opinion that they would not have looked or handled much better in a week or so, and I think Glass must have drove them well. Mr T. Charge who is here, says he believes Bulmer a very honest man but not over strong, and pays all in bills. However I am glad that you are to receive payment in country notes, and before the sheep go, and I hope you will always *do so*. I see by this day's London paper that mutton is sold in Leadenhall Market at 9d per lb. in joints, so that I do suppose that mutton is still on the advance and more than beef. But that must follow by and bye. Not but I think we have much more beef than mutton, and the wheat market is up 3s per quarter, barley a little, but oats and pulse flatter and rather lower. It is

---

12 The Welchs' son John was born on 18 Jan. 1801: DUL, BT Denton.

remarkable that we sell wheat and oats higher than you at the present. You will see by the letter I wrote by M.C. of Wark that Ralph sold wheat at 5£ 7.6 to 5£ 15 on Saturday last, and midling oats at 2£ 2s. It is very rare that the markets for corn here exceed yours, and that can only happen when the importation is in a great measure stopped, which is now the case.

The gimmers of the lot now to be sent will be in general very nice, several that I handled this morning here were thick cloven from *shoulder* to *tail*. But the ewes will be very moderate, being the shots[13] of all the rest and so uncertain or unlevel that we can't well value them by weight. Indeed the gimmers are not level here but all handsome mutton, and most of them *very capital*. Therefore perhaps it may be right of you to sort them in lots when they arrive at Denton. The ewes I would certainly sell at the price they will bring directly, but the gimmers, at least such as you approve of, may be kept a while or sold as appears to you most prudent. But I shall be disappointed if many of the gimmers don't please you and your customers also, and I have no doubt but a part of them will pay as much for keeping on as even our wethers and more than the highlanders in my opinion. But after saying this I should be very *illiberal* if I did not allow you a discretionary power to do and act as appears best to your own judgement. Because you are the best judge whether things are upon [*sic*] still upon the advance. I myself have no *doubt* of it. Then they are such *nice sellers*, such *tempting ware* that it will be in your power to sell them *over dear*. But pray do not do so to poor Bulmer, who I never saw but I think well of him, and he certainly is deserving of a refusal. Being also clear of lamb (we believe) they will pay to clip, and will clip a decent quantity of wool. However you know you are to do that as seemeth good to you. I can tell you one thing, they will look charmingly *naked*, and will drive better.

Now after this lot we shall have nothing of sheep kind worth our while sending to you, except 200 or a few more wethers and a few segs,[14] which will not drive well but will be handsome mutton, but not so nice as the shearing wethers which I expect to be very nice though not heavy owing to the dry summer, and the best sold in clipt hogs. Therefore it is your duty to consider whether they are to come to you, and if so *when*. You must always keep in mind that they will not drive well in warm weather. But that will not take place before

---

13 Ill-grown ewes, the remnants left after the best of the flock had been selected.
14 Seg: an animal that has been castrated when fully grown; male sheep more than three years old.

April, and I think we can keep them untill then. However I name this for your consideration, that you may direct your operations accordingly. Because if as I think there is no fear of mutton being considerably higher, and we able to keep our wethers untill April, you can perhaps in that case keep all or a part of these gimmers a good while on. I think we can manage also to keep our outlaid cattle till then. I offer these suggestions for your consideration, and that you may avail yourself of them and arrange your matters accordingly. But I must also observe to you that the wethers and outlaid cattle will probably come upon you with a *smack* when they do come. And therefore it may be wise in you to sell the highland wethers at or before the arrival of our own dinmonds. Because I do think that our own breed will pay more for clipping than your highlanders, because they will clip more wool and will swell to more weight. Only against this is to be set the loss ours will sustain by *driving* while yours are going on. Still in spite of all that I should lean to the side of clipping on *ours* rather than *yours*. And I must add that shearings, either males or gimmers, improve more in proportion after they are put on seeds than at any other season of the year. This we have long observed. Perhaps after recovering their teeth they feed with more avidity, and they *swell* and *sap* with fat in every part, untill fly time. But you must always keep in *recollection* that after clip time markets are very precarious and uncertain, because it is between winning and losing as it were. And when markets begin once to drop, they fall sometimes with an astonishing *rapidity*. It is nothing keeping things on now provided you have meat, because the odds are great in favor of an *advance* and often a *sudden* and considerable one. But in the *critical time*, in the end of *June* and all *July* we cannot calculate upon the sudden change that may and will take place for the worse. Therefore I hold it wise always to sell at the *prices then going*.

John I am giving you long lectures without adverting to my brother being to be with you who, when his head is *unclouded* 'can see as far into a milstone as he that picks it'. In fact though a bad salesman there are few better advisers I know. I have often profited by his advice, and hope to do so a good while yet. Perhaps if my *Matthew* had been put to the *tryals* (I won't call them *difficulties* because they are now over) which his uncle and father experienced, he would have been as active and clever. But for want of those *checks* and *rebuffs* which we met with *for our good* he is too *impetuous*. But thank God he has sense enough, and I flatter myself he will gather wisdom with experience. You must have had much more snow than us I apprehend. We had very little here. I advert to what you say about buying for you a few fresh *cows, queys or steers,* and we will think about it and do the best we can. But I assure you we never were so scarce of fodder. Indeed I never saw dry meat so scarce since 1763, the winter succeeding the dry summer 62, which you may have heard of. Harry has sent even a part of his work cattle to Wark through dire necessity, and thinks his straw will still scarcely serve. Longknow ones are all comed to the same place, and William Brown says he must send some there soon. It is happy for us that we have such a land of *Goshen*[15] as Wark I assure you. We shall become pushed here, at Westfield we shall be able to spare a piece of summer wheat I hope. But what is to *become* of the country for *corn* in summer God only knows. Because we always kept a considerable quantity of corn through summer untill harvest, but that cannot be the case within this year. At Wark perhaps they will be able to keep a little summer corn, and you know it would be *mad* and *wicked* to lie corn by at these prices. However we must defer buying you cattle at present, and when we do buy they must go to you directly. There can be no doubt of tallow advancing still more because none can come from Russia. 1s is a strange price for barley, but we exceed you far in wheat and oats.

I think I told you in the letter by Matty Culley to Durham that my brother had been poorly which has stopped him at the present, but I hope he will be going soon. I believe he goes by Mr Bates's. When our Matthew and Mr Charge go, I know not but soon I believe. I am happy to hear that you like the drills. By all means don't neglect the limestone quarry if you think it will pay anything. Pray put your money into the bank at Darling[ton] as soon as you can. If Mr Peacock don't get home you can put it in yourself well enough. I am glad that you have sold the 2 cows, and it will please my brother well. Indeed it is seldom right to keep our midling ware at any time, and as to clyered[16] and sickly cattle the sooner they are quitted the better. The soup kitchens will consume immense quantities of coarse beef, and have more effect on beef markets than we at present can calculate upon. They are exceeding general. In this parish we are to have 3, and they will consume 12 stone per week, one at Ford, one at Etall and one at Crookham. Ralph Brown goes tomorrow to contract with Marshall of Kelso if he can to supply us weekly. Now John I am sure you will not complain of quantity, however you may of the quality of *matter* from your real friend

Geo Culley

---

15 Genesis 45.10; 46.28; 47.6: the best of the land of Egypt.
16 Having a tumour or swelling.

[*M.C.*] Well John I got to Thornington today to see the gimmers drawn, and ordered the driver to buy himself a pair of new shoes for his good driving, which we will allow him for. I give you joy of a son &c. I have been a little out of order but am now better. Jane we think has got the hooping cough or king cough which is prevalent here, so dare not come south till she gets better or worse, better we hope. Stein Allen and Richard Bolton have each buried an infant, whose deaths were supposed by it. Glass is to bring you Mr Bates's colt as he comes with the sheep, or Mr Bates will send him to you and on his return he leads the 2 fillies north. Wishing your wife better I am &c

Mattw Culley

[*G.C.*] Well John   Eastfield 30th January 1801

I can now tell you that we have sent you of this day (30th) January Friday 54 ewes and 66 gimmers, 120 in all. You will find the ewes with a dot of keel or redd upon their heads or just behind them. They are very small but not bad mutton. The gimmers as I said before are very good, but so unequal in size and weight that without we had got time to put them into equal lots we could not value their weights well so must leave that to you. You will see by what my brother has wrote that he was here yesterday. His affection for his child and wife will not allow him to go to Denton at present, and when Matthew and T. Charge come I can't learn yet. The weather is once more so fine that we are sowing wheat at the Westfield. By yesterday's paper the London wheat market is up again 3*s* per quarter this week. Barley a little higher, and oats rather flatter as well as pease and beans. Yours once more Geo Culley

PS. It is my own opinion (but Ralph would say nothing) that the gimmers weigh 18 pounds per quarter and the ewes 15, making together 17 pounds per quarter *good*, which is same weight as the 100 ewes you sold. I should not wonder at your selling the gimmers for 50*s* per if you keep them a while, they are so very nice. We still have a few small gimmers left at Thornington which we shall keep on, and either send them along with a part of the shearing wethers or not as is then thought right. You know it would be wrong (perhaps) to send all the wethers together, being too many to drive cleverly at once. Especially if we shall send the few segs also, and I know not what we could do with them at Morpeth alone, I mean without any other sheep with them they would not sell so well, because the butchers rather object to them, especially if they are more than 3 years old. They are best mixed along with shearing wethers. However they always weigh well.

Berwick 31st January. Ralph has sold more wheat at 5£ 15s, but you still beat us for barley. This is a thorough snowy day, I doubt James Glass will have an awkward journey, but he must now take his chance. Mutton markets are got dearer at Edinburgh and London, so they must meet in time. I understand that last Wednesday was better also at Morpeth. The jobbers are all afloat, but many people are complaining of their turnips spoiling, and we are pretty well but I see some are going. And if that increase it may give a turn to markets. G.C.

[*M.C. jr*] As Northallerton fair is coming on we think it right for you to sell the Galloway at all events, even for some loss, and you may also try to sell your mare, because if you can sell her decently we can send you another to ride on. I expect to be in your country before the fair and have some thoughts of showing my horse. Yours

Mattw Culley

68.   *Matthew Culley to John Welch*

Wark 7th February 1801

John Welch

This will be delivered to you by your acquaintance George Winter,[17] who I wish to help the old carrier[18] which H. Rutherford made over the little Fitts.[19] It was hardly wide enough, especially at the top. I suppose John Tod has hardly watered any this season. You must let him have some assistance. It will be some days before I get to Denton, having to call at several places on my road. There will be a difficulty carrying it on where the hedge now is, but I would have it made with some solid earth before it, as when it is banked with loose earth the water often breaks it away, especially when there is a flood. Also in making the hedge have a foot or 2 for scarsements. Harry left too little room between the water and quicks, so when the earth mouldered away the quicks were ready to come away and get long necked, especially where the land was on a declivity. These things are obvious to you now. I would also widen the old carrier by taking a sod of the top and lay the sod backward a foot or so on. This may be easyly done except near the old beck, which must be made stronger by cutting some of the north side of the beck and laying to the foot of the bank before you can well widen the the [*sic*] carrier. Be sure to lay a green sod on wherever the water will run over the bank. Perhaps there should be some land watered in the carrs, especially by the well

---

17 Hind at Wark.
18 Conduit or drain for water.
19 The name of a field at Denton: DUL, DDR., Tithe plan, Denton.

strand on the north side of the carrs. I know not whether my brother wrote to you since he received a letter from Mr Deverel, saying fatt cattle had advanced 1s per stone at Nottingham fair all on a sudden, beef now worth from 7s to 8s 6d per stone, as likewise had mutton also advanced much. I was afraid of falling short of turnips, was the cause of my brother selling the best ewes to Lincoln. They are now worth 10s a head more or nearly so. Great numbers of sheep have been sold here lately on the Tweed side. Jere Clayton bought 25 score of highland wethers from [. . .],[20] 5 spayed queys of George Nicholson, and 2 fatt ones of George Hughes. John Mason bought Harry Howey's oxen. J. Clayton I heard sayed turnips were very cheap with them, so said Mr Deverel. Keep is plentifull as well as turnips, so that store stock and horses are dear with them. Indeed everything is so extra high that all judgement is at an end, God only knows the result of times. Such prices for corn were never given in our days. Matty has wrote that he got well to Mr Fleming's. He, Miss Teesdale and Nelly went to Newcastle in the chaise together. He coached to Durham and then coached to Houghton. Weather so exceedingly good, we have sown wheat all winter, also are sowing blendings in the haugh near Dry Tweed. Our crop of grain here turns out better than we expected. Ralph Brown sold 50 bolls red wheat at 107s 6d and 2s 6d referred to times, to Clunie and Home, Berwick. You can get Winter clogs, spade, shovel &c such as he needs. He is an honest good workman. I suppose you will have to find him board and bed at our expense so long as he stays with you. If you have a young labourer to set with him who may be likely to learn the business, it may be right to do so. Did your book sent by Matty get safe to Denton? He paid 8d for the carriage of it by the coach, but he says not by what coach nor whether at Newcastle or Durham. With respects to your wife, Mr Peacock &c from &c

Mattw Culley

PS. If not convenient to you to give G.W. his meat, allow him sufficient to get it. Perhaps he may lye with Harry if convenient. Tomy Laidler will not be able to keep the oxen much longer he says, but Harry may keep them a while at Thornington.

---

20 The name is left blank.

69. *George Culley to John Welch*

Eastfield 7th February 1801

John Welch

I received yours of the 4th last night. I wish I had received it before Matty went, which we might have done had we expected one yesterday, and Matty did not go from hence to Broom Park, where he was to meet T. Charge, untill 2 o'clock. However I am much obliged to you for your good letter, and I will begin this at any rate in answer, send it when we will. You were certainly right in not letting your ewes go untill you received payment. If Mr Bulmer had been fool enough to have been affronted at what you did, you were in that case not to mind it. Because as a servant you know you are to be accountable for every shilling, and consequently it is your duty to be stricter in these matters than a master, for this obvious reason that a master may indulge or give time when it would be very *unwise* and *imprudent* in you to do so. I hope you will be able to advance a little per pound in the few ewes, as Morpeth is now 8*d* per lb. sink for good sheep. Jere Clayton has bought no sheep that I have heard of, but 25 score highlanders of Joe Mills. He and George Nicholson differed for 8*d* per head for George's wethers. He bought 2 queys of George and 2 of George Hughes, which are all the cattle I have heard of his buying. People are not willing to sell yet, expecting an advance in cattle. Indeed sheep are sold full higher and readier than cattle. He would fain have engaged some good queys of Mr Jobson's but Mr J. would not. Old Mason bought a few beasts of Harry Howey, and I heard of no more. You are right, there is nothing like the south country jobber when times are brisk.

I fancy Matty of Wark must have forgot the letter to you, but I hope the books are come to your hands, because he mentioned sending them in the coach and paying 8*d* for carriage, I think to Darlington. He named this in a letter to his father the other day. Your master is quite well again himself, thank God, but Jane is poorly in the hooping cough, stops him I think. But if Jane get no worse he will go soon I dare say. Whenever grain is dearest upon the coast it is owing to a real scarcity, and the importation not adequate to the demand, which unfortunately for the country is the real case at present but scarcely happens once in ten years. 3£ per boll was offered for seed oats last Thursday, and refused by Mr Atkinson's steward of Yeavering. I dare say Matty's chesnut horse will be very unfit to shew at Northallerton, because he hunts him on Monday with Lambton's hounds. But I don't think Matthew is very keen of selling that horse, and I am myself much against selling the brown mare you have, except she can be sold for more than we can expect to sell a blemished mare for. But Matthew and I are of different opinions about that mare and I don't want to quarrel with him about trifles. But you must have one to ride, and I am of opinion that you will not easily ride a cheaper that will carry you so quickly to the market or from one part of the farm to another. And is that not very important to you? Pray who would sell such a one for 10£ or 12 pounds provided she is safe? But if dangerous, for God ['s sake][21] quit her at any price. Don't let it be said that my covetousness should persuade you to ride a hazardous mare. By no means, but I wish you to sell the Galloway, because useless in every sense. I am glad you were drilling wheat this *glorious weather*.

I now repent I did not send this by Ralph to Berwick for the post there. However in the mind I am now in, I will send it away in the morning, and it will still reach Darlington by Monday 4 or 5 o'clock, perhaps in time to catch you before you leave Darlington, or at any rate for you to get it before you go to Allerton and before Matty arrives at Denton, so that it may have some weight with him perhaps respecting the brown mare. However after all I have said I shall leave it to his own feelings and discretion. There can be no harm, if you are going to Allerton, to ride her there and shew her, but that is not a sufficient reason for giving her away. Have you any rot amongst your turnips? Ours in a part of West Flats are wasting, and many people are complaining. There can be no doubt but your wheat markets will advance in my opinion. I wish I may be mistaken, but I can't see where resources are to come from in time, except that strange fellow Paul[22] die, or go mad &c &c, which is not unlikely, and would in all probabillity be a lucky event for Europe. As long as people have grain to thrash there is no great fear. But what is to be done in the summer I know not, because in all this productive dry country upon these rivers, we are selling everything as we thrash. Indeed it would be wicked to do otherwise. But as very little corn will be left unthrashed in May, I repeat, what is to be done in summer? No doubt of it, God Almighty is sufficient for everything, and on Him we must rely. That He may enable us to subsist untill a more plentifull harvest takes place, is the sincere wish of Yours

Geo Culley

Saturday evening. Ralph Brown just returned from Berwick without doing any business, was bid much as last week without the

---

21 A word seems to have been omitted.
22 Paul, Tsar of Russia, 1754–1801, succeeded his mother Catherine the Great 1796.

2/6*d* reference which Matty will recollect. He was bid 57/5 for the potatoe oats and same for barley. Whether Ralph is right or wrong time alone will shew. But I never like to find any fault with his proceedings in the selling of corn, as I have been long convinced that he does to the best of his knowledge &c.

70.   *George Culley to John Welch*

Eastfield 13th March 1801

Well John

We got home by 11 o'clock today. My brother desires me to say that he wishes you to plow the Lammerdale hill for oats. He also desires to know what is done about Birkbeck and his tenant. I promised to send *Charles Turner, Queen's Head, Pilgrim Street, Newcastle* 20 pounds of turnip seed but I am apprehensive that we shall be scarce of that article here, having sold so much, and on thrashing our stock we find it yields worse than we expected. So be so kind as send it directly, and write to him at same time saying that you had sent him 20 lb. per Maxson's waggon[23] by your master's orders. Then let me know of it being sent, when you write to me which I hope will be soon. Direct it as above and let me know the value of the bag, as I shall charge him 1/6*d* per lb. and bag. We can say nothing at present about sending you any more sheep, untill we hear from you whether you have sold yours to Bulmer or anyone else. Morpeth was an overloaded market last Wednesday, and I should not wonder if Wakefield and Skipton be the same, because at certain times these things take place without our being able to assign a proper reason for it. But that should not discourage a man at this time, because it is not likely that things can drop at this season, but towards May I should not wonder [. . .][24] drop, provided the weather continue tolerably good. My brother [. . .] that either your Harry should be hired to board himself or another hired in his place [. . .] he get a wife. The oat seed is nearly done in this country. Ralph sold 40 bolls barley last Saturday at 10*s* per boll, which is the highest price that has been procured here as yet from the merchants. More has been got for seed barley I believe, but not above 3£ 3*s* that I have heard off. I was highly pleased John with your little mare, which I had at Bishop Auckland the other day. I like her much better than the brown one which I got from you when you were over.

---

23 George Maxson, a weekly carrier service between Newcastle, York, Hull, Nottingham etc.: *Newcastle and Gateshead Directory for 1795; Directory for the Year 1801.*

24 There is a tear in the paper here, affecting three lines of text.

Indeed I am very bad to please at my time of life, for I grow very timorous and she trips very much. Consequently I will thank you to take care of that little mare and I will have her exchanged for this at some opportunity, either when you are over or by some means or other. John please ask Mr Peacock if he has the particulars of St Helen Auckland and Thickley, which he copied from Davison's book. At least if he have them not we must have left them, as neither Mr Bailey nor I have them. The weather is turned very cold and winterish, with showers of snow and a good deal of snow upon the mountains, and yet business goes on well in plowing and sowing. As I propose sending this from Berwick tomorrow I will conclude Yours sincerely

Geo Culley

Berwick Saturday 15th March 1801. John I have just time to say that Ralph Brown has sold his wheat today at 6£ per boll without any reference, and his oats (potatoe) all at 5£ 8 per boll, but oats except for seed is not so good to sell as they have been. We have bought you 2 or 3 little dry cows for feeding, and it is said the great Mellish[25] from London has people down in the north buying fat cattle. A very cold day and some showers of snow.

Geo Culley

71. *Matthew and George Culley and Matthew Culley jr to John Welch*

Wark 21st March 1801

John Welch

George Winter got well home on the [. . .][26] Have you got the mouth of the conduit or level from the quarry opened? Should you not fill up the swallow under the bridge over the Unthank by setting a man and the lasses to throw out all the stones and gravell &c, and then fill it up with wett earth, and beat it well with a beater of wood such as you let a gale stoop with. You and Kirtley[27] should give the watered land a look now and then, as it may run too long on one place so as to eat of the grass. Have you set a parer to pare the mossy land in the carr south of the mill race? The open drains in the north side of the carrs north of where you bored so many holes in the little drain should have been all drawn across the land by the level, and

---

25 William Mellish of Mellish, Notts. He farmed scientifically, calculating the profits of cultivating different soils. Arthur Young, *The Farmer's Tour through the East of England*, London 1771, vol. 1, pp. 297–339.

26 The date is left blank.

27 Hind at Denton.

drawn according to the level by carrying it the same way as we did across the Fitts, by the water standing on the level to the top of the grass. This would have watered the small breadth best, and also drained the land best by cuting out the carriers end when you wanted to lay it dry. Cross drains dry the land best whenever you wish to drain above or below the surface in a general way. The weather is now more pleasant than when I was at Denton. I wish much for you to contrive where to set a kiln to dry oats upon made of drying bricks. I would have you consult Graves of Darlington about the bricks, and also how to make the drying kiln and where to place it and where to get the bricks. Your friend Scott said they made them at 8*d* per of pottery mettle, but no holes in them, but I suppose they might hole them properly. Jemmy Story is a nice maker of oatmeal. He is to come to you at the term. If you have a house possible it may be right then for him to come sooner. A drying kiln may be of great use in a moderate harvest, so as you could then give soft wheat a moderate drying. These things are for your consideration. Pray what wood did you buy of Mr Harrison?[28] I hope you bought a good quantity. As to a foot gang[29] perhaps there may be a willow tree or 2 small willow trees on John Todd's farm which may be cut for this purpose if suitable. You must let us know when you will have a house ready for Story. He should come 2 or 3 days before your term. Of this you must write so as he may be with you in the time that he goes away from the house. You may consult Richardson on a proper site for setting a drying kiln upon, it may be set a little way of the barn on account of the fire, if there be any danger from the drying fire. It may be set near or on the wall Richardson was building, but Graves will be a proper person to consult, with your mason. Matty Culley says young Colpitts who lets the corn tythe told him, Matty, at Darlington that he believes Lord Strathmore[30] will now let you a lease of the corn tythes, and if you will give in a proposal he, Colpitts, will give the proposal in to Lord Strathmore. This we think you should consider over and so we think. I think you have done well in geting the road over the glebe. If it cannot be hedged and made this summer it may be done partly, and finished in a future day, but it would be better still if he would sell the

---

28 John Harrison of Walworth: Robert Surtees, *History and Antiquities of the County Palatine of Durham*, vol. 3, London 1823, p. 316.

29 A long narrow chest.

30 John Bowes, 1769–1820, 10th Earl of Strathmore, succeeded to the title 1776. He leased the rectorial tithes of the parish of Gainford from Trinity College, Cambridge: see No. 17, n. 37.

principal part to buy his land tax,[31] a thing the Bishop of Durham wishes much for him to do. My brother and I think it best to leave Birkbeck's tenant to Mr Peacock to manage, but let Mr Peacock also know that if Birkbeck's tenant continue another year I am not to take him for payment to me. Birkbeck must stand to the money matter. The man is not my tenant but Birkbeck's, yet I am hardly willing Birkbeck should receive the surplus money which he pays Birkbeck. I am wishing you well. Yours &c

Mattw Culley

[*G.C.*] John Eastfield Thursday 26 March

This letter would most probably have been with you sooner by a day or 2, but I have had rather an awkward job of hiring Jemmy Story at the last, or rather his wife, for the *grey mare* is generally the best horse in this part. I could not get Jemmy to say or determine whether he would hire with corn or money,[32] so he sent up the wife and we were 'out of the frying pan into the fire'. She at last resolved upon corn, but then she had found out by her son Willy that the oatmeal had a different taste in your country. However after a great deal of backwards and forwards work we agreed at last, as you will see it stated on the other side, provided Jemmy is pleased. But I am not affraid of him, but really I thought Betty and I would have quarrelled several times. But this is a *whinging, peevish, fretful body*. However poor thing when all is said and done one can't be angry or wonder, because it is a long remove for them into a strange country. I really believe if Willy her son had not been with you and told her how well he liked the country, that she would have *taen the gee!*[33] I have advised her to sell her kale pot and all her earthen and glass vessels and heavy trumpery. My wife is to buy such as she can't sell amongst her

---

31 The land tax, originating in 1692, was levied annually by quotas allocated to counties and towns and at a rate of, in different years, 4s or less in the pound on the annual value of real property. The original assessments were never revised. The full quotas were never collected: between 1780 and 1798 no more than 30% was collected overall. Under the necessity of raising money for the war, Pitt in 1798 brought in a bill to make the tax perpetual at 4s in the pound, but giving landowners and others the right to redeem their tax by buying government stock and levying an equivalent rent charge on their tenants. Initially a number of large landowners redeemed the tax on their lands, but a substantial proportion remained, becoming regarded as a fixed charge on properties. Dowell, *History of Taxation and Taxes*, vol. 3, pp. 81–91; W.R. Ward, *The English Land Tax in the Eighteenth Century*, London 1953, pp. 132, 135–7, 142, 149.

32 I.e. to be paid wholly in cash or largely in corn.

33 Taken offence, turned sulky.

neighbors, and at any rate they are not to send more than what one cart can conveniently carry, and I have also agreed to pay her 20*s* over and above for this time only, for the loss they may sustain in selling their old things. I assure you we had a very serious piece of work of it. Now you I suppose will have to send a cart all the way to take them. I shall not be surprised if their conveyance cost 5£. However it can't be helped, and they are I truly believe honest people, and Jemmy is said to understand a mill well. I had to promise her a good house. I told her the houses were all good. I always think that the miller should have a house at or as near to the mill as possible. My brother talks of their coming a few days before the term. That you may think of and let us know, because I suppose they will have no objection. But of that I am not sure, because I neglected to put it to Betty. Matty Culley paid her from the time of Willy going to Denton to old Martinmas 22nd November 13 weeks 4 days at 6*s*/6*d* per week, £4. 8. 10*d*.

[*M.C. jr*] This was paid from the 24 August, the time of his leaving this country. She said that you had not paid Willy for that time, but we do not know that, therefore will thank you to say whether you have paid him for that time or not, as it is possible he may be twice paid. Below is James Story's conditions viz. 2 b.[34] of good wheat, 4 b. barley, 2 b. pease and 2 b. oats, to be paid 5£ for one cow and to have another cow kept along with your other servants' cows, house free, coals led, 1 measure of potatoes set, a pig and hens. James to keep a bondager.[35] If the rest of your servants do not keep bondagers you must free James of his also, and to have a 1£ of money (for this year only) allowed for the loss or breaking of their old ties.

[*G.C.*] In explanation of what Matty has said about the bondager at the bottom of the other page, you must know that we have hired James Story exactly as we hire our hinds here, and consequently we named the girl to work as a bondager does in this country, only with the difference, if none of your hinds' daughters or servants work for a certain reduced wage then she must not, or in other words to be paid the same as you pay your certain women workers. Now we neglected to name anything about hay for his cow, but you must keep his cow as you keep other hinds' cows I fancy,

---

34 It is not certain whether this abbreviation is for bushels or bolls. The latter seems more likely, as that quantity in bushels would be well below the usual payments in corn: see also No. 72.

35 A female farm worker whom the hind was bound to provide, usually his wife, sister or daughter. See p. xxvii. Bondagers were not customarily found in Co. Durham.

I have wrote to Mr Peacock about the papers for Mr Bailey, but that St Helen Auckland business is over at any rate for the present, as there are 2 different buyers for it. It seems very particular that we should have such dry weather while you have great falls of snow. Everybody here are stirring their turnip land. You know we never think of sowing barley until April at any rate, and oats are done except some little pieces that can't be conveniently plowed yet. You have managed very cleverly with the vicar and his agent about the new road. It will be still better if a part of the glebe can be bought. In regard to the wethers John, I cannot say but I now wish they had been sold, as we will be run *hard past* here I am affraid. We have all our fat cattle besides the few wethers, and not a jobber or butcher ever comes near us, and the turnips a going. However we must struggle as well as we can. We will you may be sure make every effort before we throw our stock away cowardly at this season. If I were you I would certainly submit to Bulmer rather than Houldgate. But things must come round again before May, or it will be wonderfull. Mr Sayle writes that Rotherham has been higher than ever known, and I see by this day's London paper that *mutton is now in Smithfield 1s per pound sink, and 10 ½ d per pound for naked mutton.* So that if meat keep so high in London it will not be low in the country. Do for heaven's sake let us hear from you as soon as you can again, and as often as possible. Always keep a letter in hand going. Mr Armstrong had above 200 cattle, mostly fresh steers went passed here the other day, bought in the Scotch borders I believe. Hilton Midleton I am told has bought between 30 and 40 steers and incalve young cows between here and Berwick the other day. London market for corn came down some last week, but is up again this week I see. A great deal of American flour had come in. My fear is in summer, because I am certain very little corn will be left in the growers' hands after May Day. You give a strange account of prices of barley. I am extremely sorry for my friend George Dent on account of his son. If anything we have said about James Story be not clearly explained, we will endeavour to explain more fully on your putting any questions in your answer. I am wishing you better weather and every other good wish. Yours sincerely

Geo Culley

PS. Any time I can get the little mare which I rode to Bishop Auckland in lieu of the brown one I got from you, I will be glad to exchange.

72. *Matthew Culley to John Welch*

Wark 28th March 1801

John Welch

As we think Matty may spend the vacation or holidays at Denton with you, will you enquire when they break up at Houghton, that you may know when to send horses for him. He must by no means stay longer than he ought to return to school, as most probably his mother and sisters will call to see him at school, and that will likely happen when the lasses return to Newcastle.

You must consider and write what number of wether and ewe hogs you judge your land can take, also what small cattle it may be convenient to send you and when you should have them. The sooner possibly the better if your land will take them. Milbank's farm[36] has been to a great degree winter hained. Especially when the sheep should be sent to you, as the cool weather will be best for sheep travelling with wool on. The cattle are poor and perhaps may as well be sent a little further on in the year. These are hints for your consideration, as you can best judge of your situation. How goes on your watering? I think George Winter seems now inclined to come to you. The miller Story coming to you seems to give George a greater desire. He also thinks your people keep the Sabbath better than your neibours do, and so much the better. They are no worse but better for a wish to keep the Sabbath properly, with an eye to religion. On Friday the wind got fresh, which makes the land grow pleasantly. This Sunday a fresh morning with a little rain. We have 3 fields to plow for oats, Mid Town field, Parkhill and first short side, have sown our land in good spirits and a fine season, for which we are I hope thankfull. Wheats look well, especially those sown after Martinmas. Have you eat over your wheats, which would bear eating? I am much of opinion that it fixes the land and injures the wire worm, also prevents the winter proud or the frost drawing up the corn. I think early sown wheats stand a worse chance of being a good crop than those sown later. After Christmas is better than before, as vegetation is languid until the days begin to lengthen. We also think the land is often better for being a while before sowing, and getting a bash or brash of rain and frost. It then takes very little harrowing. This can only be done on dry soils in general. I suppose

---

36 The Culleys had bought, in 1795, a piece of land in Denton (plan in ZCU 46), from Sir Ralph Milbanke, 1748–1825, 6th baronet, of Seaham (*Complete Baronetage*, vol. 3) and in 1798 his reputed lordship of Denton for £9,500: Surtees, *History and Antiquities of the County Palatine of Durham*, vol. 4, p. 2.

Mr Fleming will give up the school for a few days, at Good Friday or before, so you will have little time to spare but to send on geting this. I am with respects

Mattw Culley

On talking with George Humble today, he says George Winter is for having a cow to go with yours. Indeed his wife could hardly go to the High House or hardly to Sugar Hill to milk, and I know not where your cows are to go. It will make a grumbling among your servants if his cow is to be prefered to theirs. Therefore cannot say as yet whether he will come or not, at least at present. If you will send us as soon as you can, and time goes on, it's near May, what you give your hinds and what you think we should give him in in[sic] particular, will be obliged to you, as I will rather want the water doctor than bring any trouble or disagreement among your servants, who seem in general to be a set of clever fellows. Some of your men may attend to your works and learn in time to carry them on, or learn to water land, and we can send him Winter or someone for a season till your man gets a knowledge of the busines.

George Winter's conditions for this year is one cow kept with other servants' cows, 5 boles of oats, 2 boles of barley, one bole of peas, 1 measure of good wheat and 1 ½ measures of small wheat, 9*d* per day in summer ½ a year, 8*d* per day in winter ½ year, 1 measure of potatoes planted in the field.

73. *Matthew Culley jr and George Culley to John Welch, Matthew Culley to Matty Culley*

Eastfield 8th April 1801

Well John

We received yours of the 5th inst. this morning, and shall send the hiland wether hogs away in a few days. There are near 9 score of them, and my uncle wishes much to send a few of our own full bred wether hogs, consequently shall send a few of our own kind and as many of the hill ones as will make 9 score in all. They are now on clover at Thornington, as are all the feeding sheep we have at different places except some tups and segs, and we hope to be able to clip our wethers and segs as our grass is very good in general, but this very cold frosty weather will check it a good deal. In regard to your corn we think it advisable to sell as much as convenient, as it is likely as high now as it will be after, and we shall have occasion for the money at or about May 12th. We have been selling all along, and never sold so much at this time, consequently we shall have very little left when summer comes, and none at all near harvest, which we think right as it is high enough and it may be taken from us by the populous or burnt, as has been the case in severall places. Yours

Mattw Culley junior

[*G.C.*] Now John, after thanking you for your long letter allow me to impress you fully with the necessity of putting upon your letters wrote in a whole sheet, *Single*, because when the word (single) is not wrote upon a conspicuous part of the direction, the postmaster takes it for granted that it contains *inclosures* as bills &c &c. Therefore it is right and necessary to write the word *Single* above the direction so as they can't miss seeing it thus within hooks (Single). For want of this we had to pay double postage for your last letter, and although we can have it returned on applying to the postmaster that trouble may as well be saved. However remember also never to put the word 'Single' when you have occasion to inclose a bill, draft &c &c because I believe it is punishable, and all the postmasters have the right to open any man's letter or letters if they have suspicions of any kind. Your horses were certainly high bought at Hexham, but you were not to want horses if they had been still dearer, and I am sure you would buy them as cheap as you could. However let me recommend to you to buy good ones at all times. Not big tall horses, which tire with their own weight, but short legd thick usefull horses. You will laugh at me for giving directions about horses, although I profess to know very little about them. I approve of your deferring the drying kiln untill James Story come. There is always luck in leisure and in considering *and reconsidering* over things of importance when it can be done. It is foolish of young Story to expect higher wages than *your laborers*, and would be wrong in you to give him more. A miller is one thing and a laborer another. You may do in regard to one of James Story's daughters as a bondager or not as seemeth good to you afterwards. My brother has never said to me that he has hired Winter to go south. My brother is to be here today when he can speak for himself, and answer about the wood bought of Mr Harrison. It appears to me wise in your making the exchange for stakes with charr wood. My opinion is that it would be wrong to attempt to lease the corn tithes untill prices come to some level, which will probably take place if we get a fair crop and a peace. However there is no crime in making estimates and valuations. I have no doubt of the watering having a wonderfull effect on the low grounds at Denton in a year or 2 if properly attended to.

By a letter received this morning (*April* 9th) from Mr Peacock I find your Darlington bank allows no interest at all except the money rests 3 months in their hands, 3 per cent if 3 months and 4 if 6 months. Now our Berwick bank allows us 3 per cent if ever so few days in, which is a vast convenience to us or anybody. Therefore shall borrow at present of our Berwick bank what we need, and consider about having yours remitted at the end of 3 or 6 months. We only finished our oat seed yesterday, and they have a little to do at Wark still. But then they are only pieces of such land which we could not well do without for our turnips leading upon, and will perhaps give more oats now than earlier. Most of our other oats are above ground. But there is a strange change of weather from heat to cold. But that uncommon weather was too much, and not likely to stand. This Island is subject to great variableness in the weather. We also have received your second letter of the 6th this morning (9 April) along with Mr Peacock's, and we will forward this in answer to both yours tonight. But being now Thursday morning it has no chance to reach you before the time you see Bulmer if you meant *this Thursday*. I now am quite of opinion that the markets for both corn and *cattle* and *sheep* have seen *the height*. Not that I am as apprehensive of their coming much lower for a month or 2 except riots should prevail here as they are doing in the southern parts, which with the failure of manufactures and the forward spring are the obvious causes of the sudden drop in markets for all kinds of produce. Yet the same reasons, especially the forwardness of the grass in all parts, is the cause of the advance in the *fresh lean stocks*. However my advice is by all means to sell, both corn and cattle. The sheep you may clip, not that mutton may advance, but our sheep will improve and I don't think mutton will drop much if any for a time. I am still more affraid of beef, and if a peace should soon take place no one can say how low it may come. Hence the folly of buying in grazing cattle at these prices. I am sorry this will not reach you in time for Mr Bulmer, respecting his buying some of your cattle, but I will strongly recommend to you to sell whatever cattle he will buy of you. I shall write to Lincoln in a day or 2 to come and handle a lot of our cattle if my brother approves, because I think we shall see no more Yorkshire fat jobbers this year. Jere bought a good lot of cattle the other day of the late Mr Atkinson, never bid me for ours, and as to Shillitoe[37] &c I have done with them. You are very candid in acknowledging that you did wrong in letting Houldgate handle your sheep. You have no cause to *reject* Bulmer so long as he pays so well in country money. You were right to shew Matty the ox. I hope Bulmer will make a fortune of him.

---

37 A livestock dealer.

I think it would be right in you to take some more or *all our outlaid cattle*, and we will sell our inlaid[38] ones as well as we can, and if we can't buy things to pay a little we will buy none. You had better keep a good deal this year as you have your vicarial tithes taken, and lay in stock. Matty has recommended you selling your corn in the beginning of this letter, and I wish to enforce that doctrine. In critical times it is always wise to sell at these great prices, and who ever knew such *critical* and *eventful times*. In regard to fetching James Story, you need only send horses and gear, because some of us will spare a cart, and Ralph Brown says a *coup* cart[39] is the best, with *overings* and lightest. But you should say in your next when you mean to send, that they may be disposing of their useless things. Be so kind as tell Mr Peacock that I received his of the 5th inst. today and will answer it by and bye, but in the mean time may say to him that in the mind I am in at present I will certainly borrow what money we may want on the 9th of May, of a bank. I think I have now answered all the needfull parts of your 2 clever letters, and am yours in haste

Geo Culley

PS We will send this of tonight and it will be Darlington on Friday evening, or when tomorrow's post reaches there.

PS If you can make a slip over for 2 or 3 days it would be very well, as we can then see and consider and talk over many things. Provided you should sell any cattle &c to Bulmer and any of the money paid so as you can remit it not later than the 6th of May, so that I can receive the *bill* or *draft* by the 9th and *not later*, you must be sure to *remit it* as it will be of very great service to us if it comes at or before the *9th May* but *not later,* for our great payments. Or if John Todd or any other person should pay you any rent or other money, or to Mr Peacock, but sure that you remit it by the *9th May.* Mr Peacock must be told of this, and is the properest person to assist and advise you in getting the draft or drafts as remittances at the bank.

[*M.C. jr*] My uncle thinks that we had better send Story all the way with both horses and cart. Winter will not come unless his cow goes with yours, therefore my uncle thinks not to send him this year, and he can go over any time when wanted, and your own men can do the rest.

[*G.C.*] As to steers we can spare you none. They may be bought by you or us at a future day. There is no necessity to buy them at this dear time for *lean* or *store* stock, but we can talk and consider over

---

38　Housed.
39　A cart with closed sides and end.

these matters when you come over, which I wish for much when it is convenient and suitable to you. A few days makes no difference, but it *strikes* me that if you could come any time between the beginning and 9 of May, you could bring any money you may receive prior to that time, and if it is but 50£ it will serve us much, but a few 100£s would be very usefull. But remember that when I say this it is not now or at any time to influence you to sell stock or corn before you wish, because I make a point to give our upper servants always a discretionary power in selling, especially when they wish to do the best of their power as I am sure our servants always do. *And* I must beg you not to travel *late* or in the *dark* with cash upon you when you come *north* or any *time else.*

[*M.C.*] My dear Matt. Your mother and sisters are at Brunton and they intend to call and see you at Houghton soon. At least they intend to call on you if nothing happens to hinder them. Your cousin Matt has seen your part of John Welch's letter. I hope you behaved well at Denton. Your affectionate father

Mattw Culley

[*M.C. jr*] J.Welch. I did wrong to receive John Todd's rent, Mr Peacock must receive it as it makes a confusion. In haste Mattw Culley

74.   *Matthew and George Culley to John Welch*

Wark 9th April 1801

John Welch

I should like to know how Robert Dent is, as the doctor was not without great fears of him as I am told, although he thought that medicine might do him good. Also who bought Beaumont Hill. Mr Charles Colling intended he said to buy it. We expected a line from you in regard to Durham fair.[40] Surely you would write from Darlington on Easter Monday. George Humble wishes you to send his saddle by the horses which come for Story the miller. Your mistress and the lasses got to Brunton on Friday last Good Friday, stayed all night at Cambo, as Armstrong's horses were in the plow or harrow when they got there. You are so busy with the road over the Goose Croft and glebe that you have not time to write. What money might your friend Bulmer make on Easter Monday by shewing the ox? I hope Matty saw him there. How are your oxen doing, and sheep, and how comes on your pastures and clovers, also your lime

---

40 31 March.

kiln and watered land? Will you ask Mr Peacock if John Todd pays in the money to us he owes, or rather has he said anything to him on that score. We shall not[41] want it in May, and when he and Birkbeck pay their rents, also if or not Birkbeck's tenant keep possession or quits this term. I hope you had a good seed time. The weather has been very dry and is now very frosty.

[*G.C.*] So far my brother John some time ago, as you will see by the date above, and begun afresh on the other side today (13 April) and I am to finish this and send you per post, which I will do likely in a day or 2, as Ralph sold some wheat on Saturday to Mr Home to raise us money, which must be had and paid on the 9th May to Mr John Nisbet and Mr Howey, as I wrote you word lately. If you should be able to send me a little money, be sure you be not late of calculating the letter to be here by the 9th next month at the farthest. Consequently it should come from Darlington on the 6th or 7th at the latest. Morpeth I understand was a very heavy market on Wednesday last. Still I am not affraid of sheep, and perhaps cattle may still be better sold than we expect at present. It is said that Edinburgh was a very good market for cattle on Wednesday last. Now if Edinburgh get good, it will do some good to us. I wrote to Lincoln last Saturday but I do not expect he will be very keen of coming at present. I think we have no quantity of good cattle in this country left neither. If the York-shire men had taken away 100 more good oxen it would have done a deal for us. Jere I find bought very few when last out, a few of Mr Atkinson amd Mr Merge &c.

Geo Culley

[*M.C.*] We have got a blessed change of weather again. Yesterday was a very decent day but this is a remarkable fine day indeed. Mr Lilly of Berwick is just come to try to buy our bull, and I wish he may buy him. He is a very nice case of beef. We rather prefer selling him to Berwick as he will be to take all the way in a board by his keeper *Fox*,[42] and Sunderland or Shields is such a desperate long way. We shall ask no more than 7*s* per stone and take no less at present.

13th April

This day my brother shewed me a letter which I send you a copy of below. 'Dear Sir I have sent by the Manchester waggon a box containing the best kinds of potatoes that I could procure in this country, directed to you to the care of *Mrs Anne Thompson* (should

---

41 This is perhaps a mistake for 'We shall want it.'
42 Herd at Eastfield.

have been Miss Jane Thompson)[43] Talbot, Darlington, which I hope you will receive safe. If that should not be the case I will send you another parcell in the autumn, if you will let me know by a line directed for the Reverend Charles Mytton, Eccleston near Chester. I am sorry to say the markets are still very high notwithstanding the immense quantities of American corn and flour besides much from other countries, that has lately come into Liverpool, wheat 75£ – 25s per measure, barley 18s – 38 quarter, oats 10s – 46 ½ lb. the measure, potatoes 6s per measure of 90 lbs., and the early setts sell as high as 15s per measure. If any cause should bring you or any of your friends into this country I shall be very happy to shew any attention in my power &c. I am Sir your obedient servant Charles Mytton, Eccleston near Chester, April 6th 1801'.[44] coppy

On receipt of this you will go to the Talbot and look after the potatoes and set them in a nice manner as you can for seed for another year. This gentleman called to see Wark farm some time ago and has acted up to his word. He seemed a nice man, and so he proves, but I know not how to make him any amends. He mentioned the Ox Noble, and Champions, but has sent of different sorts which you must endeavor to keep separate, especially when you take them up.

Yours &c Mattw Culley

PS. We mean to send away 200 hogs for you in 2 or 3 days, only 40 or 50 of our own at present, 160 Thornington or Longknow ones which look well, and hope they may get safe to you. We have severe frosty weather, Saturday evening a cover of snow, but have an excellent crop of lambs. Hogs not so good as they were last season by far at Wark, and several died of a scouring[45] before they went to grass which we put them to long since, and have done well ever since. I hope you bought the horses good ones. My brother described the sort we wish for, very properly. A pound or 2£ is nothing when you meet with a thick useful one. Did Matty return yesterday for school? I wrote you a word that his mother and sisters would be to see him soon.

---

43 The words in brackets are presumably a note by Matthew Culley, not part of the Rev. Charles Mytton's letter.
44 Rev. Charles Mytton, rector of Eccleston. He died later in 1801. DUL, ASC, Hudleston Clergy Index.
45 Diarrhoea.

Eastfield April 14th 1801

John

Ralph and I have been looking over our outlaid fat and fresh cattle, and we think we can send you from 13 to 15 beasts, a few very good, some midling and a part cows we have bought, and some queys of our own. None of them unlikely things to feed, but there are 2 or 3 with some clyey[46] symptoms which we think will drive well enough with time, care and pains. However although we wish to send them in the beginning of May I still hope that you will be making a stop over in the mean time, when you could see them yourself and judge the better of them. The nice little cow is one that you wished to be sent before, and a good black and white ox that should have comed when the last came, and another good ox that was lame then, and some very fat queys &c &c. Observe that they are all outlaid ones. The 200 hogs my brother has mentioned on the other side, which should go the sooner the better, and as the Thornington ewes are now nearly done lambing James Glass can be spared, and surely it is best for them travelling this dry time and picking plenty on the road sides farther now. The other day I received a letter from a Mr Walker[47] of Woolsthorpe near Grantham in Lincolnshire, who I have long known and corresponded with, but I am not willing to engage with him in the manner he wishes. If his man chuses to buy any incalves that you may have at as good a price as other people, and pay you ready money in country notes (*no bills by any means*) I can have no objection, but on no other terms. This letter was as follows. 'Dear Sir. I have of late done some business in the cattle line (incalves), and I have several times sent the bearer, who is a Yorkshire man, into his own country to advantage. He is now going to Darlington for fifty heifers, for the purpose of keeping 2 or 3 months, and I have established a regular customer for them as they spring for calving, who pays me ready money. Now Sir I have been thinking whether it could be so managed as to be the purchaser of the cattle you may have to part with in 3 months of calving. I would be glad to do it upon a liberal plan, provided they could be sent to Darlington and a proper model could be struck out. As to fixing the price, say you fix the price of a lot to begin with, leaving a trifle to me and to the times, and if you give 2 or 3 months credit, you may depend upon a regular remittance by banker's draft on London. Mr Townend may perhaps see you or your son at Darlington, and he is prepared to talk with you on the subject.

---

46 With the appearance of a tumour.
47 William Walker, Woolsthorpe.

If not he will put this into the post office. I am &c William Walker'. Copy. This letter had the York postmark upon it, consequently he had forgot to put it in at Darlington I suppose. However I wrote to him that we did very little in the incalve way, and that what we had of that sort we sent to a man called John Welch, who had the direction of an estate of my brother's and mine at Denton near Darlington, and if his *Townend* or *Townsend* and our man could bargain, we would have no objection. Do you know this man? Was he at Darlington on Easter Monday? However if you should ever deal with him be sure you take no bills, only *gold, silver* or country notes &c &c. The master Mr Walker is a very good sort of man, but he got once wrong poor man and may do again, although I hope not. But I am determined to run as few risks in the world as possible. And you as a servant can always say that it is your master's express orders to give no credit, and that if you do, you are to be responsible for any mistakes of that sort. At any rate John you are in a very good way with Bulmer, and I hope you will continue to be so. I am in haste yours truly

Geo Culley

PS. Matty Culley is just comed from Thornington, where he has been drawing some barren ewes in good condition he says, and he tells me the hill hogs and a few of our own set of with James Glass tomorrow morning. When they arrive at your place I know not. I have instructed him to give them much time.

75. *George and Matthew Culley to John Welch*

Eastfield 18th April 1801

John Welch

My brother bids me tell you that George Winter is not to go to Denton at present. The hogs set of on Wednesday last. By a letter from Mr Lincoln he says he is full of cattle so can't come to buy ours, so we must of a bad bargain make the best. One does not know what may turn up trumps. However it is wrong to despond. As soon as Glass returns we will send him away immediately with the outlaid fat and fresh cattle, to do the best with you can. This very *wonderfull* and *astonishing* change of matters in the Baltick, and the death of the *tyrant Paul*[48] will make a strange and sudden change, not only in

---

48 Tsar Paul was assassinated on 24 March 1801. The new Tsar Alexander I (1777–1825) desired a *rapprochement* with Britain, and raised the embargo on British shipping. Britain in turn, after the victory at Copenhagen on 1 April, refrained from further naval activity in the Baltic. A definitive agreement on the rights of neutrals was concluded in June.

pollitics but agricultural matters in my opinion. The Baltick being again opened to our ships will very likely drop grain prodigiously, while it will in all probabillity advance beef &c for a while. However these are my (perhaps) weak ideas respecting it. Pray have you sent Charles Turner Newcastle his 20 pounds of turnip seed? If so tell me in your next, with price of bag, and we will set it down here as well as at your place, because we shall most likely receive the cash. I fancy we have sent you too much turnip seed this year, as our demand is so large that we are like to run short. We sell about below 2 cwt. to a hand, at 1/6d per pound. Has Pattison ever paid you yet for what he got the last year? I saw Dr Thompson the last Thursday, and I was happy to hear rather a more favorable account from Robert Dent, with the old boy's compliments to me. G.C.

[*M.C.*] 20th Well John I wish you to write often, and short letters are best. Do not mind the postage. George Winter's wife is a decent woman but very tender in constitution, and I thought it might put [off][49] some of your servants if he had his cow along with yours. May send him next autumn. Your countryman Bowlees perhaps may do some things especially in the spouting line with pare lugs, employ him in that line as much as you can, he does not seem to be an ignorant man, but little can be done before Lammas or after hay harvest. You should let us know what stock you may probably want, as we are not likely to know so well as you. M.C.

[*G.C.*] *Wednesday 22nd* and no letter from you John makes me think you are beginning to neglect us again John. *23 Thursday* and still no letter, which I rather wonder at. However I will lodge no complaint but wait untill I hear from or see you to explain and give your reasons for this *seeming neglect*. For I still hope that you will be passing over upon us some day soon, which will be very agreeable to me as I confess that I never stood in more need of advice that at this moment what to do with our fat stock. However I can't blame us for any neglect of selling, as we never had a chance to sell any of our cattle but to Jere, and at that time Shillito had wrote me such a flattering letter as made me ask Jere too much. However we must do the best we can. But I am now convinced that the corn markets have seen their highest prices, and it is very happy, as the poor and laboring part have had a long hard time of it. We have even a great reason to applaud ourselves for being so lucky as to sell 20 bolls of wheat on Saturday gone a week, when the first drop took place, at 110s per boll. Last week they would buy at no price at all, and we have nothing in

---

49 A word seems to have been omitted.

hand of wheat but what is to thrash. Ralph Brown was very unwilling to sell at a reduced price, but I began to suspect that things might drop, and the want of money had some effect perhaps with me. By a letter from Mr Walker of Woolsthorpe, he says fat stock came down 3*d* per lb. at London in a very short time, from 1*s* per lb. to 9*d*, and that he had a lot of sheep at Smithfield when the 9*d* price took place. Still I have some hope that though this Baltick or Danish business may drop corn, that it will advance beef from the trade being renewed to Russia &c &c. I would strongly recommend to you to sell corn at what it will bring, because I am convinced that grain of all kinds will drop, although it may be sold still at a fair price. We have managed the worst with oats, and we have a good few by us which might once have been sold at 50*s* or thereabouts, and now are comed down to 30*s* per. However I am going to send Ralph Brown to Kelso tomorrow to try to pop of a few as well as he can. I have all long said that in critical times such as these, it is wise to sell all commodities as they are ready for market, and we never before knew such *strangely critical eventfull times*. However I now hope that peace blessed peace is coming. If you come here John as I hope you will, be sure that you come no later than the *8th of May*, because on the 9th I have the money to pay and must get what I may be deficient in at Berwick bank on the 9th (Saturday). Indeed I do not pay Mr Howey till the 11th (Monday) as I go to Alnwick fair, but then I must get what money I want at Berwick bank on the Saturday preceding. If you bring only 50£ it will be acceptable. But remember that I am in no *distress* or *fear* of making up the money, because I have spoke to the bank folks and am prepared. Only the less we have to borrow the better. And if you do think of coming and can bring a little cash, be so kind as drop me a *line* in answer to this, be it *ever so short*. I will leave room to say how Morpeth market was yesterday, as I hope to learn that at Wooler from whence I shall send this today, and am with every good wish Yours

Geo Culley

Wooler Thursday 12 o'clock. I am sorry to hear that Morpeth has been a very bad market, both cattle and sheep too many. Write soon whether you have anything to say or not.

76.   *George and Matthew Culley to John Welch*

Eastfield 26 April 1801

Well John

Good luck it's said 'is better than early rising'. We received your well wrote letter yesterday by J. Glass, and I am rejoiced that you have sold so many of your fat cattle to your new Liverpool friend, and disappointed my old friend Crisp. I was half in despair, what with the drop in one thing and another, but as I said before, something always turns up trumps. I think you sold your cattle very well, and I wish with all my heart that you had some more from us. But John you must come over, *let what will be in the way.* Suppose you only stay one day or 2, especially as you are willing to take more of our fat cattle, I think that on that account you ought to see them, and take such as pleases and suits you best. Besides, the *more* is the *money* John, I cannot so well do without it, and these bankers of yours don't please me either *egg* or *shell*. They will remit, but how? At not less than *40 days*, which is abominable. Do my good lad *assert, regulate* and *arrange* your matters the best way you can, and bring me the money over yourself, to be here on the 8th. But the *9th* or the *10th will do*, provided it suits you better, and you acquaint me by letter on receit of this, that you will be here *to a certainty* on the *9th* or *10th*. But as I hope it may be no inconvenience to you to come at an earlier day, I trust you will contrive to be here by Friday the 8th of May. However as you now have the money ready, and 500£ is all I want, you can come at any time from this to the 8th. Besides your bankers will not allow us 1*s* interest on the money without it be in their possession 3 months. No no! They are real *Jews*! Therefore without my saying any more now on that head, be sure you lay your accounts to come off any day that is the most convenient to yourself *and business* so as to be here on the *8th* or *any other* earlier day. But be sure you also write immediately back on receiving this, saying to the best of your knowledge when you will be here, and 500£ will do my business well. Not that I have any objection to a little more if *convenient*, but not otherwise *on any account*. One thing, by the bye, you must not *neglect*, and that it to bring the money in *Darlington* or any other of your *country notes, as large* as you please. But not by any means in a *bill*, unless your bankers will give you *a bill at sight*, which I much fear they will not, otherwise it would be the best and safest for you. Indeed they ought to give you a *bill at sight*, as the money is lying with them *without interest.* However it will perhaps be the best to tell them *the truth*, as truth is *always* or generally the best, and then let them do as seemeth best to themselves, either in giving you a bill or draft *at sight* or their own notes. Admitting that the bank gives you no interest, yet it is prudent in you to put your money into the bank, because it is safer from both robbery and fire. Poor John Murray of New Ladykirk had a stable burnt and 13 horses, the best he had, all suffocated, owing to a boy who slept in the stable taking a candle to bed with him and forgetting to put it out. You are certainly right in receiving your payments in *country notes,* *gold* or *Bank of England* &c, and never in a longwinded or any *bills* or *drafts*. I am glad that you abide by that system, and my advice is for you never to depart from it. And it is a duty we owe to every respectable man to treat him with civillity and hospitallity.

We sold our bull yesterday to be killed at Hawick for 34£. Lilly of Berwick bid us only 30£ although he now says he will weigh 120 stone. Marshall of Kelso bid us 31£ 10*s*, and another man 33£. He is wonderfully fat for a bull indeed, we have not ox one so fat I dare say. We have washed our wether sheep and will be clipping a part soon. I am also glad that you still keep up your correspondence with Mr Bulmer, who I have reason to think is a respectable dealer. Mr White of Norham bought 15 fat oxen and spayed heifers of Mr Lee. I passed them going south on Thursday, their price is a little below 26£ and I think they are pretty well bought. Hutton wrote to White that Mr Bulmer was *longing to see them*. I think we can keep our cattle well untill the beginning of May, or perhaps a little later. But we will settle all these matters when you come over. You may want rain, but we never want rain here untill the middle or towards the end of May. It is most wonderfull weather indeed, although frosty. But it is cold frosty winds that do so much harm to vegetation, and we have no winds at present.

John Todd does not deceive me. It is the very character that Mr Peacock always gave of him. He is a narrow mean selfish man, which is unfortunate for himself and connections, and truly detestable. However you will not now be long plagued with him, and these things are good for us, that we see the contemptable figure men cut that behave so basely, and learn by that means to avoid such loose behavior ourselves. You certainly were right in *harrowing*, and his own interest bound him to *roll*. Perhaps it is wise to pull of the turnips when but few are left, and lie them in an oblong heap against a dike bank &c, and by that means you can get your land sooner sown with barley. In a dry time like this, it is wise to steep your barley seed I think. There has been no market for corn at Berwick these 2 weeks, and to a certainty grain has seen its highest prices and it is very lucky for the poor and laboring part of the community. We were very fortunate in selling a large quantity of wheat a fortnight ago at 5£ 10*s*, and Ralph got quit of a small quantity at Kelso on Friday last at same price. 95*s* was the highest price I heard offered at Berwick yesterday, but I heard of none sold. Ralph offered some nice oats at 36*s* and was bid 35. No wonder at large quantities of fresh and lean cattle being shewn, because there never were half so many went south so early from this country I think. I wrote to Mr Walker that if he was desirous of buying any incalves we might have, that his buyer Townend might apply to you as the person who sold our cattle at Darlington, and gave him your name. I don't see that either James Story or wife have anything to find fault with Willy's agreement and yours, as Willy was to hire himself and I think the wages fair for a young man not quite conversant in the various branches of husbandry. My brother said he would send Story, and I wrote to you and I shall tell him what you say about horses tomorrow and cart. As to the brown mare, we have drawn her lately, but you will have to ride her back when you bring me the *other*, which I hope will be very soon. Indeed you can return in the coach if thought right. My sister saw Matty very well. Poor Mr West. I am sorry to hear the account you give of Robert Dent. I wish Dr Thompson don't flatter. Suppose the days be long now I would not advise you to ride late with money in your pocket at any rate. You can always come in 2 days easily.

G. Culley

[**M.C.**] Monday 27th April. Well John I am glad that you have sold so many oxen, and so well also. John Todd is an aukward man, and very selfish. You must look to him as well as you can. Tell Mr Peacock to get John Tod's (and Birkbeck's if due) rent, at least Tod's, and send by you to my brother as you should come over with the money as the safest way. It will be inconvenient but it is necessary I think. I hope you may have a good lime season. The driver who takes the cattle to you from hence may bring the mare for my brother, may come to Mr Bates in one day and home in another. In such case you can come and go on your own mare, only it will cost you 6 or7 days on the whole. I have an unpleasant cold within this day or two, but hope it may go off with good care taking, in a little time. Give my regards to Mr Peacock and tell him he must act with Todd as seems to him most reasonable about his rent &c. Wishing you well

I am &c Mattw Culley

PS. Mrs Culley and her family saw Matt who seemed well. Mr Fleming was from home. Edinburgh [. . .][50] market keeps pretty good. One reason is there are no fatt cattle, from the distilleries, which are stopped. They furnished a many at this season. If Robert Dent does not mend this fine season I fear his life is very precarious. What are Birkbeck and his tenant doing? Proper notice should be given in time to them both to quit in time before next term.

---

50 There is a hole in the paper here.

[*G.C.*] John I beg you to write immediately on receiving this letter and say when you think you will be here, that I may be at home. Geo Culley

77. *Matthew Culley to John Welch*

[Wark, late April 1801][51]

John Welch

Well, how goes on your brick maker with carting the clay, does he approve of it? I am only afraid you may get hold of some marle or small lime stones which will injure the bricks. His [*sic*] brother George has been very ill of a sore throat but is we hope getting better as two places broke on the inside, run much matter, and so we hope he will be well soon, is now better. Andrew Bolton has been very poorly, inwardly in his stomach and bowells, is now some better but is likely not to bear much hardship. I am in great want of a young active steady man to keep accounts and carry on Wark farm. I should like to have a man of 25 within the house, unmarried, if you know of a suitable one. William Ridle assists, is a decent man but rather too soft, as Wark requires a stirring active hard fellow to look after so many turns as I have in hand. But I hope you will reach Wark soon after you receive this. You have sold your wool well, but yet never learn to sell dear. Moderate things are best and good men. We send some setts of rhubarb, should be planted in deep good soil 2 ½ to 3 ft asunder, also some seed to sow for future planting. Get your quicks transplanted by Bob in good time when the weather suits. If the weather continue frosty make a good dam or 2 where I showed you. I believe you should have a barley mill if ever you sell flower as I am told they shell barley and mix some in the flower. Am glad you have got the drying kiln finished, hope it will answer your purpose. When does George Richardson come north to see some lime kilns? Mr Bates's at Brunton at least, who has a mechanical head and who he must have conversation with, he will be glad to see him and give him all the information he can regarding their form and size. Also the line or breadth of the kiln bank so as a horse can go round them with the stones, as a single horse will best bank the stones. There are several things to talk about but they go out of my mind when at this distance. If I write at times illegible it is owing to my sight which fails me sometimes, especially with candle light. Has George George [*sic*] Richardson done anything

---

51 The address and date are missing. The letter has been labelled 'March 1801' by
   a later arranger, but from the mention of 'Wednesday 29th' it should evidently
   be late April.

at High House well (or have you covered the worst parts with thatch)? They should do it well as they go on. And do only a little piece at once. But I desire you to leave it to George R. I wish you had been here before Wood comes to see the weathers, they are got very good. They I think value them at 3£ or 3£ 3s per, 22£ per quarter, 8 ½ per lb. I wish if you have meat that they or part of them had come to you. We send Wednesday 29th we send you a dry salmon for yourself from Tom Tait, a drying machine, 2 plants rhubarb and some seed. With the compliments of the season to you and your family and inquiring friends. Your obliged

Matthw Culley

PS. I have wrote to Mr Peacock about Mally Ward's house. You should assist him, as it will be a pity to lose it and the rest of the houses and land, they are so near the mill. All may likely be sold when she dies. We hope you agreed with Grace and Co. for barley.

78.　*George Culley to John Welch and Rev. Thomas Peacock*

Eastfield 14th May 1801

Well John

I am determined not to spare pen, ink and paper to you whatever you do to me. I thought it right to acquaint you that Morpeth proved a most excellent market or rather fair yesterday. Ratcliff bought Mr Nisbet's 4 cattle at 9s per stone fully, and Ratcliff immediately went north and bought Tomy Hope's cattle, so there are scarcely any either good or bad left now in these borders except Mr Jobson's few queys. Sheep were not so dear in proportion, but were sold at a considerable advance. Jemmy Thompson said he sold his last sheep at 9d per lb. sink, but Mr James we sometimes suspect speaking a little *strongish*. However it was certainly a clearing day, and I would gladly hope that you will may have a pretty good selling time for your fat stock, but not for your corn I think. However Mr Pickering sold his wheat last Saturday at Newcastle at 30s the new boll,[52] or 4£ 10s our boll, which is much higher than we have it now. I would be glad to sell all we have now at 4£ per boll. Lean queys and steers sold strangely at Morpeth I am told, but the very fine rain had a wonderfull effect. I hope you got some at Denton. We had not quite so much here as farther south about Wooler &c &c. Bill Mills had a very good sale at Ewart today. Ewes and lambs were sold from 42 to 48s per and others

---

52　The new boll, equivalent to 2 Winchester bushels (i.e. one third of the boll used at Wooler), was commonly used in south Northumberland and Co. Durham by 1750.

in proportion. I am sorry to find by a letter from Wark today that my brother is not better yet of his cold. I hope you will not fail to send me over the little mare, as by taking away both the grey horse and brown mare I have nothing to ride but the mare which your mistress here rides double upon, as well as my Nelly. Indeed not having a spare horse to ride in case one is ill or fallen lame, we are very badly off. I believe we must send to Grindon for the mule, for Bob to ride to the post &c. Our bull got very safe to Hawick and very quietly. He died a beautiful case of beef, and a very good colour, remarkably fat, turned sulky twice on the road and lay down. *May 15th.* Very like rain this morning, has been some, glass falling. We had not so much rain as from Alnwick to Wooler.

Yours in haste Geo Culley

Mr Peacock Sir. On receiving your letter with Mr Scruton's annexed, I immediately wrote to Mr Scruton[53] that it was not in our power at this time to advance him the money wished for, but on *reconsidering* the matter we think that if you and John Welch can make up the money which is in *Darlington* and Mowbray's *Durham banks* to 1000£ or more in the course of a month or 6 weeks or such time as will suit Mr Scruton, you have our *full liberty* and *approbation* so to do. To be sure John Welch is to remit 200£ as it comes to hand, to pay Mr Bailey some money which he was so kind as lend me to make up Mr Howey's money, and some was advanced to Mr Nisbet. But if John is so fortunate as to sell the fat stock or even a part of what he has in hand, he can both remit the 200£ here and help you to make up Mr Scruton's 1000£ or more, and at any rate you can certainly get a few hundred pounds of your *bankers* for a *fortnight* or so. We don't recollect what money is in the Darlington and Durham banks, but suppose it is not less than 500£. Messrs Clunie and Home having stopped payment at Berwick has proved a little aukward to us at this particular time, because we had depended upon 800£ and rather more from them last Saturday, which will likely now never come to hand, at least not all, but I hope half may in a future day. If you think right of what I have said, you may put it in execution as soon as convenient. You must recollect to write to Mr Scruton as soon as you have made up your mind, because you will observe that I wrote to him directly yesterday that we *could not supply him* at this time, which letter he will most likely receive today (15th May). My brother thinks that John

---

53 Richard Scruton, attorney, North Bailey, Durham: *The Commercial Directory of Ireland, Scotland and the four most Northern Counties of England,* hereafter *Pigot's Directory,* Manchester 1820.

Todd should pay his rent soon, which would enable you to make up this money, or at any rate to pay bankers any part of the money you may borrow of them. I am with true esteem yours

Geo Culley

John Welch. You will see what I have said above to Mr Peacock, and I hope you will help him both with advice and money as soon as you can raise it, and I hope that will be soon, and you will not neglect to shew Mr Peacock immediately on receiving this what I have wrote to him without fail. Yours in haste once again Geo Culley

PS I hope we will hear from both you and Mr Peacock soon. Mr Peacock will explain Mr Scruton's business to you as I have not time, the letter is going to post directly by Ralph Brown.

79. *George Culley to John Welch*

Eastfield 18th May 1801

John Welch

Thomas Glass arrived last night with yours of 15th inst., and I have wrote to Mr Thomas Bates to sell the ox and make the best end he can of him. Ralph Brown, nor any of us, know anything about the ox being lame of his loins. Young Fox must have *dreamt* it. He cast himself in the place where tied, 2 or 3 times, so they put him in the loose house where lying on his own dung may have tendered his feet too much, for on the road from William Bates's to Mr T. Bates Brunton he cast several of his hoofs. If Joseph Hodgson can get the Walworth farm we will certainly help him as soon as we can, but you will see by my last letter that Scruton of Durham is to be first assisted. After that I hope you will soon be able to help Joseph, if you sell the stock as I hope you will. Tell Joseph Hodgson that my brother thinks it a very good farm, and he has our best wishes, and may assure him of our assistance as soon as you can spare him any money. I think one or two hundred pounds at most may do for us at present I think. That is as soon as you can raise the money. I cannot understand what had prevented you getting home untill Thursday. Certainly something extraordinary must have happened. We had an uncommon fine rain yesterday (Sunday) which will set all a-going.

Wednesday 20th. Yesterday we had the meeting of Messrs Clunie and Home's creditors, when everyone was unanimous in allowing them to wind up and carry on their business for their own benefit and that of creditors under the responsibillity and inspection of 3 trustees. And when I arrived from Berwick Jack Davison was just come from Denton, and I am happy in saying that I hope my brother is a little better by the account we had by a letter from himself yesterday. I shall see him today please God. I am sorry that William did not put up his appearance. I am affraid things may have taken a wrong turn at Liverpool, as well as other places. However things are certainly better here, although I understand that there is a vast of stock up to Morpeth this week. Will hear tomorrow and let you know. You are not to want cash on any account, nor horses, or whatever is necessary for your farm. But I blame you for leaving yourself too bare of cash at home. You should always keep a little money by you, because in your now extensive concern it is impossible for you to calculate to a few pounds what may be wanted, and Jackson and suchlike people are always wanting, and it is not bad policy to keep them in good humor without paying more than they want at once. Ralph Brown sold a Berwick man the hermaphrodite at 15£ and another quey at 18£, to go both away next week. Just 5£ more than canny Marshall bid him for the same two beasts. This rain will enable us to carry on awhile I hope untill you can take more stock, but it is a very cold wind today. Should not we send you a few young queys to go in your outfields for incalves? You will meet with a bull somewhere. You should go to Stagshaw Bank fair[54] on Saturday by right, it is the best place to meet with bulls. I am glad your markets are coming about a little again for corn. I should not be surprised if they should mend a little even in the ports, but I would recommend you to be selling a part. I sincerely wish Sir John Musgrave[55] luck of his bargain at St Helen Auckland &c, I think he deserves it. I have no objection to you keeping the grey horse a while. I understand that wheat was sold at Newcastle last Saturday from 30 to 35s per boll of 2 bushells. In the beginning of this letter I have named our assisting Joseph Hodgson in case he should take Mr Harrison's farm, but I think we should not lend him more than 200£ or so.

Thursday 21st May. As we have received no letter this day, must conclude that your friend William has not appeared and that you have done nothing. Had we received a letter this morning as I expected, I would have sent this away to you, but as that is not the case will keep this a day or 2 when we will certainly be receiving a letter from you. I have wrote again to Mr Bates about the ox by James Rae, who is going to Stagshaw Bank fair, for fear the former letter may not have reached him. The weather is turned windy and cold, and I think we shall get more rain. It is time our *Blue House hill* was

---

54 Stagshaw Bank fair, near Corbridge, Whitsun eve.
55 Sir John Musgrave of Edenhall, Cumberland, 1757–1806, 7th baronet. *Complete Baronetage*, vol. 1.

laid in for meadow, which makes me wish now to be quit of the fat sheep rather than the fat and feeding cows at present.

Wooler Thursday one o'clock. I have just received a very agreeable account of Morpeth market, which was still better than last week. Sheep were sold currantly at 8*d* per lb. sink, and some at the latter end of market at 8 ½, Mr Jobson tells me. He, Mr J., says that he averaged 3£ for his sheep, 22 lb. per quarter, Mr Nisbet 3£ 3*s* and Mr Robert Thompson 56 for some little ewes. Cattle were also well sold although more in than expected. Mr Jobson also says that he heard that Darlington was 5*s* per head higher than the fortnight before. Now I am a little surprised and disappointed that you did not give us a line from Darlington. However I am determined to send this letter away today to you, because the postage is nothing to the advantage of information at these critical times. There can be no doubt now but fat will be fairly sold, and I think we had better be trying a few of our fat sheep at Morpeth, as prices are so good and we want our land laid in for hay at Eastfield. It will always be making them a little less for the driving to Denton. However I must beg your opinion as soon as possible whether we must sell at Morpeth or send to you. Had you but wrote one line from Darlington, never mind quantity nor postage, a little intelligence is worth a great deal. Corn has also advanced considerably at London last Monday, that things look as though they would be reasonably got through still. I am in haste yours sincerely Geo Culley

PS. I understand that Ratcliff is still buying away, which looks very like as though he had got some contract.

80. *George Culley to John Welch*

Eastfield 26th May 1801

Well John

Yours of 22nd I was glad to receive, although it did not contain much important matter. I am sorry to say that my brother has gained no ground lately. I don't think him dangerous, but this cold and cough is severe and stays by him too long at his time of day. Matty is gone to Morpeth with 45 wethers, but I have heard a good account of Darlington today from George Hopper. So I hope you and Bulmer have made a bargain for part stock. We have had such a Whitsun Bank fair as never was in my days, for everything. Indeed we have no fat cattle, but all things were sold, it mattered not how lean. Hogs and ewes and lambs were sold strangely, milk things also wonderfully. George Hopper says that Stagshaw was much the same. He was not at Darlington but heard that it was good. A fat man bought all the bulls too at high prices *for this country*. By a letter from Mr Sayle he says sheep were 9*d* per lb. at Wakefield, and cattle 10*s* per stone, so I hope you will send me word for our Wednesday yet. Mr Sayle says long wool will be worth 1*s* per stone, wheat 16/8 per bushell. Oats 5/6 *bad ones*. By your account corn is higher with you than at Mr Sayle's country. The advance in grain depends entirely on the importation. You are an industrious man, but I am affraid you are late enough with rape in your moors. I fancy you are in no association for *prosecuting felons*.[56] But whether or not, I would have you get a couple of hundred handbills printed offering a reward of 20 guineas on conviction of the *offender* or *offenders*. Mr Peacock will help you draw one out in a proper manner for being printed off. It is the only way to scare the rogues. It is wonderfull that the rogues should have the impudence to kill the sheep in a place which is so near the turnpike road, where carts must be going constantly by night and day. I can have no objection to your staying a day at your father's, only when you do so I would rather have the information from yourself than anyone else. I am glad you want no more cows for your servants as they are strangely dear here I do assure you. I am rejoiced to hear of your good lime trade, and especially as Robert Colling is leading from you. I think Mr Thomas Charge's wethers by old George were not well sold, they should weigh all the weight you call them. I hear many sheep are up to Morpeth this week.

Wednesday morning 27th. Well John I have this moment received yours of 25th from Darlington. I have no objection to your selling to Mr Ratcliff, but I am not so satisfied about your giving him credit. Not that I have much fear of Ratcliff, but prompt payment in country notes would have suited better at this time. However I would recommend you to draw, or rather get the bankers to draw for you, at such a date as they will be willing to cash you the draft at. And not to delay drawing upon him later than he gave you leave. The cattle (if I am right) came to £747 which is a very large sum and would answer well

---

56 These were associations of property owners formed, particularly after 1760, in areas where police protection of property and the means provided for prosecution were felt to be inadequate. They generally provided rewards for information, and assistance in bringing prosecutions. There were probably at least 1000 at any one time between 1760 and 1860. Douglas Hay and Francis Snyder, eds., *Policing and Prosecution in Britain 1750–1850*, Oxford 1989, pp. 113–207. There seem to have been at least 18 in Northumberland and Co. Durham in 1801–2. John Welch's father was a member of an association at Penshaw: *Newcastle Courant*, Jan.-Feb.1802 (Penshaw, 6 Feb.). There was no association at Denton. John Welch had evidently had a sheep or sheep stolen and killed.

for Mr Peacock to lend to Mr Scruton, for I perceive now that we shall probably sell our sheep at Morpeth. We shall not need any money from you, and as I have heard no more from Mr Peacock about Mr Scruton's business I will thank you to name it to him and desire him to let me know what is done or likely to be done in that business. And if Mr Scruton agrees to take the money 1000£ or more, the sooner the money is made up the better for him. This of Ratcliff and what is in the banks will amount to more than £1000, and if he be for 1500£ you can borrow of the bank to make it up to 1500£ untill you sell your sheep, which I hope will be equal to the whole 1500£ if wanted. And as you will be selling corn to supply your own outgoings, and no money to remit *here*, you can the better spare the above money and run yourself a little nearer for a time. Besides, amongst the cattle we shall send you over there are several which will be ready to turn into money as soon as you have an opportunity. *I write this to you that you may shew or read this part to Mr Peacock as it will save both time and postage.* And although I think Ratcliff a man not very likely to go wrong, yet you can't be too cautious. Therefore would advise you in future to say to him or any other that as you are a servant you are *required* and *directed* by your masters to have *ready money*, and that in *notes &c, not bills* or *drafts*, drawn or to *be drawn*. You will excuse me writing in this strong manner, because what [with the]⁵⁷ Clunie and Home affair lately, and Paul some time ago. I think it wise and right to run as little risk as possible. Ratcliff I allow to be a very clever fellow, and I always think that he is a likely man to raise a large fortune. But still, we don't know what may happen in trade, these are most *critical* times, and Ratcliff is certainly *dashing* away at a very strange rate. He has bought a wonderfull deal of stock lately. I am highly pleased and gratified at the prices of the cattle sold to Ratcliff, and have very little fear of all being well, but I think I cannot impress your mind too strongly *with the idea of ready money*. I am very certain that it would distress you very much to lose us 100£ let alone 700£, and in your situation I have before observed to you that you can do what a master could not so well do. That is you can with a bold face say 'that your masters require ready money and therefore you must have it'. We shall not send you any more stock until we hear from you again. From the corrobarating accounts of Mr Sayle and David Green, I should not wonder at all if your markets mend for both beef and mutton, and in that case we can still send you over 100 of wethers

---

57 Two words seem to have been omitted.

more or less. But that you will acquaint us with afterwards. I am much pleased with your writing so frequently, and I am sure you will not be offended with the *well meant advice however strong* of your real friend

Geo Culley

PS. Be sure to read or let Mr Peacock see that part of the letter about Scruton's money, as I wish to hear from Mr P. or you if anything is done or likely to be done in that business. And I hope from what you have said that I shall hear from you again soon.

PS. We have Mr Boswell[58] in this country now, where Harry Rutherford went learning watering. G.C.

81.   *George and Matthew Culley to John Welch*

Eastfield 1st June 1801

John Welch

I rather expected from what you said in your last that I should have heard from you again before this, especially as I wrote to you once. I received a letter from Mr Peacock yesterday, dated 27th ultimo, saying that you had very properly shewed him my last letter, and after consulting with you he wrote to Mr Scruton who in his answer agreed to take 1000£. Therefore as soon as you can with propriety draw upon Mr Ratcliff do so. I suppose that your bankers will best direct you in regard to *drawing*, or Mr Peacock is a very proper person to consult. You can't consult a properer. I think the bankers ought to give you cash, that is *their own notes*, provided you *draw at 10 or 15 days*. At least our bankers always do so, and it is no more than fair to give Ratcliff at least 10 days. However that is entirely in the breasts of your bankers, who I think are real Jews. But I would have you shew or read this part of the letter to Mr Peacock, and he will advise you for the best. But I shall think very ill of your bankers if they will not cash your draft at 10, 12 or 15 days. And as soon as you can get the 1000£ or even more made up, give it to Mr Peacock and he will remit the money to Mr Scruton. The sooner the better. I say 1000£ or more, because if you can make it more and Mr Scruton wants it, may let him have it, tell Mr Peacock. I write this because postage is so expensive now, and you can let this part be read or shewn to Mr Peacock. But now on reading over Mr Peacock's letter I find that he is going *to London*, but makes a very short stay. However I have no doubt but you and he have arranged matters properly

---

58 George Boswell of Piddletown, Dorset, watering specialist, author of *A Treatise on Watering Meadows*, London 1779.

about Mr Scruton's money before he went. At any rate I am certain you will act right and with caution. You need not remit us any money now, nor untill you have transacted the money matter with Mr Scruton and Mr Peacock. The latter I perceive says in the course of 5 weeks or before, and Mr P. will be back from London long before that. However if Mr Scruton will take a few 100£s more at the time and you have it, and Mr Peacock approve, my brother and I will be agreeable. I say you need not to remit us any money untill you hear from us in another account, because we get plenty of money now by selling the wethers and some corn, and also cattle.

John Forster came here this morning, and will buy some beasts of us I dare say (*oxen*). The cows we will send to you when we have your orders, but you have no occasion to hurry now as pastures are growing apace, indeed I never saw corn and grass grow so fast, and I fancy fat cattle are becoming scarce everywhere now. As by a letter yesterday from Jere Clayton he is coming north soon, Ratcliff buying so many and John Forster coming unexpectedly, all shews the very increased demand. Pray take care of your wool. I find by a letter from Mr Sayle that it is selling at 1*s* per lb., and there are strangers riding in all directions, picking up any parcel that they can prevail upon people to sell. The opening of the Germany rivers has done them. However I would not have you sell rashly at even 1*s* per pound. It may be right to consult Mr Robert Colling on the wool business. Although corn came much down at London last Monday, wheat was sold at Kelso at 5£ our boll on Tuesday. There never was so little in the farmers' hands. I have not seen your master these 2 or 3 days, but am happy to say that we have good accounts from himself. John Forster and I are going there this afternoon, when I hope to give you a further good account of him.

June 2nd Tuesday. Mr Forster and I took a ride to Wark yesterday, and I am happy in saying that my brother is coming about as fast as can be expected of a person at his time of day. It has been a most extreme cold. John Forster and I have just dealt for the blind ox, and 2 more, and Richardson's cow very famous indeed at 30£ each. I think the cow near 9*s* and the oxen a little above 8 ½, perhaps 3*d* or 4*d* per stone more. He was keen of having the little white cow or quey, but I think we will reserve her for you. We have several more queys besides, and I am disposed to think that fat stock is upon the advance and not plentifull. Mr Forster talks of 2*s* per cwt. more, but I have little doubt but they will raise beef to 60*s* or 3£. If Clayton should come in the way, I will not sell him any under 5*s* per stone sink.

[*M.C.*] Well John I have had a most severe cold indeed, but have been better considerably for the last three days. I have dared to take 2 or 3 glasses of wine of diner and now find it has a good taste. Rode on my Galloway this forenoon. I will be some time before I get altogether well. We have 10 or 12 good cows and queys, some of them very good, but we have great plenty of grass, clovers exceeding good, as are most of our pastures. The country never looked better I think than it now does. Then is likely to be a good time to sell your sheep and I hope you will make the best of it, and quit what you can. If I advised you wrongly before, now you must sell, but we will not be able to supply you with stock afterwards for your pastures. Lean stock (cattle and sheep) not to be had for money, but we will be able to send you some little few lean of our own. At any rate you can make more hay land, and hay is a good article. Monday noon. Yours truly Mattw Culley

PS. I forgot to say I bought a tolerable strong useful but plain Scotch horse rising 6 at 15£. They could not be turned into money at the Bank, and there was a buyer for each cow at the Bank, and Stagshaw beyond everything for cattle. M.C.

[**G.C.**] Without your cattle markets advance it would be wrong to send you the fat ones perhaps, because we now sell them higher than you do I think. However when you write you will probably explain that matter more clearly. *June 4th*. Yesterday I received a letter from Mr Sherlock,[59] Lord Darlington's steward, inclosing what his Lordship owed us for tups, and also an order for a tup for the ensuing season, to be sent in middle of next month, *July*. So if you want any little thing that can be conveyed in the cart along with the tup may let us know, likewise anything you may have to send, that the tup cart will contain can be returned in her. Pray can you tell me if Christopher Eales (late farming steward to Lord Darlington) still lives at the farm west of Raby which he had of Lord Darlington, or at any rate you must be so kind as get me a direction enquired out where a letter will reach him as I have to write to him directly. George Dent will tell you I think, and I will be glad to know in your next how Mr Robert Dent is. You have said nothing about him lately. This is certainly a famous lime season for you. Indeed it is the finest season for everything, and every kind of land, I almost ever remember. Have you begun sowing of turnips yet? William Brown has about half done at Grindon I dare say. I wish he is not too early, but that strong land has worked so well and mellow this wonderfull season as to induce him

---

59 John Sherlock, agent to the Raby estate: Raby archives.

and many others to get forward early with their heavy soils. If our turnips be plentifull, it must be next winter from present appearances, so many will be sown But they will always be consumed. Pray have you had any trade for your turnip seed? We could have sold a good deal more by retail than we had left. Matthew had another pretty good market at Morpeth yesterday, sold from 8*d* per lb. to near 9*d* sink.

I fancy I will have to go to market next Wednesday as Matty and Nelly are both going to Edinburgh along with Mr and Miss Boswell who are here from Dorsetshire who are here at present. Mr Boswell is the gentleman who Harry Rutherford was sent to some years ago to learn to water meadows. As I have not received a letter from you this morning, and will scarcely receive one now before next week after Monday, I will send this to you from Wooler today. By a letter yesterday from Mr Deverel, wool is selling in the midland counties at a little more than 1*s* per lb. You must let us know how your prices of wool stand at Darlington. I am of opinion that if Jere Clayton do come north at this time it will [be][60] more to buy wool for his son that to buy fat for himself. There are no fat cattle left that I know of, and sheep will not pay him for driving I think, at these prices. His son can't come with any propriety before the wool time, but the old boy can. I am extremely rejoiced to inform you that my brother gets better daily, and I hope will soon now be perfectly well. I see by this morning's London paper that wheat is advanced 10*s* per quarter again last Monday at London, notwithstanding a prodigious importation. Can anything be a higher proof that corn is not in the hands of the farmer, nor even the factor. I am your

Geo Culley

PS. A report has taken place here, but I hope it is false, that James Story died soon after he got to Denton. Pray let me know if he be well. Poor old John Wright is very poorly, bedfast and not likely to get better. 'Young people *do* die, but old people *must* die'. Mr Deverill says they have had 5 days charming rain. I wish it reached you. We could take a shower now again, but we can do very well for a while. Indeed too much rain would lodge our big oats. However it has every appearance of drought at present, after some very thick hazy growing weather.

---

60 A word seems to have been omitted.

82. *George Culley to John Welch*

Eastfield 5th June 1801

John

I sent you a long letter yesterday, and today have been at Berwick fair, which was confessedly such a fair as I never saw before for lean weedy stock. Indeed the country has been so riddled and picked, that little else but *weeds* are left. Bits of midling 2 yrs old stots[61] from 10 to 12£ per, and Dunse I understand was full as dear yesterday. Milk things were also very high. The few fat things were sold as near 8*s* per stone sink as could be, which is less by much in proportion than lean stock. It is said that there were several fat cattle at Dunse, and sold below 8*s*. It happened that none of us could be there.

Saturday evening 6 o'clock 13 June 1801. John I have been at Edinburgh 3 days with your mistress to consult doctors about her health, are just returned, and I am unable to express my surprise at not having received a letter from you. What can be the reason? I hope nothing is wrong, but I really feel much disturbed. My hopes are that a letter from you must have miscarried, but letters so very seldom do miscarry to us that I know not what to think. However I will send this away in the morning, it is too late tonight, and must beg an immediate answer to *explain*, and *relieve* us from such unpleasant *suspence*. I am going to Morpeth this next week with 60 sheep and 4 cows and queys. Not that we are now in want of meat, although we are beginning to want rain, but as you have not wrote to us we are at such a loss what to do. So we are determined to be selling part, as prices may come down, and we have perhaps cows enough to send you besides these. But I really think I never was so anxious to hear from you. You must certainly be ill, or would always have wrote by this, and after Whitsun Great Monday[62] too. Matty has sold his sheep wonderfuly well every week at Morpeth, and better last Wednesday than ever. There is meeting of Paul's creditors on Wednesday at Morpeth, which takes me there. I am in haste and anxiety yours

Geo Culley

PS. We have had for some days uncommonly cold frosty weather, and still continues very cold. Please say in any corner when you write how Robert Dent is. I think if you are ill Dolly would not neglect to get Mr Peacock or somebody to write to us. Suspence is so terrible! Indeed I know of few things so unpleasant. Thank God my brother is

---

61 Oxen between one and two years old.
62 An important fair at Darlington, the Monday a fortnight after Whit Monday. In 1801 this was 8 June.

wonderfully come about and will get quite well now I trust. We dined there today on our return from Edinburgh. Corn has dropt again at London on Monday last, but Ralph sold part wheat at Kelso at 5£ and oats at 34s and 35s, barley 56 our boll. Oats are shooting as well as wheat here, and would soon be quite out were the weather warmer. Everybody are very far advanced with turnip sowing.

Sunday morning 14th June. Well John, the boy is returned from the post and no letter. I pray God you may be well. If so be sure you write by return of post, for we are in great anxiety. Wheat was the same at Berwick yesterday as Kelso the day before. We hope you have got a good deal of your wheat sold before this.

83.   *George and Matthew Cully to John Welch*

Eastfield 18 June 1801

John Welch

On my return this moment from a very good market at Morpeth I had the most pleasing satisfaction to meet with yours of the 14th. I sold my sheep from 8 ½ to 9d per pound sink, the best white cow and red quey at 5£ 10s, very nearly 9s per stone sink, and 2 little queys at about 8/3 per sink. I really think by your account that our fat butchers are better than yours. Consequently we will shew more fat next Wednesday and send the feeding young to you. I saw Mr Ratcliff who said he had got and killed most of your cattle, and was rather surprised that neither you nor I had drawn upon him. I answered that you were to draw, he said very well. So I hope you have drawn, if not pray draw directly. Ratcliff says your cattle proved very good. I thank you for your wool information. None sold here yet at 1s, I think I would take 19s, but best for you to advise with Mr Colling, make my compliments and your master's to him, and tell him we wished him to be consulted. Mr Peases's are good folks, I wish they would give me 1s or 24s our stone, and they shall have all ours, and yours too, may tell them so for me, and if they write directly they shall have it if then unsold which it is very likely to be as the buyers here hang off very much. I think you should sell your wethers, and not stand pat too hard. If hot weather come, which is very probable, prices will and must be high. Pray let George Dent have the 100£ directly, if you have not money let me know and I will send you a bill for a 100£. I am glad you are selling your corn, and my earnest wish is to sell all your corn as fast as you can. I wish you may be right in not selling your turnip seed early, but we have sown all early here. Indeed the weather is now and has been so very droughty that turnips can now hardly vegetate. Still I think it right to sow on and not wait *upon rain.* I don't think your master bought the horse for you. He is gone to Longknow. But don't you want a horse or horses, business must not be stopt for want of horses. You did well to sell your gimmers to Remmington &c. but I could have sold them better here, however it is right. We will consider about sending you the pretty ewes, and I will speak to my brother about sending you a bull. If the Todds pay their rent on Monday you will be strong enough for money. If not I will remit it if you want, by writing to me. I perceive that you have not read my distressing letter when you wrote the one I am answering. However I beg in future that you will not keep me in such suspence again, if all the draughts you have stand still for a week. I don't know that I was ever so uneasy. We will send you this tomorrow per post, and am your obliged

Geo Culley

PS. My brother's children drank tea here on their road home from Newcastle, all well, and Miss Petley with them, one of the teachers. Matt bought an ass, what could he make of an ass? Mr Edward Pease junior is this moment come here, which looks as if they were very keen, and in want.

[*M.C.*] Wark Friday 10 o'clock. Edward Pease has seen our wool, along with Matty Culley, when I was in the fields along with Tomy Bates and some gentlemen from near Hexam. I am exceedingly pleased that you sold some gimmers to Remington &c, but wish you had sold your wethers also, do sell them immediately, as times will likely fall now. I wish you may not refuse a worse price, do sell. If I was a means of your missing before, do not stand out too long. The drought is very strong here indeed and mercury high, a shew for rain and away again, in favor of our strong oats, pastures and hay going off and some corn in great want of it. I am pretty well but not altogether so. Why do you not draw on Ratcliff for the whole money? Was not Head to have a refusal of your wool? Try him, and let us hear from you. We conjectured that you were ill, and a hundred things. My brother got very uneasy indeed. We have not many queys to send you but for 4 certainly (not any cows). Possably may send you 3 or 4 spayed heifers, and 10 to 12 incalves, but not until we hear again from you. Only 4 of them are bulled yet, the backwardest of them perhaps keep them back and no rain.

M. Culley

In ten days we may spare the bull. He has had a number of cows, and Mr Bates says the ox made 24£ 2 to the butchers, leaves 22£ for us. The lameness shrunk him fast. It's very well, he might have done worse if Mr B. had not had him in charge. The horse I bought at 16£

for Jamison was not good enough for you. Are mowing the laid grass on the watered land, in places some is cooked, may mow on if weather holds dry. M.C.

[**G.C.**] My brother talks of Head, but you are under no obligation to sell wool to Head I think. G. Culley

84.   *George Culley to John Welch*

Eastfield 19th June 1801

Well John

We sent you a letter today and I am beginning another so I am determined you shall not have me to blame for neglect of information. What signifies postage if it were *double* to the satisfaction and benefit from information by letter. Mr Pease came at 9 last night. I am in great hopes that he would buy our wool, and yours which I offered at 1*s* per pound over head, but he will not give it yet, and I am resolved to take no less. He and Matty have been riding all day but have done nothing with anybody. Much wool is already sold in this country, and I am mistaken if we do not reach our market yet. By the bye, if this drought hold a while we shall have weak crops of grain where were great appearances of strong crops. However if you can sell all your wool at 20*s* I would not wish to hinder you. Mr Pease spoke at first of going home today, but he now talks of staying until Thursday next.

22nd June. We continue to have a very serious drought, God knows if it don't alter soon the finest prospect on all kinds of land will turn into desolation. Many of the oats and barley are already past redemption, and even the wheat on dry soils are burning, and everywhere falling thin. The prospect is really dreary at present. But that Being who governs existence is sufficient still to do wonders. I never remember so short a drought do so much damage, but it was the violent frosty winds from the north the week before the last, which was our ruin. Most of the potatoes are cut down to the ground and are as black as soot. Everybody are mowing who have anything to mow, and pastures dry and dead as a door nail. This has determined me to take 6 beasts to Morpeth, decent beef, 3 cows, 2 queys and a bad stot, and I only wish for as good a market this as last week. Ratcliff told me they have advanced beef to 60 or 3£ per cwt., a strange price at this season. It will make fewer to go to you, and if the drought is as bad with you as here, the fewer we send of cattle the better I fancy. It is very happy that such numbers of cattle have gone south. People began to think that cattle would not be left to eat the winter fodder, but a very few will do that now. John next May there will be 5 good horses to spare at Shotton, as we don't keep it longer. However if you want any at present, you may have one or more from any of the farms here by the man that takes over the cows.

26 June Friday. And no letter John, which surprises me the more as old Mason told me at Morpeth that you had sold your wethers to Shillito. I wrote to you yesterday by Edward Pease[63] that we sold our wool to Natt. Handasyde at 24s our stone, and requested you to give Mr Pease a refusal but not by any means to be confined to them. Only as they are good men we wish you to deal with them. I think they should give you more than 1s per lb., as yours is or ought to be particular nice wool and no ewes, that is suckling ewes in it. And they have bought it at 1s per lb. here, and would I dare say have bought ours at the price we got, but wished to buy lower so missed it. I never saw the Wooler buyers so keen. I can't understand why you have not drawn on Ratcliff. He seemed surprised himself on Wednesday, and thought you were more indulgent than he expected. I don't know that we have any reason to be so indulgent to him. I hope you are to have ready money of Shillito. Nothing like ready cash John. But I have said so much on that head to you that I can say no more, so if you will follow your own way I cannot help it. I believe the cows and queys will set forward to you on Monday next. We have a most serious drought indeed, and the face of the country changed from the greatest appearance of plenty to quite the reverse. God alone knows what is the best for us, and His will be done. It is our duty and highest interest to submit with patience to His almighty decrees. I am happy to tell you that your master is nearly quite restored, is going to Mr Thompson's tup shew today. The Yorkshire jobbers were not so fond of poor cattle on Wednesday as usual, nor would give so much by 20 or 30s in ten pounds. I will send you this tomorrow from Berwick. If you will not write is no reason why I should not, never mind postage, information is everything in business. Yours truly

Geo Culley

You will be glad to hear John that we have this year the best shew of shearing tups we ever had. Pray tell us when you have drawn of Ratcliff, and how you have done with Mr Scruton. Be sure that you write to Ratcliff when you draw, telling him for what sum and how many days &c &c, and then be prepared for the payment to the bank. Ralph Brown just returned from Kelso, sold wheat at 5£ per boll and oats at 40s per. Tomorrow is Yetholm (Kirk) fair, and Ralph says the

---

63 See No. 85.

Yorkshire jobbers were keen buying sheep as they passed through Kelso to the fair.

Berwick 27th June. I have this moment received yours of 24th and am glad you have sold your sheep at the last, and at a very good price too. We certainly sold our little nice hill gimmers at or near *9d* per lb. sink, while we could not sell our wethers for more than 8 ¼ or 8 ½. But 15 per quarter is easier to quit by either butcher or salesman than 24 or 26, and I sold 15 lb. per quarter at 45*s*. I am glad that you have at the last drawn on Ratcliff and got your 1200£ paid to Mr Scruton, and 100£ to my old friend George Dent. But the most agreeable piece of news is R. Dent being so much better. My compliments and congratulate him, and desire him from me to take great care. Tell the old boy to allow a few glasses of good port every day. What signifies him and I heaping up riches if we have none to heir them. I think this does Dr Thompson high credit. John you can't in your country avoid doing business on Sundays frequently, but the seldomer the better. However be sure you date Mr Dent's note a day before or a day after, and not on Sunday as it is not binding I believe. I am also glad that you have dealt with George Nicholson for your gimmers in part. I dare say George will kill them. I am not certain, but think we shall send you more than 20 cattle. We are so very much burnt now. They will go from us on Monday first or Tuesday. If you want money to carry on your lime trade let me know and I will remit you whatever you ask for. I am much fonder of the Peases than Head, but if you have made any prior promise before you got my letter by Mr Pease, you must say no to Mr Pease. If no promise, you may say to Head that it was your master's orders. But not to sell to Mr Pease for less than to as good a man. The cast ewes will be desperate dear I believe. However we will try and do as well as we can. We sell wheat as high as you, and oats much higher. How wonderful! However I would have you sell away your wheat in particular if you have any. I am glad you wrote to Ratcliff, that is businesslike. Matty Culley was along with his sisters. My brother well or nearly quite well, was at Mr Thompson's tup shew yesterday on horseback. My wife not quite so well as she was. Matty says we beat Mr Thompson quite for fat tups this year. As I have no more room I will conclude.

Yours Geo Culley

PS. I thank you for writing. The oftener the better too!

85. *George Culley to John Welch*

Wooler one o'clock Thursday 25 June 1801

John Welch

Having an opportunity by Mr Edward Pease, I send this to say that I have this moment sold our wool in this country for 24 per [stone][64] or 1s per pound to Mr Handasyde of Wooler. Mr Pease was not willing to give the price. However in case you have not sold your wool at Denton I have promised Mr Pease that you shall give them the first and lowest offer of it. That is I wish the Peases to have the preference, provided you are disengaged and can't sell it for a better price than they will give you. I am much surprised again that you have not wrote, because old Jack Mason told me yesterday at Morpeth that you had sold your wethers to Shillito on Sunday last at more than 3£ per head. I am glad if it is so but wonder you have not wrote us an account of it and other things. I saw Ratcliff again yesterday who says with surprise that you had never yet drawn upon him for the money. You must have your own reason for not drawing upon him yet, but I don't quite understand it. I hope you are to have ready money of Shillito. We had a failing market yesterday. I sold for less than 8/6. Little above 8s per stone I think. Lambs also dropped but mutton sheep were much the same. Pray let Mr Dent have 100£ or more. I wish you may be able to read this, but Mr Pease waits. I must conclude. Yours in haste

Geo Culley

86. *George and Matthew Culley to John Welch*

Eastfield 28th June 1801

Well John

I expect 20 cattle will set forward tomorrow for you, but I can't exactly say what kinds except the young queys for incalves. We send you none that are near fat, because we think we can sell them as well at home or Morpeth. We still have drought in the extreme, and this day appears as though it would be also windy. Respecting the ewes you name for us to buy for you, it is all right, but nobody have taken off their lambs yet, and I am persuaded they will be bought high from beginning to the end. However we will certainly be trying when opportunity offers. Kirk Yetholm fair was on Saturday last, a great shew of buyers for cattle still, but lower sold by 20s to 30s per 12£. A good many cattle still shewn. Matty sold 2 cows which was bought in

---

64 A word seems to have been omitted.

and we did not like. However they left 4£ each or better. The sheep sold very high and readily. The Yorkshire jobbers have fallen on to buy sheep more than cattle since the drought became so severe. Mr Mills Glanton Pike, a very extraordinary man, has been all the way into Perthshire and Strathmoor &c. He bought 60 beasts of Mr Proctor my friend and Sandy Hamilton's brother-in-law, agent to Lord Strathmoor at Glammis Castle. It is said he bought 150 in all fat less, not fat enough. He did not sell however 2 weeks in Morpeth, which looks as though they were not well bought. Be so kind as tell Mr Peacock as soon as you receive this, that I have wrote to Mr William Pickering to know when it will suit him to meet Mr Peacock at this place, and will suit my brother and self any time. I had a letter from Mr Peacock yesterday desiring to know what time would suit us, as he would set off either on Monday 6th, 13 or 20th July. As soon as I receive Mr Pickering's answer I will immediately inform Mr Peacock by letter, and I name this to you to save postage, which I think as little of as anybody, only it may be done when matters are not *urgent*. We and everybody are mowing to prevent the drought taking the little there is. Hay will be hay again this year. I never saw our General's Close such a pasture as this year, and I never saw anything go so rapidly. We have finished sowing turnips, which is earlier than common, and the first sown are doing better yet than could be expected. We have some fit to hoe. We have hoed our ruta baga or Sweedish. The ground is now so dry that the last sown ones can scarcely vegetate, but a very little rain will do, as the land is in such *nice, fine mellow* order. I think I told you that Ralph sold oats, good and midling together at 46s our boll on Friday. Alex Hunter bid him *40s* on Saturday for the best. Alex Hunter has taken Whinfield, a farm of 500 acres or more near Fishwick at 800£ per annum. Everyone say too dear. Cold clay, flat and damp, not the kind of land I shall ever fancy. Old Mr John Hall late of Fordhill died on Friday. I expect Clunie and Home will pay 9s in the pound in a little time and 5s about a year after. This is better than *Master Paul*, who has been in Morpeth jail long, and from whom I never expected above 1s in the pound. However we have lately got at 15s in the pound from Young and Robertson, two Scotch farmers at Nesbit for tups hired long ago, and which we had given up almost. I am rejoiced at R Dent's being so much better. Be sure you tell him to take great care now he is better, and my respects to the old boy, and tell him he must be indulged, that his son and mine can't do what their Dadys did, and we have no occasion to work them as we *worked* and *rallyed*. I would not have wrote you so much nonsense John, but this will go *scot free*. Have you not some forward steers[65] &c that you wintered, at least I never heard of them being sold? Ratcliff or the Sunderland men are the likeliest to buy such, and I would not advise you to keep them too long. What can you do with your mare but let her rest untill she be better. For God's sake don't trust those Holdgate men, none of them Ratcliff or others, and insist on notes not bills before you. We received a letter from Christopher Eales last night, he wants a tup in September. If you see him make my compliments and say that we will send him a good one. You are right in sending iron axles, they are certainly the best. Do with Head &c as seemeth good to you, only I like the Peases, they are honest punctual people. I fancy there is no long wool sold for more than 1s per pound and a grand price is it. The oftener you write the more agreeable to yours and everyone

Geo Culley

[*M.C.*] The driver Glass will bring home the grey horse, which will forward him much back. Tods should pay their rent. I am afraid that they are shabby men. We have mown and won much of hay, and it would have been better had we cut the burning land sooner, as it looses faster when burning than it gathers when growing. I am glad you quitted your sheep. I never wish you to drain selling so long again, better we selling than we keeping. Do not get the name of a hard or dear seller. Tell Mr Peacock to suit his own time to come north. I am better (but have got some more cold), hope to be better soon. Matt and sisters are tollerably well at home, your mistress at Eastfield not altogether well. Mr Thompson's tups are good this year. Ours are I believe allowed to be fatter and better by many people, especially on the top or along the shoulders, no fiddle brigs,[66] and more size. We have not shewn them so fatt and good for some years I think. You should mow especially where the land turns dry. Rake the thin hay into rows and then turn it in the wind some, never ted[67] if thin hay, as it slips through the rake teeth when dry. Hay is a light crop here, our watered land better than last season, as we did not eat it in the spring. If we had eat it, would have had none at all. The drought pains watered land more than any other where it once takes hold. Your lame mare requires 5 or 6 months to get well in. Turn her off for that time, and take off her shoes if you want her to stand riding once more. No less time will do for her after a strain.[68]

---

65 Castrated male beasts, between two and four years old.
66 Bowed hindquarters.
67 Spread for drying.
68 The signature is lacking.

**87.	*Matthew and George Culley to John Welch***

Wark 7th July 1801

John Welch

Yesterday Yetholm lamb fair, uncommonly high, some sold from 7 or 8*s* highland, to 19*s*, 20*s*, some say to 21*s* per lambs the best country ones. Although a wonderful crop of lambs and great numbers shewn, the demand was great, 3 or 4 chaps for each parcell of lambs. Your master George bought 20 barren ewes of Mr Wood, pretty stock at 38*s* per head. Your master will likely write by J. Dickinson. It seems now as though cattle were not in such numbers, as the country stands lead off, whether owing to a real scarcity or to the great quantum of grass, we know not. Perhaps to both. We shall want steers at the latter end of the year. If you can buy some good ones at any opportunity I apprehend it will be right to buy them. I mean calves, or steers turned one or 2 years old (rather 2 years old). If you can as I say buy good ones will be the sooner the better as rain is now come in time to mend pastures, and corn grows now again, and in my opinion promises to be a good crop if a good harvest succeeds. And as the turnip land was in general well wrought, there are hopes of a good crop. This is a great growing time but hardly a good hay time. We luckily began to mow early and have a good crop, a third of it in pikes. The greatest part at Wark mown and laying tedded as it is now showery. Our rape much improved, mostly hoed once down, some few turnips also hoed, all done sowing some days ago, I think too early as we keep some so farr on in the spring whenever they will keep so long. Only a midling season and for cuting the turnip seed and winnowing it. Wool mostly sold at 24*s* or 1*s* per lb. Some say they get above 24*s* but not many as I suppose except for fine wool, on which there is but little advance. I hope the lad with the 22 oxen and queys would get well up to you. I believe the rain came on here the day they set out from Thornington. The corn and grass on the dry sandy soils had given way much. Peas and beans very promising, especially the early peas, the seed of them from London last year. Beans are not long but seem well podded on those drilled. I fear pottatoes are injured by drougth, curled sore, and a sore frost which injured the tops, especially on the low-lying ground where the rind lay long. They are now begun to put away a full top, were greatly at a stand before the rains came. The check was so great I am afraid they may hardly bring a full crop. I am told Stagshaw Bank[69] was an exceeding dear fair, sheep especially out

---

69 Stagshaw Bank fair, 4 July, a large fair for sheep, especially black-faced; also cattle, horses and pigs.

of all bounds of dearness. We had an opportunity of buying hogs very low this spring, they might have paid well. We now fear that ewes will be bought very high this autumn, as will all stock probably, cart horses excepted which they say are now cheap. Yours &c

Mattw Culley

8th. Received yours last night by Glass who seems to be a good driver. My brother at Morpeth with sheep and 2 cows. For want of health I did not go to Thornington where there [are][70] some cattle which have been long yoke and full of beef, which might and should have been sent to you. We cannot spare one beast more unless we buy, and that's uncertain as they are as dear as an egg for 2*d*. I am afraid you have been eating your watered land for hay. I am as clear as 2 and 2 make 4 that 1 acre land laid in for hay winter feed will for certain grow more hay than 2 acres laid in in April or May in general. We now have a wonderful growing time, and corn comes round very fast, as do the pastures, rape and turnips promise well. The land was well wrought. Beans good, but the late peas will injure them much I fear on the parts. Where they are gross our early peas seem to load and grow well now, rather short in the straw by the drouth. Our hay a good crop if we get it well. Most all work should give way to hay when the day answers, and should then work well and late. We paid formerly 1*d* per hour for our hours, sometimes staid in the field till 7 or 8 o'clock, gave them a gill of ale and bread at a push. You do not say that you have any rape on your new or fresh moors. I go in the morning to meet Mr Peacock. I am glad your winter tares are so good. I meant that you should sell all the old corn before new comes. The crop is now promising. Am glad Holdgate has sent the bills for sheep, sometimes may oblige a friend by taking a bill or 2 when you are sure they are good, but with great caution, if backed by a good a *safe* man. I think the Tods should pay their rent now or soon if you want money. I wish Davison could get the vicar to sell glebe to buy land tax, that is I wish to buy the glebe in part. It would be very convenient, especially the Goose Croffts and the piece in the Weens,[71] also the 2 fields next the street road. We sold at 24*s* for 24 lb., Mr Colling 19*s* for 18 lb., some little less than 1*d* per lb. more than our price, will not pay carriage for our wool to Darlington. There are several people buying wool here still, yet I believe it is mostly sold. Mr Pease wanted to buy here to keep them going till they could get Lincolnshire wool home, who clip later than we do here. 20s which you ask is a high price, only yours is

---

70 A word seems to have been omitted.
71 The name of a field at Denton: DUL, DDR, Tithe plan, Denton.

all hogg or fatt sheep wool, ewe wool too short to comb in general here. We will try to buy you ewes, but they will be exceeding dear. I cannot tell what to say about your cattle. Lean are and will be sold I suppose as high as fatt now sells, therefore some of the nicest may be kept for wintering rather than buy worse pennyworths. M.C.

[*G.C.*] John I am just returned from Morpeth, a most capital market for everything. I sold some clipt hogs at 54*s* and all at 8*d* to 8 ½ per lb. I had 2 cattle, an ox and a cow. I expected 50£ and sold them at 54£. In short it appears to me that things of all kinds are and will be dear now, that turnips are likely to be a crop, being tied and half for sheep. Must conclude Yours

Geo Culley

PS. Mr Peacock met me at Morpeth and came home with me.

[*M.C.*] Your boot maker I think imposed on Matt with a pair of midling boots, and far too little, will hardly bump up for another pair as he cannot now wear them.

88. *George Culley to John Welch*

Eastfield 12th July 1801

Well John

We sent a long letter by John Dickinson, who I hope has arrived safe, and I now begin one to go by Mr Peacock who will be home some time in the week. The Bishop[72] and Mr Green &c called at Wark on their road to Berwick from Elsdon, Jedburgh &c &c, and handled a few tups which they were well pleased with or said so. This was honor but no profit say you. However these things are sometimes productive of good and no harm. Ralph sold wheat, red and white from Wark. The red very coarse yesterday at 5£ per, 30 bolls, he got 5£ 5*s* at Kelso. We have not much left now. Barley to Alex Hunter at 54, the same price at Kelso. But Morpeth was a famous market last Wednesday, which I think I told you of in the letter by John Dickinson, as well as the wonderfull market or fair at Town Yetholm for lambs. Friday and yesterday kept fair thank God, which has enabled us to get a good deal of hay piked. Last night it began to rain again, and rained most of last night I dare say. However we must mow now rain or fair. Not that I think harvest is likely to be very early now, especially if this damp dark weather continue, which is but too likely. The rains will do an abundance of good to pastures, fogs and especially turnips, will lengthen the corn much and fill it also, if

---

72 Shute Barrington 1734–1826, Bishop of Durham 1791–1826: *Oxford Dictionary of National Biography*.

mildews don't follow, which is too often the case in these damp misty seasons with great heavy laid crops, which is not the case in this fine country. I am disposed to think that grain, especially wheat, will be very scarce still before the new corn can be got in any quantity to use, which will make the new crop to be early begun with, and a wise thing it will be to thresh and sell any grain that will bear a good price, because depend upon it that government will still exert every nerve to procure grain from abroad untill they can bring grain down to something like a fair price. I don't speak thus to persuade *you* or *myself* to keep up our *bits of old corn*. By no means, I think it is a dangerous game to play, if not a wicked one, because there is no knowing what a turn things may take, and what a foolish look a man would have to keep his old corn to a falling market. We are therefore thrashing our pieces of old wheat, indeed we are nearly done, and are determined to market it as fast as we can. The consumption is small, as it is all to be sold to bakers, millers &c for home consumption.

However in regard to stock of all kinds, I am strongly inclined to think that they will be found very scarce in this part of the Island, because it is obvious to everyone that there never were such vast quantities of cattle and sheep gone south in any man's remembrance. In short the jobbers bought cattle from the oldest they could meet with till they came to stirks, and if the drought had not taken place they would have bought the calves also I fancy. Such weeds of beasts went as is surprising to think of, the very refuse of our beasts. Well then, after buying all the old sheep and hogs they could meet with, they are now buying lambs, a thing I have never seen done before, especially country lambs. What is to be or may be the result I am as yet unable to foresee. But if ever turnips are to be cheap again in this country (which I think never will be) it must be this autumn, provided we have a good crop which is highly probable now surely. Well what have you done with your wool? I think you are entitled to 19 if not 20s now, because the pieces lately and now unsold in this country is selling at more than 1s per pound. We thought that we were done going to Morpeth, but my buying a few dear ewes of Mr Wood's herd at Yetholm (20) has made us draw about 20 little weeds from the *ridlings* of our hill gimmers &c and a few coarse (hogs) and dinmonds to make out 50, and William Brown's less cow with a little quey, neither fat not such as you would shew at any rate at Darlington, but 8s is a famous price at this season, let after prices be what they will. I think it myself a very prudent game to play. So either Matthew or I will attend Morpeth again next Wednesday. Although I bought Mr Wood's ewes too dear, I considered 'that half a loaf is always better than no bread', and it is very great odds that we should ever have got any money from his but in such a like way. There are several wool men still in this country, although I believe very little wool unsold except in the hill country where it is said there is still a good deal. The demand for fine has not been equal to the demand for long wool. May that continue is the wish of your well wisher

Geo Culley

Two men they called Trumble whose father formerly lived at the Friar Law a farm adjoining Tom Nisbet's of Roddam, which is now annexed to his farm, have failed for more than 2000 pounds, and propose paying 4s in the pound. Tommy Tulip and Mr Hogarth of Fireburn Mill[73] are taken in. I am of opinion that cast ewes will be uncommonly dear this year. I certainly will be making enquiry and looking after them as soon as possible, but I am much mistaken if they will not be very high seasoned. I would wish you to consider and turn it in your thoughts whether it would not be prudent for you to breed a few lambs at Denton now you have your tithes taken.

Monday 13th July. A thick heavy morning but fair yet and has been all night. A bad turnip seed harvest, and perhaps for lime trade, but will do an immensity of good otherwise. It has continued midling fair all day, yet some slight showers in places. I have been at Grindon, and think William Brown has a fair crop, his wheat rather very good upon the whole, and his pastures much mended, but a very bad crop of hay. *Tuesday 14th July. Bolams Low Framlington.* As I am going to deliver up this letter to Mr Peacock as soon as we have taken a little dinner here, I must conclude by returning thanks to the Almighty for a very fine day, which will enable people to get up part hay into cocks &c &c, and I hope all our folks will get all or part of the turnip seed stacked which is really lying in only a very midling state. If you have a wish for any more cattle by and bye we can send you some casten little oxen and spayed queys, that are very fresh and will make nice ones to be out in winter. But we have plenty of keeping for them ourselves, only if you should have a decent crop of turnips and would like to have them we can perhaps buy more better than you. However you can turn it over in your mind and let me know afterwards. I am once more

Yours &c Geo Culley

---

73 Probably John Hogarth, son of the Rev. John Hogarth c.1740–1802, assistant curate of All Saints', Newcastle upon Tyne 1774–97, vicar of Kirknewton 1778–1802: DUL, ASC, Hudleston Clergy Index; NRO, Kirknewton Register; National Archives, Land Tax Redemption Office, Assessment 1798, IR23/65.

PS. Mr Peacock has been 2 nights with Mr Bailey, where I called upon him this morning and breakfasted, and here we dine. He goes for Newcastle tonight and I to Morpeth. Mr P. goes home by Sunderland, but will likely be home before Sunday when you will get this.

89. *George Culley to John Welch*

Eastfield 23rd July 1801

Well John Welch

I met John Dickinson today at Wooler with yours and many other letters. Yours was a good lot of wool, but I think you have sold it well. We now have charming weather to get the latter hay in. Harvest will be on soon, and as Matty Cully says he saw barley shorn today at Hungry House. All the Wark family are at Holy Island, and my wife. I shall join them on Saturday evening and shew them yours and other letters. We have not been at Morpeth this week, but understand it was a good market. We go next week with 60 dinmonds. It never can be wrong to sell them at 8*d* or 8 ½ per lb., and it may prove right. At all events it is a safe game. You do beat us for corn prices, although we sold for 18*s*/6 last Saturday at Berwick. I do suppose that grain of all kinds will sell well for a while after harvest, and I should think it wise to sell a considerable quantity of grain if it be high, say 10*s* per bushell or 3£ per boll for wheat, 30*s* for oats, 5*s* per bushell, and 6 for barley per bushell. It is certainly right to get your hay as far forward before harvest as possible. If you have any steers, fair ship beef, would it not be wise to sell them at 8*s* to 8/6*d*, or cows either?

I spoke to James Grey about his draft ewes today. We are to have a refusal of them, but depend upon it they will be at such a price as we never heard of before. People do not know what to ask. You don't say what number of draft ewes you would wish to have. We are now determined to tup you 50 or 60 ewes and send to you. It is ridiculous and absurd not to breed a few sheep with you now you have your tithe wool and lamb taken. Only think what an expence it is sending sheep all the way to Denton. And if you should breed a few it will help part, and pray what pays like sheep. We have made more of sheep than of all the things we have ever dealt in. Nor is there anything we have made our studdy so much, or understood so well as sheep. And I do really believe that we never had so feeding a kind as those we now have. Your gimmers will summer exceedingly well in your moors, at least they used to do so in our days. We let Mr William Brodie 2 tups yesterday at 45 guineas, and Fenwick Compton one at 20 guineas. I hope that will still prove a good trade this autumn. I am glad your lime trade continues so good, it is great encouragement. Be sure you pay the men regularly and keep good accounts for the lime delivered. I am sure you are very equal to it if you only have leisure. You don't say whether you have received George Nicholson's money for the gimmers you sold him before. I thought them cheap sold, and I think these 16 dear. They are above 8 ½ per pound. Your master must consider about the drying kiln when he sees your letter. It is a matter I do not understand. Mr Colling is very kind to save the pigs for us. I am glad to hear that he is better again poor man. I grant that outlaid cattle breed little manure, but they make often the nicest. However I commend you for breeding all the dung you can, it is the *primum mobile*. I hope you save all you can at the kilns. Robin White says Birkbeck is a-dying. I dare say stock of all kinds will continue dear and very dear for a time. Therefore why not breed sheep? You can't breed calves, but sheep you can. Mr Peacock and a Mr Graves who I don't know have offered 60£ for Trenham's houses for 9 or 12 years, to keep all in repair except walls, main timber, tiles and fire and tempest. I think myself it is a fair price, but don't understand it, but will be glad of your opinion. Perhaps you can inform yourself whether it be a fair rent, and let us know by return of post, because we are to give them an early answer. Perhaps you can consult some acquaintance at Darlington, but pray do it as snugly and quickly as you can, because I would not like them to know anything of it. What sort of a man is Mr Graves, and of what country? Is he said to be a man of some capital? Is he sober and of good character? Ralph Brown only poorly, and I was obliged to go to Kelso market. We can't sell corn as dear there as at Berwick. What delightfull weather, but I am affraid rain is coming again by the glass. But moderate rains will do good and little harm. I have desired Mr Peacock to give you some money if John Todd pay his rent soon, in case you are in want, as I think you have been rather pinched lately and I don't like to have you pinched. If you don't need it he will remit to me. Turnips, potatoes and everything grows apace. Stock is and will be high out of all proportion and character[74]

---

74 The signature is lacking.

90. *George Culley to John Welch*

Eastfield 14th August 1801

Well John

I would have answered your long and very intelligent letter sooner, but that I suppose my brother is with you by this time. You have harped long on about a few cast or draft ewes, and I am very sorry that we have not yet been able to buy any at a price that we thought there was the least prospect to pay. We still have James Grey's few in view, and are to look at them next Wednesday morning, and little they will be of the hop I dare say if we dare buy them. As for Mr Wood's, our people all say they would not think of giving such a price, yet they were bought at the money. But I confess things are beyond my comprehension. It was lucky that your master bought those 9 cattle of Ralph Compton, but I would have you pass over at Ninian fair if you can, and try your fortune. I fancy we are too slow of foot for these times. However I am always glad when sheep sell well. We know how to sell them long still, and I am glad that you have at the last agreed to breed a few at Denton. I will be bound for it they will pay you well. The breeding is certainly the safest game, and I have no scruple in saying that I think you never had a more feeding kind than at the present time. Why not produce them then, since I am sure we have made more of sheep than any other article we ever dealt in. Can anything be so wise then or level to common sense, as to follow or pursue good luck. Your 4 little oxen have paid you capitally, and I hope there is no fear of John Forster's. I thought you sold your piece of wool well, but by a letter from Mr Vipan[75] Lincolnshire long wool is now selling at 1s 5 per lb and advancing markets. Nevertheless like sheep it is true they may drop before your lambs become shearings. True, but they will rise again, and never will be low again if this Island keep its liberty and trade. How rejoiced I was to hear of Robert Dent's better health poor fellow. I sincerely wish it to be true. We have had a very fickle weather also, but have now got most of ours stacked. We have thrashed a few oats which Ralph is going to try to sell a few today at Kelso. They are a light crop and not a good sample, but we have a very light crop of oats through all this dry country. About Milfield the gatings[76] can scarcely be seen 500 yards

---

75 In this letter of 5 August (ZCU 24) Mr B. Vipan of Southery reported the price of wool, a fluctuating market for fat stock, smut on the wheat but good oats and barley. He was thinking of coming to Falkirk tryst in October to buy lean cattle.

76 Gaitings, sheaves of corn set up on end to dry.

of. Have you any slain or smut[77] amongst your wheat? We have none, thank God, that I can see in ours. Mr Vipan says there never was so much in Norfolk, and by a letter from Glammis Mr Proctor says there is none clear in Scotland. I am certain pickling with old chamber lye will prevent it, at least it has prevented ours these 40 years, which is a fair tryal. What a prodigious tithe is yours, yet I hope it may be worth it if corn is sold decently. Mr Bailey I find values wheat at 8*s* also, but I think it too much myself. We are to have the Grindon tithe men next week, but Mr Bell is a very reasonable man in general, has known wheat at 48*s* per boll I find. I very highly approve of your idea of buying a few Kyloes[78] if ever so young, to run in your worse land. They winter easy and will come to some time, and my advice is for you to try Brough Hill &c. Don't you pinch yourself of money by any means, but whenever you *boil over* send it to me by short bills or any safe conveyance. Perhaps you may have some to spare about your fair when your lime money comes in. But be sure you keep strong in hand, for a man is no man without money in these days. Ready money may cast a good bargain when a bare purse can't.

The plough has paid well of late certainly, especially upon good land, and without the plough you can neither have *turnips*, nor *seeds*, nor *rape*, which are all so *usefull* now in these days to be almost *indispensable*. Nevertheless I am no advocate for ploughing in the extreme, because labor is and will be excessively high. However all these things should be governed by *reason*, founded upon good *sound calculation*. Turnip soils are certainly the most profitable soils to plow in every sense of the word. However it is perhaps necessary now and then to go through the poorer lands with the plough, but not too much at once. Pray did you get through all or most of your turnip seed? I think late oats and even the potatoe oats will be a fair crop, and barley in general. But I shall be agreeably deceived if wheat is a good sample in this country. I see all the wooly eared or downy eared kinds have got a *motley, dark, dingy* cast upon it, which allways impedes or stops the filling. No wheat is so healthy, in my opinion and experience, as the red kinds, far less liable to distemper or disorder than the white tribes, especially the wooly chapped kinds, which in favorable years is undoubtedly the best sample and heaviest crop but should never be sown except on lands in high health and condition. The red wheat are not so liable to be smutted or slain or

---

77 'Slain' and 'smut' are two words for the same disease of plants.
78 Small long-horned cattle originating from the Highlands and Islands of Scotland.

mildewed or rusted or racketed or a thousand other diseases that we have not names for yet, is always the giftyest and in my mind pays more money than any other sort, admitting that it is sold even at 1s per bushell less than the best white wheat. But millers now begin to find out its merits, and will in general give as much for it as the other kinds. It in general, I may say always, weighs more than the white. We have some just now that outweighs all the wheat we grew last year at all our farms. And if red wheats were grown upon the best lands instead of white wheat, as they do grow it in all the midland counties, it would beat the white out of rights. You always have more bushells per acre of the red than the white. I own the straw is not so good fodder, but then you have more of it and I fancy it is that which saves it from winds, because by being so long strawed it gives way and yields more to the winds perhaps. However we seldom have the same loss of it on the whole. So much for a lecture upon red wheat, which next to sheep has made us more money I am disposed to think than any other grain we have grown. Many people now think of pickling red wheat because it is less liable to be smutted, but I think it very wrong to sow any wheat without pickling.

If you chuse you may inclose Mr Vanity Midleton a 20 shilling note, but I shall not because I am content with old wife's per or young girls either, and those that won't take a little trouble deserve to be punished with faulty crops. You say George Johnson would take the malting late Trenham's but you don't say what rent he would give. As to Miss Thompson it is quite out of her way to take it, I mean the house in my opinion. Her guests are never going to leave the post house to follow her there. Besides it is much better for us to have only one or 2 tenants at the most, much less trouble, I only wanted to know whether 60£ per annum is a fair rent and as good a one as any respectable tenant would give, because if Graves is not a man of property I fancy he is a sober industrious man, and as we and Mr Peacock joint with him I am satisfyed if my brother be. And as it is right for my brother to make a bargain with them or any other good tenant or tenants when in that quarter, I would recommend him to do it, and then he can condition for the manure which is certainly right although I confess it never occurred to me. I have no idea of your speaking out of any ill nature. I hope you have more sense and more liberallity. Mr Peacock remitted me John Todd's money immediately on receiving it, I fancy before he had got my letter to make you an offer of any part of it. But as you had no need it was all right, and as I said before I would rather have the money here at present than in Darlington banks, as we can make more of it here than they will allow in interest. Besides we have it ready if anything should fall out to purchase &c &c.

It is saying a great deal for your crop John on a 400 acre farm which we offered for 400£ per annum, that a 10th should be worth 100£!. However I am rejoiced at it. Go on and prosper in God's name. But you will allow me to say that nothing in my mind can pay so well in farming as combining *breeding* and *feeding* together, especially we that have so valuable a breed of sheep. For I am disposed to think that few if any have a more feeding kind. Then say you, 'We can't make these sheep so well up without turnips in winter and clover or seeds in spring'. Granted, only don't plow too much, and plow land in *general* that will grow *good turnips* and *clover* or *seeds*. Not the great sowing and little reaping, but the little sowing and great reaping that makes the profit. I suppose by this time you are tired of lectures from your real friend

Geo Culley

Friday afternoon. Tell my brother I am just come from Wark, sister well. Begun to shear Poland oats, which George says are not so good as he expected. George says they will not have constant shearing even next week at Wark. Turnips much mended, pastures good. Ralph Compton's cattle worth what they cost according to times. I left my wife well at Holy Island, comes home on Sunday. Mr Bailey intends to lay 200 [. . .]79 to Longknow house besides the other between lane and Bowmont.

91. *George Culley, Matthew Culley jr and Matthew Culley to John Welch*

Eastfield 20th August 1801

Well John

I received my brother's letter, and hope to see himself on Monday at Whitingham fair. I am very well pleased that he has bought Tod's waygoing crop even if it should prove a little too dear, because you will be clear of this plague. But I hope it may pay a little. However if new oats and barley sell well I would advise you to thrash a part, and even wheat at 10s per bushell, and then sell less in the beginning of winter when everybody are selling. If you take care of the straw you can use it very well in the beginning of winter. By a letter from Mr Sayle from his farm at Titchwell Norfolk, he says pastures are very dry in that country, and crops of corn not remarkably good, and much

---

79 The next word is illegible.

smutted or black wheat. But turnip crop good, and stock of all kinds, fat has been very dear. I am glad to hear by my brother that you have a good crop of apples at Denton. Pray watch this. As we have none in this country, they will be a *valuable prize* to us, and our carts can bring them here as they go with tups into Yorkshire. And we let 5 to go upon the Yorkshire wolds yesterday, and may probably let more still. At any rate these will employ 2 carts, which can bring home a good many apples properly packed in creels, panniers or baskets &c &c, which you must take care to provide in the mean time. And untill then you can keep them in some of your chambers. The carts will scarcely go before the end of September or beginning of October. I approve of my brother letting Trenham's houses rather than selling them at present, except he could have had ready money paid for them. Ralph has sold most of our old wheat yesterday at Kelso at 4.15*s* per, new oats at 26*s* from Wark, and ours at 28*s* per boll. I would advise you to buy some Kyloes if you can at Brough Hill or Luke fair &c &c, as they will winter easy with you and will come on. I mean young ones the best perhaps, but leave that to you. Nothing new under the sun says Solomon.[80] Newcastle used 40 years ago to be supplied from Darn[ling]ton and Teesside, and now John Davison is again making the tryal.

Well John I have bought a score of old ewes of Thomas Jeffries Crookham at 40*s* per, and 95 of James Grey at 39*s* and the 5 stots at 30 each, and I am of opinion you will not thank me for what I have done. However it is done, and now can't be helped. We think you had better not have them untill the weather be cooler, as they would take hurt by driving now and we have plenty of meat. Will you have them all, or shall we feed the oldest here? We don't shew any cattle at Whittingham as we are so full of keeping. We have cut no wheat yet but shall begin next week, but have cut all our early oats nearly, potatoes and all, and part barley. Corn ripens fast but does not win or come well in the field. I fancy we shall begin in earnest next week, and I think harvest will not be late of being done if hands can be got.

[*M.C. jr*] The carts with the tups that are going into Yorkshire will be at Denton on Friday the 9th October, and you will have to send the tups to Mr Baraugh's on the Saturday the 10th to meet the cart that comes from them on Monday 12th to Mr Baraugh's. You never informed us what you made of your turnip seeds.

[*G.C.*] *Whittingham fair Monday*. Not half the quantity of cattle

---

80 Ecclesiastes 1.9.

shewn here today that have been in former years. In truth stock might be scarce and very scarce in this country. Nevertheless the butchers are unwilling to give more than from 8 to 8/6*d* per stone sink. But many buyers from Yorkshire and Lincolnshire that would get none today. We think you had better keep the cattle you have in hand at present to winter, than run the risk of buying them perhaps dearer. And if you can meet with any steers at any time from now until Yarm fair, buy them. If not you must buy some at Yarm for your own wintering upon straw &c, and a few for us also if you can. I am affraid you may be pinched for money to do these things with, on which account you should have had John Forster's money. However as we spoke to him today, we shall now draw for it and if you are pinched let me know, and I can at any time remit you 100£ or 200£. But you have not yet received for your wool I fancy, and your lime will be a pretty large amount. And I would recommend you by all means to sell new corn if at fair prices, because corn must come down after Martinmas but may advance again towards spring. Therefore it strikes me that it will be wise to sell early and then keep of a while, and if you can only take care of the straw you may manage very well. And I would recommend to stack the straw loose, and cut it down with a hay spade or knife in the winter like hay. I am in haste your

Geo Culley

[*M.C.*] Wooler Monday 2 o'clock. Mr T. Bates has taken Haughton Castle farm of Mr Smith, 395 acres 600£ a year, tythe free of corn and hay, pays wool and lamb, within 1 ½ miles of Brunton. Mr Raws wishes you to send for Matty 3 or 4 days or more (but not to keep him after the school begins again) at Houghton feast. Hope Matt will write to you about the matter. Called on your friend Scott and break-fasted with him at Newbottle, they are a pleasant family. Harry your servant is I understand at Houghton. If tollerable wish he would come to you as he is a trusty servant and you'l have much work on your hands with Todd's crop. We wish you to thresh and sell as the prices are good. They must come down. Send one of your servants to Richmond or Barnard Castle to sell the corn, as it will keep you at home, and Matt may do well to look after your workmen at home, only he wants practice as yet and will grow better with practice.

Yours &c Mattw Culley

92. *Matthew and George Culley to John Welch*

Wark 9th September 1801

John Welch

Mr Pullen of Carlton north of Old Brough Beck and a little west of Old Brough, married Mr George Askew's sister and has been at Pallinsburn House for some time, wants much to see your threshing machine, seems to be a good young man and wants to know some things about husbandry. When he calls you will shew him anything about your farm with civility. I believe that he lives with his father at Carlton. I now begin to think that we shall be able to keep all our stock now in hand, except some of our feeding cattle. Our fogs, pastures and clovers are so good, and turnips also which have improved so very much of late. It may be prudent to sell fatt before Candlemas probably, as markets may run in, and if they run in who knows where they may stop. If plenty of corn and grass and turnips be the means of filling the markets and causing plenty such as the present year, wheat, rye, oats, barley, peas, hay, and rich pastures of clovers and grasses, such as I have seldom seen at any time or times, especially grain on all soils on a moistish bottom is an abundant crop. The crops on the lightish thin dry soils are not abundant. The red wheats on our outfield are more abundant than on our best lands. Grindon has also a luxuriant crop of grain far exceeding any crop since we farmed it. Our rapes are exceeding good, would even now be all yellow with the flower had we not eat it. We were too late of eating it, as when begun before it runs for seed the stalk parts at every joint or eye, but after it gets a long bare stalk does not do so well. What a crop of rape might you have had in Stubs's Carr had it been sown in May, but would have grown plenty sown any time in June, especially such new land, in such a season as this is. But we will learn better by and by if we live and keep our liberty &c. *Live and learn* we should do every day of our lives.

I think you should guard against a flood coming down Stubs's Carr, as it may sweep away all the soil where it may happen to run as deep as the plow has gone. Give it passage wherever you can in the upper carr in the best place and stop the way into Stubs's Carr. It is too late to lock the stable door when the steed is stolen. Also the passage at the carrs stile, and in winter we must have the bottom of the stell lowered at the carr's head and from a little above Stubs's Bridge up the carrs to above the main springs, and box in the stell. I hope the watered land in Fitts and Welsprings will flush your feeding cattle. Our watered land has flushed our stock exceedingly, more by double than ever it did before, but indeed I think we never saw such a growing time. Our turnips grow wonderfully, and our rape first eat is now a full second crop covering the ground again, sufficient moisture with a wonderfull warmth of weather. Rape should be eat as soon as it gets good, before it runs to a long stalk. It then after being eat and the stock taken away to a later sown piece, puts out at every eye again. I think it may answer very well drilled on the poorish clay soils as it feeds much faster than turnips, and may be eat over, and over, and over on rich land soon sown and well attended to, then will bring a crop of wheat to almost a certainty. Land I think was intended by the God of nature to bear a crop every year, *and so the gardens do*, and may do for ever by a return of the dung, husbanded properly. The run from the dung hill should be led on to the land, or conveyed by some means on it. I am afraid you would not send the peas till spoiled, but it is a thing of no moment. Yours in haste

Mattw Culley

PS. Mr Peacock has to send a pair of globes to me, hope you will charge the driver to take care of them, but do not send but by a carefull cart of ours, and let him know the value of them.

[*G.C.*] John Welch   Eastfield 11th September 1801

I beg you will not neglect to see Christopher Eales, and let me know when he wishes to have his tup. These are things we should not neglect, otherwise we may offend our customers, and I think Mr Eales wished to have his tup in this month. I did hope that you would have answered our last on that head before this, because I am affraid that Christopher Eales may want the tup. We get done our shearing tomorrow except a few beans at the Westfield, which are not quite fit to cut yet. I fancy the Wark folks are done all nearly, black corn and all. And William Brown is very near being done if not quite so. But we all have much to lead, owing to the weather being so calm and warm. This has been the first winning day of a long time. Harry has a good deal to do yet, both at home and Longknow. Shotton near done. My brother has said plenty about turnips, clover fogs &c &c and turnips [*sic*]. But in truth I never knew these things make such growth as this year. Indeed I never knew such a warm close growing time, with enough and not too much rain, and the warm growing time still continues. In a letter I have wrote to Mr Peacock, I have requested him to get the globes cased with wood, or otherwise so managed that they will not take any damage by coming in a cart. If you have a clever carpenter I should think he would contrive it, with your advice and assistance, as well as Mr Peacock's. I have heard nothing concerning Morpeth market of some time, as I have not been at

Wooler for some weeks but have stuck to the harvest. I am wishing a continuance of good [. . .][81] Yours truly

Geo Culley

September 12th a fine morning. John I said a good deal in my last about your selling *any kind of grain* as soon as possible, because I was apprehensive that corn would drop very soon after harvest, and by a letter from Mr Peacock dated the 7th I find that on the Saturday preceding (5th inst.) corn came down considerably at Richmond, which I do not wonder at. 'The first ripe cherries are generally worth the most money'. However I have no apprehension of corn coming very low untill about or after Martinmas, when everybody begin to thresh. And even then I should not suppose that it will continue long very low. These are my opinions and my brother's, which are founded on sound principles and experience. Nevertheless we find old observations and experience give way sometimes to new ideas (and young commanders such as *Bonaparte*[82] and *Nelson*[83] defeating the more experienced!). Witness the present advance in stock of all kinds, and beef and mutton advancing at a season when we always knew it to drop before. Yet Solomon says 'There is nothing new under the sun', and in general it will hold good. That is, what has been the case in former days, will in general be the same in time to come. This advance in stock, and consequently in butchers or *shambles meat* is *certainly* an unprecedented matter, that never happened before in our or our forefathers days. But remember an old remark 'that foreseen *dearths* never happen', and the reason is obvious, because everybody expect *the dearth*, and consequently by guarding against it *prevents* it. And I shall not be surprised if that should be the case in some degree *this winter* or in *the spring*. Especially if a peace should take place, which is neither *impossible* nor *unlikely*. And should that happen it will in my judgement make a strange alteration. The immense *stores* or magazines of *corn meal, flour, beef, pork* &c &c will all have to come to the common market, and the very noise and rumour will have a sudden and evident effect. This is a long lecture and may not take place. But it is sound reasoning, and all I want you and me to avail ourselves of *is this*, that if beef get to 10*s* per stone or near it about Christmas or thereabouts I would recommend to sell a part. Same

---

81 There is a tear in the paper here.

82 Napoleon Bonaparte 1769–1821, French general, First Consul 1799, Emperor of the French 1804–15.

83 Horatio Nelson 1758–1805, British admiral: *Oxford Dictionay of National Biography.*

thing of mutton. But you will answer, what are we to make of our winter meat, turnips hay &c &c. I reply, that hay will keep, and as to turnips people like us who have young stock had better be kinder to them than common. At any rate I would advise to eat turnips and straw, and save hay at all events, and then you need not make so much ground hay another season. For depend upon it that hay is the most expensive produce that you raise, in every sense of the word, pays the least money in general, especially when you can't sell it, and takes upon the land more than any other crop I think, and I have long thought so without having any cause to alter my ways of thinking. Hence the very great merit and benefit arising from *water meadows*, which if well managed get better every year, and instead of taking *muck*, helps you with *muck* for the rest of the farm, while otherwise you must either muck old land every *now* and *then*, or grow your hay on *your seeds*, the most valuable *spring pastures* to a certainty you have. And I have often said that were I a farmer *for myself*, or to begin the world again, I would try to do with as little hay as possible. And it is surprising how little [*sic* ?much] may be dispensed with, by giving to your horses Sweedish turnips or even common ones, potatoes, beans or pease, straw &c &c, and we verified this last winter at this place. We had a very little to begin with indeed, yet by a strict economy we spared 15£ worth to Mr Askew. But our Ralph is the best economist I ever knew, and I think George Humble the worst, although one of the cleverest servants I know. But they use more hay at Wark than any place I know considering what bean and pease straw &c &c they have. G. C.

Besides turnips sometimes rot in winter, and should that happen it may make ugly work. Then hay becomes *really valuable*, but still only pays a little. I have given you a long lecture and could have said more, but have put it in [as] few words as possible. I hope you will read it more than once, and I have no doubt but you will profit by it, because I know you are a good *economist*, which in these days and at all times is an excellent thing.

93. *George and Matthew Culley to John Welch*

Eastfield 21st September 1801

John Welch

I was really mortifyed that you don't let us know when Mr Eales is for his tup, and it hurts me to think that we should have so often to remind you of a matter which is so level to common sense. If you don't meet with Mr Eales at the markets, can't you send a boy afoot with a letter to him? I await of knowing the precise time that Mr E.

wants his tup. We not only are kept in a very unpleasant suspence but run the risk of offending him without any intention, but in a way that he may easily construe into a neglect on our side. It is true, we might have wrote to Mr Eales ourselves, but I thought much to put him to the unnecessary expence of postage when you could do so easily, and as he wished to have him in the month of September we could send him any time, although it would be less expence to us to send him with other Yorkshire tups about the tenth of next month October.

22 [*September*].[84] A dark morning, wind north north east, but it has been dry all night although many showers fell yesterday from north east. We are pulling our beans, a very nice crop indeed, especially where the harvest pease are amongst them. Too many pease prevent the beans from loading. They are podded so near the bottom that we find it hard to pull them although a good length, and 3 pullers act together whenever their handfulls are ready, set together as they return like a 3 footed stack. If few pease they stand well, but if they fall down soon dry if good weather and not too drought sets. Pease are all in *well* and covered, a very nice crop, and all our white corn except 2 pieces of late barley which will lead in a day or 2 if weather be fair. Some of it mows where standing, which is best mode I think, the rest reaped or shorn. We are all nearly in upon every farm but Thornington, Longknow and Shotton Hill and all about covered. I now think that grain of all kinds will keep the present prices until Martinmas or thereabouts when the general thrashing takes place. But it must then come down to a level. But what that level may be I am not competent to say. Prices advanced a little last Saturday. Ralph sold his wheat mostly red and not in nice condition, for 36s per boll. Angus oats were sold at 19s, Polands at 24s. Ralph thought he could have sold for more but we have none to sell. Good barley I was told might have been sold to the maltsters at 31/6 a capital price I do think. I could wish to hear of you selling some at that or more, as I suppose it will be more in your country. Barley is a bad sample here, *brown, small* and *coarse*. Oats are better but not a very nice sample. Wheat upon deep or cold lands a good sample, on dry lands there is much small in it. But certainly take it upon the whole, the crop may be called the most abundant that we have had of many years of all kinds, and in all the country well harvested and very full yards in general. But in the cold country towards the sea especially, the yards will not contain them. Many people have cut a 2nd crop of clover,

---

84 The date has been left incomplete.

better than the first and well got in that fine weather lately. Turnips a most wonderfull crop, and stock will be wanted to consume them. The sheep buyers I am told are all out again, but I apprehend that there is little for them to get. It is not improbable but Mr Nisbet and self may be in your quarter about the 10th of next month when we may happen to leave our horses a few days with you and take a chaise. But this depends upon circumstances, especially on a letter which I look for in 2 or 3 days time. We are doing very well in the tup way this season and look forward to do a good deal. Sheep pay so well for breeding, gives great encouragement.

Well John, yours of 20th came to hand yesterday (this is the 24th), and I am glad that Mr Eales will take the tup along with the Yorkshire ones. Mr John Taylor of Salton Yorkshire came the other day in the coach and hired one. We are very likely to have a good tup season. Indeed we have already done a great deal, and many good customers expected. Notwithstanding that you had put (single) upon your letter it is charged double. I apprehend it is your manner of folding your letters. Be so kind as observe how we fold ours, and do you the same, and write *single* in large letters. We shall get the wrong charge allowed again by sending the letter down to the Berwick post office, only it is attended with a little trouble to Ralph Brown. I can have no doubt of your industry and attention, but you have had more rain by a great deal than we had. Many hundreds of uncovered stacks in this quarter, which took no harm that I heard of. Ours was not a heavy rain at all. We have had some driving cold showery weather lately, which has impeded the latter harvest a good deal. Very little corn led lately, but no harm was caused to it. We have two little pieces of late barley unled only, which we shall lead the first fine day, and a little few beans. All the rest is covered and coped out. It is a wise thing to reserve a piece of old wheat or maslin[85] straw to cover with, but few people could do that the last year. I remember my father always proceeded upon that doctrine, and said what lesson he had known by stacks being wet down through. I am glad to hear that your corn markets have come about again. You beat us in everything but oats. But I perfectly agree with you that the drop in corn was owing to people being under the necessity of thrashing for covering. It has every appearance of being well sold now until the general thrashing takes place about Martinmas, and I think you are extremely wise in thrashing a good deal of John Todd's crop before that impends. It will

---

85 Mixed corn, usually wheat and rye.

enable you to answer better whether you make anything of it or not, and it will measure farther now than after Candlemas a good deal, take my word for it, and I shall be much mistaken if it don't sell worse after Candlemas than at present. You don't say how your different grains yield, do tell us. I don't think ours yields so well as I have known it, nay not so well as the last year. However I can explain that upon the strong cold lands, it yields as usual about one bushell per stack. I see by the day's paper that the London market for wheat on Monday last dropped 8s per quarter, all other prices rather advancing. Much wheat had arrived from abroad and [. . .][86] off the demand be considerable into Westmoreland and Cumberland.[87] I should not be surprised if Barnard Castle be the best market for you. It is very clear that you are luckily situated as to be able to go to either Richmond or Barnard Castle. I am affraid you have kept too long of your moor rape, but you must be the best judge and I do perfectly agree with you that it [is][88] a strange price &c and [. . .][89] to lye on cattle.

30th September. I had been saying the same to my brother, but who is competent to judge of times? I pretend not to be able, only I incline to think that if the war continue, which is now doubtfull, the flesh markets will be very dear, much more so than corn. Corn can be had from abroad, butchers meat can not so well, and it is clear that stock of all kinds are scarce still if beef be at 24 or 20s and so by Christmas or Candlemas. It may be wise to sell a part and I think it will be the case if the war go on. We think you have done right in selling the little brown mare. The money will go far to buy you a stronger. If the blood mare carries you safe and well why part with her? Can you be carried so cheap and well? But I leave this matter to the two Matthews. Observe John that rape will grow where turnips will not, at least with half the dung upon midling lands, eat of with sheep it affords a crop of red wheat. I leave Mr Buston's steers to my brother to answer, but I approve much what you say about buying queys if to be got. I should suppose in your situation queys will pay better than steers. I hope you will get a score of Kyloes at Brough Hill fair. We can winter Grey's ewes if you can do without them, and I assure you I am affraid that they will leave very little plunder. Not but they are doing very well, only they are little for money. Your master and I both wish you

---

86 Two words are illegible.
87 The reading of this sentence is not clear.
88 A word seems to have been omitted.
89 Two words are illegible.

to attend fairs in general, fairs in general are best to buy at. There are always chances, and you are young, active and sharp as most people. Besides it perfects your judgement, no cleverness without practice. I am yours in haste

Geo Culley

[**M.C.**] My brother has left me nothing to say, he has answered your letter so well. I am going to weigh of Grindon wool in the morning. It's the last of weighing wool, although light will come to a good sum at such a high price as 1s per lb. Beaumont was a strong flood yesterday morning. Not much rain here. I apprehend we have more feeding cattle and sheep than we can carry through with turnips, cattle very good as are our turnips. They are an extra crop indeed. What can we say, times are wonderfull, past men's comprehension. The children got your pears and apples post paid for them. I am obliged to you.

20 nice oxen for turnips and very good
<u>13</u> smaller
33 at Wark besides small invalids
<u>4 or 5</u> cows old queys also now fatt for selling at present
Harry has
15
<u>8</u> gone to Eastfield
<u>23</u> good oxen and spayed queys
56. These we can winter on ours if sent

I can't tell how many weathers and ewes we have, but may in another letter as I am going to Grindon this morning to weigh the wool. Yours in haste

Matthew Culley

Friday morning 7 o'clock

94. *George Culley to John Welch*

Eastfield 29th September 1801

Well John Welch

Times are so wonderful and seasons so very critical that I think we should not spare pen, ink, paper and pains to carry information back and forwards. I did not suppose or expect that my brother would have sent away our last letter to you untill after yesterday, which was Ninian fair, that we might have conveyed you the account of that fair in the same paper. There was one of the Bulmers at the fair, he and George Ascough and George Hopper were going to Brough Hill fair, and we sent our remembrances to you by them as we believe you will be there. There were full as many cattle and sheep at St Ninian as could be expected, considering the many thousands sold before, but the stock there was were very indifferent, and half a dozen buyers for every lot of either cattle or sheep. However I contrived to buy one score of very good ewes of a Mr Dunn of Buckton at 48s per. We could have had 56s per for them again, however I also bought 6 pretty cattle, 5 steers and one quey rising 4 only at 20£ per, I think them and some others 56 stone, but Mr William Spours said only 54 and he is a knowing man. They are fruitfull and well bred, and such beef as the butchers are glad to kill. We rather wished to buy some steers to work but could not. I think they were still dearer in proportion. Tom Nisbet of Rodding bought a lot of Bill Snowdon the same age at 18£ per and very poor, and one of them is declining I think. Some ewes sold as high as 58s yesterday but I believe they were for breeding. John Forster of South Shields and his *master Bowling* as I call him were at the fair, buying sheep for turnips. They bought about 100 I understand. Forster came home with us and we went up to Wark to look at 2 or 3 cattle. A large cow was William Brown's a [. . .]90 stot and a niceish red quey which was also bad all winter. Ralph asked him 20 guineas each but he did not buy them. Ralph sold some wheat to Mr Walker yesterday at no more than 50 per boll. It was all or mostly red, and rather soft. We are going to thrash a few oats because the merchants are very fond of buying them at 19 to 24s our boll, common oats at 20s, Polands or Churches at 23 or 24, which are certainly undesirable prices at this early season and when oats are confessedly the greatest crop of any other grain throughout both Great Britain and Ireland. Barley seems also to be in greater demand than wheat as well as oats. Mr Barf, Mr Meakin and others from Yorkshire very keen of buying cargoes to send into Yorkshire. Still I think it will be wise in you to sell some of your barley, especially off Todd's farm. I was not able to learn what these new merchants from Yorkshire were bidding for barley, but can learn perhaps on Saturday when I will inform you.

Mr Nisbet and I set of for Alnwick fair on this day week on horseback, and if we can reach Newcastle that night we can see Mr Mason's and the two Collings's stocks, and perhaps reach Thornton Mr Peacock's on Wednesday evening. If not it will be some time on Thursday, on which day we will be at Denton. Because on Friday evening we propose taking Mr Peacock with us in a chaise to Northallerton so as we can the next day come at Tadcaster in good time, when we meet a Mr Claridge to try to purchase Easington

---

90 There is a blot on the page here.

Grange, a part of the Belford property. The price I believe will be 13000£, 323 acres of freehold and tithe free, lambs excepted, and perhaps might be let in this country at near 40s per acre. There is a good deal of strong wheat land, but no clay near the surface, with a considerable portion of turnip soil and a small piece of old grazing land, and much of the deeper strong land would perhaps be best employed in that way after being well laid down. One field very boggy, but does for young cattle, and can't well be drained effectually on account of mill race coming through the middle of it. A nice situation for a mill and a thrashing machine with a good supply of water. I dare say it will be dearer than what we have bought, but is in my opinion a most desirable little property. John Forster has just bought the 3 cattle at 27£ per, he returned again and left us, he takes them away in ten days. This is by much the finest harvest day we have had of a long time, a fine south west wind and a good deal of sun will save our beans fast, and the late corn upon the moors. We have two carts helping to lead at Harry's. We have put the 6 cattle and a cow my brother bought yesterday upon the wheat stubble clover, where they seem to enjoy their place, and if this weather continue the other stubbles will carry cattle also and nothing will grow thereafter, and we have such clover where our wheat grew as I have seldom seen before.

Wednesday 30th. Another very fine morning. Carts have gone from Thornington, have got through most of his leading, a full stackyard. We are all going to thrash oats as they sell well. Natt Handasyde is going to weigh the Grindon wool. I don't care to go as the smallpox are so prevalent in that village, but thank God wonderfully favorable so far, above 30 have had them, Mary Brown amongst the rest had a great load but was said yesterday to be out of danger. I would recommend to you and your wife to inoculate yours with the cow pox, if not already done, which is by much the safest way and least sickness to the poor children. Many have been inoculated with the cow pox in this part, and have all done well.[91] This has turned out a soft day, not

---

91 The practice of rubbing pustular matter from a patient suffering a mild attack of smallpox on to the skin of a person to be protected was introduced into England in the early eighteenth century by Lady Mary Wortley Montagu, who had seen it used in Turkey and had her own children inoculated. Vaccination was put on a safer and sounder footing by Edward Jenner (1749–1823) who, learning that cowpox was said to immunise dairy maids against smallpox, conducted experiments and published a report in 1798. Vaccination was soon taken up in Europe and America.

much rain neither but kind of soft haze and showers every now and then.

Friday 2 October. I fancy I shall do with this as the foolish man did with his letter to his sweetheart. He took it to her himself, and I fancy I may as well do so with this. We are going to put some of our largest cattle to turnips on Monday. I was at Wark and Thornington today, my brother was along with me. Harry will get all or mostly in tomorrow I fancy, except a few beans and pease just below the stackyard, if such weather as this day only continue. Indeed this is the best harvest day we have had. We never had so brisk and warm a fresh wind. London market has come down 8 or 10s per quarter this week again for wheat, whilst oats and barley keep up. So we are going to sell a quantity of oats tomorrow. Harry never had so full a yard, I think we never had so good a stackyard since we farmed Thornington. But crops are very midling upon the whole. Neverthe-less wheat is said not to yield well in general, may be small in it and not quite so pretty a sample as the last years, and barley is a very ill coloured low sample in general. Oats are good but not so well coloured as last year neither. We are putting our ewes to the best tups for tup lambs, and part of them have already taken the tup. We are also going to have a better tup season than we have had of a long time. We have let to the number of nearly 500£, and many customers informed and not comed yet. I hope you have bought a few Kyloes at Brough Hill, as I think all the world are going to Falkirk.[92] The carts set off with the tups for Yorkshire Mr Sayle's on Tuesday next the 6th inst. and will be at your house on Thursday evening, and William Brown junior who goes with them will go all the way to Mr Barugh's himself we think, as your carts may not be so properly quitted to carry many tups. William Brown must go forward on the Friday, and will return to your place on the Saturday evening, stay all day Sunday and return on the Monday of the week following with such loading as you have for him. You will be so kind as to send Mr Eales his tup immediately on his arrival at your place. As this letter gives an account of both our travels and the carts coming to you with the tups, I think I shall send it per post tomorrow from Berwick. I am in the mean time Yours truly

Geo Culley

---

92 There were three trysts a year at Falkirk, in August, September and October, mainly for black cattle. The one in October was the largest: R. Belches, *General View of the Agriculture of the County of Stirling*, Edinburgh 1796, p. 45.

Ralph Brown is just returned from Kelso. Corn markets running on a good deal. If we should buy the place we are going to Tadcaster about, you must make a calculation what money you may be able to raise us by May next, when the first payment will I expect take place. But be sure you always take care to calculate a good deal within, so as not to disappoint either us or yourself. *Saturday the 3rd Oct.* By a letter this moment from a Mr Thacker,[93] a most respectable miller and correspondent of mine by the Trent side, and one of the cleverest men in his line I ever knew, he says they have the most abundant crops of all kinds he almost ever knew, and he is convinced that prices must still come down considerably. If that should be the case, which indeed is the universal opinion of all reasonable men I have conversed with or corresponded with, as well as my own sentiments, how prudent it is then to sell as much grain as we conveniently can at this season before the general thrashing comes on. Wheat is good and fine with them he says (which is not the case here in general), but the most smutted or black wheat he ever saw. Barley and oats rather coarse samples. All kinds of grain rather on the decline, with full markets, whilst butchers meat is scarce and unusually high. But he is disposed to think it will come down as well as grain when the general thrashing takes place. This is another fine morning, which will crown the latter harvest and suit the wheat seed on all cold lands. By a letter from Mr Claridge this morning I find the bargain is to take place. If you see Mr Peacock be so kind as tell him so, and that we will be with him at the time fixed. William Brown, who goes with the tups on Tuesday, will leave a parcel at Mrs West's if she still keeps the White Hart at Chester, if not at John West's, for George Thompson. It will be directed to your father, and I must intreat you not to forget to tell your father or brother to make enquiry for it there, or it may lay long enough.

Berwick, Saturday 12 o'clock. Ralph has only had 23*s* bid for his oats, and is not willing to take less than 24*s* per boll at any rate. Barley and oats are both in demand, and wheat flat.

Yours again Geo Culley

---

93 In this letter of 25 September (ZCU 24) Mr S. Thacker, a miller of Wilm Mill, reported on grain crops and prices, and meat prices. Potatoes and apples were abundant, pigs scarce.

95. *George Culley to John Welch*

Eastfield 14th October 1801

Well John

I got home yesterday about ½ past 3 o'clock, and thanks to the Almighty found my family full better than I expected. I trust in God that Matty is now in the way of getting well again, and most assuredly he will endeavor to guard against such an imprudency in future, or he will lose the sight of common sense, the best of all sense. I have been at Wark this morning, and I think my brother and George both think they can do without steers untill spring. If so it would be unwise to buy at such strange prices at this critical time. It is only buying in the spring even at higher prices provided things keep up, and I wish they may because we always sell more than we buy. However notwithstanding what I have said above, if you can buy 6 or 8 or 10 or more or less, at what one would think a reasonable price, I would advise you to buy, either at Yarm or the Martinmas fairs as you fall in, especially if they are well bred ones. I would even go a little further for well bred ones at any time. I think I have prevailed upon our folks not to take any horses to Newcastle, not only as times are critical and likely to come down, from so many of the soldiers' horses being to come to market, but because we shall want 6, 8 or 10 horses in the spring for this estate we have bought and going to take into our own hands, Consequently it may be right in you to buy 2 or 4 colts, or good fillies, at Yarm or Newcastle. I am at a loss who to make the manager of the Grange. My brother thinks George Jameson now at Longknow. I know so little of George that I can't give an opinion. Somebody we must have. I quite forgot in my hurry to ask you about James Grey's ewes, whether we must send them to you or sell them here when fat. Do you think you can get us a few acorns for Matty? I should think by the look as one comes along the road on this side Sunderland Bridge, that they would come to full maturity this autumn. I don't well know where you would succeed best. It should be where the oaks see the sun. Every tree appeared to be heavily loaden as we came along the turnpike on this side Sunderland Bridge. Don't by any means get any without first asking leave from the person where they grow. It might be right for you to make a slip over here after Martinmas, when you are not very throng.[94] Some haws also you should get gathered, as they are a very scarce commodity here this year, and a few quicks will always sell if you

---

94 Full of work.

have not occasion for them. You will most likely have to come over here after the 22nd November with your account books and your account of stock and crops &c.

16th October. As we have little more to communicate now, will send this away today. Matty says we shall have no horses at Newcastle fair now except one he may ride up, so that you need not come until the day before the fair without you chuse. I think half the farmers from this district are gone to the trysts at Falkirk, where Kyloes will be dear bought I dare say. So many have come south before. I am sorry you have missed sowing your naked fallows, it can't be helped, but wet and moor should never be neglected. I highly approve of your drilling in general, but wet moor land should not be wasted on any account. A season may come, but it is very uncertain at this season of the year after the weather once break. However it will make you wiser in future I hope. We have had a wonderfull tup season. We let 3 yesterday exceedingly well to near customers, Mr Weddell of Mousen and Mr Dinning's brother &c. Indeed few days pass without some being here. We have not had so good a tup season of many years. The London market for corn is now completely down. I am only sorry you have missed selling of the barley, but it can't be helped now. In critical times like these, an opportunity should never be slipped, because there is no saying how low grain may fall. It was obviously clear that corn would come down as soon as people began to thrash, peace or war, from the wonderfull crop and immense importation. The peace has certainly hastened the fall in that commodity, however I am inclined to think that may not be the case with stock, especially fat stock, because I think there is a real scarcity. But it is very different with corn, and I would not be surprised if these matters make many people run upon stock and neglect corn. Yet my doctrine is to have as many strings to your bow as you prudently can, and some will answer although others may fail. I am no advocate for growing too much corn in these dear times of labor, but it is impossible to get turnips, seeds or rape &c without the plow. Therefore it is wise to plow a reasonable quantity, because there is no farming without turnips, rape and clover &c &c. With every good wish I am yours

Geo Culley

PS. William Brown got all his naked fallows sown on his wet cold land at Grindon long since. Indeed there is no trusting wet lands. The best way is to keep your fallows well gripped[95] so as to have no water

---

95 Dug with trenches.

standing upon it, and then if a season come you can the better attack it. Be sure to have chamber lye always ready to pickle your seed. If you have not tubs enough buy more. You may sow barley upon your low grounds, but the moors will not do well for it although I have seen Joseph Hodgson's father grow very decent crops of barley upon their moors formerly. Indeed it was once a very common practice to sow naked fallow barley upon moor lands in the County of Durham, but I prefer red wheat very much. Only of a bad bargain you must make the best. You should use long swingletrees and each horse to go in the wet furrow, it is much the best plan for wet land. William Brown uses them to all his land, and we use them here to our wet land north of the house. It should be exactly the length of the breadth of the ridge, and fastened to each horse by a common short swingle-tree at each end. It saves the land from being poached, and you cover the land much wetter this way than the common method.

96. *George Culley to Matthew Culley and John Welch*
Eastfield Saturday 7th November 1801

Dear Brother

We sent you a longish letter today by Ralph to Berwick, and I begin another to say Matty and I have been up at Wark where they are busy batting the hogs, tup hogs new done or taken from look badly, but George is going to put them to turnips directly. Things are looking well, even Collinson is walking about poor animal, with a very thick looking leg. *Sunday 8th* has been a most remarkable fine day. Ralph sold 10 bolls white and 20 bolls red wheat all from Wark yesterday at Berwick for 45s per boll. I understand none was sold for more than 38s to 42s except the above. Barley rather brisker, but little if any at 24s per. Oats 15 to 18 rather flat. Ralph brought 150£ from John Graham, and we want 100 or nearly of him still. John is very slow but very civil. Brown bought the wheat. I had been walking alone as far as the General's Close this morning, and this moment on my return I found a letter upon our table directed to M.C. Wark Esqr. So as that gentleman is in the *south* I ventured to unfold it. It is dated Jedburgh 3 inst. November and signed William Oliver, and relates 'that Thomas Bowcroft be liable in the maintenance of the child of Ann Donaldson'. I fancy George Humble will understand it, and he is to be here tomorrow. Your friend Sowers was arrested the other day by somebody in Berwick, and must have gone to jail if George Purvis had not become bond for him, and George silly man will have to pay the debt I will answer for it. Sowers makes his brags that he earns 50 guineas per annum by Coldstream weddings, which added to his salary might seem to keep well now, but somebody at Coldstream has set up against him and marrys at 10/6d. Douglass with his 2 artificial legs was married by this new man on Friday last to one of our eminent ladies at Westfield. Mrs William was brought to bed of a daughter a few days ago. More news you see.

Monday 9th November 6 o'clock in the morning. Mistress just going away to Weetwood[96] to meet her brother there. Poor Mrs Peterson is very poorly. No immediate danger, but they think she is much like what Mrs Bailey was, and is now I believe taking the same drops (lignum) I think that cured Mrs B. and many others. It is a very nice morning although a little frosty. George Humble just come (9 o'clock) and I have received your 3rd letter of the 5th inst. this moment, and although you say you are for leaving Denton on Tuesday (tomorrow) or Wednesday I will hazard this by sending it away today or tomorrow at the farthest, and I hope John will have sense to open it even if you are come away. However I will now direct to John only for fear of the event. You don't say why my letter was so long of coming to hand. It must have laid at Piercebridge or somewhere, because I am certain it would be at Darlington on Saturday the 31st October, or John Welch had forgot to ask for it at Darlington. I do like investigation and the saddle to be put upon the right horse. If a man acts right and means well he need not be affraid of having his actions scrutinized, although he may make mistakes he will not do it willingly, and will not be affraid of those mistakes being known. Now my 2nd letter away for you should be comed to your hands, as I am certain without some mistake it would be at Dar[ling]ton on Sunday last the 7th inst. I think very highly of Mr Darnot's judgement, but I can't say but I am disposed to think at present that grain will be well sold. Not but I recommend selling as long as prices are this good, indeed John's prices for wheat and barley are such as by no means to be missed in my judgement.

I perfectly agree with you that John Welch should have an under steward, a married man by all means, and if he be a gardner or half gardner so much the better, and it is not at all unlikely but we may be able to send him one of that sort by and bye, or he may try to meet with one, or if Mr Thomas Bates can recommend one of that description to him it will be a shorter distance than from this country. And as you will likely come home by Mr T. Bates there can be nothing wrong

---

96 John Nesbit leased this farm from John Orde of Weetwood, 1739–1818: Land Tax Assessments, IR23/65; *Burke's Landed Gentry.*

in applying to him on that head. It is impossible but John Welch must have turnips to serve more stock than he has at present, or his turnips are worse than I suppose them. However I don't wish him to take a tail more from us without it suit him, because we can be selling them at an earlier period. However one thing I must remark to John Welch, that he should not have solicited us so earnestly in many letters to buy James Grey's or other ewes, although I repeatedly told him in answer that they could not be bought to make a probable return in profit. I mean no other reflection in this remark than that John Welch and everybody should think and consider a matter twice over before they give such positive directions as John's then was. And as to the 60 ewes that are tupping for Denton I preached long about them to everybody *before I could be heard*, much less prevail upon the privy council to adopt the idea with a *great deal to common sense.* And in my humble opinion you will not act so wisely as I think right, except you keep 20 or 25 ewes at Denton to breed from, because nothing we can deal in pays in so well as sheep, and because upon every farm there is a certain description of land that is fitter to keep store than fat stocks, and because we always kept 60–70 or even 80 breeding ewes upon my brother's estate at Denton formerly, and fed all their produce, and more because surely Mr or rather Sir Ralph Milbank's estate is very competent to keep as many or more than my brother's part did at that time when we had as much plowing as there is upon it at present I think.

The wine will be attended to according to your instructions. I hope you will make an end of Shotton one way or other. I am highly obliged to my brother for answering my letters word for word or sentence after sentence, which is always the right way. My brother's idea of your meeting the fat ewes at Chester to sell to such butchers as you could approve of, is to me absurd in the highest degree! Either take them all the way, or let us sell them at this place as well as we can. The thought had struck my brother at the moment without reflecting upon it or reconsidering it. I recomend to you John to sow the wet wild part of the Church field with barley in the spring by all means. It is everything to get land made clean in the first instance, especially when it is in the proprietor's own hand. The moors &c I again repeat will not do for barley so well, but if you can't make a better of it you can't now help it. However here is now very promising weather again. The moors will not on any account do for spring wheat. The infield lands may. My brother speaks about single horse carts. I have certainly long been convinced of the propriety and advantage of using single horse carts, but never could get it brought about here yet although many other people have adopted them in this country now, and they will one day become general in this Island, and I commend you more than I have time to say at present for adopting them at Denton so much.[97] Be so kind as look about and enquire for a 2nd steward for yourself. You cannot do well at all without one. I am in real haste. Yours truly whether M.C. or John Welch

Geo Culley

97. *George Culley to John Welch*

Eastfield 11th November 1801

Well John

I shall address this to you, and the rather because in the first place I wrote (the last) in a great hurry in hope that there might be a possibillity of its reaching my brother before he left you, and secondly because I am pretty certain that if I don't write to you first you will not write to me, because you will improperly conjecture that my brother will tell us everything about you and Denton. Now unfortunately my brother will tell us next to nothing at all, and as he will probably come by Mr T. Bates's it may be a good while before he reach Wark, and then I may not see him for some days. Now I don't nor can't desire you to write such long letters to me as I do to you, but I do request that you will write as often as you can, and about such things as are needfull to communicate. I was rather in ill humour about my worthy brother's weak remarks of you meeting the ewes at Chester, and so I got nothing said about when they should go if they do go. Therefore I beg you will be so kind as say when we are to send you a 100 fat ewes, and the 60 tupped ones for you. But if you are affraid of falling short of turnips, we had better keep them here altogether, as we can probably buy turnips cheaper than you can, and if they are not sent this month the snows may come on and make it very bad driving of them, particularly over the moors. Yet they can't go Morpeth road except they go by Alnwick or at least Felton,

---

97 In north Northumberland farm carts were on the whole heavy and clumsy, drawn by two horses. Single horse carts were coming in in some parts of the county, especially the Tyne valley, probably in imitation of Cumberland where such carts were universally used. In Co. Durham long carts, drawn by two or three horses, were used for most farm purposes but single horse carts were becoming more general for carrying goods on the public roads. Bailey and Culley, *General View, Northumberland*, p. 38; Bailey, *General View, Durham*, pp. 82–3.

because Weldon Bridge is fallen down, and it is a very awkward river to cross with sheep without a bridge as it is a bad ford, at any rate just below the bridge, and besides the Coquet and most rivers are liable to floods at this time of the year. I hear that Hallow fair was [. . .][98] for both wintering cattle and horses than Newcastle was. [. . .] be of thinking that fat stock will be pretty well sold by and bye, but at any rate we must be selling a part of both cattle and sheep about Christmas, or buy a good many turnips which I am not very fond of doing. 'But necessity has no law'. By yesterday's paper (this is Friday 12th [13th]) I see that London market is advancing for all kinds of corn, rye excepted, and people have now got it into their heads that wheat will be well sold, because first it was begun very early with, and 2ndly we sell more or can grow so much wheat now as serves this populous island, and 3rdly if barley is high sold wheat will, and perhaps most other grains. I hope John you will be able this fine weather to sow some more wheat, and I am sure you will not slip the opportunity. We are busy sowing, and William Brown and Harry, after beans, pea stubbles and rape &c &c.

Berwick 14th November 1801. Ralph Brown has been bid 48*s* for wheat this day, and both barley and wheat have advanced today considerably. I have heard nothing about my brother and sister being upon the road home yet. We have wonderfull good weather here still. Mutton 8*d* per lb. A strange dear market at Hallow fair Edinburgh I am told, near 10*s* per stone for lean cattle. I wish we could see 9 for fat and I would sell a few. Write soon and you will oblige Yours

Geo Culley

98.   *George and Matthew Culley to John Welch*

Eastfield 19th November 1801

John Welch

After we had sent away the last letter to you, Matty luckily found that the ewes we had put to the tup for you were not missed tupping as we had been led to understand. Consequently sent you along with 105 fat ewes and gimmers, 80 ewes tupped for keeping and a tup along with them in case any of them should come in season, which tup you can let stay with the ewes as long as you think right. But perhaps a barren ewe or 2 are better than over late lambs, and that you can best judge off. The tup sent with these ewes you must return to us when the rest come home, as Matty thinks pretty well of him,

---

98 There is a tear in the paper here, affecting two lines of text.

and we can send a tup or tups against another season for your own use. My brother and sister got well home thank God on Tuesday evening. Don't neglect to send your stock account. John my wife begs you will in your next letter to me after receiving this say how much George Thompson's board &c amount to for one year, and we will immediately send you the money. Pray don't neglect this. You had better write it upon a slip of paper and put it in the inside of the letter. I do wish you could get a decent married man to act under you as a 2nd steward, to go with the work people and have a charge of everything in your absence at markets, or any time when you are from home. If you can't meet with a proper person in your own country, perhaps we had better try to hire one here, and if he could act as shepherd as well as under steward it would be better. But I have two objections to this. The first is the expence of sending a married man so far, the 2nd is that shepherds in general are so averse to any other charge except their own, and are in general indolent. My brother says a man that knows something of gardening and watering &c would be best. Now I could wish for your own opinion and thoughts on this subject. Bob Deepy would answer well but he is unmarried, and perhaps he would not like to go. He is a famous clipper too, and would easy learn to look after your stocks I think. To be sure it is not easy to meet with a person capable of doing all these different things, and yet such people are to be met with at times. However you must turn it over in your mind and make enquiry in your own neighborhood, as it is highly important, nay necessary that you should have a person behind your hand, in whom you could confide. Whatever money you can spare without pinching yourself, put it into the Darlington (Mowbray's) bank, but remember you are not to pinch yourself. Mr Peacock will remit the money, whatever it is, at or before May next to London, and the remainder, after I know from Mr Peacock what it is, I must take care to remit from hence. This was a very fine white frosty morning, but about one o'clock it began to snow the largest flakes I ever remember to have seen, and it now has a close cover of thin snow. *Friday 20th November*. A very severe frost indeed, but will not prevent the ewes from tripping along that are going to you.

Wark 20th November

[*M.C.*] Well John

Mr Bates will send you some thorn quicks, and take back some apples, a hamper or so on. Let him have as many as he writes for by all means, they will be in lieu of quicks. My brother and Matty were here today, assorting the feeding cattle. Many good ones, and in good order except the few new castoffs from one of our drafts. Would Bob Dippie do for a 2nd overseer? He has been accustomed to water land, could do little things in the gardening way, but has not been accustomed to overlook workmen except in the watering land, but is unmarried. You may trust to his honesty, only he knows nothing of plowing. Then I have never said a word to him about his going to you. He would be soft at first, is not without a good understanding as I think. As we now are determined to use single horse carts in general, shall be in want of iron arms or iron axels with whole length bushes.[99] Ask and give us your opinion which are best, arms or pull through, with bushes or one bush only to one wheel. At Sheffield they had a long bush to each wheel only, and the axel ran for a number of years (20) without any mending or repairs, I have been told by Mr Sayle of Wentbridge, whose brother in law Mr Booth used them. I think we should send you 6 or 8 pigs, *shots*, and perhaps a sow in pig. Do let me have your opinion of this matter. I have not yet enquired how they sell here, nor the price of pork, butter 10 to 11*d* at Berwick, dressed goose 3*s*, ducks 9*d* to 13*d*, wheat near 50*s* per boll, all grain advanced at London except barley.

Sat. 21st. I think we should send you some young oaks and crabs to put in your hedges and other places where convenient to be putting, keep and sow this winter some acorns for your own or Denton planting. We should plant the Lime Kiln carr, quarry hole and a strip also above the quarry next winter if all be well, and I rather wish for acorns or young oaks in hedgerow with tap roots than other trees that have collateral roots running near the surface. The ash &c injure land more yearly than they improve in value yearly. Do not pull down houses or hedges until you have leisure to build others up. A bad hedge or house may be better than none. The cabbin beams fallen in so as to want new spars and roofing. Fine wheat 5 to 6*d* advance per qr, rye 4 ditto, barley 2 lower, oats 3 higher, peas 1 ditto, white ditto at a stand, London Monday. Newcastle on the advance for grain, wheat near 19 to 20. Do not forget to renew the hedge on Mr Bowses's planting, cut the osiers on your hedges and along the race on the carrs. Nat and another should cut them and use them in such things and on such days as he finds opportunity without neglecting your business. With esteem

Mattw Culley

---

99 Metal boxes or linings for the cylinder in which an axle works.

We did not get anything settled about your byers or places for your feeding cattle. Single room stands are best where they have liberty, but should have a trough to eat out of raised so high as to hinder them from dirtying their meat, an old shallow tub or trough. Use some of your coarse wood or trees for railing the partitions. They should be sawn strong, so as to prevent the cattle breaking them down. I paid Mr Bates 15£ for the colt, he is likely to be a good slave. If you could slip over he has a good mare 1 yr older, a little fidgey in giers asks 20£. Perhaps he could buy you or us a strong colt or mare 2, 3, or 4 to 6 yrs old by the help of Rowell of Whitington who bought our stone horses.[100] Ralph Brown was at Dunse fair on Wednesday, thought he could buy a horse or 2 but was disapointed, made Matty and I think better of our Newcastle ones. Matty found much fault with those we bought, and with Hare's who fell and broke his knees on the road home. Also Orwin's they say is a little one and a bad middle. He had probably been made up with softness, and his flesh is gone off with a cough or cold so that he looks badly. However they had kept him at Eastfield, sent Lyne's mare and the next mare we bought to Harry, Hare's horse and lambs to Wark. They say the 20£ mare is an old one. Usefull horses will always be wanted, they pay no more cess than midling or bad ones. Our people had kept the tup lambs on the 2nd growing of the rape so as to injure them much. They are fond of the first growth but not of the latter. Matty they say desired them to be kept on. They want of ½ the day to good grass, when put to turnips fell greedily to them at first, and 3 or 4 of them died. Two or 3 wethers kept on the rich clovers have died of the yellows, a thing I am always afraid of. They piss bloody urine, the bladder is affected always before they dye. Saltpetre may be useful but they are not seen till past cure, except the knife. We don't find that any *Cape wheat* came north last year. We should have some sent by some means for tryal whenever it be had from London or what port cheapest by sea carriage to here. William Shotton would he answer for you as a 2nd grieve or overseer? He did wish to be under us. Has T. Hutchison comed to you or not? Or is Kirtley gone into one of the rooms? Is it certain that Lindsay Birkbeck's tenant has taken a farm, and where is it at? How come they on with the water land? Does the conduit take all the water off the foot of the upper well spring except in a flood? Have you had any floods since I came away? You must

---

100 Stallions.

give us an account of old Martinmas day's market turned out. It falls on Monday. York and Newcastle are on the same day as Darlington.

99. *Matthew and George Culley to John Welch*

Wark 1st December 1801

John Welch

Well, we received your letter of the 26th on Monday by Harry. It was our blame, and Glass's. As the ewe was lame he ought to have left her at Milfield. I met him on the south side of it, and he said she was lame and I did not consider it well. The weather pleasant although frosty and some snow. I am glad the good weather has allowed you to sow, although it's too late, leave it to you to sow or not the high Church field. The south side is very clayey and soft in wet weather. Do not sow wheat in the spring but on fresh good soily land as a general rule unless soon in the spring. The newspaper does not say Newcastle stores fair was so good or high as you speak of for fat or lean cattle. If the markets for fat be good, should you not sell some fatt stock? Would not some of your ewes be suitable for Christmas? Kyloes &c are now said to be as high as ever, by Mr Proctor of Glahams. Your Church field will produce good wheat, but not an abundant crop of wheat. Barnard Castle 3 guineas, Berwick 52s per boll to 52s 6d. 7s per stone for lumbring steers is an extra good price. We have sold wheat as fast as we could get it threshed, Why? We think the prices are likely to be lower, and time and good weather prevent threshing. Cattle have plenty of grass yet, no occasion to house them this mild season. Prices of barley and oats I leave to my brother to answer, but in general people suppose the distilleries will hardly be suffered to go on as they have done formerly. Tom Wright's cow (of ours) is a great loss but cannot be helped, only it appears to me to be a very dear way to find a hind a cow. Yet we sometimes do so. I hope you have bought the house of Nicholson's. Do not stop at 5£ extra or even more for a good house, although I believe my brother gave George Nicholson the front. I suppose it is worth 40£ or more, although no land annexed to it I believe. Tell them I am obliged to them for the offer to me. I am glad George Winter gets the land so well watered. I have ordered a spirit level from Mr Miller, South Bridge, Edinburgh, with a screw and a spring, very good and simple, and low priced about 17s without feet, which I can get made here. Mr Cradock should not see so expensive a matter until he sees the hay &c produced by water. Bob Dippie has not been spoken to as yet. I suppose ½ of the acorns are a new to send here. We should or you should have had some crab pips this season. I wish you to put in acorns in your hedges, and to sow some of acorns for planting in future. You were to enquire for 20 or 40,000 seedling quicks and send by carriage, if could be bought at a low price, as they are very scarce here and those sown here last season have not grown 1 in 10, I may perhaps say not one in 100, 10 or so on a large bed. I had it much to heart to send you from 12 to 16 or twenty pigs, shots, only our people think they will not pay for driving. Had thought of sending them by Newcastle, but I am told that one arch of Weldon Bridge is down, which will make a bad passage for small stock such as sheep and swine. William Brown sent a nice young sow to Thornington to be sent with the other pigs for you. We have 28 small sucking ones on the sows, but apprehend they will not be able to travel for several weeks so far as Denton House. Glanton and Co. sent you the 2 large or small casks of varnish to King's Head, Darlington as I ordered on coming by Newcastle. I suppose cart horses will be dear in the spring. Mr Bates's mare is fidgey, so I cannot tell how to advise about her. So far

Mattw Culley

PS. Mrs Culley desires to be remembered to Dolly and Johny, Miss Middleton &c. Tell Winter his wife is well but thinks he stays long away, but he must do what is necessary and then come away when you and he think necessary. I think a hind on reconsidering things might be indulged with a Kyloe cow if he is a valuable man and an excellent hind. Of this my brother may consider. In such case he should have 2 or 3 Kyloes to fat a little before the term. Also I have a wish (but it goes no further) for the 2 queys to be sent back here out of Andrew Bolton's cows, as Andrew gets a quey of me that I wish not for him to have, also William Brown wants a quey in exchange for the one which he got last year, but these things and the swine I leave to my brother and his son entirely now, for I have got my thoughts examined over, and if it be a loss or gain to me it is the same to him. Therefore I leave swine and the heifers to what they say and determine upon as I have made up my mind to be perfectly easy about the matter whether they go and come or not. M.C.

Eastfield 2 December 1801

[*G.C.*] Well John my brother sent me this just now. He took your letter by Glass home with him on Monday, has answered thus far and sent it to me to finish. Pork as well as wheat is for the most part dearer by 1s per stone with you than in this district, and wheat generally is 1s per bushell more. Our pork markets will only give 7s although they can afford full as good a price as yours, as their shipping is readier and more certain, but I suppose will not deal for so little profit.

However Ralph Brown by great industry has found out a good man at Kelso who will give 7s/6d per, head and feet on, which is nearly as good as 8s without. It was most certainly a wonderfull seed time for so late a season, but I am affraid it may answer badly upon your late, cold weak moor land, as these months of December have only a very *weak vegetation*, especially as the frost (although mild as yet) has set in very like a lying storm. I consider if it only *vegetates* even within the ground it will do, provided it be not very wet land, for it has pleased the Almighty that *wheat*, the means of the *staff of life*, should be able to bear great hardship. However it is the *duty* and the *interest* of the husbandman to make use of the properest seasons, and although we may not know to a certainty, yet I am disposed to think that a late season is very often the best or produces the giftiest crops provided it do not founder. I mean this upon clays or wheat soils properly so called, but upon all dry light turnip soils I would prefer February to any other month. But because we can't impose a February season it is wise to take a January one. But really from some late experiments I should prefer the first week in March to any time in January for light early ground, especially if in high culture or fresh *young land*. You understand me I think. But I often say that we have a deal to learn yet, and every wise humble man will learn every and every day, but a conceited self wise *animal*, I will not call him a man, will not nor ever can learn.

Although fat cattle have dropped in value I am glad that sheep are still sold with you at 8d per lb. sink. It is a famous price. And do you know that I would recommend to you in strong terms to be selling some of your fat ewes at that price, because prices may be weak in future. But at any rate we have a good many ewes still which we could send you, James Grey's &c very nice I believe, although neither so heavy nor so fat as our own. Besides we have upwards of 200 wethers or dinmonds which, if we can, should be clipped. Wool will be so valuable in all probabillity. Now do you turn these matters in your mind, consider well and then give me your sentiments. Can you take any wethers towards spring, and clip them if you can so long keep them? I am rather affraid we can not, without injuring our hogs, tups &c &c. I do not find that any such price as 8d per lb. sink is given at Morpeth, indeed Morpeth is held out to be very bad at present, and I am sure that I shall be very unwilling to sell William Lincoln the lot of ewes which we have promised him an offer of at less than 8d or even so little, as I expect they are particularly good. Do give me an early opinion of these matters. I would also like well to sell a lot or 2 of our turnip cattle pretty early, say between Christmas and

Candlemas, as we have a large quantity and they are forwarder than we have often had them. Besides we are not over plentifull I find in turnips. We shall have some very nice outlaid cattle I hope also. Will you be able to take all or a part of them think you? I rather think we shall have 20 or better, but you may be sure we shall keep them as late as we can. I hope to receive your account of stock tomorrow morning or very soon, because it stops me from getting our annual accounts done. As we have not seen your books or any summary account from there, would [it][101] not be proper for you to make a slip over about Christmas or sooner if convenient? You could then see what stock we have and help Matty and I to consider what is best to be done. For as to my brother honest man, I do assure you we have very little help from him. I truly and sincerely wish that you had a decent man behind you [. . .][102] next to impossible for you to do as you ought to do without an under steward [. . .] saw that in so strong a light as I now do, or I would have had [. . .] conversation with you upon it when I saw you at your own house, and [. . .] consulting upon a matter of that importance it would be best to have you here. For it is not easy to say everything one wishes upon paper. However you have never said by letter or otherwise what you think is best to be done about a man under you. I never knew until my sister of Wark told me that Harry turned out so badly, methodistically bad! When you [are][103] at markets or fairs it is terrible to have no one behind your hand that has authority and that you can confide in. My brother says very justly that you have far too few pigs for so large a farm, and that pigs are not to be got for money scarcely in your neighborhood, on which account he proposed sending a man all the way with 7 pigs, which I objected to on account of the expence running away with all the benefit. I said we had better send 20 more or less, and you could sell such as you could not conveniently keep. So the matter stands. Therefore be so kind as give us your opinion of the matter, and we will endeavor to act to the best of our knowledge.

Thursday morning 3 December 1801. We have this moment received yours with the account of stock, for which I thank you, as I can now get forward. It is very likely to continue a storm I think, at least it has every appearance. Matty desires you will send for the 2 tups at Lord Darlington's and Christopher Eales's directly if done. You can put them either amongst the wethers or some little field by

---

101 A word seems to have been omitted.
102 There is a tear in the paper here, affecting four lines of text.
103 A word seems to have been omitted.

themselves, and you may hold yourself in readiness to send for the Yorkshire tups to Mr Barugh's in a few days, which we will take care to let you know as we are expecting letters every day. As it will require two carts you must take care to have 2 in readiness properly prepared. I am glad you are for trying to get an overseer or 2nd steward in your own country. Will not Tommy Hutchison answer, go with the women workers if no other? He was brought up to husband labor and was a very nice farmer since I knew him more than 40 years ago, and I always thought he had more *spirit* and *firmness* than any other of the brothers. Be so kind as try him. I have a great notion that he will answer better than you are aware of, and perhaps may be trusted poor man with authority over all the workers, As a relation it may give him some weight. To be sure he is too old to be active, yet I think I can manage shearers or hoers &c as well as other folks still, and I am not many years younger than he is, and he was once as clever as ever I was. But misfortunes perhaps break our spirits. John you must buy George Nicholson's house at all events, or you will offend my brother. He thinks it well worth 40£. My brother certainly gave George the ground to build on, but that will not make any weight with these people, and I would not have you mind a 5£ matter. By this day's London paper corn is coming down apace again, and I will be much mistaken if it do not, because nothing but the fine weather which kept people from thrashing and made less come to market, was the means of the advance in the price lately. We have thrashed and delivered all we could, as I was never more convinced that prices will drop as people begin thrashing, Nevertheless I do not imagine that grain will come very low. I don't know the Cape wheat. Thank you for your remarks about bushes for wheels. I am glad you think the same about pigs that we all do except my brother, and I have no doubt but he will send some to you, and I would have you say some more about them in your next. It is too far to send crab quicks and we have few. But I would have you write directly to Mr Barugh about procuring you a few crab pips which you can bring home with the tups. I hope Birkbeck's tenant will go quietly away. Write again soon and let us know how the Great Monday[104] market or fair proves, which will oblige

Yours truly Geo Culley

---

104 7 December.

100.   *Matthew Culley jr to John Welch*

Eastfield 8th December 1801

John Welch

We have this moment received a letter from Yorkshire saying that our tups will be at Mr Henry Barugh's of Birdforth near Carlton on Friday next the 10th instant at night, and you must be so good as send there for them as soon as you receive this letter. You will have to send two carts as there will be eight tups to come from there. Also if you have any potatoe oats you must send one quarter for Mr Taylor for seed, directed to Mr John Taylor of Salton near Malton, Yorkshire, and desire Mr Barugh to forward them if Mr Taylor's cart be gone. If you have none ready we must send them some other way, and let us know in your next whether you sent them or not. You must take care to get Lord Darlington's and Kit Eales tups to Denton by Wednesday the 16th inst. as our carts will be at Denton on that day to bring home all these tups and yours, eleven in number, which will take three carts.

Corn has been advancing here lately. Ralph sold wheat last Friday at 55*s* per boll at Kelso, and it is worth 52*s* at Berwick. Barley and oats keep much the same. We wish you by all means to sell your fat ewes at 8*d* per lb. sink, and as much more as you can get. I am in haste

Yours &c Mattw Culley

101.   *Matthew Culley jr, George and Matthew Culley to John Welch*

Eastfield 11 December 1801

Well John

I am sorry you should have had so much trouble about these tups, but it cannot be helped. I have this moment received a letter from Mr John Taylor of Salton saying that his tup is not to be at Mr Barugh's at the same time as the rest, and not until Monday week the 28th instant, when you will have to send for him, and if you have not sent the quarter of potatoe oats for him you might send them then at all events as his cart will meet yours there on purpose to take the oats home, as he wants them for seed, and I have this day written to Mr Taylor to this effect.

Yours &c Mattw Culley

[*G.C*] John Welch. It used to be the custom in the country where you now live, to give over plowing from a little before this time of the year until Candlemas. For what reason I know not, nor perhaps the people themselves, only it was the fashion before we left that country, and I only name it to guard you against it, because I think all winter plowing the most advantageous besides the forwarding of business, as the frosts and changeable weather *pulverizes* and *ameliorates* the land and is a more certain preparative for crops or even for fallow, than all other means which the art of man can contrive. Even if land is plowed a little wet, the frost heals and puts all right again. But plowing wet after the frosts are gone is extremely wrong, because the dry weather which generally succeeds in the spring bakes wet land especially into *bricks*. I have no doubt but you are aware of all this, but will communicate our course for good manners. That is to say, we are often led away by habits and customs of countries to do what we ought not to do, without reflecting that we are doing wrong. But a wise man will and ought to learn every day. At that period the sowing of spring wheat would have been accounted madness, and perhaps may not succeed so well with you as in this country, because there seems more life and vigor in the land here than with you. Whether this (if it is so) is owing to the difference in the lime, or the land here has not been so much trashed, I know not, but I think there is a strange difference between your lime and the lime here. I am disposed to think that you will find the lime burnt from the white stone in the east side of your quarry much better lime for land than the other side, and on that account I would recommend you to make some tryals with the two sorts of lime as soon as you can, because if I am right it will be an important matter.

G. Culley

Saturday 12th December 12 o'clock at noon. I had got thus far John when your kind letter of 10th came to hand, and one from Mr Sayle which I will speak of first. Mr Sayle says 'lean cattle at Pomfret last Saturday (a fair I fancy)[105] dearer than was ever known. Fat also from 8 to 9s per stone sink and very scarce. As to real fat I saw none. Fat sheep 9d to 9 ½ per lb. sink'. So you see Mr Sayle's accounts are very high indeed, and accounts from all the Wakefield jobbers being at Darlington on Great Monday. Yet your prices were low certainly compared, but I attribute that to the badness of your fat stock shewn at this time. When looking over your letter I find the Wakefield men were at Richmond, not Darlington Great Monday. It is very extraordinary that William Story was never at this place, at least none of this family ever saw him. I would certainly have wrote by him if I had seen him, and it seems very odd to me that the young man never came near me. I don't recollect ever behaving to him improperly. I am

---

105 These words are presumably an interpolation by George Culley.

rejoiced to hear that you have sold so much barley at 6s 3d, although I now think that barley will not be lower as the distilleries are now to go on. Still it is a safe game and a famous price. I am also happy to hear of your selling your red wheat at these good prices. Whatever the prices may be afterwards, I am sure you will not grumble nor repent what is now done. I doubt you have picked the ewes for George Nicholson. Otherwise it appears a very good price and a ready taking away, and I hope the money will be good. However from Mr Sayle's account, of which I can have no doubt, and from your own ideas and reasoning, sheep will be higher, and cannot be lower well. Turnips are so plentifull and large cattle seldom advance in price much until at or after Candlemas. From your account of Sunderland being so dull I think we shall make little out with William Lincoln, and the 100 ewes we intend for him ought to be worth much more than those you have sold to George Nicholson even if you did pick them. However we shall scarcely write to Mr Lincoln until you come north, when we can consider things over with some attention. We can certainly buy turnips, plenty yet. Indeed we wish rather to eat our own nearish before we buy, as turnips must be plentifull without a frost destroy them, and eating our own enables us to sow more spring wheat, a very important thing here where we can grow nearly as many bushells of wheat per acre as barley, and barley is always ill sold here. All my fear is that the lot of ewes we intend for Mr Lincoln will be over fat to travel to your place should we send them, which perhaps may be wise as by Mr Sayle's account your markets will and must get worse before ours. Still I think by giving of them time they should make it, except in rainy or snowy weather. We have some very nice wethers, which I am so covetous as wish to clip, as wool has a chance to be so dear. But without you can do it, it will not be in our power as it will interfere with our hogs and tups which pay too well at this time not to pay every attention to them.

I am extremely obliged to you for your sensible suggestions and remarks, and assure you that I quite agree with you, and we will certainly try to keep everything for a time. Only old fat and heavy ewes will be obliged to be sold by and bye. Such as Jemmy Grey's we can drive anywhere, and nice fat they will be, 18 lb. per quarter or better I hope. But when you come over we will consider about things. After all it appears a little odd to me that the Wakefield and Skipton jobbers should not be at Darlington, but perhaps they like better to buy at home. What has become of your friend's business? You don't name him amongst them. I am very glad that you have bought George Nicholson's house. My brother will be highly pleased with that transaction. What you suggest about pork pigs is extremely right, but we have sold all ours now at 7s/6d per stone head and feet on. The weather has been very changeable here of late. Wednesday and Thursday great rain and heavy floods, now again frosty and snow on Cheviot again. It will be well done if you can procure 30 or 40 thousand of seedling quicks. I can't conceive who would send us horse chestnuts, besides they are an ornamental but most useless trees. I hope they will prove acorns but if so we know nothing who has sent them. I am very much pleased that you propose coming north. Do you chuse your own time, but I would wish you not to hurry when you do come. If that was a real good horse, why did you not buy him? I think 2 or 3£ is nothing in a good horse, especially for you who have strongish land to plow, and single horse carts. I am glad that you have sold Monkhouse more wheat. 9s is a capital price.

Geo Culley

[*M.C.*] Saturday night. I had wrote ½ a sheet to you when J. Davison brought this from my brother. Your mistress and I are greatly pleased that you bought the house at 40£. It gives you upputting for 2 more servants, and that is a great acquisition. I hope we may some future buy Birkbeck's and Wade's &c if we get a fair offer of them, and gives you a command of labourers, a thing your active husbandry cannot do without. You should have 2 families also if you can, and think it right to have, at my moor or Hutton's Close house when Linsey quits. I think the house has 1 or 2 chambers, stable and barn, but you should have a clever active overlooker, a trusty fellow to assist you, not only when you are absent but at all times, as your businesses are such you must see markets and not be hard tyed. It is of vast importance for you to be at liberty to buy and sell, and you cannot do that and be confined at home. You seem to like Mr. Croft. It is right to keep to good men such as he and Monkhouse. I agree you ought as my brother says to buy a good cart horse when one falls in your way. 2 or 3£ are of little consequence between a thick useful one and a crab. Try to buy him if unsold. I suppose G.W. has only carried the water over the north side of Howlet flatt. If we live to another year we may possibly water part of the Gilla field, but perhaps we should plow that part first, and take some crops of corn of it. In that or either case it should be hedged off the dry land, by where the water could be carried along, the ditch or gutter to be the carrier in the same way as over the 2 Fitts. If you wish to plow and hedge back part I must send you the new level to take a proper level with. More of this when you come over, but mature it over in your mind. You do not say Mr Bates has come for some apples with some quicks in exchange. If you cannot overset us with the seedling quicks, give some extra price if delivered at Newcastle. From there we can get them by Orwin to Eastfield. They should be put within a matt or bags to keep them from frost or drought. You say the flood was over all the watered land. How does the carr flood, is anything on the north side, and was the oflet properly vented by the side of the old water course on the south south [*sic*] side above where you pared the coarse land? That should be one to prevent a flood injuring your new sown wheat in Stubbs carr. I shall tire your patience soon and end with respects to you all.

Mattw Culley.

I forgot to say Matty was happy to meet us at Chester and foot it to school next morning.

May it not be right to get 2 or 4 gallons of the best brandy and keep for particular company, as some people cannot drink rum. Wine cannot be allowed off except when we are at Denton. It is now so excessive dear, it stands to 3*s* a bottle. Whenever you have company on our account you must make us answerable for the expence. Also should have some spirits of an inferior kind for drams, or for horses when we intend, but I will send them by J. Davison.

102.   *Matthew Culley to John Welch*

Wark 12th December 1801

John Welch

As Jack Davison can hardly drive the 3 carts home without help, we think G. Winter may as well return with him. I suppose Kirtley with your assistance may finish what's necessary when G.W. is away. I have got a spirit level which may be sent to you next year, cost 16*s* of a good construction, will I hope answer well for the purpose. We had a great fall of rain on Wednesday, the burns were very full, the river only high. The weather turns to frost although the wind is from the west often. How was your market on Great Monday? Have you opened the road over Unthank and given the land back to the glebe in lieu of what you got? Does Mr Peacock think that any of the glebe will be sold on account of buying their land tax? I wish you may not have missed them horses of the Nicholsons, what does a 5£ matter make in buying them, let us hear by return of Jack, as I wish much to have them. Have you done anything at the drying kiln? Days are now short to build in, will be longer after the new year. John Dickinson is a good workman, you may get part of your work done by him, but this I leave to you. I suppose Birkbeck will have to keep the tenement, the rent can be settled after. Should he not also have the garth below the dam, and the wastes when the new road is done, and the part of the cow pasture laid to it? Altogether this will be a large field, but much of it little worth, only he has set his heart on it. Tomy Hutchison and Kirton may be usefull to you, but do not expect too much from the Hutchisons, are rather indolent people. Can he, Tomy, learn to let on the water, or off as occasion requires, and look in the field sometimes, or after the women working &c? But I am afraid slow and lazy. We send 300 crabs and 300 oaks to lay in your hedges at about 7 or 8 yards distance as you choose or think proper

From Mattw Culley

Give your mistress's respects to Miss Middleton, also to Miss Nicholson &c, and ours to all enquiring friends.

PS. If Mr Bates has not got the apples we are afraid they may spoil. Had you not better dispose of the most of them before they spoil? May keep the winter greens such as will keep, and the hard kind of apples for Mr B. if ever he send. It will not be worth his sending if he does not send soon. You have never told us how Misses Dent are lately. I do not know how to advise you about my Moor farm,[106] as to plowing any of the lee. It is so very much run out, but as it is annexed to Tod's farm and the 2 fields which you got from Mr Birkbeck, better perhaps plow some of the good land and lay off the moor with white clover and ray grass so as to fleshen a little. Little profitt can arise from poor worn out moor lands. This you will be able to judge off better when you are better acquainted with the land. You talked of ploughing Limekiln field and the cow pasture, perhaps it may be right to plow them but I am hardly master of the subject.

Story seems to want ideas. I wrote to Nelly by him, in which he saw me inclose 10£ for her to be left with her at Newcastle. We had a letter since from her dated the 8th in which she takes no notice of him or the money. Do ask him properly about the matter, as we are in suspence, as he said he would be in Newcastle soon but had to call at Rothbury. M.C.

---

106 An area of some 55 acres adjacent to Denton Grange farm: DUL, DDR, Tithe
     plan, Denton.

1802

103. *George Culley and Matthew Culley jr to John Welch*

Eastfield 26 January 1802

Well John

You have disappointed us in not writing according to your
promise. I suppose the Panshaw man won't buy the butter, but you
should have acquainted us because we must sell it somewhere if he
will not buy it. Indeed I was in hopes that you would have wrote
from Morpeth and told us how the market was. However we have
had a busy week I do assure you. Mr Lincoln came and bought 100
ewes at 3£ 3 per, and 15 pigs at 50s per. A Berwick butcher also bought
a cow and an ox of us at 8s per stone sink, but he would not touch the
shot ewes at 8 ¾d. We have also let the Grange farm to Joe Mole
junior, so you see we have got through a good deal of work. And to
crown all, it has pleased the Almighty to send us some charming
fresh drying weather, which will set us sowing wheat if it continues
any time. But I am sorry to tell you one piece of melancholly news.
Poor John Potts[1] was drowned in the river Bowmont the last
Thursday night, in consequence of having got too much liquor. He
had stopped at Downham after coming from Wooler market, and
although Mr Forster sent a servant with him, he when near his own
mill rode away at full gallop and told Mr Forster's lad that he was
going to Paston. The lad had presence of mind to go directly to Mr
Potts's steward, but before the steward got away Mr Potts's horse
came home. The river was so much out that none except a madman or
a drunken man would have attempted to ride her. He was found the
next morning opposite to Downham House where the river begins to
run broad and shallower.

I know not how your corn markets are John, but ours are very flatt
and dropping except wheat which stands its ground the best. Oats
are rather lower, and barley is *done*. We have missed our market
entirely with that grain for our little we have, and I would really
advise you to sell a part of yours if you can, if even you should take
less than 6s, for I am apprehensive that barley has seen its highest
price. Perhaps your market for barley may not be so flat as ours, I
wish it may not, and in that case let me advise you to sell by all

---

1 John Potts of Mindrum Mill, 1763–1802, DUL, BT Carham.

means, you may depend upon it wise and prudent so to do. Perhaps oats, if the weather keep dry, may hold their price, but I have the best opinion of wheat of any grain, because I find it only yields midlingly in most countries. I don't intend to send away this letter untill we receive one from you except you put of longer than I can either suppose or wish. I think turnips are damaged a little with the frost and changeable weather, especially the old sown ones, but I hope this fine fresh weather will set them a-budding in the tops, which will prevent them rotting any further. At any rate I hope the late sown ones pretty safe, and able to serve until seeds begin to spring up for the sheep.

[*M.C. jr*] We reckoned the ewes on turnips 25 lb. per quarter, but Mr Lincoln said they would not weigh more than 20 lb. The pigs are sold for near 9s per stone, or between 8/6d and 9s. The ewes were not so good as they were last year, and are 13s per head now. They all go away on Monday week. We intend sending your books by Matty to Houghton, and he must forward them to you if he can.

[*G.C.*] *Friday 29th January.* I heard yesterday at Wooler that Mr Nisbet had sold part of his ewes on Wednesday at Morpeth at 3£ 6s per head, that the market was good for both sheep and cattle. The latter 8/6d per stone sink, and sheep about 9 ½ d per lb. sink. These are strange prices for mutton, and there is little doubt but cattle will follow. Sheep are really falling scarce in this district, and no wonder when such numbers were taken from us the last autumn. The thing that surprises me most is that Mr Sayle should say there are fewer sheep at turnips in his country than usual. What in all the world is becoming of them? By yesterday's London paper all grain but wheat dropped in price last Monday. Wheat advanced 2s per quarter. I now begin to think that we have missed our barley market compleatly, and I must repeat that my advice to you is to be selling, even provided you abate more than you once expected. But you are better able to judge of your own country than I am. However when barley drops at this season, and continues to do so for some weeks which it has done lately in London, I consider it a bad sign, and London is the main standard to judge the whole Island by. Clem Stephenson was at Wooler yesterday, very keen of either cattle or sheep. We have had very nice weather ever since the thaw came on, and the snow is now very nearly all gone by the hedge sides. I dare say you would have a good deal to do on your return home, but you had better have wrote a little letter than none, because as you said you would write from Panshaw at all events it keeps us in suspence. But I hope all is well with you. I have deferred sending this away because I still am expecting a letter from you every day. However I will certainly send it from Berwick tomorrow after market, that I may give you what information I can. Ralph Brown is just come from Kelso. Barley quite done, nobody will buy any. I saw Tommy Bates who is comed to Wark to see my brother, and says nobody will buy any barley at Newcastle. How things run into extremes. Ralph says a man of Kelso was bidding 9/3*d* per stone for pork today, head and feet off! I never knew that price before in this country. Don't neglect to write about the butter by any means, because we will send it to London directly if you have heard nothing from the Panshaw man. I am yours in haste

Geo Culley

Berwick 30th January 1802. Well John I met with yours of 29th as I was coming here this morning, which I think must be a mistake because wrote yesterday could hardly be here this day without being put in at Darlington yesterday. John I could have wished that you had sold a part barley even at 5/6*d*. because I am persuaded that you will not meet with the same offer again these 2 years. However you did it for the best and I will not say one word more, but that I think it is always right to take a good price when going for a part. I have seen a man today who says barley will still be much down, therefore I would still advise you to be doing if possible even below 5/6 or perhaps 5*s*. However you will remember that I only advise, I don't insist, and you must be better able to judge of your country than I can. Wheat as I said before is the only grain at present that is selling, and is very likely to keep its price. I think you have a right to expect 9 ½ *d* per lb. sink for your wethers. There is no fear of them coming down, and they may advance. At least I think you should be stiff at 9 ½ till you try and hear more about it. At any rate people here are talking about 9 ½ and even 10*d* with the same confidence as they did a while ago of 8*d* sink. But every cord may be too much stretched, and I would have you content with 9 ½ to good men. But whatever you do, have ready money. I am glad the Whitby butchers are looking out. They will drive the Sunderland chaps this road. Lincoln wishes to handle the Wark cattle, and I will write him word. We have sold Lincoln our pigs far too low, but I wish him luck. I am very well satisfied with 46*s* for the butter, but you must ascertain the time and the manner how they are to be sent to Chester. As soon as you are certain of the quicks, be so kind as write and we will send immediately. I dare say Matty will have no objection to send the horses you speak off. I will speak about the Mazagan beans.[2] But you surprise me about the

---

2 A small early variety of broad bean. Bailey and Culley (*General View,*

ewes being in lamb that Rimmington bought of you. I will have it investigated. It is quite unknown to us! I will add about market by and bye.

Berwick market I have just time to say is very dull for everything but wheat, which is a little advanced. Ralph sold for 52s which is 2s more than last week. Geo Culley. John Nisbet says he got no more than 9d per lb. sink for his ewes.

104.   *George Culley and Matthew Culley jr to John Welch*

Eastfield 7th February 1802

John Welch

We received yours of 5th this morning, and the cart and firkins[3] shall be at Chester by Thursday 11th as you request., if all go well. John I don't say it to blame or mortify you or us, but to make us wiser and more attentive if we are well informed. Butter firkins are worth 55s at Newcastle this moment. We had this from Mr Bates of Brunton who was at Wark lately, only went away the other day. By Mr Nisbet's account cattle are better sold at Morpeth than Darlington. He says he sold as near to 9s per stone as could be, and he seldom gives too much weight. But he says he only got 9d per lb. for his ewes and very good, and perhaps you ought to have taken 3£ 3s for your 150 wethers, it is a famous price. Not that I am affraid of prices dropping at present, but I am affraid of these Wakefield or rather Leeds jobbers paying you in two months or long-winded bills. I know Jere pays nothing else. Pray don't take them except bankers will take them without your *indorsement*, because if you *indorse* or put your name under them you are liable if they fail before the time runs out that the bill is drawn for. If old Clayton was there I would rather have you deal with him than his son Ben who I know little of, but I have heard that he was a wild young man. Indeed if Jere Clayton's name be upon the drafts I would have very little fear, because I believe Jere is able enough. You must not take it amiss, my repeating these matters about money payments so frequently, because I wish certainly to impress you with the greatest caution in receiving your payments in *notes*, not *bills* or *drafts*. We have so many failures nowadays. I am much of your opinion that mutton goes far, and will not get higher than 9d or 9 ½ per lb. sink at any rate. One thing must be kept in mind, that *offals* of

---

Northumberland, p. 89) said that it ripened earlier than the large horse-bean but yielded less well.

3   A small cask for holding liquids, butter etc. Butter was usually sold by the firkin.

sheep *never* were worth so much money in my days. Well after saying this much to you, I leave it entirely to yourself to take such prices for your wethers as you think right. I am certain you will do the best you can, and let us be content, the prices are such as we can afford, and such as we hardly ever knew before. Ratcliff must be very quick in his motions. He was in this quarter the other day, and bought 8 steers of Laird Burns, which saved the laird from going to Jack[4] at present, but it must be his fate soon if he don't alter his ways. I agree with you that it is right to quit our remaining ewes, only the hill ewes are not quite so fat as they should be. However we will think about it, and I hope with you that the fat cattle will get up, as the ships must be supplyed. As to barley we must be content, both you and we did for the best. It may stir a little but I think all grain will come down towards spring. You have sold your last wheat well, and I would advise you to be doing, for depend upon it it will be no higher but may be lower. Indeed the very high price of butchers meat, which precludes the labouring classes from buying it, will make them use more wheat in dumplings, bread &c &c. We have a man from Scarborough who bought 199 bolls of pottatoe oats yesterday of Mr Nisbet and some of John Mole at 23 and 24s per boll. He will buy a ship load, and many hog buyers are out already from Yorkshire. I am glad you have got Church field sown with wheat. We have sown a good deal, but the weather is turning frosty again. In regard to John Todd I am for being clear of him to a certainty, but will say more when my brother comes tomorrow. You were right in paying for your corn tithe. I do assure you that I think George Nicholson's ewes cheaper than Lincoln's, but I wish them both luck. George Nicholson sent a very little fine boy for his ewes last week, which looks as though he needed them. We sent him to Mr Nisbet's, and a boy of Mr Nisbet's sent him on to the low turnpike and hope he has got well away poor fellow. We will order the man to call at Mr Charmley's for your day book. Matty Culley put your shirt &c along with your great coat into a box which will come by cart, but you should write to Houghton if Matty has not sent your books. Your account of small pigs is astonishing. I have no doubt of Birkbeck's old lime kiln field proving a nice plow field for turnips &c. We will speak to my brother about Mazagan beans. I would blame you much indeed if you don't go to Allerton fair, business or none. How are people to learn information but at fairs and markets? I forgot to say that it appears that Jere's son has given Mr Davison above 9d

---

4   Going to ruin.

per lb. I hope you will sell your few ewes at any rate, although it does not seem as though Rimmington was affraid by coming again, of their being in lamb. We can't find that any tups were amongst them, except when the lad Glass drove the tup with them to your place. We have killed many ourselves, and never one in lamb. But if they were by the tup driven with them they could not be many weeks gone in lamb. How it happens they are with lamb I know not, nor can we find out yet. It may be known afterwards. It is said there are the fewest farrow or dry cows this year ever known. It is owing to the last fine summer and autumn, made them all lean and bold. Consequently they will be scarce and dear, and will be a means of making fat cattle scarce and dear through summer. It is said wool is to be uncommonly dear at clip. I would like to clip our own wethers, but I am of opinion they will be so fat as to be scarce able to travel to you in their coats. Perhaps after all best to sell them, but you will give me your opinion when you write again.

Monday 8th February. I have wrote to Mr Lincoln this morning per post, as he expressed a wish when here to buy James Grey's remaining ewes about 60, to come and handle them and my brother's oxen at Wark, which he also wished to buy. I have given him until 7th March, and we will be selling remainder of our ewes at Morpeth in the mean time. I am thinking of going to Morpeth on Wednesday week with a lot. John I can't help thinking that there is a probabillity of fat sheep being as high sold here this spring as with you, if not higher, owing to the scarcity of sheep in this district. Such numbers went south the last autumn, and if that should prove to be the case there is no inducement to our wethers going to you, but the clipping them. To be sure dinmonds, or rather young wethers, do increase in bulk and fat upon clovers in the spring of the year. But of this you can consider and give your opinion. I am of opinion that the 100 wethers we have at this place are the fattest at this moment we ever had any, and those are our middle lot. They have 50 at Wark that should be much heavier, as they were picked ones. The offal of these weathers I consider to be worth 15 or 16s, which is 2d per lb. for offal. Therefore butchers can afford 9 ½ at 8d per lb. for naked mutton, and 10d where they sell at 8 ½, which they now do in many places. George Nicholson's little boy told me that they sell at 8 ½, consequently George Nicholson's ewes will leave him 12 or 13s a head at that rate. Don't think by this that I begrudge him his profit. God forbid! I wish him great luck, it happens on the right side. Besides the loss in weight is against him, in cutting out some little especially where they make so many cuts in a quarter. It is necessary that you should know that my brother gave his son Matty Culley an order on George Nicholson for some money to pay his school master, but my brother has forgot how much as he did not make a memorandum of it at the time. But George Nicholson will tell you. It will be somewhere about 30£. But I think John that you should get settled with George Nicholson and Rimmington &c. There is nothing like a short account in your way, and in everyone's. It prevents keeping so many accounts and does not oppress your mind so much.

[*M.C. jr*] In regard to the paying of John Todd we wish to be entirely clear of him, and would not wish you to pay him any money untill the final settling. Only if he is much in want of money at present you may let him have 100£ but not more at any rate. We are not tyed to pay the half at Candlemas, neither are we liable to pay interest from that time forward.

[*G.C.*] *Tuesday 8th*. We had our rent day[5] yesterday and received in full. Mr Lincoln's man came last night, and Ralph is going to set him of at Grindon this day. He will not acknowledge that they sell for any more than 8*d* per lb., but says that beef is 3£ per cwt. took place ten days or a fortnight ago. I hope you will understand what Matty has wrote above concerning John Todd. If any difficulty occur to you, let us know in your next. We don't mind any trifling benefit we might gain by the money continuing in their hands at interest, and we want to be perfectly quit of John Todd. The less one has to do with troublesome people the better, after you know such folks. It is impossible to avoid meeting with such in the course of business, but the sooner you can shake them of the better. The horses and everything will be sent per cart by Jack Davison I hope. William Wright has hired himself to go to beside London to Lord Tyrconnel,[6] son in law to Lord Delaval,[7] an infamous set, all for not more wages than I would [have][8] given him except about 4£ per annum. A fine slow sober lad, and we all were fond of him, but young lads will take *freaks* as well as lasses. Now John *observe* what follows and don't *neglect*. You will receive 4 bushells of Churches oats in a bag by Jack Davison, and you will observe how they are directed, because you are to send 8 bushells of your best *potatoe oats* along with them and to be directed in the self same manner. If you can put them all into one bag (*I mean your 8 bushells*) it will be best. If not you must take two bags, and let them be

---

5   For the Akeld and Humbleton estate, bought in 1795.
6   George Carpenter, 5th Earl of Tyrconnell 1750–1805: *Complete Peerage*, vol. 12/2.
7   John Hay Delaval, Baron Delaval, 1728–1802. *Complete Peerage*, vol. 4.
8   A word seems to have been omitted.

good ones. The sooner they and the 4 bushells of ours go the better, and be sure you give Maxim or whoever they go by a very serious charge about them. If 2 bags both must be directed, and when you write to me say what the bags are worth as I will value the oats. And I shall write to Mr Deverill about them directly. We have more to send by sea if an opportunity offer, both to him and Mr Stubbins. You must write to Mr Deverill the day you send them, or the day after at the furthest, per post, saying when sent from *Darlington* how many *bushells*, and what kind of *oats* and how *directed*. Your letter must be directed in the same manner as the bags, and say that you were directed by your master to write. I think you will understand perfectly what I have said above about the oats to Mr Deverill. John I have been thinking that we have 37 outlaid cattle that will all have to come to you at any rate, and perhaps may come earlier than you may wish, at least one half of them. Therefore I think it right to prepare you, that you may have early grass laid in for them. You may be sure that we will keep them as long as we can upon turnips, but grass we shall have none. You have one comfort, that I hope most of them will not have to stay long in hand without times should run counter, which is not very likely. I am yours truly

Geo Culley

105.　*Matthew and George Culley to John Welch*

[Wark n.d.]

John Welch. Wrote before you were here.

I think it wrong to give coals to Cape and other paupers, as it's a means of making them burn too many. If you have as hard frost as is here and changeable weather, it may endanger the turnips. This I apprehend the hardest winter since 1740, only the country is well provided with fodder of all kinds. Hay is the safest article for winter of any food I apprehend in general, as it's not lyable to be destroyed by frost as turnips are, but turnips, pottatoes and Swedish turnips &c are all necessary with hay and straw for stock. If hay be most valuable, then you see the necessity of watering land wherever there is an oportunity of using it. Then the farmers on the mountains are only in an unenlightened state, much behind some of their neighbors. It shews that we are only beginning to learn farming yet. If Sir John Sinclair[9] gets forward with his designs in Caithness, what a blessing

---

9　Sir John Sinclair of Ulbster, 1774–1835, first president of the Board of Agriculture, did much to improve his large estates in Caithness. Matthew Culley may have seen Sinclair's newly published *Hints regarding certain Measures calculated*

will he be to his country in general. He who makes 2 grains or ears of
corn grow where only 1 grew before, does more good to his fellow
creatures than he who conquers a kingdom. I am glad to hear that
you are making the new road over from Croffts. You have great need
for an overseer to assist you, especially when you go from home and
have many irons in the fire. I fear you may injure your health by too
great attention and going about. You must have people at the head of
some of your workers. Bailey suits the quarry men but then who rules
your labourers? How goes on your miller? I was thinking, as
Monkhouse is not a miller, that Story might grind the wheat he buys
of you, to the purposes which he wants it for, and allow him for the
bran. But the bran must go with the flour. You may begin with a small
tryal, and so get on by degrees with him once Story gets into the
method with him. Monkhouse must have some trouble in geting the
corn grinded properly. When you have plenty of water it may employ
Story to advantage. This Story should consider about himself. How
does Kirtley manage the watering of the land, or is any other of your
men seemingly more attentive or knowing of the business? This
weather suits watering better than hedging. The storm makes us
unable to keep our men properly employed. Wishing you and yours
many happy returns of the season, and good health

Mattw Culley

I think the Richardsons may be best with some of your men to
make a bridge below the box in the carr (a conduit). Lay your foundation sufficiently low as there wants a graff[10] up the carr from the old
conduit to where the old water course was, in the foot of the corn carr.
I was thinking, if you could bore flatt ways in to the draw well at
High House when you get the open drifft as far up as you can, so as to
save mining. But if you have to mine, get a pitman or sinker to do it,
who have more caution and knowledge about such things, and will
run less risk and have proper tools to mine with.

1802 February 8th. We are watering some land at Thornington in a
proper manner, under H. Rutherford's direction, which you must see

---

*to Improve an Extensive Property, more especially applicable to an Estate in the North-
ern Parts of Scotland*, London 1802, in which he set out what he intended to do
by agricultural improvement and fostering trade and new manufactures, fish-
eries and harbours. See also Sinclair's account in the following year in *Annals of
Agriculture*, 39(1803), pp. 465–78; also *Account of the Improvements carried on by
Sir J. Sinclair . . . on his Estates in Scotland*, London 1812, extracted from John
Henderson, *General View of the Agriculture of the County of Caithness*, London
1812.

10 A trench.

next time you come here. My brother fears much they will be short of turnips. We can relieve them and other places I hope. Pay some attention to your watered lands, and boring near the tub in the carrs. My brother and Matty let Easington Grange to Mr Mole's son of Barmoor Mill (the youngest sons at home) for 500£ the 2 first years, 570 the 3 last years, with easy covenants to them.

Mattw Culley

[*G.C.*] My brother brought this down yesterday the 8th inst., but most of it except a little at the last, was wrote long ago. However we will send it as it will cost no postage.

106.   *George Culley to John Welch*

Eastfield 26 February 1802

John Welch

I have been absent from home 10 days in the Tyneside, but was glad to meet with yours of 18th inst. on my return yesterday, which I will answer after telling you what a strange market we had at Morpeth on Wednesday where 60 ewes met me. There were not more than 400 or 450 sheep in Morpeth. I don't believe that I ever saw so few by 400 in my life. Nay I have often seen 1000 more. Nevertheless we had a bad market compared. There were plenty of butchers also, consequently we every right to expect a particular good market, but it proved quite the contrary. The only way it can be accounted for is that the butchers declare that they have lost so much that they had better do nothing than lose on. And it is a fact that few sheep as there were, 40 or 50 remained unsold at 2 o'clock when I came from Morpeth, and I understood at Wooler yesterday that a part were not sold at last. To be sure I sold some lots at the first to some men from Hilton ferry boat, and others from Wearside, at 9*d* or better, but any who sell at Newcastle would not give that price. They say that much mutton was sold at 6*d* per lb. in the evenings lately at Newcastle and much salted, and I sold my last lot or 2 at little more than 8*d* rather than set up. Yet I might make about 9*d* or nearly over head. A strange quantity of midling cattle, stots, cows and queys, 4 decent steers of the Teeswater, 3 or 5 yrs old were sold at 8*s* or a very little more, but many at a deal less and a great number unsold when I came away. Your last letter has made my brother keen of buying more cattle for you, but you know I told you in my last that we have near 40 outlaid cattle of different kinds which must go to you, and I am affraid that if we buy more we shall scarcely be able to keep *them* and *those* we have, until you can take them. Now Morpeth is the best place to buy in at present, especially such days as Wednesday was. But then you can't take them at present, and it would be aukward to drive them north and then south again. However it is the only place, because you can buy some outlaid steers there, but no such thing in this country, and naked cattle out of a warm house are long of turning. At any rate from present appearances cattle are sure to be lower *now* for some time than was expected, and in all reasonable probabillity very high in May and June &c, because by the Tyneside where I have been the turnips are entirely rotted. Pray say how yours are. I was rejoiced when I came north yesterday to see the turnips green and tollerable, but not any alive to the south of the county. They look like a dead half eaten pasture, scarce a green top to see, and fall out of the cart with a *plash* like soft muck! Mr Nisbet who came with me said they might *sup*, but they could not eat them! I have not seen them so bad these 30 years. Mr Armstrong and another man have been buying some Kyloes &c in Berwickshire and East Lothian. The other man Carross (I think is his name) has bought inlaid ones, which must drive some in and look sadly before he get them to York. However it will do good, and keep up people's spirits a little, and I cannot have a remaining doubt but fat will be very scarce and very dear afterwards with them who can weather the storm. There never were so many and such good turnips in no man's time, also very reasonable to buy, but now can't be had for money. I don't suppose Mr Lincoln will now buy either sheep or cattle from us, consequently we shall have enough to do to weather the storm with what we have until you can take them, which can't be till grass rise. Ralph Brown and my brother are gone to look at 8 cattle of the late John Potts's. They are housed, and I can't say but I wish them not to buy. I should have gone, but my wife was taken so poorly that I could not think of leaving her. Besides we should clip what weathers we have now, both you and us in my opinion, because wool will not be below 15*d* per lb., perhaps more. I really would advise you to clip your few shot wethers, put them in your *carrs* or any of the worst of your watered land. They will do very well, and will not rot at this season I believe, but if a little tainted it will be of no consequence as you will sell when clipped. A very midling fleece must be worth from 7*s* to 9*s*, and if you can only put them of any how until your grass rise, so as not to sink in condition, it will do. As you have not many, you can put a few in a place. I have known the Mr Taylors and Joe Clark put fat sheep upon Killingworth moor in the spring a while to gain time, and upon heath or ling do *midlingly* a while. I am very glad that you keep selling corn. I am still of opinion that it will be lower, certainly not *higher*. You were right in selling your horse. We had heard that there is a demand abroad for light gay horses, which is very lucky.

Let us hear from you, pray, as soon as your first Monday in March is over. We shall be anxious to hear how it is, and I am affraid it will be flat. However, whatever it is, I am glad you sold your sheep. We will certainly buy cows if we can, but they are very scarce indeed. Be sure that George Nicholson, C. Charge, Shillito, Rimmington &c &c all pay you before April, as the money will be to remit south soon after that. But don't let that hurry you in selling any other stock, because we can raise the money many ways by banks &c, which money made by the stock in hand will repay again. I only wish you to collect all arrears. You sell maslin higher than we do wheat! I am highly pleased with what you have paid Mr Peacock, which I shall write to him about in a few days. Only be sure and take a memorandum of him for every sum paid, and keep a regular cash account of it. It was right to let Mr Harrison have the piece barley, near hand and safe money. I wish you had it all sold at that price. I would have you put of your steers as long as you can, as markets will and must advance by and bye. Give them turnips for water if new pinched, and part straw and a little hay so as to keep them where they are. I am glad the roaned steer proves so well. Good ones always pay the most for keeping. Give yourself no pain about the butter, only let us remember to be better informed in future. However be sure you get your money by April or sooner. I wish you would remember to write to Mr Deverill when the oats went. If not write when you receive this, saying when *sent*, how *directed*, and what *quantity*, and let me know the value of sacks and what price you think of charging for each sent, but you need not name that to him. By all means get the 20,000 quicks if you can, and run them at Denton. O pray John, don't neglect to procure and send us 2 or 3 measures or bushells of the best rape seed you can meet with. We forgot to name it before. Perhaps you will hear of some in Yorkshire. Ralph Brown is returned from looking at the cattle of Mr Potts, and my brother and he have not bought them as they asked far too much for such midling stuff. As I propose sending this to post per Ralph in the morning I shall conclude. Yours sincerely

Geo Culley

PS. Don't neglect to write from Darlington on Monday when you will probably receive this letter, or rather the next day or as soon after as possible, as you will be much busied on the fair day. This is a most wonderfull fine drying day, which if it holds by a day or 2 will set us all to work sowing spring wheat, beans &c. *Saturday 27th.* I have been thinking John, as cattle are so hard to come at, and likely so to be, whether we had not better buy a few hogs forward in condition, as wool is likely to be so well sold, and you can either sell them again in the autumn, or winter them upon turnips provided you are well laid in. However you can turn this over in your mind and give us an answer about it when you write again. We are inclined to think that they may be bought at present with a greater likelihood of paying money than cattle. Be so kind as let us know what money is in the Darlington bank and how much you have paid Mr Peacock when you write again, also how much you think you may be able to raise by the 1st of April exclusive of the Durham bank, as we must then think of remitting our money to London for the purchase of the Grange.

107. *George Culley and Matthew Culley jr to John Welch*

Eastfield 3 March 1802

Well John

I expect a letter from you tomorrow in answer to the one I wrote to you on Saturday last, and giving an account of the 1st Monday in March. We could not wait longer about buying you some hogs, although I wished to have your opinion about it. Ralph saw a lot of good hogs of Mr Wood's of Preston at Thornton, so Matty and I looked at them yesterday morning early, went to Wark, took my brother with us to Mr Wood's and bought them, 150 at 2£ 2s 6d each and 20 at 30s each, a strange price and but little hogs, but I never handled any so fat before. If they can be kept going on (but they always sink when losing their teeth) they will make very nice sheep in the autumn, either to sell then or keep as circumstances suit. He is to keep them until the 1st April, when they will go directly for you, so you must be prepared. I did not wish them on any account to be longer kept on turnips, and I am certain your moors at that time are better for them than turnips. But seeds or good grass is certainly better. I only speak thus strong to shew you my opinion, that moderate grass for hogs in April or even perhaps sooner is better than turnips, on account of their *teeth*. I also name this on another account, because if our fat wethers are to go to you to be clipped and sold, the same driver (or some other) must immediately on his return north take away the wethers to you, and I have repeatedly told you that we have 37 outlaid cattle which must go to you, besides 7 or 8 cows which will turn out to grass very well I think, and are too backward to be sold before grass time. We can at present meet with no cows or forward cattle for you, which was one great inducement to buy Mr W.'s hogs. Perhaps they may pay as well, and I think what

[with][11] them, the wethers, and 45 cattle will not be a bad spring stock for you. Besides you will have to take our hill hogs I suppose as usual. Now I have been thinking that it may be right for some other driver to keep meeting the other upon the road. But this is an afterthought which you as well as us can consider of. There will be 3 droves of sheep and I fancy as many of cattle in the spring. You may answer that these cattle should all be sold by the end of June. I admit it, and perhaps you had better keep your land uneaten untill the autumn, that other stock can be promised, or a dry time may happen. Fat stock are always better than lean, because fat can always be sold at some price. Besides all this stock you have your little Kyloes, and it is great odds that either *you* or *us*, before you want, may fall in with a lot of 2nd hand cattle of some kind. If you can meet with any at Darlington or in your neighborhood, don't baulk yourself. Only always keep in mind that it is better to understock than overdo it. As you have your tithes taken, you can lay in a field ot 2 for hay later on or do many ways.

X[12] In regard to money matters, I wrote to you in my last to get the butter money, George Nicholson's &c &c all gathered together, and if Mr Peacock will not take it, put it into the bank. But tell Mr Peacock when you see him that I am come to a *determination* not to remit the money to London for the Grange estate untill the 1st of May in bills at 40 days. But I will write to Mr Peacock in a little time, only I expect a letter from him in the mean time. But I would wish you to read this part of my letter to Mr Peacock the first time you see him, and tell him that I will give him my reasons when I write to him, which indeed are these, that we can only pay discount at any rate, and it will suit me better then than sooner. X

John be making enquiry about wool, how it is to be sold. We have got it into our heads here that it is to be from 15*d* to 18*d* per pound, and that is an inducement for our buying these hogs, for although they are fine wooled they will clip a good deal, I hope 7 and 8 pounds a fleece. John yours will be a valuable clip in quality, and not a very small quantity. I saw Joe Pease junior[13] on Wednesday at Morpeth, who desired I would enquire for any old clips and write to him, but there are no old clips here. He seems uncommonly keen, said 'it would be dearer than *thou* ever knew it', but did not say what. Since

---

11 A word seems to have been omitted.

12 The following passage, to the next 'X' is circled.

13 Joseph Pease 1772–1826, brother of Edward Pease, known as Joseph Pease of Feethams.

Wednesday's rain we have had excellent weather, and this week I hope will finish our wheat seed, and then we will get to work with our beans next week if the weather hold decent. T. Tulip sowed a large field of oats yesterday. Now I cannot help repeating that 600 hogs and wethers, and 45 head of cattle will fall upon you pretty full at once. To be sure the 200 wethers and most of the cattle should not stay long with you, and you have the Kyloes you bought. The wethers are so very fat and full of wool that they should at any rate set forward from hence in the beginning of April. Will your clovers be fit for them then? They will take 10 days at least I believe to go from hence to your place. We would keep them longer, but if the weather should come warm it would do them much harm. Do think of this and give us your opinion. Perhaps we had better keep a part of the cattle later than injure the wethers. Now John this leads me to another question. Where do you wash your sheep? You should certainly pave a washing place of your own, and the water ought to go upon the grass land when let off. However at all events it would punish these fat sheep far too much to drive them to the Teese and back. I find my brother and sister are going to Denton this month, and my brother is a proper person to advise you about a sheep washing pool. It is highly necessary for you to have one now, you have and will have so many sheep I hope to wash every year. My brother and sister talk of being at Newcastle about the 17th or 18th inst., and after that will be with you.

Friday 5th March 1802. John we fully expected a letter from you yesterday morning but got none, and this morning I sent to the post again and am still disappointed. I know not how to account for this neglect, after so important a day as first Monday in March, but will say nothing until I hear your own reasons, because many things may have prevented you that I cannot foresee. However I have not often been more disappointed. Perhaps I feel more at this moment because I have now got into my mind the payment of the 6000£ at London, and as the money is to be paid on May 12, the bills must be remitted on 1st April so as the 40 days are run out before *12 May*. I am under no pain but I can accomplish the business with your help and Mr Peacock's, and therefore you must be sure to pay *every shilling* you can possibly raise to Mr Peacock (without straitening yourself) on or before the 1st of April next, as he is to remit whatever money you or he can come at *of ours*, either in the *banks* or other places on that day to London, which I expect both you and he will give me an account of when so done. I have told Mr Peacock in a letter which I have wrote to him this day, 'that I would give you orders in this letter to pay him all the money you can possibly collect together on or before the 1st of

April' *without straitening yourself*, which you are never to do because your agricultural operation must be carried on smoothly or it will not do. However I would advise you to sell what grain you can receive for before that period, as corn still sells very well and may sell worse, while fat cannot at present be turned into money. But they that can weather the storm will I hope still sell high enough, and we must now do the best we can. These sudden changes are good for us. It awakens our latent powers and gives a fresh spring to *our actions*. Mr Peacock luckily said in the bottom of his letter which we received yesterday 'that the market was full of fat stock at Darlington and they went of heavily'. That needed no prophetick powers to forsee, as the turnips have failed so much in the south I understand, but pray is that the case because we have never heard you say so. But I know that to be the case in the south part of this county. Now John, as you will still be raising money by corn you can run yourself the nearer to assist Mr Peacock on the 1st of April, and I am certain that you will leave no stone *unturned* to help *him* and your *masters*. I have desired you in a former part of this letter to get all your money together that is due to you for *butter*, for *sheep* and for grain &c &c. Therefore pray set about it directly. It is very important I do assure you at this moment. When my brother comes to you, make him steward while you ride about collecting.

X[14] Since writing to you in a former part of this letter to *shew* or *read* that part to Mr Peacock I received a letter from him which has decided *my opinion*. That is to pay the money in bills on the first of April, which I have said a good deal about above. Therefore you don't need to read that part to Mr Peacock without you chuse, as your *instructions* now are to get all the money collected you can, and give to Mr Peacock who will remit it to London on 1st April. X

A very hard frost came on yesterday, and today a slight cover of snow which will rather impede the sowing a little. I will not finish this until tomorrow at Berwick, and hope to meet with yours by the road. Could I only prevail upon you to have a letter wrote *in part at different times*, I should be glad and you would find your account in it. Besides, you can *explain* and *remember* things much better by writing at different times than all at once, and much *easier* to you, and *simpler*. Only try this for once, and oblige yourself and real friend

Geo Culley

---

14 The following passage, to the next X, is circled.

[*M.C. jr*] PS. We do not wish you to meddle with the 340£ of yours which my uncle put into the Durham bank in November, as we mean to let it lie till 12th May when we shall receive 4 per cent for it. If you have any money in the Darlington bank take care to let Mr Peacock have it to remit, along with your other monies. Be sure you send us an account of what money you can raise by 1st April.

[*G.C.*] I wrote to you last week, I think the letter was put into the post at Berwick by Ralph last Saturday. I hope you received it last Monday and will receive this next Monday. G.C. *Berwick Saturday morning 6th March*. Well John I had the pleasure of meeting with yours of the 4th this morning as I came here. I am glad you have embraced the good weather to sow wheat in, and are sowing beans as we have a change but pretty dry frost yet. I am not affraid but markets will mend for cattle and sheep both. I would have been as well pleased if Shillito had bought your cast wethers, but sell them if you can but not below 57s. You know they will not improve like tups or better sheep. You talk of rye grass fit to keep sheep. Had we not better send you our 200 wethers while the weather is cold and roads dry? I am terrified that they will not drive without much hazard in warm weather. Think of this, and give us an early answer. Besides it will enable us by that means to keep more fat cattle to a good market I hope, at least it will give us a good chance. The hogs will drive any time. I really do not know how many hill hogs we have, but you need to take no more than is agreeable to you. If Mr Wood's hogs 170 only keep going on and get well up, I think they will please you. It is certainly a great price, but I saw the Yorkshire man on Thursday that bid 42s for them, and he said it was true, and that they were the fattest hogs he ever handled, and so say I. I am sorry that James Kerr says that they are not clear of scab, but they dressed them all over and since that none have broke, and this is honester than saying they are and had been clean. We will take care while with us and James will, and you can always mend a scabbed sheep. But they appeared quite clean at present. Expect no fresh cattle as we can't meet with them in this country. I don't wonder at George Hopper and J. Bates selling even at 8 or 8 ½ per lb. rather than give them corn. They have nothing else. I am affraid you will not be able to sell your few shot wethers now for a while. If you can do, or rather let me say what seemeth best to you, I have no notion of limiting any man of sense. We will not omitt buying Kyloes or cows if to be had, you may depend upon it. I do wish you John, to raise what cash you can by first of April to give Mr Peacock. You must sell corn now, cattle and sheep can't be sold at present, and you are selling corn wonderfully well. Depend upon it that there is a deal of grain in the farmers' yards still, and let it be as it will, these are prices not to be refused nor repented of. But if corn markets should also drop, we might then repent. I have ordered Ralph Brown to sell wheat, which is the only grain that sells here at present. I am very glad that you have so good a market for your potatoe oats, sell away if you can. We have sent Mr Deverel some more by sea from this place, but I am affraid we will not get all ours sold for seed which we intended. But we will sow the more, I like them well. Respecting John Todd I think you can still raise money for him after 1st April, but my brother will be with you very soon, when he and you can talk over and settle these matters. Notwithstanding the advance in barley in London, let me advise you to sell when you can decently. Not that I think it will drop any more. Mr Maynard's must have been capital oxen. Ralph Benson's hogs would be much bigger than Mr Wood's, but not fatter I think. Do collect all your bills for money to give Mr Peacock, and let me hear from you soon. Don't mind whether the letter be long or short, but I want much to have you keep one on the stocks. Have you any money in Darlington bank? If so how much? Let me know what cash you can raise by April 1st.

Geo Culley

108.   *George Culley to John Welch*

Eastfield 8th March 1802

Well John

   I think I gave you the fullest letter from Berwick on Saturday I ever wrote to you, but I forgot to say how you were to send the rape seed, as it will cost so much by hired carriage and will not be wanted until the middle or end of May. You had better rest it a while, as it is not impossible but some cart or some means may be hit upon. It is only sending it by the carriers at the last. If it could be got to Newcastle there are ships constantly coming and going from Berwick to Newcastle. What have you done or are likely to do with your potatoes? If you were to sell them to your butter merchant or Newcastle &c, it might be got to Newcastle. If so a direction to us must be put upon the bag or poke. *Wednesday 10th.* Such charming weather now, as will greatly forward the seed and vegetation, and perhaps *enliven* the markets again a little, and it strikes me strongly as being a nice dry time for sending you our wethers, which would enable us to weather the storm with our housed cattle. I am not affraid of the outlaid ones. This charming weather has very much the appearance of standing a while, and as we have had so much severe weather during the winter, it would seem as though we had a kind of reason to expect it. However His will be done that gave the weather. *12th March Friday.* I was glad to hear yesterday that there was a good sheep market at Morpeth on Wednesday, *9d* and better per lb. sink, cattle not so good by far, hardly 8s per stone. I have been thinking that perhaps we had better clip all or a part of the wethers here still, and you take the mere hogs, because the flood land will be good and forward at Wark and it is safer keeping wethers than hogs upon it. Wheat I see is advanced *5s* per quarter at London Monday, oats a little, barley sinking. But there is too much appearance of our going to war again with the French. However if it should be so, which God forbid, it will advance all kinds of provisions. Indeed I have little doubt but we shall have very good both mutton and beef markets by and bye. This letter will most likely come by my brother and sister, who it seems are talking of going south in 2 or 3 days. I never saw a sweeter morning. We are still sowing spring wheat, and will finish all or most of our beans today I hope. I am just going to Tweedmouth to meet Mr Bailey to finish that division business. Ralph is going to Kelso. Tolerable horses not capital were sold at Kelso the last Friday at 30£ per. Mr Vardy[15] bought a mare at very near 30£.

Monday 15th. As your master and mistress talk of going of about Wednesday for Denton I shall not send this until they go, without something happen which I do not at present foresee. We have had and still have very severe frosts, so as to prevent harrowing until 10 o'clock often. But being dry is fortunate, and upon the whole it has been pretty calm except a part of Friday and all of Saturday, when it was fit to *skin a flint*. We have sown no oats yet, but Wark folks have sown a good deal. Horses were exceeding high on Friday again at Kelso. Ralph sold wheat at 56 on Saturday, which is a great jump on 50*s*. But what with seed wheat and one thing or another, wheat is fallen very scarce and people begin to think that it is not plentifull at any rate. Barley much the same. I don't know whether I told you that we sold 100 bolls barley to Mr Gregson the other week, and he would gladly have us promise 50 more last Saturday at the same price 25*s* per boll. But we are too busy to do that. Oats still flat except the potatoe oats for seed. We are still selling at 25*s* and have disposed of a

---

15 George Vardy, Fenton, had been a pupil of the Culleys. He developed the 'Vardy horse', a medium-sized working horse, based on mares (probably black Shires) bought from Bakewell and crossed with local breeds such as the Cleveland Bay and Fell pony: John Gall, 'The search for the Vardy horse', *The Ark*, Feb. 1993, pp. 59–60.

good many, and Ralph sold potatoe or Churches at 18*s* per boll on Saturday last.

Tuesday 16. By a letter this moment from Ben Sayle he says it was reported at Doncaster 'that a man was gone north to buy *stores* for government, which has made a sudden advance on wheat, and barley rather better.' If this be true it will soon be known, and an advance will take place in all kinds of provisions. If not true it will fall to the ground like the story of our having received 20,000£ for old Pate Thompson of Wark and never gave it to him. It is amazing what a noise that has made. You know old Pate, and may remember that we tried to render about 1000£ or 1300£ at most, belonging to a son he had murdered in the Grenadas when poor Mr Hume of Paxton was killed by an insurrection of negroes. But unfortunately the money never came to hand, and we are thus rewarded with *evil for good*. But people who have got so fast forward as we have will have *enemies* and *envyers*. But we had better be envied than pitied. I need not say any more to you about getting all the money you can, and given to Mr Peacock by the 1st April, or at any later day he may name to you, because I am certain that you will leave no stone unturned in an honest way to do so. My brother will likely be wanting a little money from you when he returns to the north again, but you will always be able to do that I hope. Is the barley any mended in price with you? Mr Charge and Mr Cradock are on their way south again, but don't go directly home or we would have sent this by them. Pray write soon about what stock you will take from us, and whether hogs or wethers or both, as I have altered my mind rather about the wethers, but either may be done still as you most approve. Perhaps fat sheep may be as well sold at Morpeth, and there is some danger in driving the wethers but none of the hogs. If we were quit of part hogs we could keep our fat cattle a little longer perhaps. Mr Wood's hogs will come to you at all events in beginning of April. I am yours &c

Geo Culley

109. *George Culley to Matthew Culley*

Eastfield 30th March 1802

Well Mr Matt

We paid our rent[16] yesterday, and I hope we shall be able to pay the 6000£ to Mr Claridge in proper time also, and consequently we have no need to sell our fat stock through necessity, but only want of meat,

---

16 For Wark, Thornington and Longknowe, to the Chillingham estate.

and I would gladly hope that we can strugle with our housed cattle untill the 12th of May upon turnips, and if there is no change then for the better we must be content to take such prices as are going. We sent you a letter on Sunday in answer to your 2 kind letters which I hope you would receive at Darlington by 3 or 4 o'clock on Monday last. George Humble and the other stewards were here this morning as we were at Chillingham yesterday, and we are going (Matty and I) to Wark and Thornington tomorrow which is the first day I have had to spare of a long time. Except Sunday and this day I have not dined at home since you went away. Harry's wife and child have been both very ill, but we hope they are now in a fair way of doing well. Expect to hear from you again after Durham fair. This day and yesterday have been cold and frosty. We hope you will not neglect to pay Mr Thomas Charge (with our thanks) for advertizing our contradiction of the report in the York paper. Also you should be sure to pay Mr Walker and Mrs Hodgson of Newcastle for the same thing, and if you can recollect to thank Mrs Hodgson who prints the Chronicle in Newcastle it will be well done, for the following paragraph in her paper besides the advertisement for which we are very much obliged to her certainly. 'A more gross and nefarious attack upon the reputations of two respectable men was perhaps never made than that so long in circulation in this neighbourhood on the Messrs Culley of Eastfield, Northumberland. But the purpose of the monster who invented it is effectually defeated by the advertisement in this paper.'[17] This is in the part of the Chronicle where the Newcastle news is put, and must be Mrs Hodgson's own doing, or some of her

---

17 *Newcastle Chronicle*, 27 March 1802. The advertisement (which appeared also in the *Newcastle Courant* and the *York Herald* of 27 March) read: 'Eastfield 22nd March 1802. Whereas a base and scurrilous Falsehood has, by some Person or Persons unknown, been propagated throughout the Country, that Messrs CULLEYS have defrauded PETER THOMPSON, a poor Man, of Wark, of a large Sum of Money; which Report they have hitherto treated with *Silent Contempt*; but by Letters from their Friends in different Parts of the Island, finding that it has been spread far and wide, they think it necessary to contradict so base a Report in this Public Manner, and also to offer a Reward of FIFTY POUNDS, to the Person or Persons, who shall detect and convict the Propagator or Propagators of such an atrocious Calumny. MATTHEW CULLEY, GEORGE CULLEY

I, PETER THOMPSON, of Wark, do hereby assert that there never was any Foundation for the above Report of Messrs CULLEYS having defrauded me of any Sum of Money whatever, said to be due to me, on Account of my late Son in the West Indies. PETER THOMPSON X his mark. Witness, GEO HUMBLE.'

people, because none of the Newcastle people (our friends) could well know of it until the Chronicle and Courant were distributed.

Wednesday 31st March. We are just returned from Wark, the seeds are promising, and very capable of keeping sheep, so the tup hogs are to be put on today. The fat oxen look well, I never saw them out before. I will have ours turned out. Your cattle will be fit to be turned into a grazing field, or drive anywhere, but George will have turnips to keep them until May Day if not sold before. Harry Potts flecked ox weighs 96 stone I think. How stiff and stubborn the field plows below the Gallow Hill, I prophesy a midling crop. I could not have supposed it would have turned up so *sterile*. The flatt land at Wark has been plowed for too long and cropped too much. I am of opinion that most of your outfield has paid you more for cultivation than the *haughs*, and takes far less labor. Ewes half done lambing. We have still more orders for turnip seed, and George is thrashing all yours. We came home by Harry's, his wife much better. William Surtees, Mr Thomas Bates's mason, came here the other day to look and enquire about lime killns, in consequence of what Mr Nisbet said to him when we were there. We wrote to Mr James Bell, and sent Bob along with him. Whether he has gained any knowledge I know not, but he did not seem quite satisfyed with what he had seen. Mr Thomas Bates did not write by him, and I might have been from home. However I wrote to him and scolded him, but I despair of mending him. Mr Wood's hogs come to Grindon tomorrow Thursday. Friday will be employed in clodding[18] them, Saturday to Ford Hill, Sunday to Weetwood, Monday Glanton, Tuesday to Rimside Moor House. These are nice short stages. They will go more miles per day after that, but as I don't know their stages and lodgments I leave John and you to calculate. But I think it will employ them all the next week well. If John would have sent and met them it might have been very well, as the lad Tom Glass will be to return with the hill hogs directly, and after that the outlaid cattle will have to go in part to a certainty, not less than 20.

Thursday 1st April. Well Sir, I congratulate you on the definitive treaty[19] being signed at the last, but I cannot congratulate you on the drop of wheat, 14s per quarter this week, and 8s the last, must bring it to a level by and bye surely. I don't like these great and sudden rises and falls, they are often productive of mischief to corn dealers, witness poor Clunie and Home. However we have nothing to blame

---

18 Removing clods of earth.
19 The peace treaty of Amiens between Britain, France, Spain and Holland, signed on 25 March 1802.

ourselves for in this matter, because I believe we never sold so much wheat by this season in any year I recollect since we grew so much. And not only us but most people in this district, led by the same idea, viz. that wheat could not hold the price it began with, sold more than usual. Besides it raised so much money. The only thing I regret is that we did not sell more oats in harvest and immediately after, because it was totally out of question that they could keep at such high prices. I said all I could at that period, but was overruled. Indeed I had been wrong 2 years before, but often right prior to that time. And indeed if we will take it for a series of years we will find that *grain, oats* in particular, make more money at and immediately after harvest than at any other period. But remember that I don't repeat these as grievances, by no means, but only to impress my friends with a very important observation, in my judgement, and that we may profit by our mistakes. Oats are now no more than 12 to 15s per boll here, and scarcely saleable to the merchants at those prices. And now that wheat is so much dropped we can hardly calculate how low they may come, because there certainly are more oats in this country than all the other grains put put [sic] together. Indeed I am persuaded that there is less wheat and barley than usual. To be sure the time for selling the latter is nearly over, but I am disposed to think that wheat will not come, or at the least continue long very low, except a great importation take place, which is not very likely now from the drop, and more particularly if it be true that corn is scarce in France. However one thing is very certain, that is that if wheat come much below 40s, oats below 12, &c &c, that it will break or at least be very hard upon most of the late taken farmers. And more land will [be][20] appropriated to pasture and stock than plowing. We have luckily sold a very great quantity first and last of the potato oats, which have proved a benefit article to us. If prices of fat stock should come down, and horn and corn often go together, we shall be in a worse hole, but I think stock is still really scarce. However it may have more effect than we are aware off, and we must of a bad bargain make the best. I am now anxious to hear how Darlington and Durham fair was, and I am sure you will not be long of giving that information to your affectionate brother

Geo Culley

Friday 2nd April. Mr Robert Thompson told me at Wooler yesterday that he saw both you and your son at Durham fair.

---

20 A word seems to have been omitted.

Morpeth was a worse market on Wednesday than ever for both cattle and sheep. If our consumption don't increase there will be too much stock in this country now. However I was glad to hear that Darlington and Wakefield were both pretty good. I should suppose that the signing of the definitive treaty will encourage manufacturers. John Pratt bought Adderstone at 13850£, and Mr Bailey says Shotton will go to more than 6000£,[21] so I wish them luck that gets it. I dare say land that is property will increase in value much now.

Well Sir I have this moment received yours of 30th ult. last from Durham, and am sorry to find that you have not received mine last wrote, which to a certainty would or could have been at Darlington last Monday the 29th ult. in the afternoon, as it would go from Berwick on that morning if your friend *James Forster* and his clerk acquit themselves properly. But I do not know what time the post reaches Darlington, but I well remember that we were at York by 10 o'clock in the evening when Mr Bailey and I went in the mail to London. I shall send this, at least I shall *put it* into the Berwick post office with my own hands tomorrow Saturday, and it ought to be at Darlington on the Sunday afternoon. You must have left Darlington the last Monday before the post *arrived*, or before the letters were delivered, or you had neglected to make enquiry. And being at the post house one would have thought either you or John would have recollected. I have been thus diffuse not only to awaken your *recollection* respecting the arrival of letters at Darlington but John Welch's *recollection* also, because I have often found that John Welch has not received my letters regularly. Consequently it may be right in you to make enquiry (or John if you forget) at the post office respecting the letter I wrote last, and which to a certainty would or *should* have been at Darlington last Monday 29th ulto. I rather suspect that the Darlington people put your letters as soon as they arrive into the *Piercebridge bag*, or the *bag* that goes west, and then they lie ever so long at Piercebridge. At all events I am certain there are mistakes about letters to Denton. But none to this place, because you will observe that we have received yours from Durham this very *morning the 2nd April Friday*, and you could only put it in at Durham on Wednesday 31st.

---

21 Shotton was advertised for sale at the end of March 1802 (*Newcastle Courant*, 27 Mar., 3 and 10 Apr.). It was bought by the Selbys of Paston, who already owned West Shotton: *History of Northumberland*, vol. 11, pp. 186–7. The Culleys' lease, for six years from 1795, was not renewed. See No. 114.

Well Sir I am much obliged for your attention in writing indeed so often, and am glad that the sheep markets are better to the south. It may mend them here by and bye. But there were none above 8*d* sink I understand at Morpeth, so that it may tempt the Wakefield jobbers to come north still. Mr Lox you say, I suppose you mean *Maynard* was quite right to refuse bills, I wish everyone would do the same. I am sorry my quondam friend Jerry 'fed on Pate's story'. It argues a bad heart. We are much obliged to Tommy Charge and his father and all *our numerous friends*, which I am glad you *have realized*. Long may we have friends, and long may we deserve them. Pray don't forget to pay Tommy Charge the expence of advertising at York. It was in the *County Chronicle* yesterday, so that I hope we have checked the flames everywhere now. I hope you have had good weather lately. Ours is uncommonly good. Perhaps we had better not get John Welch's queys bulled in future. I have no recollection who *young Pattison* is, and yet I think John was naming some young man to me once. I hope John will have something handsome for his own trouble and board. It is said young Lorrain is away already from Robert Thompson. Harry Rutherford's wife and *daughter* both doing well. I am glad John has bought horses, he must not want horses or anything proper to promote business. Mr Mason got the premium at Durham. You see I know before you! This 2nd April has been a wonderfull fine day. We get done sowing oats tomorrow I hope, but we have our hastening pease to sow yet. Sheep might have been washed today perhaps better than when we wish to wash them. *3 April Saturday.* Quite a change for the worse, a cold north east wind, like the season. Indeed yesterday was too fine a day to continue. You must mind that there is scab amongst those hogs of Mr Wood's which I am sorry for, especially as John Welch has no shepherd. They must not be put where other sheep go. Berwick 2 o'clock. Nothing to be done in the corn way today I think. I wonder I don't hear from Mr Ben Sayle. I have heard nothing, and his money was due the 29th ultimo.

110. *George Culley to Matthew Culley and John Welch; Matthew Culley to John Welch*

Eastfield Monday 5th April 1802

Dear Brother

I sent you a long letter on Saturday from Berwick, which should certainly reach your hands today at any rate. I am dissatisfyed with letters getting so late to Denton. The cause should be investigated, it must be owing to the Darlington post people I do suppose. This is the 5th April, and Ben Sayle's money was due upon the 30th March, and no letter or account of it, which is the more extraordinary as Mr Sayle is so punctual a man. However I wrote to him yesterday for fear any mistake or misunderstanding has taken place, because he is a very accurate man in business. But there is time enough yet for it to come before Saturday, the day when we shall remit 4000£ from Berwick, which with 2000 remitted from Darlington by Mr Peacock will do the business. I have heard nothing from Mr Peacock lately, but as he promised to remit 2000£ I can have no doubt of his punctuallity. Mr Nisbet was here yesterday and says he can only let me have 250£ instead of 600£ the next Saturday, which with Mr Sayle's 400 will oblige us to borrow a good deal at the bank for the present. But Mr Nisbet had mistook me and thought we would not want the money until 8th of May, which he says shall be then ready. Ralph sold a little wheat on Saturday at 49s, a very good price still. It seems there was a mistake in our London paper last Thursday. It sayd 14s drop, but we have since been told it should only have been 4s. Ralph had 40 bolls barley from Wark which could be only bid 21s for so did not sell, and oats are scarcely marketable at any price. And the potatoe oat seed work is done. This is rather a cool morning, yesterday was a remarkable fine day, and upon the whole the weather has been very favorable lately although dry. George Humble has thrashed most of his turnip seed at Wark, very good but not quite so gifty as he expected. We only finished our oat seed on Saturday, and not begun our barley yet. We are plowing the land the first time over, and have not much to sow this year. We have a few early rape to sow at the Westfield, which will be done this week. Lambs do well and we have more twins than common this year. Also two cows calved twins, one of the 4 died. I understand you have had bad luck with your calves at Wark. I am told it is a bad thing to give young calves too much milk at the first.

John Welch I shall address this to you for fear my brother be come from Denton. But let us be thankfull that no human lives are lost. Last night about 10 o'clock an alarming fire took place at Wark in one of the stables where 10 or 11 of their best work horses were burnt to death, but no cattle thank God nor no corn. All the hay or nearly so, and all the range of stables and byers where the cows stood. The big stable escaped. Upon the whole although a *terrible thing*, not near so bad as it might have been surely. It is covered by an ample insurance. John you will have to take many of our fat cattle and some earlier on this account as the hay is all burnt. And I think you should try to buy 2, 4 or 6 horses if you light on in your neighborhood, usefull draft ones, never mind if they cost 20 to near 30£ per, 4 or 5 years old may do and some older just as you can light on. The neighbors all were kind. Mr Hogarth was there from the first. The Comptons and he are sending carts to lead away the burning rubbish to a safe place, and all their sons in general behaved well indeed and we must make them our amends. I have always said that no candles should be used in fodering after the middle of March, and never in windy nights at any rate. But thank God it is no worse says (7th [6th] April Tuesday) yours sincerely

Geo Culley

I had forgot to mention the butter money, which was not sent to me. I had to borrow 40£ at the bank.

[**M.C.**] John I opened your letter.[22] Mr Bates will try to buy us some horses, and write to you and us by the driver Glass. We will try to get home tonight if we can, as this misfortune has happened at Wark. Thank God it's not so bad as might have been. Jack Davidson complains of fatigue from exertion and anxiety, he is going to bed, came from Newcastle this morning.

In haste Matthw Culley

111. *George and Matthew Culley to John Welch*

Eastfield 11th April 1802

Well John

Your master and mistress got home about 11 o'clock on Wednesday night, and were truly thankfull for our wonderfull deliverance. Indeed when I went up again in the daylight I was more astonished than ever. If the poor horses had only been saved, all the rest are but trifles compared. But that was not possible without the fire had been discovered sooner by our own people. But the greatest conduct had been displayed, and every effort used that could be suggested, and all was safe or nearly so before we got there. As it certainly was occasioned by the negligence of the men when foddering, I shall only wish to impress you and everyone with the utter necessity of being carefull with candles at those times. I believe every stable should have a large broad-bottomed lanthern to slide along a cord from one end of the stable to the other, as you have seen them at gentlemen's houses. We have one in our hackney stable here.

Well John the markets seem now to be going against us in almost everything but wool. But never mind, we have had admirable times and ought to submit to a change of measures with *philosophick calmness* and resignation. And let us take the prices that are going and be

---

22 Matthew Culley and his wife were at Halton on their way home. This letter was brought to him there by Jack Davison. See. No. 111.

thankfull. At the same time I think I would certainly put of a while still, for I think things may be a little better although they may be worse, but I think it not *very likely* so long as fresh lean stock sells so high. I certainly would buy nothing of cattle or sheep kind until we are likely to get through all or at least a part of what we have. The sheep market was not so bad I am told on Wednesday, 8 ½ per lb. or rather better for good sheep. Cattle not more than 7/6 per stone sink even the best. Now it is very likely that we shall send a great many of our cattle to you, and yet I want to put of so long as we have turnips, especially as the weather has turned extremely cold and stormy which may hold a while. And I think you should continue if possible to come over or meet us at Alnwick fair, as we must shew a lot of capital cattle there and I think Ratcliff the likeliest man to buy them. And I can't sell him those you know, and if not sold there they must go forward to Denton or somewhere. John I am sincerely sorry for poor William Story's death,[23] particularly as I understood that he was become exceedingly usefull to you. I forgot to say that you need not buy any horses as we can do with fewer than we at first thought of, and my brother had ordered Thomas and William Bates to buy some, so that we may get too many. If you can meet with one or 2 good draft horses or mares you cannot be wrong, but not more I think until you hear further about it. Remember that any money you can spare must be given to Mr Peacock to put into the bank at Darlington to lie there until Martinmas, when our other payment is to be made for Easington Grange estate. I have wrote to Mr Peacock on that head yesterday, without you chuse to place it there yourself. I fancy they will give 3 per cent, which is as much as we can make of it here until that time, and it may as well be there as here, and can be as safely remitted from Darlington to London as from hence. We remitted our money yesterday from Berwick, which with what Mr Peacock has sent will make up the sum wanted at this time, and now we must provide for Martinmas.

Sunday morning same date as before. This is as cold a day as ever blew, but still not unlike March weather. And this is still old March. No kind of corn market yesterday at Berwick. Things are in a kind of stagnation at present, and will be until matters take their proper level again. Some wheat sold at 44 to 46s our boll, barley 20 to 22s, and oats 12 to 15s. We did nothing, nor have we any reason to blame ourselves about corn at all, except it be not selling some oats immediately after

---

23 William Story died on 6 April, aged 23: DUL, BT Denton.

harvest, for which I am always very keen, because in general they not only sell well then but measure more than ever after. However we have all along sold our corn as we thrashed it, which I believe is always wise and prudent, except when grain is very low indeed. I do believe that there are strange quantities of fat cattle still in this country, notwithstanding there are seldom less than 100 shown weekly at Morpeth. We yet have many large lots in this neighborhood. I don't apprehend that sheep are so plentifull by far. But the worst is that our consumption is so small. I don't know that I ever remember more dust in the roads at this season than now, nor such a continuance of winds, yet I never saw the spring wheat and oats come faster nor better away, and the seeds grew amazingly before these 3 or 4 cold days. Consequently it shews that we don't want *moisture* but *warmth* or at least *fresh weather*. Indeed we never want rain in this climate until near the end of May. John I must beg of you to exert yourself a little in writing, as I shall expect to hear from you frequently now. You know I have always recommended to you to have a letter *begun* as soon as another is *sent away*, and then you can write a kind of *journal* of your proceedings. And by writing a very little every *evening* or any part of the day when you can find a little *repose*, it will be easier and less trouble to you and not oppress your memory so much to recollect anything you may have to write about. Mrs Grey has taken Milfield again at *900£* per annum, her present rent *250£*! It is not ascertained yet who are to get Midleton Hall farm, at present *Mr Hughes's* near Wooler. Its present rent is 260£, and houses in Wooler let for 20£ per annum, then say 280£ per annum in all. And I know to a certainty that *1125£* per annum is bid for it by Messrs Robert Thompson of Fenham and his brother William. Mr Nisbet bid, and gave in a proposal for 1066£. Some say there is more bid than Messrs Thompson, but that is pretty well I think. It is rather more than 4 times the present rent I think, and certainly shews that people are not affraid of the times in this country. Indeed the farmers have made much money in the district of late years, and are still willing to sport a card. I don't know if 20 different sets have not viewed and given in proposals for Midleton Hall, whilst at the last letting 21 years ago no one opposed George Hughes. He *walked the course*.[24] I am yours sincerely

Geo Culley

---

24 George Hughes kept the lease of Middleton Hall. He bought the farm from
   Greenwich Hospital in 1832.

[*M.C.*] Sunday evening. Well John. Your mistress and I got well and found the lasses all well. M.C. But the butter money not come. I think this was mentioned to you in my brother's letter from Halton by me. You must look to it as I could not open your friend's letter and see the sum, as I think you expected it to come by the boys Bolson and Pattison. The latter seems strong. I wish that he may answer your expectations. I have spoken to Bob Dippie, who seems not very keen of coming. He will hardly get from here before the latter end of this week. George is not fond of sending him as he is so useful here. You should be cautious to treat him well, is of a shy temper and is a very quiet valuable lad, will improve in your hand daily. We got to Halton on Tuesday night. In the morning after Davison brought my brother's letter we set out as soon as we got breakfast, reached Wark past 10 that night. We made no stop, eat a little on the road as we travailed. A disagreeable sight next morning of bare walls and 10 sick scorched cows. They are still sick, most of them may recover, only they give little milk, their paps are so burnt. They may if they live feed tolerable well. George Humble's quey and B. Dickinson's cow are sore burnt. Bob Tait's seems to give most milk. The other 2 Hs Ds [*sic*] will hardly come to milk again, at least not likely from their udders appearance at present. We may be truly thankfull to the Almighty so little damage is done. Great exertions by men and women, whose debt we can never repay. Great exertions have been made indeed, and great numbers of people who wrought beyond credibility. The house safe and cattle and corn. The more it is considered, the more wonderfull it is that so little damage is done. The committee come on Tuesday to settle the damage. Now my brother says you need only buy 2 horses, but 2 will hardly do as Mr Bates may not meet with any and we lost 10, 7 of them very valuable, 1 young and 2 old. We have 4 sent from Eastfield and 2 from Grindon for the present. Buy 3 or 4 good aged, or strong colts or fillies fit to work. We intend Dippie to take a Galloway to Denton, then the driver may ride north on it. If you have any horses to send here he may ride on the Galloway and lead the horse or horses. He should on his return call on Mr Bates, who if he has bought any the lad may bring also. Perhaps if he wants help may hire one to assist for the first day's driving. You will write by Glass what grass you have. Our clovers and Thornington's are so good we have put tups, wethers and hogs to grass or wheat, and hope our turnips &c will keep fatt cattle until you can take some. We are very strong here of turnips, and if weather keeps tollerable may enable the clovers to get fit for cattle and sheep, and then we must sell fast, cattle and sheep when clipped. My brother is afraid of giving pottatoes to cattle as they hove[25] them, but I hope they will not hove fatt cattle, only lean or weak ones may be too greedy of them. Do let us know how the fever is in your town and how your neighbours are. I hope the rest of the Storys are got well, but they met with a sore loss and we fear they would be injured by his death.

I have not said anything to my brother about a smith's house and shop, but on considering it over think right to build both together, as it's right to have a smith near his work. Should there be a door between the shop and house, but perhaps not, as the door between the stables at Wark was the cause of the cows being so very sore scorched and the byers being set on fire. They should be filled up to the tyles on the mid gabel, so as no communication be of wood touching. The rife and rig tree should be separated by the gabel, and the wood not to touch. If no hole in the gabel the fire would not have communicated to the tyled byers, unless the fire had burnt through the solid wall which is not the case. With extra exertions they saved the machine barn although the fire was through a hole in west end, but the hole and door were continually thrown water on. Yet the door was burnt nearly, so as the boards are fallen in pieces. They with extra force pushed the timber when burning of the byer westward from the barn end. Hoping you are all well is the wish of

Matthw Culley.

[G.C.] PS. As there is now a probabillity of our parting with Shotton this first term, which I have learnt today by a letter from Mr Bailey, pray in that case you had better buy no horses until we write to you again. It will be known by Friday first whether we keep Shotton again or not. I am in haste Yours

Geo Culley

We have most terrible cold frosty weather at present as I have often known. Cheviot covered with snow at nights and is taken of by the sun in the day

112.   George Culley and Matthew Culley jr to John Welch

Eastfield 14th April 1802

John

We sent you a long letter the other day, and I make a point immediately to begin another to you, But I am concerned to hear by a letter this moment from Mr Peacock that you are and have been poorly for some time, but that thank God you are now recovering apace.

---

25 Cause to swell up.

However I must beg and entreat you to take care of yourself, and if a bottle or 2 of wine will do you service, which I am persuaded it will if it be anything of that nasty slow fever which the Storys have had, I beg you will procure some good wine and I shall certainly pay for it should my brother refuse to join me. But I am certain he will not. I am also sorry to hear by Mr Peacock that his wife is very poorly. I sincerely wish her better. However I am rejoiced to hear that the Storys are doing better poor creatures, although I am sorry that they have not been so *prudent* and *discreet* in taking proper care of themselves, or they might have been further forward in their health. Now do let me entreat you not to want anything or any assistance when either you, your wife, or any of your family are out of health, but if you need a doctor get who you most approve of. But not that Mr Hodgson of Darlington, who although I do not know personally yet all the family were extremely weak people, and by the account my brother gives me I am affraid that he had quite misunderstood poor William Story's case. We have had cold weather in the extreme lately and the hills covered with snow. But this afternoon it has rained a little and turned warmer, and the mercury is coming down, so that I hope a change is approaching. Times seem at present John to run against us, but let us be thankfull that we have had our share of wonderfull good times, and what is more, we made use of them, so as we can bear a brush and many a one, God knows, can not. But no one knows still how things may turn before Whitsuntide. However, be as it will, we have done for the best, let us be contented and thankfull that it is no worse. Sheep will be well or decently sold still I think, and if we can get 8s or even 7/6 per stone for our cattle it will do. We have all known things worse a deal, and remember we have another comfort. We buy few of our cattle or sheep, consequently can afford them the lower, at all events let us keep up our spirits and do the best we can. We have been successfull beyond most people in our ways and in our time. We have had many escapes, but none more extraordinary than the one lately at Wark. May we be gratefully thankfull to the Father of mercies. I hardly expect to hear from you until after Easter Monday, more especially as you have not been well. Do take care of yourself, and want for nothing you can get that is necessary.

15 April. John we have this moment received yours from Wark, which I suppose came by Tom Glass. I find you had not got the letter which I sent you the other day. I am sorry that you did not sell your cattle even upon the terms that Holdgate and Shillito bid you with the keeping of half a fortnight. I am certain that it is 1s 6d per stone more than the best price yesterday at Morpeth. However I know that you did it for the best, and will not blame you as it cannot now be helped. But I am doubtfull that you will not have the same offer again of some time. Besides we have such a large number of cattle to come to you as quite alarms me. However we will certainly put them of coming as long as we can, but come they must as they are litterally not more than 7 to 7 ½ per stone sink at the utmost here I do assure you, and a wonderfull deal of good cattle still in hand here, such swarms are at Morpeth every week as never was known before. And the Shields butchers all are full, nay said to have more than they can kill in 4 or 5 weeks, so that there is no chance for Shields butchers coming to buy any at Darlington. For heaven's sake, if *ever you* can come at that price again, take it. We have more than 50 fat cattle that must either go to Denton to be sold, or half given away here. You can form no kind of idea how bad times are here and at Morpeth for fat cattle. If you had, you would have sold to a certainty. And these Wakefield jobbers must learn that it is so soon. The worst fault you have is being too hard a seller. I don't like to tell you so, but you must excuse me at this time, as I don't think I ever was so affraid of times as now in my life. We have such a strange quantity of cattle, and really do not know what way to turn ourselves. For God's sake write and advise us. The Wark cattle should come the first I do suppose, and they were a charming lot of cattle, but by the misfortune of the fire were obliged to be turned out both night and day. However their hair is coming again now upon most of them and they begin to look like outlaid ones. We must shew our Westfield ones at Alnwick if worth while, but I really believe it will mean nothing, and if they come to you also it will make 70 and more. But I will say no more now, as I shall only mortify both yourself and me. We must just do the best we can. We have one comfort, we can afford them as low as anybody, as few of them are bought in asking as good a price as you have refused, and 8*s* for ours. I am yours truly

Geo Culley

PS. I forgot to say that the sheep were well sold at Morpeth in proportion, that is from 8 ½ to 9*d* nearly. But then the fear is that too many sheep may be clipped and fill the markets too full then. But we must run hazards of that also. We got a fine rain yesterday after a most severe frosty morning, and it has kept pretty fresh ever since the rain. I fancy Tom Glass will set of directly with the hill hogs for your place again. John you don't say one word whether the hogs Tom brought you please you or not. You must remember there is a scab amongst them, therefore they should go by themselves, it is an unfortunate event but can't now be helped. Perhaps it may be wise to sell them in the autumn or as soon as you can. You will think I am joking, but I declare I believe it is not wrong to sell cheap now and then. It makes your customers come again. Now if you had lost ten pounds by selling your steers at this time, I am of opinion that it would have been 20£ in our way. I only name this that you may benefit by it, and I am sure that over hard selling is unwise, I mean to do it always. I have often been called a dear seller, and perhaps justly so in a market where I think a man is right to sell for all he can get, because the butchers and jobbers will buy as cheap as they can of you, and this is all fair. In a dear selling day you have the advantage, in a bad selling day they have the advantage. But at home to certain customers like Holdgate &c I think different, that is I would not be too tight at all times. We must and should give and take, at least that is my way of thinking. John you say nothing of being ill yourself, and Mr Peacock says in his letter to me that you are poorly. Pray do take care of yourself and want for nothing. I would certainly sell lime for profit or not burn at all. Why plague yourself without some advantage? You have enough to do besides. I need not recomend to you to be gentle with Bob Dippie. He is a fine lad but of a reserved and shy temper, and may be easily offended. I understand beside that he is not over and above willing to come, but perhaps he may like better by degrees.

[*M.C. jr*] You need not buy any more horses untill you hear further from us again, as we are likely (indeed almost sure) to lose Shotton this first 12 May, and if we do we shall have 4 or 5 horses to spare from there for Wark. Wheat is up this week at London from 5 to 8*s*, also oats. Barley is rather down. We think wheat will be pretty well sold still and hold its price.

[*G.C.*] *16th April.* A little frosty but a fine morning upon the whole. We will send this today to you per post. I still think that you get our letters very badly. I wish you would send oftener to Piercebridge, and speak to old Dicky Thompson at Darlington or whoever is post-master. It is of the greatest importance for you to receive our letters regularly, and I am sure we neither pinch for number nor quantity of stuff. By all means sell if you can, if you should lose 5 or 10£ by your lot. I wish you could draw these men back. And write us a few lines often as possible. Things are most critical at this time indeed in the fat way, and we have such a vast quantity in hand. Not but I have known times take a strange turn for the better even after Easter Monday. One thing, people have bought lean stock in so high that they can't sell them very low without much loss. Indeed many people here are selling at Ninian fair prices. Nor am I certain that Wark cattle are or will be sold for more than what they were worth at Ninian fair. John I fancy you will need no turnip seed this season. If you do may let us know. We sell wholesale at 1*s* per pound or £5. 12*s* per cwt., and by retail at 15*d* per pound

113. *Matthew Culley to John Welch*

[Wark Thursday 15 April 1802][26]

This will be given you by Bob [. . .][27] who sets of if all be well on Good Friday. Dissenters do not mind our [. . .] fast days, yet I think we ought to reverence both them and the sabbath for fear we forget our God and our Redeemer. But I do not say anything against them, as they are a set of decent people. The lad is an honest worthy man and so I hope you will find him, also diligent and active. I hope things may settle soon, so as some judgement may be formed of the price of cattle and corn. It is believed that this country is full of fatt cattle. A few weeks will shew that. The fine rain last night *Wednesday* will mellow the ground so as to bring vegetation on apace, especially if we get warm weather after it. One of the queys died, the rest we now hope may recover, but they are an unpleasant sight and very much maimed all over, especially legs and udders and paps, are likely to loose the milk if they do mend. George's quey, Bob Tait's cow and R. Dickinson's cow are of the burnt ones, which makes it hard on them, but as we were insured it will help to lessen the loss. The cows and horses are the most unpleasant part of it, houses and hay may be over got. Indeed the poor people were many of them sore pained after they cooled, but some amends will be made them. This morning will I hope enable us to carry forward our sheep as the clovers may grow again. The grasses were sore hurt with the sheep feet yesterday morning by the frost on the clover. I hope you have planned the smith's house and shop properly. Fir wood is now supposed to come down in price. It is only 18*d* Norway logs here and believed to lower more soon. Bob D. will sow your seeds, look after the water in all places, plant you some cabbages &c &c, take out the stops of water coming to the mill &c. I hope the miller's family are in a degree got better in their health, but they will long feel the loss of their son poor things. They are to be pitied, but unpleasant things belong to this world. We forget that we are to go hence and be no more seen, but please God to a better life, especially if we submit to, and act according to the Scriptures. I received your letter from Glass today

---

26 The address is lacking. The date is inserted at the end of the letter.
27 The paper is torn here, affecting two lines of text.

which I sent to my brother. Bob Dippie will bring my brother's letter and this. We have washed our weathers this afternoon, and also the segs. The wethers seem very good and fatt. I cannot say but in these dubious times I wish you had sold the fatt queys and steers, but I fear now you may be hardly able to reach the same price. It was a good value, and there is an old adage, better me sell than me keep. As you did for the best, there can little more be said. You may shew some of them in the market on Monday, and do what seems best in your own opinion. You have got them made good, then take the market price for what you have to sell. We cannot hope for such prices as have been going in the war, unless it be until August. But I cannot form any accurate opinion now when things are got so high. We shall know tomorrow Friday likely whether we keep Shotton a year more or not. In such case we shall get some midling horses 4 from there but not till May Day. We must have some good ones in the mean time, and the one you have sent here is cheaper than we can buy here. If you can buy us 2 or 3 more good ones, along with Mr Bates who you say is to write to you what he can do, as we must have our land wrought for turnips &c. The rape seed should be sent over by some means. Thursday 15th April 1802 Yours &c

Matthw Culley

I have wrote till my fingers are sore. I send the level by Bob. The feet shall be sent in a future day. In the mean time Bob and you must use a stick with a sharp iron point and a flatt top with a hole for the iron pin to run about upon and turn round, but you have little to [do]28 with a level at this season. Bob should cut some old wood out of the trees, but that season is over for this year. I do not hear that Birkbeck has paid his ½ year's rent to Mr Peacock.

114. *George and Matthew Culley and Matthew Culley jr to John Welch*

[Eastfield 17 April 1802]29

[. . .]bridge that same evening [. . .] and down to Piercebridge on the Sunday evening late, provided the Barnard Castle post go by Piercebridge every night especially Sunday nights. I name this to you in particular because I find your letters come badly to you, especially I perceive those we send from Berwick on the Saturdays to reach you by or before Monday, seldom are received by you at the time they

---

28 A word seems to have been omitted.
29 The top of the sheet is torn off. The starting date is probably 17 April.

ought to reach you, and I think this is owing either to a neglect of the post to Piercebridge or the postmaster at Darlington, and it is worth your while to take some pains to investigate this business as letters are of great importance and especially at this *critical season* Perhaps by consulting with Mr Peacock he could help you, or it may be that your letters come along with Mr Harrison's and Mr Peacock's. That you must know and I do not. I dare say we shall have to send you a lot of fat cattle before Tom come back, otherwise they will be coming upon you in such a *clash*, and we can't keep them long, although we will certainly put of as long as we possibly can, you may be sure. Had you ever reflected upon what a *shower* must come upon you in a little time, you would have been glad to have taken 30£ per head of [*sic ?or*] 30 guineas each. I really think you should either come over before Alnwick fair 12th May, or meet me there with our Westfield cattle, because I am such a novice in selling cattle now. However if you can neither come nor meet I will do my best, and if I can't make 7/6*d* I shall send them forward to you. John I never heard one word before today of Tom Glass being going to be hired to you as a shepherd, but I am very glad of it provided you can hire him reasonably. I am sure you must now have much need of a shepherd. Tom told me today that he had now taken it into his head to live with you.

Sunday Easter Day 18th. Wheat rather better yesterday at Berwick but oats and barley low indeed. Everybody saying what will be done with fat cattle. I may forget, but I think I do not recollect so very critical a time. But sometimes things turn out better, as well as worse than expected. The weather thank God is very good for the season since the rain on Wednesday last and the preceding severe frost. *Monday 19th*. What a blessed day! I never saw a finer milder sweeter morning in my life at this season. We will get grass however if we never sell. They have washed the best wethers at Wark, which we shall begin to market about Wednesday week or so. Do let us know how your sheep markets are, ours are 8 ½ to near 9d per lb. for rough sheep. I am affraid we shall not get 8*d* for clipped sheep, as skins are so high it takes of the price greatly. I understand skins are sold from 9 to 12*s*. Our wethers are particularly good. I could wish Holdgate would come and buy them. Mr Selby of Swainfield by Alnwick, who is proprietor of Paston, has bought Shotton and we are to quit everything this 12th May. They take the servants and perhaps it is as well as we shall get the horses for Wark. It was to be sold by auction at Kelso last Friday but he bought it after the sale by private contract 6000 some 100s better, I know not how much, very dear indeed. We are under no necessity to sell things too dear, and it only gains a man too hard a name. Nay I have known some butchers and jobbers, by way of setting a hard seller fast, when they have found him *positive*, bid him even more than things were really *worth*. Every man naturally wishes to sell his commodity for a times price, and he has a right to do so in general certainly. But in serious and falling times like these, a man should not be too stiff. I know your cattle are good, indeed my brother says that he seldom ever saw any as good not to have got oil cake. But I know your *motives well*, and it is a commendable *ambition* for you to *wish* to make the most of your *own* cattle. But you perhaps forget at the moment that Mr Colling's cattle were *oil caked*, and if I remember right Mr Colling did not exceed 9s much if any in brisker times. And there is nothing so foolish as for a man to fight against times. Let his stock be so ever good, he must submit to times, and I would gladly hope that this is the last time you will part with good men for 10s per head and a fortnight's keeping for half of them. I freely forgive you because I am satisfyed you did it for the best. It strikes me now that we had better send you the Wark housed cattle first, with a few more large outlaid ones to make up a lot of 20 cattle or so, because the large cattle are best sold before warm weather sets in. Then we will send all the next heaviest cattle and keep the *smallest steers, spayed queys* and *cows* until the last, both because they are the fittest to sell in summer but because we can keep them longest here, as we can put a part of them upon our seeds &c &c. Tell us as soon as you can how the wool is likely to start. Your own should be a valuable clip, I mean because you are all hogs and only a small lot of ewes. G.C.

[*M.C.*] *Monday 19th Easter.* Well John you never said anything about your own illness. You should have said how you were. I am much afraid of Story and his daughter. As we loose Shotton, at least it is 10 to one, we get 3 or 4 horses from there but they are not such as Mr Walker's mare you sent us at 25£. She is more like the money than any can be bought here, and we can take 2 or 3 more if Mr Bates and you fall in with 2 or 3 good ones. If you meet my brother at Alnwick I suppose that's all you can do, as Bob Dippie will hardly be able yet to fill up your business. We got a letter today from Matty from Houghton. The 9 scorched cows seem to get better ever since we gave them guide fat ale to drink, 1 or 2 quarts at a time once a day. They seemed to [be][30] doing better and cooler, but most of them are gone

---

30 A word seems to have been omitted.

dry. 2 of ours give a little milk, but they are very sore and stiff of their limbs, unable to walk back from the courtain[31] to the house. M.C.

I received an order upon the overseers of East Ord from Mr Peacock or someone (when at Denton) for 2£ 2s 0, was presented by Ralph Brown but they refuse to pay the money, alledging the men are not married &c. Shall return the order as soon as an oportunity offers. It will not pay postage or I would have sent it in your letter. You must learn to sell at moderate prices. I am a bad salesman on that account. Hope this day Easter Monday may be better than expected. I have seen it a bad, exceeding bad market yet an excellent one succeed, but I have lost my judgement now for want of custom in selling, and I am now over old to learn. If the miller's family get better [it] will I hope be a relief to you. I hope they will be better soon and please God in His blessings to us and that family poor things. I hope you have got the money for the butter. How goes on the smith's shop and house? Has Birkbeck paid his quarter year's rent to Mr Peacock? Let us know how you and your family are in your next.

Matthw Culley

[**M.C. jr**] *Tuesday 20.* Rather colder, wind a little to the north, but a good morning for the time of the year, very inclined to be dry weather I still think. You have some rape seed for us. It is drawing near the time of sowing, and there will hardly be an opportunity of sending it free of charge, therefore you had better sent it by the waggon, properly directed. M.C. junior

[**G.C.**] *Thursday 22nd.* John I am glad to hear that you have sold your cattle. Mr Vardy told Matty Culley so at Wooler market today, and was to have brought a letter from you but by some mistake did not. However we are much disappointed at not receiving a letter from you per post, which I think may be the best way of sending letters. Mr Vardy is a reasonable man and might have brought it up safe enough, but depend upon it that nothing is so safe and proper as the post. The expence is not worth naming in business. Morpeth was a much better market yesterday, good cattle near 8s per stone sink and mutton sheep 8 ½ to near 9d per lb. sink. I don't by any means expect such long letters as I write to you, it would be quite unreasonable to drive you to write such, but I should be happy indeed to receive letters oftener, did they only contain half a score of lines. And now when we have heard that you are poorly yourself, and the Storys poor things so very ill, one wishes to hear of both you and them

---

31 Enclosing wall, not supporting a roof.

certainly. If we don't hear from you tomorrow we will send this away. I think some mistake must have taken place, or we should have had a letter from you since Easter Monday. However you can hardly conceive how glad I am to hear that you were at market and sold your cattle. I had begun to think you were very ill.

Yours in haste Geo Culley

Friday morning 23rd. We have this moment received yours of 21st and was never better pleased than at you selling your cattle. You don't say anything about the money, but wish you to lodge it in Darlington banks if they will allow 3 per cent. If not, to remit it here. But they will certainly allow 3 per cent if it be more than a month, and this will be till near Martinmas. However I would like to have you consult Mr Peacock about your money matters as he is more conversant with these things than you are, and it is aukward to send it here because it will be as well and safely remitted to London at Martinmas as from here. We were thinking of sending you the Wark oxen as we could better keep the small cattle, and the large ones should be first sold before the hot weather. But I will go to Wark and have a consideration today. I am glad Mr Wood's hogs please, be sure you look well to the scab in them, but Tom Glass will do that if he be with you. We have plenty of capital turnip seed. It may be right to send a cart with 6 or 8 cwt. to you if you can sell it at 1*s* per lb. or better, but not less than 1*s*. We sell wholesale at 1*s*, and 1/5*d* retail. Say directly if you wish us to send and how much, and we can bring the rape seed back. Your best way is to mix the old and new together, we always do. We have more turnip seed by a deal than we can sell this year. We are rejoiced to hear that Mrs Peacock is better, and Story's family, and particularly yourself. Tell Mr Colling with my compliments that I think Mr Jobson, Mr Vardy and Mr R. Thompson the properest to send a young man to, but they have all 2 each at the present. Mr Nisbet has only one and is a steady boy. He being my brother in law I don't care to speak of him. He makes his work moderately and I think it right. The fire happened to a certainty by negligence at foddering. If you can get 2 good horses at or before Allerton fair, it may be right I think to buy them. Matty Walker's mare pleases my brother well. You have not said how many cattle you sold. Mr Vardy says 17, you see we had heard before your letter reached us if we are rightly informed. Yours in haste

Geo Culley

As we shall mostly likely set of 20 cattle today or tomorrow, Tom Glass had best come along the Corbridge road to meet them. I think you should sell your wheat as 20*s* or near it is a famous price. We can't get 16s scarcely. To be sure if you buy horses and Mr Bates also it will not be so well. You and Mr Bates should correspond and know what you are each about. However if you can meet with a particular good horse or mare don't trouble yourself. Notwithstanding what I say about putting money in the bank be sure you always keep enough in hand to pay your way. To be sure when you are constantly selling corn and cattle &c you can't be in want, but it is wrong to pinch yourself. If you should come to Alnwick might bring money with you, especially if the bankers will not allow 3 per cent. But I do think they will, and in that case it is safer and better in the bank. If you can't conveniently come to Alnwick you need not, as Matty says he will sell the cattle and if not they will come to you. (Wark) We have sent of 20 oxen, 14 inlaid cattle which must be sold to Sunderland I believe. However there are some very capital oxen and some only midling, one of Harry Potts yellow with mustard near 100 stone, another reddy one being about 90 stone. The 6 are outlaid and will improve with grass if you chuse to keep them a while. One of these is a nice red ox that laid by himself and weighs 90 stone I hope. These 6 we have tied a piece tape about their tails, that you may not mistake any of them. Now John do write directly and say whether we should send you any fat sheep. They are very good if one thought the Wakefield jobbers would give us more than Morpeth. They are nearer to the Yorkshire market than here by 100 miles nearly. Besides they will travel well when clipped. Do write and give us your opinion. I suppose we shall send another lot of spayed queys and cows very soon, or the remainder of the oxen, but would wish to know which you would like to come first. So do say. There will be 20 or more of cows and spayed queys and about 30 oxen still, besides what we shall shew at Alnwick. I must beg you to write on receipt of this. Yours

Geo Culley

PS. Also write about the turnip seed directly, as no time should be lost about sending it, and the rape coming. If you can't sell 5 or 6 cwt. it will not be worth sending so far.

115. *George and Matthew Culley and Matthew Culley jr to John Welch*

Eastfield 23 April 1802

Well John

We have been sending away 20 oxen to you, and a letter at same time per post, and I immediately begin another which I think is much the best way. Had you had a letter in your pocket on Monday at Darlington, written so far as you had matter to fill it, how soon you could have said in it that it was a good or a bad market &c &c, and then sent it per post from Darlington and we would have had it on Tuesday or Wednesday at the furthest, which would have been 2 days sooner at the least than the one we received today, a matter of much moment surely in business. But you now promise to write *journalwise* so I will say no more at present. But I thank you for selling your cattle, and will thank you to say next time how many you sell at once. Mr Vardy says 17, you say nothing. I hope it was 17. I had not time to examine the cattle today as I had to finish the letter, but I should think the inlaid 14 oxen will weigh about 80 or better one with another. You know I have said in my letter sent today that one ox, Harry Pott's yellow flecked with curled in horns, weighs 96 stone, and another that was an Eastfield one, lyery[32] and very leeddy[33] in his handling, quite a ship ox, weighs 90. He is also unhorned. There are several more good oxen and some only midling. Still I think they should be better than 80 stone over head. Perhaps near 84 stone each. But they will not look so immediately after the journey, and it must be left to you either to sell directly or give potatoes to them as you think right. The 6 outlaid ones will be better I should think for giving a little grass to, which will bloom them and sap them. However we leave that entirely to your discretion. The pretty little nice ox of my brother's which stood loose at Wark weighs 90 good, many think. I can't say but I think it plenty for him. Another 2 weigh 80, so that we thought the whole score about 80 over head. But I own when I have said this that I am perhaps mistaken, and more likely to weigh them over heavy than over light, and can only say that you must examine them well yourself and sell them as well as you can. I am affraid that the inlaid ones will not improve, and if you can get 8*s* per stone sink I would let them go. If you can't get so much you must take less. Not but I do believe they have a chance to die well if the turning out has not hurt them, and if my brother would have let William Lincoln have had them 2 months ago when he was in the mind, I am clearly of opinion that they would have been sold for as much money then as now, and I suppose more. But we can't always hit the nail on the head, and let them be sold for what the times will allow, I will be content. The outlaid ones will perhaps not *prove* so well in tallow but are certainly worth more per stone, and will *improve* by keeping if you can't come at your price, as I am certain they will grow *like trouts*. And

---

32 Having a superabundance of lean flesh.
33 Hard working.

we have some more outlaid ones, both oxen and queys, which will pay well for a little meat. Indeed many of them are so far back as to require meat or keeping a while, and will pay better in my judgement than buying lean or fresh off stock.

Sunday 25th. Blustering winds with driving rains, but yet much wet. Corn very flat at Berwick. Best price for wheat 46 to 47, barley 20*s* to 21*s*, oats not marketable. They really will not buy them. Now John I think it wrong not to sell your wheat at 20*s* which is 3£ our boll. What a difference to our prices! I can't help thinking it very wrong not to sell. No doubt you have your own reasons for not selling, but I can't think it wise to refuse 20 or 18*s* either. Times are certainly uncertain and critical, consequently every wise man will sell at 3£ for wheat. Now John you asked how the fire happened at Wark. I told you to a certainty, by negligence when foddering. Better so than if done *maliciously surely*. But does it not strike you with the necessity for *sure thought* and *attention* when foddering. Young men and even old unthinking men say 'O never fear I will take care', when very likely he never thinks nearer about it. But without any intention to do harm, tops the candle with his finger and thumb and even puts his foot upon it, but looks the other way at the moment, his foot misses a part which kindles, and burns *man* and *beast, perhaps besides houses, grain &c &c* Thus I have no doubt the Wark fire took place. I do not attend foddering myself, it can hardly be expected indeed, but I am always preaching up to take care of fire, indeed a sudden smoke in the day time anywhere makes me start. But that will wear of, and we are with much reason truly sincerely and gratefully thankfull I am sure. But the poor cows are a melancholly sight, breaking out in different parts of their udders and skins, and running matter, lee &c. Indeed I am affraid many of them will never do much good. But John I forgot to remark one very important thing. It has always been a rule with me to order our people to fodder after the middle of March without candles. And if those orders had been obeyed, we had had no fire at Wark. When the sun goes to six o'clock I am certain people may fodder without candles, and had not one better fodder a little earlier than run such risks. Another rule with me was never at any time to use a candle in a very windy night. Better grope and fodder aukwardly than run such *risks* surely. I believe William Brown has adhered to these rules strictly, but I am affraid no one else. Willy and I you know were long together, and these rules become habbits and excellent habbits too. William Brown always paid the greatest attention to my orders, rules and *observations*

Monday 26. Still cold windy and dry, yet we might have worse weather I think. It will be odd if we have not rain by the time we need it. But God's will be done. It is always better to have fat stock than lean, because you can turn the first into money while the last may starve. I may forget, but I do not think I recollect so windy a time as we have had for 2 months now. But what a blessing it was that the night the fire happened at Wark it was quite calm, which has scarcely been the case before or since of many weeks. Now that is sufficient to convince us that it was not done by design, because they would have chose a windy night. But by insuring you defeat the *diabolical intentions* of such unfortunate persons. How depraved must a human being be become, who can cooly and deliberately set fire to their neighbor's house, yards &c when dead asleep, and not only burn him but many others to whom they could scarcely have any ill will. They must be in the last stages of depravity! How unenviable, rather how pitiable are such forlorn wretches! We can hardly conceive that such abandoned creatures exist. It can't be in the country, it must be in the sinks of vice in some great and wicked town. Pray God keep me and mine from such places, and turn the hearts of these unfortunates to better ways of thinking. Another of the Wark cows is dead, and indeed it is not easy to say what may be the result with more of them. Poor things, they are shocking objects to look upon. This day is strangely cold indeed, and the wind violent. Was not for the strength of the sun it would be hard to bear. Such continual storms are remarkable. *Tuesday 27.* Wind violent all day, but being stuck to the south I had hopes of rain, however only some slight showers in the evening came on with a tempestuous wind. Snow had fallen upon Cheviot at the same time. *Wednesday 28th.* A calm frosty morning, the white hoary frost lying in the garden walks. You will have nothing said about wool yet. Ask Mr Colling's opinion. It is right to have the best information one can upon that subject. Mr Stubbins writes that it is selling at 35 and 36s the todd of 28 pounds. Now 35s is just 15d per pound, or 30s our stone of 24 pounds, a famous price to be sure, which help to make sheep still a valuable animal.

[*M.C. jr*] Mr Bates of Brunton sent my uncle yesterday two good strong horses, so you need not buy any now for Wark unless you have already bought, as they have now got 9 horses, that is one from you, two from Mr Bates, six from Shotton, three of them very good horses the other three only midling. We have a steady[34] and pair of smith's

---

34 A device for holding steady something being fashioned; in this case presumably in a smithy.

bellows to spare from Shotton. The bellows are very good. You may
have them if you think them worth carrying so far. They could come
in a cart along with turnip seed if you want any. You can consider of it
and let us know in your next.

M.C. jr

[*G.C.*] John by a letter received this moment from Mr Peacock he
says he will allow us 4 per cent for what money we may let him and
Mr Graves have, until wanted at Martinmas, as we did for the money
we let them have before our last payments to London. Therefore I
must request you to let Mr Peacock have what money you can spare
all along, until you have my brother's and my orders to the contrary.
You know I had ordered you to put your money as it came to hand
into the Darlington bank. But this is *more* than we can make by
putting it *into the bank.* You must let them have it whenever you can
spare it, and if you have already put any cash into the Darlington
bank you must take it out again and give it to Messrs Peacock and
Graves, except the bank will allow you 4 per cent for it until we want
it, which will be on the *1st* of *October* next. Whether we receive a letter
from you tomorrow morning or not, I will send you this to you per
post, that we may have your answer about turnip seed and the *steady*
and *bellows* now at Shotton. And I *beg* and *intreat* that you will write
about these things immediately on receiving this letter from us, even
if you only write *half a dozen lines.* Never mind postage on matters of
importance. Along with the bellows and steady, 2 or 3 cwt. of turnip
seed will be as much as a horse can take perhaps. Do write and oblige
yours sincerely

Geo Culley

PS. Upon second thoughts John, perhaps you had better send
away a single horse cart to fetch the bellows, steady, and as much
turnip seed as will make up the load, particularly if you are in want of
or wish for any turnip seed. I am sure if you do want any to sell it is
more than time that you had it, and you may have a *shrewd guess*
surely by this time whether you have a chance to sell any or not this
season, because if you can not sell it at 1s per pound, don't have it by
any means. Besides the *steady* and bellows can be got to you afterwards. What I mean by your sending immediately on receiving this
letter is to save time, because before you write an answer to this and
we receive it, your cart would be here. But of course I don't desire you
to do this if not agreeable to you.

[*M.C.*] John on speaking to Bob Tait today he says the steady and
bellows were new when we got to Shotton, and the steady is a decent
one. Only if you have bought a steady and bellows, then you will not

want those at Shotton, which are now of no use to us at all as we have neither plow nor horses there now. So we think they may be usefull to you. Mr Bates sent us a very strong mare and a horse also very stout, but 7 and 8 years old. I paid 28 for 4 through bushes charged to you by Mr Winfield of Gateside,[35] have an idea that you got them from there. We will hear how you come on with smith's house and shop in your next, and that the Storys and you are well. Adieu

Matthw Culley

PS. As we have plenty of turnip seed you may as well send a cart and have some. Also we want the rape seed from you

[*G.C.*] John you ordered us a digester[36] about 2 years ago, and Whinfield got pay of my brother at Wark today, only 8 6*d*. Did you ever pay for it? *Thursday 29th* As there is no letter from you I will send this per post to you today. It is a most severe frost. John I have received a long letter from Mr Faulder today wherein he says 'I have wrote to your man at Denton to pick me up if *he can make it convenient* without interfering with your business about 30 or 40 cows to value, any time from September to Christmas'. Your master and I could have no objection to your doing a thing of this sort with advantage to yourself, but I am persuaded that you have quite enough to do for us without working for Faulder or any other person, except it was to assist him or his son in a market provided they were present. Besides 30 or 40 cows will amount to a large sum. He may give you leave to draw on him for the amount, but how do you or I know what his situation is? We believe him a *good man* and I hope he is, but we can't be sure of that, and at any rate I think you already have more business than you can or will get through without an assistant behind your hand when you are at market &c &c, and I sincerely wish Bob Dippie may answer for you in that way. If not you should look about for one. I am sorry to find that Faulder says that fat stock comes down and will be down at Smithfield. If that be the case it will be further down with you and us. I must therefore once more beg of you to sell if you can, the 20 oxen at 8*s* or less if you can't make 8*s*. We will send some spayed queys &c soon, which with a little grass will make nice ones. The worst is that the consumption is nothing from Smithfield to

---

35 Richard Whinfield, iron founders, Pipewellgate, Gateshead: *Newcastle and Gateshead Directory for 1795; Mackenzie and Dent's Triennial Directory*, Newcastle upon Tyne 1811.

36 A strong vessel in which bones or other substances could be boiled at high temperatures so as to be dissolved.

Morpeth. However we can bear it as well as other folks. Wheat has comed down 8s per quarter at London last Monday. I wish you and us all to be quitting part, although I am not so affraid of wheat in stack. But one thing brings down another.

Geo Culley

Wooler one o'clock. I have only room to say that I hear Morpeth was the worst market ever known yesterday. 3000 sheep some say, and such numbers set up. I don't know what the price per lb. was, but John if you think you can sell our sheep at 7 ½ per lb. naked, we will send them to you. They will not be sold for more than 7d per pound at Morpeth now. Do say your opinion. G.C.

116. *George and Matthew Culley and Matthew Culley jr to John Welch*

Eastfield 30th April 1802

Well John Welch

I sent you another long letter yesterday from Wooler, which I hope will reach you tomorrow. After violent and cold frost, we have again got milder weather. I mentioned Morpeth market. My second account that I heard was not quite so bad. A very full market, but nearly all quitted at last except 100 wethers which little Tommy Bates[37] turned out directly, and said he would take them and 300 more away to Skipton where he had met with a good sale before. As far as I can learn the clipped sheep were not sold for much above 7d per lb. sink offal. But it is evident to me that both corn and stock is coming gradually down in price, only the high prices things to feed were and still are bought at makes people fight as hard as they can. But down they will be to a certain level I think. Should they come down to such prices as we have known not in these few years (which I do not expect) it will be hard upon new taken farms. I have agreed with Tom Glass to be your shepherd, to have 5 ewes, 3 hogs (2 of them wether hogs and one a ewe hog) and 2 fat sheep with 5s per week and a cow kept, if you approve of it. Thomas Glass.[38]

John I believe I have given Tom Glass more wages than perhaps I ought to have done, but I think you are so badly of for want of a shep-

---

37 A livestock dealer. His relationship to the Bates family of Aydon White House (*History of Northumberland*, vol. 10, pp. 377–9) is uncertain. He is thought not to have been Thomas Bates later of Kirklevington, 1775–1849, the shorthorn breeder (see No. 25, n. 49).

38 Thomas Glass's signature.

herd. He is to find his own cow mind. By a letter received this morning from Mr Peacock he says 'John Welch has paid me in bills, good ones but rather long winded ones. Three hundred and eighty pounds on your account. But I have been obliged to use part of the money to make up my last remittance to London. When I have your directions I will apply the balance as you chuse'. So I have wrote to him this very day that I had directed you to pay him what money you could spare. Only do you mind to keep an exact account of what money you do pay to him and the dates *when*, as they, viz. Messrs Graves and Peacock, are to allow us 4 per cent for the money given them untill a certain day. And I hope you will have it in your power to give them a good deal in the course of this business, as most of our fat cattle are likely to pass through your hands. As Tom Glass said 'when he came away from Denton that you promised to write per post to us', we have anxiously been expecting a letter from you, but none is arrived yet. I am glad that Tom got all the hogs well up, and the one left upon the road also, and says that honest Jemmy with the 20 oxen all well about Wallington. John I have said a great deal to you about bills. It may be wise nevertheless to take bills rather than nothing at a particular time, but you ought to know that you are never safe with bills until the time they are dated is run out, and your payer may be broke or fled long before it runs out. However if you are under the disagreeable necessity of taking bills, and *long winded ones* too. I would certainly go to a bank if at Darlington along with the *payer*, and see if the bank will cash them. If not why take them? In short bills are a very unsafe way of doing business you may depend upon it. But I have said so much on that head that if it has not had an effect upon you it is vain to say any more, and I am extremely unwilling to give you any uneasiness. But I am certain that it would give you very great uneasiness to lose us a few hundred pounds by bills. It would be much better to take a less price for your goods of good people, at least those who will pay you in good money of any sort that you are acquainted with. Can't you tell these jobbers that it is my express *command* that you are to take no bills?

Saturday 1st May. A very nice morning although rather frosty. We have just been drawing 42 fat sheep for Morpeth which Matty is going up with, as I cannot attend upon account of Wooler fair on Tuesday and Wark court on Wednesday. The sheep are very fat, from 22 to 27 lb. per quarter, 6 segs 30 pounds per quarter. We are expecting 7 ½ per pound sink, but it will be well if we don't come of with 7*d*. But whatever is the price we must be content. Ours will pay as much as other people's for their meat I am certain. A Mr Wood from some part of Yorkshire has been buying hogs in this neighborhood, some say he comes from near Wetherby. However he has given better prices than any others. He bid George Purvis 50*s* per for 76, not one to be cast, money paid, 8 taken away directly. But George happened not to take it, and I am doubtfull that he may not soon have the same offer. They are large coarse hogs with much wool. George says they are fat, but I never knew those kind of hogs so fat as kinder ones. It is not impossible John that things may be better sold towards the end of the summer and autumn again provided the turnip crop proves promising. In that case it may be safe to sell Bill Wood's hogs in the autumn. To be sure it is long until then, but there is no harm in speaking of it. You will have a good few dinmonds besides. *Saturday 11 o'clock*. Well John we have received yours of 29th this moment. I really think it would be wise to send you our wethers, at least the best of them. However we will see how Matty comes on at Morpeth on Wednesday, and if he can get no more than 7*d* per lb. we will send you the remainder away directly. I have no other objection but giving you too much to do, but if they are sold at your own home it is not so much trouble as selling them in the market. We cannot take them now to Morpeth under 1*s* per head, reckoning all expences, and I don't believe 150 sheep (which will be about the number) will cost much more than 1*s* per head to Denton. Then if you can get ½ penny per pound more they will pay plenty for sending, and save us much trouble. You will have time to answer this before we can send them of. I believe *I am wrong* and will therefore be guided by Wednesday market all together. So you may partly expect them I think. Mr Peacock and you agree in the sum of £380. You will get pay for butter firkins some time I hope, but the man should have paid early as he got a *hit in the eye* which is all fair now, nor would I have named it but for George Nicholson who our wethers would suit well a part of them as they are very fat, and it might be right to let him know. We have received 29£ for my brother from Mr Peacock. We have no want of rain thank God yet, nor is there any appearance of rain, but I trust it will come in due time. I am determined to sell if possible at Alnwick fair. Nevertheless we will send you no more cattle until we hear from you again. Not but the queys should be going to you soon if required, because they will require a little time at grass, most of them, and you never reflect that cattle are at least 8 days on the road to you when set off. I hope you would receive my letter in time to stop you buying any horses at *Northallerton fair*. We have got plenty at the present. Mr Bates ought to have wrote to you that he had bought and sent 2 to my brother. He is a good man but a *most indolent writer*. You will see that I have hired you Tom Glass, and perhaps at too much wages, but you are in want and Tom is a willing decent herd. He is to drive, or come here for stock, I mean whenever you can spare him, which is any time except in *lambing time* almost. He is such a famous driver and knows the roads so well. To be sure it is a benefit to him to drive, which should have been considered in his hiring, but I am not so good a bargain maker as I used to be. I am very glad that you have hired Bob Dippie, as I hope you will find him usefull many ways to you. He may be too quiet for an *overseer*, but he will improve with practice. I am sorry we have not your answer about *steady* and *bellows* &c, because time burns about turnip seed. It is more *than time* you had it, but I hope you will either write again on receipt of my last or send away a cart *of your own*, or both. We are rejoiced to hear of Jemmy Story and family recovering poor things. We shall most likely send away the cows and spayed queys first, and likely before it be long. Shotton is very dear sold to a certainty, but land will be very high sold now that stocks are so high. We are going to Wark this afternoon when we will send this letter away per post, which is the quickest and safest for letters. You see this of yours is only dated 29th (Thursday) and we have got it on the 1st May (Saturday), only 2 days, and you should receive this tomorrow evening or Monday at the furthest. Yours in haste

Geo Culley

[**M.C. jr**] If any of your customers enquire for turnip seed of you, tell them they will get it as there is some coming, you fixing the price according to the rate in your country. Also if you have any old seed be sure you mix the new and old together, as they will sell better that way. Mr Wright of Cleasby will be a proper person to enquire the price of as he sells a good deal.

[**M.C.**] John Welch I received of Joseph Hodgson an order on East Ord for 2 milititia [*sic*] men, John Kirtley and James Lilly, from the township of Darlington. They, the East Ord people, refuse to pay it. It has been offered to them. It shall be sent by the first man who comes to Denton, to save postage to them. We likely will send to you about a score of cows and spayed queys in a little time, only I am afraid of over stocking you which will be wrong, as your pastures are not great and full of grass. Yours in haste

Mattw Culley

117.   *George and Matthew Culley and Matthew Culley jr to John Welch*

Eastfield 2nd May 1802

Well John

We sent you a letter yesterday directly by post in answer to yours of 29th April, which came to hand yesterday. So you see my method is to begin a fresh letter directly after sending the last one away, and by writing every day a little of what occurs to me, nothing almost *escapes* and thus the letter is ready to *finish* and *send* away to you whenever one *arrives* from you. This is a fine sunny morning but rather droughty to what yesterday was. A great many fat sheep have gone past from Scotland yesterday to Morpeth, which denotes a full market. There are such improvements now on the Scotch side of the Tweed as well as the English, that immense quantities of stock, and turnips in such quantities, are now produced where none were grown 20 or 30 years ago, as well as corn that really surprises one. But it is a most productive country all through the Merse and Roxboroughshire, quite up to above Hawick now, where they are getting more and more into the Dishley breed of sheep, so that inexhaustible supplies are produced and fed now where none were fed 30 years ago. And this wonderfull expected advance in wool, and particularly long wool, will encourage and diffuse that breed more than ever. So that I think it more than probable that you may expect the principal part of our wethers by and bye, only Harry's are not nor will not be clipped until the end of this week, so that tomorrow week is as soon as they can possibly set of from here. Only you can name it to your customers and feel what they say on the occasion. They will take more than 8 days going up, but I know well that by giving them plenty of time they will lose very little when clipped. Be sure to pay attention to the scab among Mr Wood's hogs. It will be an aukward business if it be catched by your flock, ewes and lambs or gimmer hogs. That makes me rather wish for you to part with them in the autumn or at any rate about Christmas. Scab is so very unpleasant a companion. I wish you had 8*s* per stone for the cattle comed to you. Ours are now full as good, perhaps better on the whole, and I will take 7/6 or even less than 7/6 before I will send them south from Alnwick. I think you will have enough to do without these inlaid cattle. It was different with Wark cattle because they had been turned out in the day for some time before the fire, and then altogether. I am glad that you have got settled with the Todds. We will most likely send the spayed queys and cows first, although there are some very nice fat oxen still. I hope you would understand me about the inlaid and outlaid oxen in the 20 comed to you, because the outlaid ones will be worth more per stone, or at any rate will improve more for keeping. If you are able to sell so much turnip seed as a cart will carry, I think you are very wrong not to let us know that at an earlyer period. We can furnish you with plenty I am sure this year and next also, and I hope in future so long as my brother and I are concerned together. Our [. . .]39 well afford it at 1s per lb. surely. It will pay at that price beyond anything else we produce. Berwick was very flat for grain yesterday, wheat 42 to 44s, barley 19s to 20/6, oats nothing. Very little done in anything. But if this drought continue a month or 6 weeks longer, oats will become more valuable or I am deceived. At all events it cannot be wrong to keep oats now, when people have granaries or conveniences for keeping them. However Ralph sold 20 bolls wheat at Kelso on Friday at 51s, he could have asked 55s the week before. But Ralph is like, I had almost said John Welch, and perhaps George Culley also. He likes a little of his own way. *Monday 3 May.* A very fine morning, although a *little frosty*, with a *north east wind* and a *new moon* yesterday and a *rising glass*. But I never saw a greater appearance of dry weather in my life. And if the wind only keep down we can do without rain a good while yet, at all events it is our duty to *await the decrees* of that Almighty Being who best knows what is most expedient for us. And let His will be done. 7 o'clock. After a very fine warm morning at 5, it has turned cold and as ill natured as can be.

Geo. Culley

[*M.C.*] Well John I am not going to blame you, Matty Walker's mare was dead in her stall this morning when Andrew Bolton looked at her, never supposed but she was well, as her legs were laying under her, had died seemingly without a struggle. It is possable pottatoes were the cause, but that's all conjecture, altogether conjecture. But our own horses are accustomed to such things. She might be an ailing mare or not, she brought well and I thought her cheap (there is a pair of vice also at Shotton, as well as *new bellows* and steady). The vice are not capital and want a spring which Tod Hunter will repair for you, the steady midling but the bellows capital.

Mattw Culley

[*G.C.*] *Tuesday 4th May.* Another fine sunny morning without a cloud almost although the wind is south west. However I think this will prove a fine warm day for Wooler fair. We went to the post this

---

39 There is a blot on the page here.

morning hoping a letter might be from you about the said steady &c and particularly the turnip seed, which it is more than time you had. But I hope you are sending a cart yourself which will be at this place tonight. I wish it may be so. But if so you should have wrote a line per post that we might have the things here ready. The steady &c are all at Shotton and we don't care to remove them until we are certain whether you will have them or not. Some were saying that perhaps Mr Morton will buy them as he gets the farm. Well John just returned from Wooler fair, where there was a number of people and many more cattle than I ever saw there. A vast of clever 3 year old steers in high condition, 13£ to 14£ a head. Day very warm indeed. *Wednesday 5th May.* A very fine mild morning, glass very high, and no appearance of rain although a little clouded so as to conceal the sun. Nice weather for working turnip land &c. Indeed dry weather is always favorable to the operations of husbandry. *Thursday 6th May* 5 o'clock in the morning. At 9 o'clock last night Matty Culley brought me yours of 3rd inst. from Morpeth, where he experienced, as we expected, a terrible bad market, 6 ½ per lb. the outside price. So you may lay your accounts for 150 or more of the best sheep we have. 55*s* was the best price he got for sheep, 27*s* per quarter, and he was forced to trust George Nicholson and John West 8 sheep at the last. The price he will tell you in this letter when he gets up poor fellow, but he went to bed so fatigued. It is likely George and John will pay you the money, I hope there is no fear. No cattle sold above 7*s* per stone. We shall have a devil of a market at Alnwick to a certainty as the butchers at Shields &c are all full. However I am determined to sell if possible. Nevertheless if we can't they must come forward to you. Only as you have and must have some large lots from us besides, we will certainly try to sell at Alnwick. I am only affraid that you will be too stiff with the Wark cattle that are comed to you. I must beg that you will be doing, even at 7/6 per stone or less rather than keep or not part with them. We had a surprising shew of 3 years old steers at Wooler on Tuesday, very full of condition, nay many of them decent beef. Bill Pickering bought 6 of James Darling at 5*s* per stone pretty good beef. What can we expect now after this for some time but *down the hill* until things find a level more. Turnips you see having been so plentifull in this quarter that everybody had their 3 years old steers in high condition, so as to be marketable in a few weeks to the butchers, and a vast of these are in hand unsold in this district. I am glad Glass got the 20 ewes well up. I hope the next account will be that you have sold them. But I am sorry that your ewes are so many of them barren, but as you say we must try to manage better the next year. We will certainly send you the turnip seed next week, and the smith's tools we must dispose of to the best end we can. I have no doubt of your being strong and busy, which is good for you, and I am glad that your lime trade is brisk. Only I will blame you very much if you don't sell for a fair profit. Why run over my brother's pretty land without a fair compensation for so doing? You forget that in all trade there is a loss by bad debts, and in none more than the lime trade, or it is much altered for the better since we learnt lime at Denton. Besides it gives you a great deal of trouble in keeping accounts, collecting &c &c, which I know you don't mind as it is your duty and pleasure to do all you can for your masters, I am convinced. But I hope you will make Bob Dippie write at nights and assist you all he can. It will ease you and do him good. But whatever you do, don't work *for nothing*. By all means keep the horse you have bought, and even buy another rather than want. It is impossible to carry on business without *men, money* and *horses* all *usefully* employed, and then I will be bound for the consequence or result being advantageous. You know that neither you nor we could know that Shotton smith's tools would be to spare so soon. But I hope Harry Morton will buy them.

My brother is busy roofing his burnt houses. 2 more of the cows are dead, and there is no saying what may be the consequences of the rest. If it had been properly considered they would have been best killed at the first, but nobody can be blamed because we could not foresee the result and it is well it is no worse. Only the poor animals have had a sad time of it. Whatever they sell mutton at, you must know better than me, but I know that at the prices sold above they can have a handsome profit at $6d$ per pound in joints. But to a certainty butchers have had a most beneficial time this spring. They make more of one sheep than they were accustomed to do of 3, and same of cattle. Well well, we have had our good times, 'and every dog has his day'. I will be only glad if you can get $7d$ per lb. for the clipt sheep we shall send, let alone 7 ½. You may well, but I think it will come to that at Morpeth that they can't be sold scarcely. You see we have such crowds of lambs all likely to be very fat in this country, that will be ready and many already in the market, besides steers and queys in abundance. This is a land of plenty and well it is, so let us be thankfull. We must be obliged to send you another lot of cattle very soon John, as well as the wethers. God bless me you can sell if you will, and I hope you will, but we cannot hardly sell here at any rate. Ought we not then to be thankfull that you are so situated as to sell for us, and why not avail ourselves of that advantage. I am only affraid that you will be too stiff. However if you be, don't blame me.

Only recollect what an escape you had with your fat cattle. O! it is unwise in times like these for a man to be too stiff with a good customer. I am happy in saying that I believe few men have had or have better servants, but both you and Ralph Brown are too *stiff*, too *hard* at times. I need say no more to you about paying money to Mr Peacock. Continue to do so until you have any contrary orders or see cause to deposit yourself. You buy iron and wood cheaper than we do, but I believe your wood measure is not so good as ours. Have nothing to do with Faulder. Make your apology. Surely you have work enough without his employment. I don't wish to blame you and Ralph Brown for not selling your wheat when prices were better, because you both do it for the best, but nothing like being content with good and fair profits. Wheat is down 5s per quarter now this week at London. So it is. But never mind that, if it will only make you and Ralph wiser. But I have my doubts. My brother must answer you about the repairs at the Moor House in another letter as this is done and I will send it away today. But your master had no business with Birkbeck. I am in real haste yours truly

Geo Culley

PS. Certainly the best way for two men to value the premises at Moor House, but my brother will write in next what he thinks right. Mr Peacock says he will explain to you about the East Ord paper. I am sorry to hear by a letter from Mr P. that Joseph Hodgson's eldest daughter is very ill.

[*M.C. jr*] I had a terrible bad market at Morpeth yesterday. I sold your friend George Nicholson and John West of Chester 8 wethers at 50s per, 20£ which you are to receive of them. They bought them because they were so very cheap. Indeed I am sure there was nothing sold at 7d per lb., beef hardly 7s.

118.   *George Culley to John Welch*

Eastfield 7th May 1802

Well John

We sent you another pretty well filled letter yesterday, and Ralph Brown and Matty went to Thornington and Wark to see about what wether sheep to send you. We can't ascertain the quantity yet, but will write by the cart on Monday which goes with turnip seed &c how many are coming. George Purvis sold his hogs after all for something above 50s per, but many lots are sold at 44, 45, 46 and some upwards, which in my opinion is still dearer than fat wethers by much. I have heard nothing this long time from my friend Mr Sayle about the Yorkshire markets, so that I know nothing how Wakefield is. But whenever it comes across my mind about your selling your 17 cattle I am rejoiced above measure. I only wish you may get near 8s for those that are come to you, and I wish that you may not have laid a *lance* on Holdgate and Shillito so as they dare not come again to you. The Shields and Sunderland men will never be so mad as come to Darlington when they can buy so much cheaper at Morpeth, and this is all against you. However let me advise you to get through them and not be *too stiff* on any account. You may depend upon it that you have a good deal of work to go through. This is a wonderfull fine mild morning. But say you, 'We want rain so much'. True, but then if we had not some wants we would not know, or be truly *sensible* of the *blessings* of the Almighty Being! I look upon it, that all these *difficulties* and *hardships* or rather more frequently only *seeming difficulties* and *hardships* which we meet with are real *blessings* If we had nothing but *plain sailing* as the sailors call it, we should forget ourselves. If we did not know what *pain* is, we would not nor could not be so sensible of pleasure when it arrives. In short many things which we consider as *hardships* at the time often turn out *blessings* and *comforts*. So little do we know ourselves and are led by our *passions* and not *reason*! Indeed a great deal more than half the troubles of this life are of our own bringing on or creating. We become often *fretfull* and *uneasy*, and we don't know *what about*! All these seeming difficulties bring us to a sense of *our duty*, and when real troubles come upon us it is generally for our good although at the moment we don't or can't think it so. And how many *blessings* and *comforts* do we *enjoy* that we are not truly sensible of or do not see *at the time*. If we compare our *situation* with many of those around us, how much reason have we to *rejoice* and be *thankfull*. And when misfortunes befall us, it is very *seldom* indeed, *perhaps* never, but it might have been worse. Wark fire was one of the most alarming things I have ever met with to my remembrance. To be called up in your *first sleep, in the dead of night* with the cry of *Fire*! A fire in the most valuable farm you had! And when we got to the north side of these buildings to see the *immense light* in the atmosphere occasioned by the fire, was to me the most *tremendous* and *sickening sight* I had ever seen. I could have turned back and gone *to bed*, for I was *really sickened* at my *heart*. But we were to go on then, and I may truly say that I was perhaps never more truly thankfull to the *God of mercies* than when we arrived at Wark and saw such a *wonderfull deliverance* in the *midst* of *so many* and *great dangers*. And would not I have been a *brute*, an *ingrate* and an *unfeeling being* if I had not been truly and sincerely thankfull to the *Father of mercies*! Now John I have given you a larger piece of morality than I intended at the beginning, but when one gets into a serious train of thinking one thing brings on another. But I hope what I have said will neither do you nor me any harm.

Saturday 8th May. Ralph Brown was at Kelso yesterday and sold only 15 bolls of wheat at 50*s*, which he thinks right as wheat will not at Berwick give above 42*s* per boll. This is a very droughty morning indeed, and very cold. The wind was south east yesterday a good part of the day, and the glass although very high fell a little. But this morning it is stationary and the wind a little south west. It has so much the appearance of drought as I ever remember seeing it in my life. Yet we don't know how soon it may please God to send a change. I am going to Thornington today to draw such lame wethers and segs as are unfit to do much, and those which are fit are to set of for Denton on Monday morning, as well as 11 cattle for Alnwick at same time, and the cart also with turnip seed and chests &c for Bob Dippie and Tom Glass, and by it I will send this letter if not sooner per post. Ralph Brown sold some midling wheat at 46*s* per boll today at Berwick. John I am glad to tell you that by a letter dated 5 May (Wednesday last) Wakefield from Mr Tom Sayle, saying that clipped sheep were sold at 9*d* per lb., which is 3*d* per lb. almost more than we got at Morpeth. However if you get 7 ½ I will be well pleased, but 8*d* per pound will be better. Cattle not quite so brisk but sold from 8 to 9/6 per stone sink. The lowest price is more by 1*s* per stone than we can get in this frozen country. However John don't let this make you too stiff, but get those sold you have, to make room for more which will soon be with you. I have been looking at our Westfield cattle which really look well, and this Wakefield account will perhaps tempt me to send them from Alnwick to you rather than take 7*s* at Alnwick. Otherwise I am determined to sell them at all events, but this has given me fresh spirits. Still I dare say I will sell them if I possibly can, because although a very fat lot they are too heavy cattle for the south country markets at this season of the year.

(Sunday 9th May). A fine mild morning but not the least appearance of rain. So we must be content and thankfull with what it pleases the Almighty Being to send us, 'and let His will be done'. Now in my judgement you should have had another score of cattle (the queys and cows) sent directly, because they would be better many of them with a little grass before selling if you have it. If you have it not, they would be better for a little rest, if it was only in midling meat, nay even *your moors*. I speak this *strong* because I can't help thinking it very unwise not to have one lot ready as another goes away, and you certainly will or must sell at some price. Otherwise take them forward to Wakefield. Besides I always think that you forget that it takes the cattle 8 or 9 days in going. Then they ought to have as many more to refresh in. This to me is *so level* to *common sense* that I am amazed you don't see it, and supposing that they should eat up your pastures, which I will also suppose rather dry and not growing owing to the drought, yet they may as well take the grass as let the sun have it. To me it is the height of absurdity, because if our information from Mr Sayle is just, you can sell at a fair price, nay prices we did not expect from late appearances. But at any rate they *are* and *must be* sold now or soon, let prices be what they will, because they cannot be kept though summer. But I am scolding when it is too late, for I dare say my brother would distress himself ever so much rather than send them to you before you wish to have them, and now our turnips are done and I would suppose full less grass than you have. Well I have only one vote in the Senate House, and perhaps it strikes me in a *stronger* manner than other people. But it can't be helped now, only it appears to me and always appeared to me that it was absurd to wait until we saw whether we could sell at Alnwick or not, because as I have repeated, sell we must. Well after Alnwick fair on Thursday next, whether we hear from you in the mean time or not, if I can have any weight in the *counsel* we will certainly set of 20 more or less of cows and queys, therefore I believe you may lay your account for them to be set of at that time. Mr Tom Sayle says 'Wakefield was better for sheep by 5s per head than the day before'. 'Mr Booth of Killerby and one of the Todds were there. Mr Booth had some cattle'. I have known and been there with Booth, who is about as heavy a man and midling a salesman as I know. I have no doubt but you could do as well again. In markets of that sort they are always fond of buying of a stranger. He, Mr T. Sayle also says that 'he understands that Skipton was much brisker'. He also says that he was at Huddersfield (a very populous place) where they had sold some nice Scots, which the butcher was selling at 10d or 1s per pound for choice pieces. These are very encouraging accounts, and you may depend upon it that we have some very nice queys indeed to send you, and some clever cows. But I really think they should have been upon the road now, and within 2 days journey of you. It is not one year in 10, nay not in 20, but fat stock is sold for more by 1s per pound with you than here. I mean sheep, and cattle full 6d to 1s per stone and often more. Ought we not to be thankfull then that have it in our power to *avail* ourselves of that *opportunity* of selling with you and by you, who have no fault as a salesman but that of selling too dear, which arises from the *fair principle* of *making* the most you can for your master. But don't forget that a hard seller is not the *best salesman*. But I hope I have said enough on that head before. I am not joking about going to Wakefield. If you had a clever fellow behind your hand, that you would be spared, I will be bound for it you would sell at Wakefield well. The only misfortune would be that you would be turning jobber.

Well John I think I will send this letter per post tonight and write by Jack Davison again in the morning, as the letter should be at Darlington tomorrow afternoon some time. I don't know when the mail passes through Darlington, and then the different bags have to be sorted and regulated before you can get a letter from the post office, which will require half an hour more or less I suppose. The number of sheep we send will be 200 as far as I can learn at present, or it may happen a few more, but that few will not be more than 10, likely less. Big and little, all wethers and segs, but more of the latter but young ones, 2 and 3 years old. Some of them are not very fat, but handsome mutton, and worth more I fancy with you than us by 1*d* per lb. at least. Many of the wethers appear to me very fat and many strong weights. Some little ones from Wark not very fat but all good mutton. They have not cooked them well at Wark. This winter ours were well managed. We have taken out 2 or 3 to keep over year, which are likely to make remarkably fat I hope. And if Tom gives them as much time as he is requested, and I hope he will, they will come up fat and clever. Tom wished them to have a mark, as he said one might be stole missed in the road. So we gave them a slight tar mark behind the head. They would not *boost yet*. I am yours wishing you a good market

Geo Culley

PS. Let Tom come back immediately to meet the cows and queys.
PS. If I possibly can I will drop you a line from Alnwick.

119.   *Matthew Culley to John Welch*

Wark 9th May 1802

John Welch

If we send you stock so as to be over stocked, you must sell away to good and safe men ready money people or good bills. They are good and we must take times as they run. Sometimes seeing too far for high prices is a loss. Take times as they run. Human judgement errs. I probably have been the cause of our lott of cattle being kept too long on. They might have been sold, and now we could have bought plenty of fresh stock &c cheap. They were at Wooler fair. I bought 2 dear ones, but Mr B. Pickering bought of Mr Darling a lott of beef and beefish steers at 13£ per if 52 stone, only 5s per stone. When times are at the worst they may take a turn. I hope Bob Dippie will now be serviceable to you, and if rain comes not very soon your old pastures that are not covered with grass will soon fail, as Denton will not bear drought without a cover to keep it out. We are now able with good looking to keep our stock if rain comes soon, but it seems very like drought. G. Humble's wife has brought him a daughter.[40] You should not sell too dear to Shillito and his friends, as you must have affronted Jery Clayton. I thought the Thornington wethers (they were wintered at Eastfield) were very good when we clipped them on Friday last. H. Rutherford's wife is now seemingly beginning to mend. Pray how is Mrs Peacock? If you think it right to say anything to Birkbeck about the tenant which is gone likely before you get this, may get Johnson to write to him and say I will prosecute him, Birkbeck, for not leaving things in repair and for his tenant taking away the straw and also laying his dung on his wheat after all reasonable time was past. I am ill treated by them. After all much depends on the lease, which Peacock made, and I think he leans to Birkbeck much. I think P. leans the way he likes too much. You will probably have to pinch your ewes and lambs on the moors, or the backwardest hill hogs. You I hope have not eat your watered land. If you have, beware of a [. . .][41] crop, will have to water the edges by hand. How does Pattison come on? Does he like to harrow or plow? He may as the weather is so dry and pleasant. We are glad the miller and family get forward again. The son had been injured by working that morning I apprehend and not taking proper care to keep from cold. Can you get your smith's house and shop soon? Surely you cannot sell lime at last year's prices without being out of pocket. Ought to advance according to the advance in coals and labour. I hope to hear from you by Jack on his return. Glass will give you the order on East Ord, but I forget who gave it to me, Joseph Hodgson or Groves. Let us hear how Mrs Peacock does, mending we hope. It will be after turnip sowing be over before we can get to Denton. Respects to all your neighbours. Hoping this will find you all well. Adieu

Mattw Culley

PS. You must let us know how the wool trade goes on. Mr Vardy

---

40 Jane, daughter of George Humble and his wife Jane Oliver, was baptised at Carham on 12 May 1802: NRO, Carham register.

41 There is a tear in the paper here.

saw Matty on his return north, also Charles Selby and Anderson of Glanton.

120. *George Culley to John Welch*

Eastfield 9th May 1802 Sunday

Well John

I have this moment sent you a long letter per post, and like a man of business and who by my actions I *think* proves that I don't wish to spare pains to do what is *my duty* to do. I can't exactly tell the number of sheep still, but there will be between 200 and 210. You see Jack Davison goes early away in the morning with 8 cwts. turnip seed for you from Wark, very good I believe, 1 cwt. for Mr E. Potts to leave him at Glanton, with two chests or boxes for Bob Dippie and Tom Glass. Tom's sister will not go south on foot, and I think it fair for her to ride. So I will endeavor to prevail on my sister at Wark to let her have roany Galloway and it can be returned by the drivers. I will if possible drop you a line per post from Alnwick, but I think you may rely on a lot of cows and queys, many of them very good, to set of from here on Thursday first to you, which I would wish you to send Tom back directly to meet, as the remainder of what are good of the outlaid steers should follow soon for fear your markets may drop. Many of these you will I hope find very good also. I am with every good wish Yours

Geo Culley

PS. I think I said to you in a former letter that the selling of hogs was the wisest way, and I really believe that if you could sell Mr Wood's hogs over head for 47*s* per with their wool on, it would be famous pay and perhaps the best end they may ever make. However I leave it entirely to your own judgement and opinion, only we could send you more wether hogs in their room if you should want. Indeed from the condition of Mr Wood's hogs when we bought them. they may pay for clipping and keeping until the autumn, because they are not a large hog. But really hogs are selling for more here by ½ *d* per lb. than fat sheep, at least in my judgement. For what can pay like hogs at 40 to 50*s* per? It is better than fat lamb a deal. However don't forget to look near to the scab in Mr Wood's hogs. That is one reason for selling them. But by keeping them from other sheep and having a shepherd I hope you will keep clear of scab, or else it is an awkward disorder. John I shall be sadly mortifyed if you don't sell all or most of your fat cattle when you have an opportunity, because we shall be hard fast here if we don't send you away the rest we have here. So pray sell and make room. G.C.

121.   *George Culley to John Welch*

                    Eastfield Monday 5 o'clock 10th May 1802

Well John

This is as dry like morning as ever with a rising glass. Jack Davison just setting off. I have sent a short letter by him, and I will send you this from Alnwick if I don't forget, and I hope you will receive the letter I sent you yesterday per post today or tomorrow at the furthest, as it contains a good deal of important matter and will prepare you for Jack Davison's arrival in case you have anything to send back with him, because he can only stay all night with you on Wednesday night and return directly on Thursday morning as it takes him 3 days each way. The sheep also set of today, to Humbleton first night, Tom Allison's Glanton Bank 2nd night, and Rimside Moor House 3rd night, to get a good bellyfull of fodder. After that I know nothing of their places of stopping. The 11 cattle likewise set of for Alnwick fair today. 11 o'clock. We have this moment received yours of 8th inst., for which I thank you for writing so readily. I hope you will learn to write journalwise by and bye. I am also glad you are selling your wheat and maslin so well. It is a very good congratulation John to have good markets after so many bad ones. It is a pity but Matty had sent his 42 wethers from Morpeth. By your account they would have sold for 10*s* per head more with you, which is 21£ However we have sent you this morning 206 wethers and segs, mostly very good, a part only small and some not fat, also some young segs not very fat, but they will mend if you keep them a while. I am glad you have engaged your cattle to Holdgate, and I trust the next account will be that they are sold. By all means let him have the whole. If you get 8*d* per lb. sink for clipt sheep it will rejoice me much. But John the Claytons are usefull as prickers or pushers of the other jobbers, and although I don't admire them, yet it may be right to sell them a lot sometimes. However I certainly leave that to your own discretion. I did not know that Holdgate ever stood Skipton. However I am rejoiced that they got a good profit by them. It will make them keener of buying of you again. I should not wonder if Manchester &c make your Yorkshire merchants be at London now. It was the case sometimes before the war, and by a letter from Mr Sayle from Norfolk the London markets are not so good as Wakefield. We may be to blame for not buying steers at Wooler, but what were we to have made of them? Let us first my good lord get quit of our fat stock. If this drought continue, which is too likely, we can buy you plenty of steers, or if you would come to Morpeth fair the Wednesday before Whitsuntide you will see a great shew of fat and fresh stock. John I am and all of us much obliged for offering to sell for us. It will be mean and shabby if your master and I don't allow you for dinners and drinks to jobbers and all men on business on our account. Indeed it is ridiculous not to avail ourselves of the advantage of your selling our fat stock. You only stand out for too good prices, which no doubt of it arises from your zeal and wish to serve us in the best way you can. And if Holdgate &c have got 5£ per head by your cattle I shall rejoice. We have in consequence of your letter altered our minds about Alnwick. We will send none there now, but will tomorrow send away the cows and queys to you, and they will be with you before the sheep, perhaps in 6 days time. And I would wish you to say in answer to this whether you would wish us to send the large oxen, viz. those we meant to send to Alnwick, *next* or the lighter steers and oxen. And we will immediately send this per post and write again from Alnwick. Never let us mind postage in these matters of importance. I am going to a meeting and am in haste Yours truly

Geo Culley

122.   *George Culley and Matthew Culley jr to John Welch*
                            Eastfield Monday 10th May 1802

Well John

If I don't pepper you of with letters this week I shall wonder at it. This is the 2nd today, one yesterday, and one this morning by Jack Davison, and I intend to send this from Alnwick. Indeed the one I sent per post today was not to have gone until after Alnwick fair, but your letter today altering our resolution of sending the cattle to Alnwick, I then determined to send it to you per post today, particularly as it was proper to acquaint you that we meant to send you a lot of cows and queys tomorrow. And Matty tells me just now that they have picked out 19 to go, viz 8 open cows fat, a black Kyloe also open, and 2 mottled open queys, one of which with a blind eye is unbulled, the other they think is, which you must look to for fear she get too forward and in that case should and will be well sold for an incalve. She is a little clyered but not much. The other 8 all spayed ones, [**M.C. jr**] 6 or 7 of which are very nice, and with a little grass will be made very capital indeed. We have 6 or 8 more she beasts that are far back, will likely come to you afterwards for summer feeding, the last lot we shall send, but we have a great many very capital outlaid oxen and also inlaid ones to come first.

[**G.C.**] John my daughter is come home today, and I have to thank you for your attention to her in taking pidgeons &c to her. She begs her thanks to both you and Dolly. *Tuesday 11 May*. A cold, windy, dry unpleasant morning, wind a little to the north, yet the mercury has sunk a little. But I am affraid it is only an act of wind. But what we wish, we are apt to hope for. It is said that wool is rather lower at present, but there can be no good reason for its sinking now peace is ratified. I believe I told you in my letter of yesterday that by a letter from Mr Sayle from Norfolk he says 'Corn and cattle both fat and lean, upon the decline. The merchants are very shy of buying corn as the London market falls every week. Fatt cattle also lower in Smithfield this week than last by one penny per pound. Mutton keeps its price. It's very dry and cold, little grass. Wheat sickly in some parts upon strong land. We finished plowing up near 20 acres yesterday to sow with barley, being destroyed by what they call wire worm'. However John nothing is equal to paring and burning for preventing the wire worm or any other worms or insects, and you have certainly had a nice time for burning the Sugar hill. It strikes me very *clearly* John as the wisest way, to send you all the *fattest* cattle in the next drove, *big* or *little*, as they will be the fittest to sell, and perhaps the sooner we do this the better, but not until we have your answer to this observation or rather *recommendation*. I wish we may be able 'to keep the wheel back in the *nick*' if we keep sending all our live stock to you this way. In fact I am not quite certain that we can make all our payments, *rents* &c except we can sell our wool to be paid for by October rent time, which we must continue to do. But at any rate we will take care to let you know in time if we should want. In the mean time *keep paying* your money to Mr Peacock except you have notice from us to the contrary.

Alnwick fair 12 May. I have this moment received yours, and am glad you have sold. I hope you will make better of the 6 outlaids and those coming. You would have been very wrong indeed not to have sold those inlaid cattle of my brother's, but before I would take 7/6 for the outlaid ones I would try Skipton and take it there. We came past the sheep all well last night, 150 or 160 capital ones, the soft ones but nice mutton. I fear a bad market here but will tell you before I send this. I would take both a lot of cattle and sheep to Skipton or Wakefield before I would sell one at 7/6 and the other at 7*d* per lb. sink. I see George Hopper and little Tommy Bates here. I should not wonder, if things are very low, that they will buy a lot to take to Skipton as Tommy was keen to buy. Alnwick 12 o'clock. Well John how glad you may be that have sold near 7/6. No cattle sold this day at more than 6*s* sink, and many selling lower now. Ralph Fenwick indeed sold a lot of very uncommon steers and some say nearly 6/6. Little Tommy wanted to have bought them and taken them south, but

I missed them and it is as well I hope. I can only say that you must keep doing the best way you can. If you can't get 7/6 take 7s or less, or else take them to the Yorkshire markets. However if you can keep Holdgate and Shillito in tow, I think it may be as well as going south. Not that I will restrict you one way or other. You have either managed so well or been in luck that I wish you to back good luck. I have seen many changes in my time but never one like this! A lot of cattle worth 100£ today were this day 2 years worth 200£. How wonderful! Just 100 per cent difference. Lean cattle are selling still lower, but I am determined to buy none until we quit some more fat ones. Sheep were rather better sold at Morpeth this morning than last Wednesday by more than ½ d per lb. I am told James Thompson and some others are comed from Morpeth and now buying steers. Let us hear from you again as soon as possible, and don't sell your sheep below 7d if you can help it, as I think sheep are not very plentifull. Yours in haste

Geo Culley

PS. There are a deal of nice fat oxen here this day, and many Jeff said were not sold. The report will reach Wakefield. David Green is here but not buying cattle.

123.   *George and Matthew Culley to John Welch*

Eastfield 12 May Alnwick fair night [1802][42]

Well John

We have got safely home thank God from Alnwick, which I told you in my letter from there was a terrible bad fair indeed. But I would gladly hope that things have seen the worst. Just after I had sealed and sent away the letter to you today came David Green in the coach, and bought several little lots, which enhanced the spirits of people a little although he gave very little more than what had been given, and picked his cattle, cows and queys chiefly. I fancy he might give 6/6d for a part of them. Tommy Meek also bought a few. However one way and another the greatest part of the greatest shew of good cattle ever seen in the town of Alnwick were got through, which makes me hope that things may be better but cannot be worse. But come of our cattle what will, I was very glad that we shewed none at Alnwick, and I am glad that you sold the Wark inlaid cattle (14) to Holdgate. I am certain that they would not have been sold at Alnwick for more than 26 or 27£ per at most. Ralph Fenwick I told you shewed 7 very capital 4 years old steers as ripe as cherries, outlaid and fruitfull, more than 80

---

42 The year has been omitted, but is certainly 1802.

stone each, I fancy 84 stone, at 26£ 10 per. Little Tommy Bates and George Hopper repented much that they did not get them. They had a good opportunity but did not embrace it. Bill Moffat bought them. I don't think Messrs Hopper and Bates bought any. But Tomy Bates, as soon as ever he saw David Green, set of immediately for the Tweedside to buy all the fat sheep he could. So you see that these *little* fellows will put some *little* spunk into us again it is to be hoped. Indeed from the account of Morpeth market today there is great reason to suppose that there will not be too many fat sheep. Mr Andrew Scot says there were plenty of butchers very keen of sheep, the Sunderland butchers especially, and thinks that fat sheep will be well sold still. To be sure the noise of the Alnwick fair will go far and do some harm, but I do prognosticate that things will be better and not worse after this. However, do you be doing by all means, and say whether we had not better send you the fattest cattle we have whether big or little. It is clearly the wisest way, but we will be directed by you, only do not forget to say in your next which you would wish us to send and when. I have no doubt but Matty Walker's mare was killed and poisoned with potatoes. It was a pity, for she was a valuable mare. Cheviot was white with snow this morning, and is quite white again tonight, and I hope it will be rain by and bye in the vallies. I assure you we begin to want it pretty much.

Thursday 13th May. Snow to the door cheeks, which will do an abundance of good, especially as it has come on calm and mild. Nevertheless it has an uncouth appearance at this season. John did I tell you that William Mills of Howtell has bailiffs in the house at present, on account of arrears of rent I understand. The poor naked sheep will tremble this morning on Rimside Moor, but the cattle will enjoy themselves and both will do very well when the sun gets up, as the dust is laid. No alteration in the London markets respecting corn this week. Oats rather brisker as well as barley, wheat the same as last week. But I perceive that mutton is advanced and that will help beef. I am very much disposed to think that mutton sheep will fall scarce in a few weeks, and if we should have another promising crop of turnips we shall have ewes and *dinmonds* alias *shearings* as dear or nearly so as the last autumn. I fancy there are still a good many fat cattle both in the Scotch and English sides of Tweed. When I reflect on your selling of the Wark inlaid cattle you can't imagine how pleased I am. God bless your soul! What must we have done? No grass, and the greatest part of our cattle still inlaid! We are not to *regard* or *repent* now, that those said cattle might have been sold in February for 3 or 4£ per head more than now, and all *keeping expences* &c &c saved. No, but we are to consider, and do what is best to be done under the present *existing circumstances*. That is true *philosophy* in my opinion. What signifies a man making himself *unhappy* for what is past, and what he can't *now help*. But let him make all the amends he can by *present conduct*. And although he may again *mistake* and get *wrong, still*, so long as he has *done* and *acted* to the best of his *knowledge* and *judgement*, let him be content and *thankfull*. He has committed no real *crime or mistake*. He has only *failed* in his *judgement* and *opinion of things*, and that is *human* and what every man, be he ever so *clever*, is *liable to*. We have had good times, and we have benefitted by them, thanks be to the Almighty Being. Had we like some thoughtless inconsiderate people who I could name, lived up to those good times, and when bad times came had made no provision, then God knows we should have had *great reason* 'to have repented in *sackcloth and ashes*'.[43] Let us then be thankfull to that kind Being who has instilled these good ideas into our understandings, and pray to Him to continue to *enlighten* our *understandings*, and *acquit* ourselves as *conformable beings* who are truly *thankfull* for those good things which it has pleased God to bestow upon us and not despond as men *without hope*. John I am no *Methodist* but I trust a true believer in the *God of Righteousness*, and I hope these reflections will neither hurt you nor myself.

Friday 14th May. It is really a very curious thing to see snow lying to the very doors every morning in the middle of the month of May. A hare may be traced and that is about all, and I think the quantity is rather more this morning than yesterday. Indeed it never went of the north side of Cheviot all the day yesterday, which shews that the air must be very cold indeed. It will moisten the ground a little, but in a very unpleasant way. The poor naked sheep will be very cold indeed in crossing the high grounds. However they must take their chance. *Saturday 15th*. What another dreadfull morning! The snow to be sure does not come to the doors today, but it is upon Mr Reid's farm, and Cheviot is lying as thick as winter and the snow has not melted upon it these 3 or 4 days. God knows what will be the consequence. The poor naked sheep will all be starved as they come to you, and their skins will be checked and chapped I doubt. It will thicken their skins vastly and swell them. To be sure a part of them has been clipped a fortnight, and will stand it better. I have heard nothing yet what little Tommy Bates has done since Alnwick fair, but David Green is buying

---

43 Matthew 11.21.

all before him in this vicinity. He bought Frank Peacock's yesterday rough. I could not learn the price, but I suppose cheap, as Frank had missed selling before 80 good sheep. He was to handle Mr Carr's of Ford last night or this morning. These and what little Tommy buys will all go to Skipton I suppose. It will help Morpeth much I do suppose, and I hope you will have your customers to handle yours before they know of the coming of Green &c. And if you can get 7 ½ per pound take it in my opinion. Not that I think sheep will be worse, but better. What is very extraordinary is they have had a wonderfull good market at Edinburgh the last Wednesday, when Alnwick was so bad, good it is said for both cattle and sheep. Ralph Brown heard so at Kelso yesterday, and that Jemmy Bruce of Kelso had since Wednesday been buying up all the fat sheep he could lay his hands on. I understand Jemmy is become a very great jobber. David Green was at Ned Smith's at Cornhill yesterday, going to buy his steers very good ones, and his wethers clipt, also very fat. He has some thoughts of clipping Frank Peacock's sheep before he sends them south. It is an extraordinary circumstance that fat cattle were 100£ per cent dearer two years ago at Alnwick than last Wednesday. What was worth 100£ last Wednesday was worth 200£ 2 years ago. Indeed things were then at the highest, and I think it is highly probable that things were at the lowest now at Alnwick fair again. I am confident that fat stock will advance here, whatever may be the case with you.

Saturday evening, and Jack Davison just arrived with your letter of 12th. I am sorry that Mr Harrison has behaved so badly to you John, but I am glad that you have kept the man. You have no occasion to regard Mr Harrison otherwise than as a neighbor that it is your interest to live peaceably with if possible. But and if Mr Harrison will behave improperly to you and your servants it is your business to do what is right and just, and never mind any more. Never mind if Mr Harrison do write to your master. Your master will certainly not attend to Mr Harrison in preference to you. It is very bad neighbor-hood for one man to interfere with another about his servants, and I would wish to guard you against any such practice. Mr Harrison being a richer or greater man is no excuse for improper behavior. If the man is a clever fellow, as by your account he is, you would have been much to blame indeed to have lost him for 6d per week or even more, because it is good and valuable servants that are best worth keeping. There is no washing or clipping sheep either, such weather as this. What signifies a few unwashed sheep? You will get them washed and clipped and sold some time I warrant you. I am glad that you have got fruit about the Moor house, and very happy to hear that

Mrs Peacock is mending. I dare say it will be quite right if wood come to 1s per foot to do your granaries at the mill &c. But that and the house roofing I shall leave to my brother to answer, and will send this to you tomorrow per post. I always hope John that you will find the benefit of writing a little every day, or whenever you have anything to say, day by day journalwise in the manner I do in every letter almost. I am certain that it is the best way. However I am much obliged to you for writing so often at any rate. *Sunday 16 May.* No more snow but still lying close upon Cheviot and other hills, with a clear sky, not a speck of cloud, and frost very hard. God's will be done. But should this severe frost and drought continue, what will be the result? This is too like the year 1762 which was the most fatal drought ever remembered. It began early, and no rain until 4th September! We had spring wheat that year in your Weans and Gill Moor which hardly shot. Nothing but the grass of the carrs. The few heads which did appear never came to any perfection. No hay anywhere. My business was all the summer to cut ash branches to feed our cattle with, and they followed and roared after me. The gooseberries are gone in the garden. A strange time indeed. Well John the day still continues to shew the greatest appearance of a lasting drought that can be, and as I am going to send this per post to you I will conclude wishing better weather. Yours

Geo Culley

PS. I forgot to say that Ralph Brown thought the market at Berwick yesterday better for all kinds of grain, and you may depend upon it that if this weather continue a little longer oats and pease in particular, and all other grains perhaps, will advance. The hay crop must be bad in general. G.C.

[**M.C.**] Well John, Ned Smith sold Green his oxen and sheep. Sunday 1 o'clock a very strong shower of hail from the east. I hope it may turn the wind. Can you not stop the swallows in the carrier in the Fitts and also tread it wherever it gets into the carrier, so as to get the water over the land? It is worth while to attend to your meadows by every means, even to have a scoop and throw it onto the land with the scoop. As to the mill, run all the water you can at once so as to get the water on the land, as it is best where you can run the greatest quantity at once. Also when you water a place, cut out a single place as deep as the bottom of the carrier so as to get out all the water you possibly can on to as much land as you can at once. I suppose at the same time that you have very little water to get on the land. M.C. As to Mr Harrison, my brother's ideas are properly expressed. Do not kick against the pricks. At same time be steady and do your duty quietly without complaining to anyone except a friend you can confide in, and to your masters who will hear you as well as commend you for just action, also stand by you for doing your duty. Pray how is Robert Dent and the old man? Let us hear from you. We must think of doing the mill granaries now when wood comes to 12*d* per foot. Your mistress wishes to have the front wall raised a foot so as to make the garrets into lodging rooms. This to be considered of when I come over. The time when you can have me best when your turnips are sown. Of this let me know in your letter when it will be best to do them, I suppose the latter end of July or beginning of August between finishing turnips &c and the harvests of hay and corn, I am glad your young mare seems to do so well. Your M. Walker's was I believe hurt by too many pottatoes. Yours &c

Mattw Culley

124.   *George and Matthew Culley to John Welch*

Eastfield Sunday 16th May 1802

John

We have this day sent you a long letter in answer to the one by J. Davison &c &c. Your master and mistress from Wark dined here and are just gone. We have had several showers of snow today, and the atmosphere is become very thick and cloudy, which makes me hope for a change. But what we wish we hope. *Monday 17th.* Snow again to the door cheeks! What is to be the end of this God only knows! But such severe weather never happened within my recollection before. 'But God's will be done'. He best knows what is most expedient for us. A more severe [. . .]⁴⁴ there never was. Had it been in February or even March it would have been most severe, but at this season it seems so unnatural. In the evening yesterday the wind got south west, and I would fain have flattered myself that it was turning fresh. It fell mild also before bed time a little, but how different this morning!. Wind dead north. *Tuesday 18th May.* Surely this is the most extraordinary weather ever known. No snow this last night, but the frost in the greatest extreme. All the vegetables in the garden seem to be totally destroyed! The forward pease lying flat upon the ground, as well as brocola and all the large plants. In short things have a most dismal appearance, and what will be the result God only knows. I don't know how you feel, but I assure you I am glad we have bought no cattle to graze. I beg that you may take the greatest care of your straw of all kinds. Nobody knows how much it may be wanted. In the

---

44 A word seems to be missing here.

year 1762 the thatch upon stacks and even some houses was eat in the following winter! We can tell nothing what may happen if this most severe drought continue, and it has every appearance. However there never can be any crime in taking care, and possibly for the worst. It is in the power of the Almighty to send us immediately genial showers and fresh weather so as to recruit everything and still send plenty. But if the drought should continue a few weeks longer, we know not what may be the result. God forbid that it should! 'But His will be done'. It is our duty to submit without murmuring. *Wednesday 19th.* Thank God is much milder frost than we have had lately, but the snow is still lying close upon Cheviot. However the wind is south east, and we will hope for fresh by and bye. David Green has bought a few cattle of Mr Smith of Hay Farm, 4 I believe 74 stone each 26£ per, of Ned Smith 6, 72 stone at 26£ each very good and outlaid, Mr Smith's Hay Farm inlaid. Those with the 13 at Alnwick are all the cattle I have heard of his buying, but he might buy some of Mr Jobson as he went south on Sunday (but hardly). He bought only 80 wethers of Frank Peacock and 40 of Ned Smith, at least we have not heard of any more. And as to Tomy Bates I think it has been false information, as we have heard nothing of him since Alnwick, when it was said he was gone in haste to the Tweedside as soon as he saw David Green. But if that had been the case Ralph Brown would have heard of him either at Berwick or Kelso. *Thursday 20th May.* A clear frost but not so severe as usual, but Cheviot is still close covered. We heard yesterday that little Tomy had been with Mr Oliphant of Eckford but know not if he bought anything. We have had one very heavy shower of snow and sleet yesterday which wet the ground and laid the dust for the present. By this day's paper wheat is dearer in London, and indeed all other grain. I do imagine that both grain and fat cattle as well as sheep have seen their lowest for this year, and if I am right it is very extraordinary that *fat* should be *lowest* at a season when it has usually been the *highest*. G.C.

[**M.C.**] Thursday afternoon. We have this moment heard that Morpeth has been a good market yesterday, which looks as if things would mend. There was not much stock in, and a great many butchers. Sheep clipt worth more than 7 ½ but not quite *8d*, and cattle were much better sold at *7s* per stone. Indeed we hear there were very few to call fatt cattle in. This has been a much milder day, but still frosty. The wind has got to the south or rather south west tonight, which I hope will bring about a change. This country looks very bad[l]y, and stands great need of fresh weather. Yours

Mattw Culley

125. *George Culley to John Welch*

Eastfield Sunday 23rd May 1802

This afternoon John in consequence of a letter by post from you received this morning, we sent you an answer because I consider the expence of postage as nothing compared with important communications. If it reach you tomorrow it will be well, if later you will still get it in decent time I hope. This evening has turned calm but quite clear and frosty, scarce a cloud to be seen and the glass rising, with every appearance of continued drought. But we are always murmuring. So long as we are in the hands of a Beneficent Being let us be content, as He best knows what is most expedient for us, lett us in every respect do those things which we ought to do, and then we need not fear what man can do to us. If we were not now and then to meet with checks, we should perhaps forget ourselves altogether. However it is always wise to look before us, and consider what is best to be done under the present existing circumstances, and leave the rest to the great Governor of the world. We ought to be thankfull that we have bought so little fresh stock. The fat can always be sold at some rate, will keep their condition better than lean ones, and when parted with will always enable us to support the keeping stock better than if we had no fat ones but all lean or keeping stock. *Monday 24.* A very fine calm mild morning, such a one, if we had moisture, as vegetation would go on rapidly, and it is amazing how wheat is growing, especially where drilled and hoed, even on Pond Close where you would think no moisture is. The oats and barley also look well this morning where there is any cover. But it is amazing how hoed ground retains moisture compared with unhoed land. So much for cultivation. Mr Askew's man just here asking for hay. They have got a few tons, but I am determined to part with no more let the result be what it will. Indeed we don't know when we shall have any grass fit to eat for horses. *Tuesday 25th.* Another clear frosty morning with every appearance of a continued and severe drought. We heard that James Glass was got home on Sunday only. It will suit your lime trade I fancy John. We have sown our ruta baga in part, and they have sown some rape at Wark, and this weather will make people begin to sow turnips also in part, and wise too to sow a few to eat early off. *Wednesday 26th.* A very mild morning with more dew than usual, and a considerable deal of mist flying about the mountains for the first time of late, which may in time engender thunder perhaps. Would it begin to ascend the hills I would think it a better chance for rain.

Thursday 27th. Mist to the doors but no drop in the mercury, nor any appearance of rain at all. John you promised to write from

Darlington on Monday last, but no letter has comed yet. Indeed we neither got letters nor papers today, owing to some mistake I fancy in the post man or boy. *Evening 8 o'clock.* The son of a *gun post* boy took your letter past the place where he should have left it and some others and our newspaper, and we have only got them this moment. However we will send this away in the morning, and it will be at Darlington on Saturday afternoon. I am very well content with your going to Skipton, and will not complain if you lose 10 or even 20£ by the 10 cattle, because I hope you will get as much knowledge by going as 20£. I was surprised to see Holdgate along with David Green today at Wooler. They asked me what we had. I said we had some usefull steers and oxen. What would we take? I said as near 8*s* as I could get. They asked no more questions but went to Chillingham Barns to see some cattle of Robert Thompson, and I suppose we shall see no more of them. I am told that they have bought very little yet although they have been into Scotland. But Mr Green missed it. He should have bought in Alnwick fair week. I will be bound for it that what he bought then would do him much good, but nobody will take less here now than 7/6 per stone sink for cattle, and 8*d* per lb. sink at least for sheep. There has been another famous market for sheep at Morpeth yesterday, full 8*s per* lb. or better, and John Nisbet sold his cattle for near 8*s*, prices from that down to 7*s* I understand. I believe there are still a good few cattle in this quarter and Scotland, but none very fat I am persuaded. They bought some queys of Mr Jobson at 8*s* he says. Sheep will fall scarce here and dearer I am certain. I am under no pain about your sheep, and we will put of another 10 days with our cattle, or even a fortnight before we will send, although it is a serious time indeed and I would advise you to sell your cattle by all means as they come to hand, as grass will scarcely keep cattle going on. But sheep will do well enough and are sure to be sold well. But however, don't be over stiff. Holdgate complains of you. But that I don't mind, because that is their way. Only it is wrong of you to be too stiff. Green told me you had asked him 50 for your hogs. I said I could not wonder at it, as I was certain they were fat. He allowed they were, but were little. Never mind said I, they will be nice ones out of their coats. I was told some clipped hogs were sold at Morpeth yesterday for 50*s* per and more. Holdgate said he came to *buy hogs.* It may be so, but I rather think they expected to buy fat sheep but are disappointed, and when fat sheep are gone, fat cattle will follow. I am glad you have sold some more barley at so very good a price. Not but grain of all kinds must advance, oats and pease in particular, and wheat and barley soon.

Matty Culley in his road south will be at Denton about Wednesday evening late I hope, when you will be returned from Skipton. If market is good then you will either bring a customer home for your sheep or take them there in part against the next fortnight. That day Matty was at Morpeth was a very bad day indeed. I do believe he sold some hogs and good ones at 6*d* and his others at 6 ½. However a strange change has happened since, and will now hold I think for a while, and I don't doubt but many clipt hogs will be sold. But that will make young sheep sell well after, if any crop of turnips take place. I am affraid wool is on the decline, but we will know better by and bye. I am very well pleased with the prices for your 3 cattle at Darnton, but I had rather it had been the 3 worst. But never mind. Green says you are a good servant but too *hard*. I told him I believed you would do the best for your masters you could. He then said would I sell him the sheep and hogs at Denton. I answered that whatever was at Denton *John Welch* had a power to sell as he thought proper, and whatever it was I should be satisfyed but that I would not interfere. I do believe they were much disappointed about getting few or no sheep here, and hope you will sell well enough. However if you can deal to your mind with Holdgate do, I like that man and you must have somebody to be a customer. However you are a better judge than I am. Only be sure to deal safe. Yours

Geo Culley

By a letter from Mr Deverel they foresee a severe drought. Corn advancing and fat pastures gone, wheat advanced at London last Monday, but no barley near the price you got. Be doing at that price I pray you.

126.   *George Culley to John Welch*

Eastfield 28th May 1802

Well John

You are an enterprising man. I do admire your *firmness* and *resources*, and sincerely wish you a good market at Skipton. I have sent you an answer to yours of Tuesday last the 25th inst, which I hope will reach you before you go to Skipton, and this will go by Matty post free. We have been looking over the cattle at Wark and Thornington today, and think them in general much improved since they went to grass. If you have a tolerable market at Skipton I think it would be wisest to send you 20 or 24 of our fattest cattle directly after we hear from you on your return, because we have a good many more still to come to you. John I do think if Holdgate would give you full 56 to 58*s* per for our wethers it would be a very famous price.

They can hardly weigh more than 21 or at most 22 lb. per quarter. 21 lb. per quarter at *8d* lb. is *56s*, at 22 lb. per quarter it is *58s*. *8d* is a capital price. When I asked Holdgate if he would give you *8d* per lb. sink he said he never meant to bid you so little. Therefore if he was speaking truth you must have been too stiff. However you must know best, but I am under no pain about the sheep although I think the Sunderland &c men must come to you soon as the Morpeth supply will begin to fall short in my judgement and then cattle will be the quicker confirmed. I like Holdgate, at least he always appears to me a steady man. Not but they will all take advantage of times, and no wonder. But I consider David Green, although a *sharp fellow*, a man that buys by *fits* and *starts*. He wants resolution. If he had been *awake* at all, he might have seen that he could not have got wrong in buying in Alnwick fair week. I hope you will make more of your 6 oxen than they bid you, and they would fain have persuaded me that you began at the wrong end of your *web*, and perhaps you did. At least it is in general the best to stick to the *good ones*, But as you could have sold a 4th ox at 31£ per, it brings the 2 remaining ones very low to 49£. I cannot conceive how you manage to sell barley higher than in any other part of the kingdom. Your price is *44s* per quarter, London best price is *35* and Nottingham *38*, where their measure is *above par* considerably. How do you account for this? Do they mix barley meal in their bread? Or what way do they employ it at this season? It is now too late for malting I suppose. Certainly barley, beans, pease and oats will and must advance this dry season, but I have no idea of oats coming to *4s* per bushell again. Ralph Brown just come from Kelso and heard that little Tomy Bates had bid Mr Oliphant *44s* for his wethers, and Mr Clement Stephenson came and bought them at *54s* per! Ralph sold his wheat today at 54 and 55 per boll.

Saturday 28 May. Well John, long looked for is *comed* at last. It came a little drizzly rain in the evening of last night, but the wind being north by east (where it still is) I wanted faith to think it would be rain. But by the land being wet a good deal it must have drove on all or most of the night from the north east. The glass is still coming down although still high, but as it is coming down with a north east wind it may come a good deal of rain before we have done with it. And if we do but get it, it matters not from what *quarter*, or whether *cold* or *warm*, as the land is warm enough I dare say. *Monday 31st* 5 o'clock. A very cold frosty morning, the rime lying white. Cheviot was white with snow yesterday morning and may be this, but can't see it for clouds. The glass going up, and much appearance of frosty droughty weather again. However the rain will and must do a vast of good. The market for corn at Berwick not very good, not so much advance as we expected. Ralph sold nothing. Mr Nisbet tells me Morpeth was very good for sheep but not cattle. He got near 7/6 per stone sink, but were particularly good. 7s and even less were common prices. But there are a great number of fat cattle in this quarter still, but sheep are scarce. Mr Nisbet sold your Mr George Nicholson his best 10 on Wednesday at 66s per. The Sunderland and Wearside men were the best customers. By a letter from Mr Peech of Sheffield dated 28th he says both beef and mutton very dear. Beef 8/6 to 9s per stone, mutton 8d to 9d per lb. sink, and all on the advance. Both cattle and sheep come sparingly to market and are rather dearer, and notwithstanding we are in very great want of rain, lean cattle are dearer than ever known. Hay advances from 3/10 to 5s per ton, oats 16/6 to 22/6, beans new 33 to 35s, wheat 70 to 84s all per quarter.

Monday evening 7 o'clock. Just returned from Mr Robertson's tup shew, where were all the world and part of Gateshead. But if we have no more dangerous opposition than Mr Robertson this year we shall have little to fear. I think his shearings very indifferent indeed. His old sheep were in great plight to be sure, but nothing else to recommend them. His *bulls* and *boars* I did not stay to see, but came across the Tweed to a sale at Wester Newbigging, where ewes and lambs were sold from 50 to 70s per couple, and indeed everything else were well sold while I staid. But I came away before the horses were sold. David Green's little brother was there, and has bought several fat cattle I find at 7s to 7/6 per stone sink, 12 of Thomas Smith of Berryhill, a lot of William Smith of Shedlaw, some of White of Norham &c &c. William Smith's went of directly, but I did not hear when the others go. However they can't injure markets tomorrow at Skipton, and you will scarcely go the next fortnight I think except you go with sheep. This day is very frosty but much sun and very mild, and corn already looks greatly better.[45]

127.   *George and Matthew Culley to John Welch and Matthew Culley jr; Eleanor Culley to Matthew Culley jr*

Eastfield 1st June 1802

John

I shall direct this to you as usual, but it will be wrote to both you and Matty who went this day for Morpeth in his road to Denton and then into Leicestershire, and you must be sure to shew him this when

---

45 The signature is lacking.

he comes down or any other letters which I may send you before he returns to Denton again, and then he will know all we have done, for I shall continue to write every day as usual. *Wednesday 2 June*. A clear frosty morning, no appearance of rain, glass stationary. Grass a little better but corn very much mended indeed. *Thursday 3 June*. Wind north east and the hills thick. At any other time we should expect rain from that appearance, but the mercury is so very high and scarcely comed any down. I attended David Lee's tup shew yesterday, when he made a better shew of dinmonds than the gentlemen on the other side of the water. In general pretty good forward, but large boned and rather hollow or lowish in the middle of the back, and yet the old sheep their sire is very fine boned. He was got by a sheep of Kitty Mason's, and shews a great propensity to feed. There are 2 dinmonds out of 14 that are handsome sheep, the last but one the best or will be best in my judgement. Another small sized but very smart and finer of the bone than most of them, and very light in his offals. Mr Robertson there, very *civil* indeed, but tells me he has no thoughts of going south to the tup shews at all. The 2 Murrays, old and young, George Logan. The latter and Major Johnson went away before dinner. Tom Alder junior, Robert Thompson and Mr Jobling. I staid till near 8 o'clock as there was no drinking, indeed the wine glasses were no larger than good thimbles. I drank gin and water. Your mother and sister had been at Berwick and I overtook them and we came home together. We have still great orders for turnip seed in the retail way. I know not if we shall do without thrashing some more at Wark. Sir John Swinburn[46] seemed highly pleased with what he saw, an amiable baronet I think, says he will visit us again. Mr Robertson declared that the dinmond which young Harriet bought at his shew at 51 guineas was not bought in, but he did not intend him to be let under 100 guineas but had not apprized Harriet of it, but was just going to speak to him when he was struck of to Mr H. junior. That young man Harriet it seems has everything in the farming way turned over to him by his father, and is highly spoke of by the gentlemen yesterday I do assure you, as a sober, steady, sensible young man. I have received another letter this morning from Mr Thompson London, telling me that the Grange business was put of until yesterday. Wheat dearer at London, other grain much the same. Mr Thompson says 'We have had a fine fall of rain these last two days

---

46 Sir John Edward Swinburne of Capheaton, 7th baronet, 1762–1860: *Complete Baronetage*, vol. 3.

but the wind has got to the eastward and is very cold'. He dates on the 31st May. So their rain had likely fallen a little later than ours here, but it is possible that they may have got more of it. Old Tomy Chisholm late of Moorside Law died of a few days illness lately poor old boy. Ralph Brown very poorly in his old complaint, Mally bad of her headaches, and little Willy in the toothache. Old Sanders supposed a dying. But I hope Ralph will be better *now* a fine rain has comed and is now running in a fine stream out of the courtain towards the pool. It has spitt on all the forenoon, but since one it has rained pretty drealy[47] now 3 o'clock and past, and as it comes from the south east I hope that it will not only bring rain enough but warmer weather which would do a great deal still. Straw may get to a fair length, but I think hay will scarcely be a good crop, perhaps hardly a midling one. Not but warm weather and seasonable rain will do wonders.

Friday 4th June. Well John the old adage at Denton was 'When the Unthank is down in May then Denton is stocked with hay'. But say you this is June. Granted, but still it is only old May! However, if you have got such a steep as we have and are likely to have from appearances, if you are not stocked with hay it will do you an abundance of good. We had a good deal of rain yesterday from about the middle of the day, but it has rained all night or most of it I understand, and now half past 5 o'clock it is raining very drealy indeed, and as the wind continues in the south east I see, and the mercury much comed down, it is more than probable that we shall have a great deal of rain still. If it please the Giver of all good to send warmness after it, a wonderfull change will take place. But at any rate either your land in general or ours in this country will be greatly benefitted, and can bear with a deal of moisture at this season. The land parched before, and the days nearly at the longest. Our first sown Sweedish turnips are comed away clanking and very mily indeed in the Planting Close, and we have sown near half a score acres of common turnips which will do well I hope. If a good crop of turnips should happen this year again, which is now very likely, I am disposed to think that sheep will be as dear as they were the last autumn. If David Green and you have not agreed for Mr Wood's hogs you need not fear selling them well now I trust rough or clipped. Perhaps you will now wish that we had bought you some steers &c &c to feed, but I don't repent at all I do assure you, because in the first place they would have distressed us before this rain came and I will promise you that we have been sore

---

47 Heavily, persistently.

enough put about. But if you be over lightly stocked, lye in more land *for hay*. Besides it will enable you to keep some Kyloes &c the better. I should not wonder if it throws a damp upon oats and barley, and indeed it may be on all grain. But particularly oats I think, because oats will now have a chance of being a very plentifull crop and have plenty of length of straw too. Barley will come up with greens and may not do so well, and even wheat may be too big. But although we may guess at things we can tell nothing of what may happen. After so long a drought we may naturally expect a good deal of rain, which may be too much for the cold lands, and starve and yellow the corn and grass both. To be sure this comes at a time of the year when it must suit every soil. but by a continuance it may overdo the deep lands, while it may make plentiful crops upon the dry soils. But God's will be done, let us always be thankfull. The country I believe is full of old oats, at least I know it to be the case here. But still a few good old oats in the granaries is no bad thing, especially with you where a few clever sweet old oats always sell to hunters or innkeepers &c &c. It is very different in this country where the consumption in oats is entirely in meal, except what the farmers eat with their horses. Indeed I understand that the consumption in oatmeal is not near so much, in neither this *district* nor *Scotland* as it used to be formerly. The poor eat as much more wheat bread as they used to do even here and in Scotland. However that makes a greater consumption of wheat, 'So that you have it not in the *singing psalms* you will have it in the *reading psalms'*. I hope you will get the water turned over your watered land now, and give it a good drenching. If Bob Dippie approves of it you may tell him that your limestone land can hardly be over done. Only the water should not be run too long in a place, especially at this season, because it brings a *scum* upon the land in a very short time at this season of the year.

[*E.C.*] Saturday morning 5th. Dear brother. I was at Milfield yesterday seeing the troops exercise. Shocking bad charges they made! The meeting you know was to determine whether to continue or no. It was proposed first to have it put to vote, and they all agreed to abide by a majority except Tom Smith. Now the question being put there was a large majority for continuing. Consequently you are still a troop but I think you must leave out the word *loyal* in your title as it's no notion of loyalty that now binds you. You I am sure would have wished to dissolve but I believe as an officer you would not have had a vote. E.C.

[*M.C.*] Well John Wark 4th June. We had a full fatt market at Dunse. Tommy Hope and 2 good ones, James Bruce bought about 20 fatt oxen and was the only buyer except 3 or 4 country butchers. Not one south country butcher or buyer (not even John Mason). George Logan sold 2 nice fatt cows (and Mr Wilson 2), Logan 30£ per 5£ returned. The rest in general went home unsold. *A fine rain.* Morpeth full as full of oxen and sheep, sheep 4 to 5s per head lower but fatt cattle 7s per stone. But this is a wonderfull country for stock of all kinds. We run the water over our meadows this morning and the grass is all alive. Corn looks admirably well here and on the road to Dunse, said to look very ill all the way to Morpeth. The land here fresh new limed, and farmers have liberty to plow fresh land to Morpeth, and with you tyed down to a slavish system of white crops. Oatmeal fell 5s per load at Dalkeith &c last week. Our fatt cattle are got fresh of their legs, and growing apace. Many Sunderland butchers at Morpeth saved the market or it could have been miserably bad, Crisp at their head. After all as rain has fallen in plenty it will keep up fatt stock in my opinion, for when people have no grass they take a half fatt cattle to market. When pastures are good or likely to be so, they will not sell without a profitt, when grazing stock are bought at extravagant high prices. The rains we had some days before had done good, and now I hope our clovers will rise and hay got on the meadow old lands. From yours &c

Mattw Culley

[*G.C.*] Well Matthew I have just returned from Wark where your uncle wrote the above, and I hear that your troop is to be continued. They dined upon the plain in a tent, the dinner by McCloud's people. Poor Lady Errol[48] a dying. Lord Errol was there, but no one sees her except the housekeeper and her own maid. It has been fair since about 10 o'clock and a sweet warm day indeed. *Saturday 5th June 5 o'clock. A sweet, pleasant, charming* morning indeed, with a gentle warm shower. If vegetation do not come forward now we must not blame the weather. I hope we shall meet with a letter from either you or Matty on our road to Berwick this morning or I shall be much disappointed indeed. If we do I shall send this per post, if not keep it until we do hear from you. *Grindon 10 o'clock.* We received yours and Matty Culley's letters at Twizell as we came along. Am glad that John sold his cows and queys so well at Skipton, and approve of his taking more stock thither, either cattle or sheep or both, as he thinks right. I

---

48 Alicia Eliott, second wife of William Hay Carr, 17th Earl of Erroll, 1787–1819. He inherited the estate of Etal from his maternal grandfather. Lady Erroll lived until 1812. *Complete Peerage*, vol. 5.

am only sorry that he has not an able hand behind when from home. We will send of a lot of our fattest cattle on Monday morning. We hear that Morpeth was worse by 4 or 5*s* per head for sheep, and cattle 6/6 to 7*s*. I did not know that Tom Glass's sister was gone at all. I am sorry to hear so bad an account of your pastures John by Matty and meadows, but hope these rains and warm weather will bring them about again. I will be glad to hear from you again John as soon as convenient, and be sure you let Matty see these letters on his return. I am yours affectionately

George Culley

PS. How far is Skipton from Denton? What road do you go? By Laburn[49] and up Coverdale &c? By a letter from Mr Barf of Wakefield who married Miss Spours this day he says 'The market at Wakefield today was very full of both cattle and sheep, clipt sheep from 7 ½ to 8 ½ , cattle 7*s* to 8/6 sink. Corn of all kinds advancing wheat 10*s*, oats 3*s*, barley 4*s* per bushell, but people are buying in speculation', and in my opinion oats will come down again. He dates the 2nd instant last Wednesday, says wool is coming down owing to cotton so low. Berwick one o'clock. Much thunder and rain but none to the west north west. I have just seen Mr Nisbet, who says Morpeth came down to 6 ¼ per lb. mutton, or nearly as bad as the day Matty Culley sold so badly at Morpeth, and cattle low indeed 6 to 7*s*. Mr Counsellor Alder sold some famous oxen at 27£ for which he had been bid 36£ for. In short John my advice is to send cattle and sheep both to Skipton, as the shipping trade is so bad at Shields and Sunderland that the butchers are affraid to buy. Cross of Sunderland bought Alder's cattle. You must be the best judge, but really I should think it wise to quit, let prices be as they will, and in my opinion H. and Shillito will not buy of you but at bad prices now as your queys are gone, and let prices be as they will at Skipton you will quit and no more about it. That I advise but you are better able to judge than I am. But depend upon it that Morpeth will be overdone for cattle compleatly, and that lambs and hot weather will hurt sheep. Lord have mercy what a rain! It comes down whole water. We shall perhaps get too much now, but I shall not complain yet. Nor you for your low ground. I think some trade will be hurt. Berwick stuck in one compleat sheet of water. We may think ourselves very lucky that have you to sell for us these very critical times, and now you have broke the ice at Skipton don't fear if another day you should lose 15£. The wheat market was rather more

---

49 Leyburn.

than 5s but Ralph sold none today. He sold Mr Walker some oats at 15s/6d per boll. Let me hear from you again I desire you.

Geo Culley

128. *George Culley and Eleanor Culley to John Welch and Matthew Culley jr*

Eastfield Sunday 8th June 1802

Well John and Mattw Culley (if not come past)

I sent you a well filled journal-like letter yesterday, I am sure, from Berwick. The rain had fallen partially yesterday like thunder showers, for notwithstanding that there was a sea of rain at Berwick our people had been sowing turnips all day, and this is a bright sunny but fresh morning with a fine breeze of wind from the south west. A fine rain fell between one and two o'clock, and everything is now reviving and looking chearfull. If Matthew shall see this, may let him know that Mr Lilburn, nor none of the publicans and *sinners* of Holy Island would take in one woman, being only a single person at once. So Bob went down today to the island and has engaged Mrs Thompson's rooms, the steward's wife, at 10s per week, and our Nelly is to find her own victuals. She will take a maid along with her. John I think you sold your cows and queys well, very well indeed at Skipton. It is great encouragement for you to try again. Little Green, brother to David, is still in this quarter, John Batters tells me. Indeed Mr Batters is his head quarters. He has been buying hogs lately in Bamboroughshire. It is amazing how both corn and grass have recovered since the rains and warm weather have come. I hope your pastures are also recovering. But this light free land comes faster about than yours. Let us know how Mr Wood's hogs clip and look after being shorn. If you can sell them well this autumn perhaps it may be wise to sell them as we can supply you with more dinmonds if you should want any. However when I say so, remember that I only wish to advise, and suggest hints. But you must be much better able to judge than we can.

Matthew, I have received many satisfactory letters indeed from Mr William Thompson this morning. All the business is now settled for this half year. A person along with Mr A. George Onslow stood out sore for interest for the time past 12 May. But Mr Trebeck, although their attorney, behaved with great *liberallity*, Mr Thompson says, said we had paid our money better than any of the other *purchasers*, so that was got over on condition that we pay the remaining 6000£ and interest from 12 May in due time against Martinmas next. So it will be our interest and duty to do so. Your mother, sister and self were at church today, received the sacrament and brought Mr and Mrs Thompson home and two of the little ones who dined here, and Bob is gone home with them in the chaise. Mr Thompson sent a statement of the account between us, the balance 70£ odd, but says 'as a dividend will soon be due we need not remit'. This I do assure you is very agreeable, as I have not a 5£ note in the house at present. I gave Harry 60£ and 7£ for James Glass to bear expences with the cattle, and I gave William Brown 70£ the last Tuesday, and he will want some more tomorrow which I can get of your mother at present, and we are to receive some money on Tuesday at the bank, and Ralph expects 70 or 80£ on Friday at Kelso. And we have as much to receive of Brown on Berwick fair as will pay Tommy Bryce. So that I hope we will still be able to make all ends meet, although we shall be sharp enough kept I do believe. But it is good for us to be pinched a little now and then. We owe Bill Pickering and James Rae each 50£ still, one is 50 guineas, and if anything should fall in our way to tempt us to buy at the Bank[50] it will be a little aukward to have nothing in our pocket. However we can make many shifts. But we have always many payments at these term times, and no wonder in our extensive dealings. But May is always the heaviest.

Monday 7th June. Rather windy and coldish, but the wind is from a fresh quarter, the west, and everything is looking alive and well. I don't know if I told you in my letter from Berwick that Counsellor Alder sold his cattle 9£ per less than he had been bid long ago, and his sheep 10s per head less! I think this will compleatly knock him up for farming any more. It is amazing how people rejoice over that poor creature's ill luck. We should learn from this to avoid that low *meanness* which has sunk that *animal* so much below the level of other men. It hurts me though when I recollect the respectabillity of his father. God grant that my son may always behave himself like a man. We had much better err on the side of *liberallity* than *meanness*.

Well John we have this day set of 23 oxen to you by James Glass, over large for the time of year but it can't be helped. Several of them weigh very great weights. One ox with only one horn is a very great weight but not so fat as I could wish, nor near so fat as some of the others. My brother was much against sending him, but you have two chances and we only have one, that is to Shields, where they will not give above 6/6 at the outside. Now you have a chance of Sunderland, which is generally 2s per hundredweight more than Shields. Besides

---

50 Whitsun Bank fair, Wooler, Tuesday of Whitsun week.

you may by chance get him of south, although he is a very unfat beast I am affraid for that part. There is another very large red and white flecked ox, rather plain of his shapes but very fat, and several more very good oxen. If you can only make a little more than 7s per stone sink, it is more than we can make here. I cannot advise you which way to sort them so well as you know yourself, but I should think it best to put the large ones by themselves and the lesser ones by themselves. But perhaps you will put the fattest ones by themselves, big or little. We have one famous ox between 80 and 90 stone left on account of being a little lame, and we have kept a handsome dark red ox besides that is a little backward. He was late of being turned of by William Brown. And we have a considerable lot of *half* and *3 quarter* fat ones still, which can either come to you afterwards or be kept here, just as is most advisable. Be sure that you send back Harry Rutherford's white Galloway by James Glass. I will send this letter away tomorrow or the next day at any rate, that you may be apprized of the cattle coming and the number.

[**E.C.**] Eleanor Culley will be much obliged to John Welch if he will let the Miss Greatheads know, any day this week, that he has a man coming north next week, as they have a parcel for me which James Glass might bring.

[**G.C.**] Whitsun Bank Tuesday. Having an opportunity by Mr H. Midleton the bearer. I send this. He will tell you the market uncommonly high for little small queys and steers, and freshish Kyloes. Hogs flat but very dear for all that. They had a heavy market at York it seems on Thursday last. Still good hogs sold at and near 50s per. Your hogs (Mr Wood's) would have sold at 50 today, while fat cattle nearly 6/6 sink. Tom Lumsden sold 8 to Ratcliff at 6/4d per and no more very good beef. I heard by young Kettle the jobber that you had a bad fat market at Darlington on Monday. Let me hear from you soon I beg. I believe I shall want 6 or 700£ from you, which I expect 5 per cent for to a safe hand and the money paid in before we want it again. But I will write again in a day or 2 to catch Matty who could bring it, or you could send it safely by Mr Hodgson Midleton if I want it, which I will let you know in a day or 2, it depends upon a man speaking to me today. I am in haste yours

Geo Culley

129. *George Culley to John Welch and Matthew Culley jr*

Eastfield 8th June 1802

Well John

This day I sent you by Mr Hodgson Middleton a letter which will I hope be delivered to you tomorrow evening, and in that letter I said that I would likely want 6 or 800£ from you by Matty or Mr H. Middleton. But I saw Captain Orde[51] after I had sealed my letter, and I think the money will not be wanted now as the business is not likely to take place for which he wanted it. However if it is, I will immediately let you know. Lean poor queys and steers, and midling fresh ones, were strangely dear today indeed, and at Stagshaw Bank, but fat low and very low, 6/6 the outside. Oats were lower today too, the Berwick merchants would not buy today at all hardly. Ralph contrived to pass off a few 10 bolls to millers &c. What a sweet mild evening, a wonderfull change indeed in both grass and corn already. I understand that Mr Ratcliff bought Thomas Hope's oxen at 27£. They were not at the fair but he had seen them at home. The cattle were all sold I dare say fat and lean, and sheep too, and prices very good for all except fat cattle. A bad shew of horses, and rather lower. *Wednesday 9 June.* A charming fresh morning, likely for wind and showers. I still think that the fat cattle will begin to wear scarce now, and perhaps the sheep too. But then lamb at this season comes more in vogue, and is more used than any other shambles meat. Still lambs are, will and must be high sold on account of the high price of hogs, and I do look forward to good autumnal prices for sheep again. John Davison was out to buy sheep yesterday and other Newcastle butchers, and they would not come out to buy if they did not foresee a scarcity.

Well John I have this moment received Matty's and yours of the 5, 6 and 7th last from your bad market at Darlington, which I think a very good one, at least I think you sold your 2 cows very well and the old cow well enough, and I think it was not a very bad price for your 3 oxen. However at 28£ per they would have made far more with the others sold at Darlington the fortnight before than Holdgate bid you for the 6 &c. I am sorry that you did not get my letter at Darlington on Monday that was sent from Berwick on Saturday, but you should always send to Piercebridge on the Sunday nights in my opinion at

---

51 William Orde (see No. 40, n. 28) was a captain in the Northumberland Militia. The reference is presumably to him, but may possibly be to his third son Charles Ward Orde, 1775–1810, who was in the army.

this time of the year, because we frequently send letters to you from Berwick on Saturdays and they reach Darlington on the Sunday and so go to Piercebridge the same evening I do suppose by the post that goes to Barnard Castle. However you would get it on Monday evening or Tuesday surely. And this day Wednesday you will receive another safe by Hodgson Midleton I have no doubt which will be nearly as soon as the other per post from Berwick. 3/5s [*sic*] per for the 100 wethers was not a bad price, but it was not quite 8*d* per lb. yet you may do worse at Skipton. However if you do I will not blame you at all, because I am certain you do it for the best. I will be bound for it that little Tommy Bates had no other view but his own interest in what he said to you. I fancy you should try *Wakefield* with your large oxen, rather than Skipton. But that you will know better than I can pretend to. Only I have heard it said that Wakefield was the likelier place for large cattle. You are right, we have very great reason indeed to be thankfull for this uncommon fine weather. I think I never saw such a change for the weather in so short a time. Thank God for it. What wheat we have sold lately is sold at 55*s* per boll. But I could have wished that we had sold more oats before the rain came, as the oat market is quite down again, and likely to be so as this country is fuller of oats than ever was known perhaps at this time of the year, and now such an appearance of a plentifull crop of oats and everything. Nothing to be feared now but too much straw to lodge the crops which will probably be the case if the rains hold on a while. A great deal of land is too wet to sow turnips on in this country now. We have too much wet for our Mile Knows &c, and William Brown is quite knocked down at Grindon with wet. We have had a good deal of rain this evening here. Our Sweedish turnips are in rough leaf. It is possible that Green has found these Leeds jobbers, and if so they will be a strong force. Wilson is said to be strong in hand. It is certainly right to go forward to Skipton or Wakefield with your stock in these critical times, even if you should lose a little of the reward. David Green's little comical looking brother is still in this country I believe. He bought some nice Scots yesterday at the Bank, but they were for Tomy Dreck, at least he paid for them

My dear Matty. I should be wanting in both reason and affection if I did not commend you for coming back to Denton to do your duty there rather than attend *hops* &c &c at Richmond or anywhere else. I did not expect Mr Wood's hogs to clip above 6 lb., they are bred from a fine light wooled kind. But I hope they will make fat, and be sold for profit. But depend upon it that all wool especially ours will be light this season. We shall be very glad to see you back in good health on

Wednesday my dear, but I think we will send no stock to meet you at this time. We wrote about our *little ewes* and the barren *ewes* going to John, and I think well of it myself, and your uncle, and shall be glad of your opinion and John Welch's on it. The sheep just cost *6d* per head going to Denton with a very trifle over, which is cheaper by a deal than sending to Morpeth, and John can quit the barren ewes better and safer than we can here. You are right to see Mr Colling's stock. I do hope to hear from you from Loughborough if it be only 3 lines and a half, but I am happy in saying that you are a good lad, and may you long continue is the earnest prayer of your truly affectionate father

Geo Culley

PS. We have so many payments, and are at present so pinched for money, that perhaps it would be right for you to bring a hundred pounds from John. But John will be on his road to Skipton, so that you may miss him or not be able to procure it except from Mr Peacock. Or John can send it by Hodgson Midleton. Or at all events we can do without it as we shall receive the insurance money early in July I hope. I got William Pickering paid yesterday by borrowing of your mother, aunt &c, but Jemy Rae is still unpaid and Thomas Bryce and many servants, and your uncle is in want of money which I don't like to have been pinched by all folks. William Brown and Harry have taken so much cash lately as well as our own payments. But Ralph hopes to receive a decent sum tomorrow at Kelso, and I received a letter from *Suncarty* with 17£ 2 the other day for turnip seed. I am of opinion we shall still have to thrash all or part of the turnip seed stack at Wark. That is become a most *lucrative* and wonderfull trade. To be sure there is much sowed in this country this year. I wish Tomy Hutton and Mr Thackerton don't deal upon it. I fancy John was right in sending the largest first to Skipton. They will travel better by themselves than along with the light small sheep, and these will go best by themselves. This is Thursday morning the 10th and a fine fresh morning thank God. I am not for Wooler but going to dine at Wark along with Mr Sitwell[52] and his steward Mr James Bell.

John I wish you would mind where you put your wafer and leave a bare place for it always. I could not make out a word or two in your last. We have plenty of grass now to keep the rest of the cattle and

---

52 Francis Sitwell inherited Barmoor Castle from a cousin in 1791. He had the house rebuilt, and took an interest in sheep. *History of Northumberland*, vol. 14, pp. 109–10, 117; Pevsner, *Buildings of England: Northumberland*, pp. 158–9.

ewes until you give orders to send them. I have a great notion that Capt. Orde's business will not take place now, and I shall be as well pleased upon the whole. We might be disappointed of the cash when to be sent to London. I see wheat has advanced this week at London, and without an importation so considerable it will advance. I am disposed to think that it is not plentifull in this quarter. Oats keep their price 3*s* per bushell and rather better owing it is said to the dearness of *hay*. If the drought had continued one week longer, hay would have been done. At any rate the early crops can't be good, and that will still help oats although the country is full of oats. John I have just been over at Hay Farm to ask about Morpeth market, and was disappointed to hear that it was little or no better than the week before, owing to the Newcastle butchers *glutting* themselves the week before, and 300 sheep left over days, and a cloud of lambs over. Mr Adam Smith says that he was bid in the morning more than he could come at afterwards, and I would advise you John to be doing with a part at Skipton in the morning at any rate. It is right to sell in a morning 6 times out of 7, perhaps 9 out of 10, but especially in a market where you are a stranger. When you know your market, the quantity of sheep and the butchers, as I used to do at Morpeth, you can then calculate pretty nearly. But in a strange market it is unwise to speculate *too far* or *too fine*. However I hope you will be better acquainted with both *Skipton* and *Wakefield* by and bye, at least I wish it. However Mr Adam Smith says that the fat cattle market was the best that he has seen of late, and the butchers confessed themselves that they had given more than 7s per stone sink. So there is some consolation still. Indeed I begin to think that fat cattle will be full as scarce as sheep presently. Such numbers of lambs come into Morpeth at this season, and highland sheep, and they are fond of thinish mutton at this warm time of the year. But I hope your Manchester friends will still stick to the fat mutton. If your [*sic*] in any way in luck this time John, it will encourage you to take a part of your fattest cattle next fortnight to either Skipton or Wakefield, but perhaps to the latter. However of that you will take care to inform yourself. If you agree to our sending you over the gimmers and barren ewes, which I think you will, they will perhaps be best sold at Skipton, but you can inform yourself of that also. Besides at Skipton they will never think of keeping any of your best looking handsome ewes to breed from. It is quite out of the world for breeding these kind of sheep, and if they ever should it will not hurt us nor any other breeders I think. I suppose mutton was not sold at much above 6 ½ per lb. sink this week at Morpeth, but I am thinking that Hodgson Midleton would see a part of Morpeth in the morning, and you can compare his account and this together. But Adam Smith says it was best for sheep in the morning. What a sweet day! *warm, mild and fresh*. God be praised for all His mercies. I see by a passage in this day's London paper that corn is confessedly dearer in France. That being the case it will make wheat dearer here without a large importation indeed, and perhaps it will advance barley also. But if other parts of England be as full of oats as this, it is not likely that they will advance. Indeed if hay be a bad crop in the south it may affect them. I am yours

Geo Culley

Wark one o'clock Thursday. Mr Sitwell is just come to dine with my brother so I must send this to Coldstream. I am glad to hear that the keelmen[53] &c are got settled, and I see by a paragraph in our newspaper today that a person is come from France to London to look to settle and regulate a commercial treaty with England, and that it is already in great forwardness. As that is the case the shipping will all go right again, and I should not wonder if *mutton, beef* &c get up again and wool also. At any rate they will all then find their true level. G.C. My brother is gone to take a ride with Mr Sitwell while I finish this, as he could not find time himself. We are to look at some steers of Ralph Compton this evening if we have time. My brother is busy covering in his burnt houses I see. What a rich beautifull appearance Wark has at present. The only fear is too much upon the ground, and only a few days ago we thought we should have both a shabby crop of corn and hay as well as pastures.

130. *George and Matthew Culley and Matthew Culley jr to John Welch*

Eastfield 11th June 1802

Yesterday John I finished a letter at Wark for you and George Humble sent it over the water to Coldstream to go per post. This is a dry hardlike morning. People will be able like William Brown to sow turnips on their damp lands. The Tweed was past fording yesterday which has not been the case of many weeks. Wark farm has a rich luxuriant appearance indeed. They have cut clover for the horses a week past. We only began yesterday. *12 June Friday*. This morning is also dry and rather cold, with a rising glass, and looks rather like dry weather. People are a going and sowing turnips everywhere, which

---

53 Crews of the keels, flat-bottomed boats carrying coal from the colliery staiths to shipping on the Tyne and Wear.

certainly have a good chance, but there is no saying what may be the case. However I never saw the land in better plight, not a quicking[54] to be seen scarcely, but I think I never saw so many small or top weeds amongst corn and other grain, and especially yellows, what you call *kale* and we *masicks*, and *cress* and wild *mustard*. At the time I am writing here I can see Tommy Reid's fields as yellow as go land! We have few ourselves but a strange quantity of other annuals. I know not a county where so much *corn, beans, pease* and *turnips* &c are drilled neither as this, by which mode one should think that they would almost be extirpated. But these bare frosty winters pulverized the land so much as to break all the small clods where they lie quiet until they be broke. However nothing will do if the drill husbandry will not, and no doubt but they will be in a great measure extirpated in time. We were looking at 13 steers and queys yesterday of Ralph Compton '*high low* Jack and the game', 3 years old which is the general age now sold at. Ralph asked 18 guineas and we desired him to set them at 16 per, so we parted for this time. One quey (spayed) than which I perhaps never saw many better of her age. But then he would not part with her. I told him he should have kept her out of the way then, because she hurts the others exceedingly. Nothing like shewing stock *as equal* as possible, and always the *worst* first, and consequently the *best* last, as we do our tups. By the bye, I hope our dinmond or shearing tups will cut a pretty good figure this season. They were clipping them yesterday at Wark, and what I saw were well formed and full of mutton 20lb. or *24* and upwards I am certain most of them. I really would not liked [*sic*] to have bought Ralph Compton's queys and steers 10 of them at 16£ per. They don't weigh above 50 stone each and about half beef, and many of them weedy and ill bred ones. Nothing like such cattle as used to be bred at Learmouth. However Ralph has improved his sheep much. Not a ewe now but what has been bred from our tups, and his cattle will improve again soon I think. He owes my brother 30£ for cesses. I know not how these people do. But one thing is certain, he cannot help getting rich upon so nice a farm, his crops very promising indeed. Our Nelly went to Holy Island to bathe a while yesterday. Thank God she keeps pretty well. The sale of oats is quite over for the present. Ralph could sell none yesterday at Kelso at all, was scarcely bid 14*s* per boll. He still sells wheat at 55*s* which is a famous price. On Wednesday week at Morpeth I am told Counsellor Alder asked 4

---

54 Quickens, couch grass and other creeping weeds with long roots.

guineas for his best lot of dinmonds clipped, and sold at 3£ 3s. The butchers had nice sport with him and soaked him well. He will scarcely fight them again I think. The steward William Embleton was there the last Wednesday, another bad market, but I know not how he came on. But he had sold none when Adam Smith left the market. They are famous sheep too. He was bid a deal more at home than he got at Morpeth, but he is such a *mean* man that everyone rejoices at his bad luck poor *animal*. He made a most piteous complaint to me last Saturday, which I could hardly keep from *laughing*. Very few of the corn merchants will buy from him because he does not make the stock equal to the sample. I am sorry to say for his father's memory, who was certainly a respectable man, that he is the most *unpopular* man of his age in this part of the world I think. John whenever an opportunity falls in your way of buying any steers at Darlington, either stirks or 2 or 3 years old which will suit either to run in your moors or for us to draw, that you should not miss them. Now we are likely to have a plenty of straw I hope. I dare say all these kind of things will be high sold from this time forward, but opportunities may offer. I went as far as Grindon along with Ralph on his way to Berwick market. William Brown's pastures are now mending as well as his corn, but his pastures were and still are very bad, and his cold land does not recover as soon as our freer soils. But we have and will take part stock from him which will bring him about again. This day has sent us several heavy showers from the west although the glass is rising.

Monday 14th June is a little frosty but a fine clear pleasant morning with the mercury still rising very gradually, which in my opinion very much denotes dry weather, which if of not too long continuance will prove extremely serviceable as we have plenty of moisture at present and everything appears healthfull and thriving. We had the misfortune to lose my old brown Nisbet's mare in foaling with the foal also to Tickle Toby.[55] I lament my poor old mare, as I have rode her many thousands of miles with great pleasure and safety. We have been very lucky with Toby. I think he has cost us about 15£ and only one swing a year old I believe. The[y] had a full horse[56] at Wark bought by you, got his leg broke the other day by the careless lad overturning him in a gateway. They have been exceedingly unfortu-

---

55 A stallion standing at Hutton Hall, Berwickshire. *Newcastle Courant*, 13 Mar. 1802.
56 A stallion.

nate at Wark with horse flesh of late. William Brown is just come and no letter from Matty, although he promised to write on Thursday from Loughborough, and I am much disappointed. But no news is good news is the old proverb. If he do but come home well poor fellow I shall be thankfull, and at all events may God's will be done. I thought he would scarce have leisure to write, they have such labours and hurrys. Our tups I understand have clipped *very fat out* of their wool, so I hope we shall make a decent exhibition. *Tuesday 15th June.* This is a fine fresh morning, wind westerly, and we had a very little chance of rain in the evening yesterday. Indeed the weather though cool is very seasonable, but it strikes me that we are going to have dryish weather as the glass has gone up gradually for some time and is now pretty high and seems stationary. Well we have received a letter from Matty at the last. Thank God he says he is well, and I suppose he will have told you before you receive this that he has hired a tup from Mr Buckley. He poor fellow 'says he has done what he is affraid his uncle and father may blame him for'. I hope neither will blame him as we are certain he has acted to the best of his knowledge, and the rest must be left to futurity. Ralph Brown tells me that Jemmy Thompson of Bogend told him yesterday that a man they call Wright, a neighbor of J. Thompson, sold a large lot of clipped hogs or shearings last Wednesday morning at Morpeth at 48s per over head and all together. It is a strange good price, and the thing is not impossible, but we sometimes think that Jemmy Thompson talks a little fastish. But 18 lb. per quarter at 8d is just the money, only 8d was beyond the last week's price, and 18 lb. per quarter for a strong lot of 70 or 80 hogs is a great weight. The young butchers might take them to be a year older too. I have known them deceived sometimes, even Billy Dunn who thinks himself very knowing. 75 very nice barren ewes comed from Grindon on their road to Thornington haugh, where they are to stay until you think of having them and the little winter fed gimmers or not. The gelt or barren ewes are unsorting by being of so many different ages, but they appear to be all pretty mutton. We have taken out 4 of the fattest to keep as a shew. It is a fancy of mine and it may do good and no harm I trust. It leaves 71 now, and how many there are of the little fat ones at Thornington I know not, but perhaps Tom Glass can tell you. They will amount to more than 100 in all I should think, and will be very pleasant ones to sell I should think in any market. Only we dare not sell them at Morpeth, I mean the barren ones, they are many of them such *beauties*. But you will consider about them and say whether we must send them to you or not. This has been a sweet day indeed, and the weather seems settled in my opinion, as the mercury is now stationary after rising gradually.

[**M.C.**] Well John these rains will I hope bring you a crop of hay on the watered land, at least ours now promises to be a crop at least where the land is good, it now improves exceedingly, only your land has only been part of 2 winters under water. Whenever the hay on spots begins to lay down, then we begin to mow the laid spots if it be but a rood of land in a place, with short scythes, and before we get all the laid places cut then other places are ready for cutting. I am afraid 23 oxen which we sent last may overstock you, and some of the very largest are not quite so ripe. I think Morpeth was better for cattle last Wednesday. Ratcliff bought so very cheap at the back of Mrs Grey's 21£ servant from Kimmerston, also bought 12 of Tomy Hope I believe 21£ per. They were it's said under marts prices. Buying these 2 lotts so low may stop him, otherwise you would have had him ere now for your largest oxen as the keelmen are now under way again after a stop off work. We have near 20 almost ripe, either should go to Alnwick fair the last Monday in July or come to you provided you be clear handed and have meat, and your markets answer better than here. They are mostly 4 years old steers. Perhaps if you sell away you may take them sooner. We have them going very well.

Mattw Culley

[**G.C.**] *Wednesday 16th June* is a very fine grey morning. We are clipping at every farm except this, and the hogs clip in good condition I am told. Ewes are begun with. Was there no wool market on Whitsun Monday? Let us know the prices up on Great Monday. There will certainly be some sale on that day. Turnips come clanking away this fine season, and the work far forward with many people. *Thursday 17 June* is a very good morning although the wind is to the northward. The weather seems settled and seasonable at present. By the London paper this morning I see that wheat still advances a little, other grain pretty stationary. But what is rather remarkable, mutton is highest sold in joints at 9*d* per lb. This confirms me in the opinion that sheep will be high in the autumn, perhaps as high as last year, especially as appearances are very favorable for a good crop of turnips. Indeed things seem turning *topsy turvy*, because beef and mutton used to be much the highest in May and beginning of June, and always fell rapidly after lamb time and warm weather commenced. The last year it continued dear all through July and August, and this year has the appearance of being something like it.

Geo Culley

[**M.C. jr**] Well John. I am this moment arrived, and have found all well. I wrote you a short letter according to promises which I hope you will have got before this one. I heard after writing that the market yesterday was full as good as the week before for sheep, and rather better for cattle although there was an uncommon number in. Still I advise you to sell both cattle and sheep if you have a fair offer, and even rather submit a little than not sell and then have call to take to Skipton and Wakefield. Crops here and grass looking much better than with you. Yours

M. Culley

[**G.C.**] *Friday 18th June.* Berwick fair where we are just going. A clear frosty dry morning but very like drought again. However everything is growing apace here, even young turnips. I would advise you John if William comes to you, to take 7/6 for your big cattle and be thankfull, or to anyone else. Berwick 12 o'clock. This is the dearest market for both fat and lean that I have seen these several months. Fat or rather half fat full 7s per stone sink, some near 7/6. I really think that if we can get that for what we have left we shall not trouble you with them. But I am affraid that this day only proceeds from our *great graziers* and the *Scotch graziers* buying in opposition to each other. Then the little Yorkshire jobbers are buying the veriest little weeds at from 12 to 14£ such as I never saw before at the money. I do assure you I think fattest to strong 8d to 1s per stone higher than at Whitsun Bank. But Ralph could only be bid 14/6 per for oats. Ralph has not sold neither oats nor wheat yet. He expects 54 for the wheat, all red and leanish such as will not sell at Kelso. I will now conclude Yours

Geo Culley

PS. Let us hear from you whenever convenient. I was very uneasy about Matty as he had staid a day longer than he said and did not write to me. Tell Mr Peacock when you see him that I have wrote to Mr Pickering, and when I have his answer I will write to Mr Peacock.

131. *George Culley to John Welch*

Eastfield 19th June 1802

Well John

I should have sent you a letter yesterday the last date Berwick, with an account of that day's fair, but I came away without recollecting to leave it at the post. However it went last night to Twizel Bridge by a lad going with some turnip seed to Grindon, so that I hops you will get it at Piercebridge by sending there on Sunday evening. This is a cold raw morning, wind north west. One thing is certain, that at the rate lean cattle are now bought, they cannot leave a fair profit without they are sold again when fat at 8s per stone or thereabouts. Never did I see such weeds sold at such prices! Ralph sold his wheat at 53s and oats at 15s per boll. *Sunday 20 June.* A very dry fine sunny morning, but we begin to wish for rain again for our pastures. Our General's Close grew a while but is now going back again as fast as can be. However we should and I trust are truly thankfull for the rains that fell. I know not what might have been the result if the drought had continued until this time. The corn in general will be able to stand it now I hope, except the late sown barley and some very dry land oats. Wheat is past danger and has a chance to be a very gifty crop at any rate. *Monday 21st June.* A thick raw morning, wind north west. Just going to Wark tup shew. *Tuesday 22nd.* A cold blowing morning, wind still north west. The mercury has dropped considerably too, yet I think it is more like wind than rain. We had a large and respectable company at Wark yesterday, and our sheep made a particular good appearance. The young sheep in much better condition than we expected. Wheat at Haddington market, Mr William Brodie says, is 2£ 2 or 10/6 per bushell. This is all owing to the foreign wheat being sent to France. Nevertheless I think it wise to be selling, because we are very likely to have an early harvest and a most plentifull crop of wheat in this Island, and the country full of old corn. Indeed I never saw so many stacks in my life I think at this time of the year, and although it is mostly oats and barley yet that will always have an effect. Besides in France they have a very early harvest in general, always before ours, and if they once had enough of corn then it will come back upon us directly. So that I really would advise to be doing. *Wednesday morning 23 June.* Severely droughty, and as cold as to 'skin a flint' as the old proverb says. But as the glasses are and have been sinking these 2 days I would gladly hope a change is approaching. Indeed we are now in much need of rain for our pastures and meadow lands, especially those that were late eaten. Wark stands it better. I was there again yesterday and their crops have a chance to be hurt with any quantity of rain. Harry's haugh also at Thornington is full of grass, especially where he watered it. The barren &c ewes and a score of midling outcast hill wether hogs by way of experiment are all upon it.

Thursday morning 24th June. A clear dry sunny frosty morning, wind still north west. We had a few very cold drops of rain, rather sleety rain yesterday, and as the glasses had dropped for 2 days my hopes for rain were very sanguine, but the mercury rises again this morning and all over for rain at present. Not a cloud scarce to see. We begin to hoe our earliest Sweedish turnips this morning. I never knew them grow so fast. The rain came in nice time for them, and we gave them a piece of good old muck. They are in the Planting field top which is only midling soil. I see by this day's paper that wheat has fallen from 8 to 10s per quarter last Monday at London. I did expect a drop as there is so much grain in the country, but not so very great a fall. I never was better pleased with our selling, for we have sold nearer than common and have been delivering all this busy turnip seed time, as it frequently drops after that season. People can't leave the turnip sowing and the market is badly supplied at that time generally. Mr Pickering in a letter received this moment says 'he will meet Mr Peacock at the Eastfield on the 12 or 13th July'. John I have this day received yours of 23rd, wherein you tell me the agreeable news of having sold the cattle at Boroughbridge for 30£, 10£ 10s returned. I heard yesterday that the drought is very severe in your country by old George Ascough. It is bad enough here for grass, corn stands it yet. Nothing done here about wool. We shall certainly leave you to sell your own wool, as you can sell it better than we can, but would advise you not to be in a hurry. I hope it will be sold well enough by and bye. Wooler High Market contained plenty of wool buyers, but they are only willing to give 24s per stone, the last year's price, while we expected 28s at least or 30s per stone. How is it with you? Mr Nisbet yesterday at Morpeth sold to George Nicholson Panshaw clipt hogs at 49s per head. A number of cattle at Morpeth and sold at no more than 6/6 to 7s sink. Old Mason was at Wakefield last market which was a good one, and sold his cattle at 8/6 over head or from 8 to 9s per stone. He bought several beasts at Morpeth, and 4 capital ones 4 years old steers at near 100 stone each of Mr John Lawson. I don't know him.

Yours Geo Culley

Friday 25th June. Cold and frosty, glasses rising and every appearance of drought. This day I propose please God to set of southwards. Joe Mills bought Aby Hogarth's cattle at Berwick fair and sold them last Wednesday to Ratcliff at 20£ 10 or less. That is the way to lose not gain money I think. G.C.

132.   *Matthew Culley jr and George Culley to John Welch*

Eastfield Tuesday morning 6th July 1802

Well John

I hope you have got plenty of rain now for your pastures, at least if you have got as much as we have you will have had a good deal, for I think it has rained every day more or less for several days, and sometimes it has fallen a great deal, but not in heavy plannet[57] showers but fine mild soaking rains for 2, 3 or 4 hours at a time. It rained a good deal last night, and it has been raining again this morning, the mercury falling, and we shall likely get more rain now than the strong soils can bear. Willy Brown's lands are already almost past working, and I think they have had more than us, which is often the case down the Tweed. I went to Thornington on Saturday morning and drew 102 ewes for you which went off that afternoon. There are part of them very good to be barren ewes, and a part not so fat, but all decent mutton and many of them very capital. You must remember to sell all the oldest clipt ones as they have been fed all winter and will make no better now. They are the shots of James Grey's and our own hill ewes. Indeed they should all be sold soon, as they have been fed on Thornington haugh which is now flooded land, and there is a hazard of their being injured, and therefore not safe to keep long on, that is not to turnip any of them. I got my father's last letter on Sunday. I went to Wark that afternoon and sent 12 oxen from there yesterday morning and two from here to Thornington last night, which with one there was at Thornington makes 15 in all, which the lad will set of with this morning, and you may expect them at Denton on Monday first now, but it would be very well if you could send to meet them at Cold Rowley or Witton[58] as the lad is quite a stranger, and send him back as we have great need of him. The ewes will likely be forward about the same time or sooner. Our corn and grass are very much improved and I hope may be decent enough crops still, excepting on the very dry soils which were burnt past recovery. Spring wheat still very thin and will be a poor crop. Turnips look remarkably well and have the prospect of being a very good crop. If so fat and feeding stock will be higher this autumn than ever known. We have been hoeing common turnips these two days and some of our neighbors part, perhaps they are too early.[59]

[G.C.] *Tuesday 2 o'clock.* Well John I am just arrived safe thank God and found all well. I met the sheep this morning a little after 5 o'clock at the Forrest Burn Gate, but did not see Jemmy as he was in bed asleep the woman said, that had been up 2 or 3 hours looking at them and all were well she believed, so I did not disturb him. It was then a charming hazy morning. The cattle I never met, but they are to be at

---

57 Heavy but local.
58 Witton le Wear.
59 The signature is lacking.

the Haugh Head tonight, I understand 15 as Matty has said on the other side, and has requested you to send to meet the cattle which would be very well done, but Tom will not be returned from Skipton. However you may have some man that knows the road to Witton or so, and I think it would be right to bring forward Jemmy's sheep, or a man and boy as your men (hinds) will not know so much of driving I am affraid, or else Jemmy would be best to turn back for the cattle, because the Wark lad knows nothing of the road, is ignorant and much wanted at home. Or if you approve of it the man who goes to meet Jemmy may go on to the cattle if he knows the road, and he cannot well miss the road, because I am affraid that none of them will be able to drive the sheep so well and safely as Jemmy Glass, and whenever the sheep arrive at Toft Hill the road is quite alive with coal carts which are dangerous for hurting the sheep. However these are my suggestions, and I leave you to act as you think right. The sheep I think will be ahead of the cattle all the way as they have got such a start of them, and being mettled ewes will go nearly as far on a day as the steers. Not much rain today but damp and mist and will be more rain so I think.

Yetholm lamb fair yesterday was beyond anything ever known. Country lambs from 20s to 27s! and since we came north you might have picked Yetholm fair at 5s to 6s per head. What a change! Part wool sold at 24 and 25s and 1s more, if times will afford it. We have done nothing yet. Your clip is so good that I think you should not sell under nearly 21s the 18 lb. or 14d per pound. At least I would wait a while. Ralph Brown I find has sold most of the oats we had in hand at 15s or 2/6 per bushell. Tommy Bates told me that he had bought a good few oats the last market day at Hexham at 2s per bushell. He did not want them, but thought it a very bad price and so it is, below the present price of land. Wheat is 8s to 8/4d at Berwick. Wonderfull weather for turnips indeed, and all folks are hoeing almost in this district. Well you will hear of many people about the Stagshaw Bank fair. Sheep were very high and will be so in my judgement. Therefore I think you will do right in keeping your shearings, at least until it be further considered of. Some black-faced hogs sold at 20s per. Very few country cattle shewn, I fancy they are become scarce. But the most Scotch and Irish cattle ever known, and I dare say many would be left unsold. They were asking strange prices for them. How that fair, like all others, is altered. I can remember since you might have bought a quantity of good aged oxen. Now none to be seen. Indeed few English cattle of any kind, but many more Scotch and particularly Irish than was ever known. Several fat cattle shewn, and two men from Lincolnshire bought many of them. Large cattle 7/6, Kyloes 8s per stone. They cut out the Sunderland and Shields butchers compleatly, which I was not sorry for. Lincoln had bid William Bates of Chollerton for 7 winter fed oxen 29£ per, and 19£ for 5 clever cows which Mr Bates offered to him at 21£. These men met the cattle and bought the cows at 22£ 10s and still very cheap. I would have bought the cows for you had I seen them before they were bought. They then came and bought the oxen at 30£ 3 guineas again, about 7/8d per stone I thought. They call one of the men *Farmer*, and on my demand he said gentlemen had a lot of bullocks sold at Boroughbridge cheaper than these. I said perhaps they might, but that was a heavy selling day. He replyed that they were dear enough and did no good. I could say nothing to that. Where do these men take them to, think you? I should imagine Wakefield. Do write immediately on your return from Skipton. If Mr Peacock is not comed away he may bring your letter, but the post is the quicker in general, and never pay any regard to postage for letters of consequence. I shall perhaps write again soon after Wednesday that I have heard how Morpeth is, and I do expect a better trade now at Morpeth. For who will take 18s for fat lambs and can sell for 24 for keeping. I am in haste yours

Geo Culley

[*M.C. jr*] Morpeth was very full last week of sheep and lambs, but were sold much the same as some weeks before, only 20 cattle in all the town, and high sold, 8s per stone for some. If Mr Thomas Charge comes to Denton to look at your ewes that are come up to you this week, let him pick them as he chuses, as he wants to buy some of them for keepers. Only he is not to take less than 10, and as many more as he chuses. My father had some talk [with][60] him and we shall fix the price. If he fixes on any he will take them away immediately. Highland lambs at Yetholm sold as high as country lambs in proportion. We shall have to pay Longknow hind 18s per for his lambs this year. We never paid above 15s before.

133. *George Culley to John Welch*

Eastfield 6th July 1802

John

I sent you a letter today per post, and my brother came down after, on my asking him why he would not allow you to sow hop or yellow clover, he said he knew nothing at all about it. My brother forgets as

---

60 A word seems to have been omitted.

well as myself, but you will recollect to sow 2 pounds per acre next year, or more perhaps on your moors or outfield land where red clover will hardly grow. We get showers every day and our land begins to look greener, and turnips grow apace. *Wednesday 7th July.* A fine fresh morning and very like more showers. I am going to the funeral of Mrs Air of Coldstream. She kept a large shop, died very suddenly the other day of a kind of palsy or parralitick fit. *Thursday 8th July.* Yesterday was by much the finest, mildest and hottest day we have had this summer, but this morning is showery and very like rain again. Not much alteration in the London corn prices this week. Several nice showers today and very likely to be more. Pastures turning green, and turnips growing so fast we can hardly keep up with hoeing. Morpeth a very good market yesterday I am told, but not seeing anyone that was there I could not learn particulars. But it must be better because the uncommon advance in lambs must send fewer to market at Morpeth, and that will affect sheep which are not now very plentifull. I mean 2 years olds, and people will not be very fond of selling shearings cowardly, because it is now clear that sheep must and will be as dear or dearer than the last autumn. There never was in my recollection a finer lambing time or a better crop of lambs, and still selling 30 per cent hig[h]er than the last year. How are lambs to pay from such prices, without being high and very high sold again, both in hogs and dinmonds &c. It is the jobbers from Yorkshire that raised the prices so unaccountable high. Nobody had any intention of asking such prices untill they were met by the way and taken from them at any price, and the rumour soon spread. Is it not wise then to breed both lambs and calves? I will be bound for it that the breeding is the safest game, the least expensive, and the *most profitable.*

Friday 9th July. Mr and Mrs Peacock arrived last night. My wife and I are going to Holy Island today, but I shall return on Monday when we expect Mr Pickering. This is a very fine fresh morning, wind west south west, the first time that I have seen the dew worm[61] out this summer I think. But very likely for showers still in my judgement. We cannot get our wool sold. Many have now sold at 24*s*, 25*s*, 25*s*/6 and Mr Vardy at 26*s* the only one that has sold at that price I have heard off. I think we should take that price 26*s* and no less. It is not perhaps wise or prudent to keep it if you can get 26*s*, but less we will not take, and I still think it will advance higher. Remember we leave you to sell as you think right, and you must remember that

---

61 Common earth worm.

yours is worth 1s per stone at least more than ours or most people's, by having so few ewes in proportion to fat sheep and hogs. But don't sell without being certain of your money before or at Newcastle Luke fair, because we shall want all the cash we can raise then. Matty tells me that Mr Mason would gladly have bought Mr Wood's dinmonds (they are now) of him or his uncle the other day, but they told him that whatever was with you they would not interfere with, that you were able and accountable to dispose of your own stock at Denton. This shews though how keen they are, and that they must expect things strangely high. John I beg you will attend to your breeding sheep, ewes and gimmers to keep them on sound land, for it is very dangerous weather for rotting. I hope you have eased your clovers and things will soon recover, so as to enable you to keep your sheep pretty well now. It is said that cattle were worth nearly 8s per stone sink at Morpeth last Wednesday, and sheep 8d per lb. sink. I should not wonder if our markets get to be as good as yours now. If so we may have done wrong in sending our fat stock to you. However they must take their chance now. Indeed the wonderfull price that lambs brought at Yetholm must advance Morpeth both in lambs and sheep, and that will affect cattle by degrees. Ralph Brown could not sell his wheat on Saturday at more than 50s so he sold none, but got 15/6 for oats. People begin now to wish for dry weather for getting hay, but I am glad of the wet. We never got a good steeping until now. However it is too much for a deal of cold lands. I rather think we shall not get our wool sold at present. We have determined to take 26s and no less for 24 pounds. John I would wish you to agree with Wilson about the road scrapings, it will be good policy.

Tuesday 13th. A very fine clear frosty morning. I have been at Holy Island a few days, where I left your mistress, and shall go to her again on Thursday or Friday. I would have sent this by Mr Peacock, but he and mistress don't return for more than a week. *July 15th.* Yesterday received your letter by Glass. I think that John Forster will deserve a bottle or two of wine and you a few guineas if these sheep be as cheap as you say and all fair play I hope. We have begun to mow and the weather seems settled. Light crops I assure you. I highly approve of your going with the large cattle to Wakefield, at least you will see both markets. Wakefield is the further from you. I think with you that cattle and sheep will both fall scarce, but can't think that beef will come to 10s neither. I won't say but mutton may get to 9d again. The Sunderland butchers will be obliged to advance beef and mutton. I hope Tommy Charge will buy a few of your ewes. Do let us know all you can about wool. I have just been writing to John Forster to come and buy the famous lame ox and an invalid or two more of us. He can come from Alnwick fair which will be on Monday 26th inst. By a letter from B. Sayle this morning he says (12th July) a poor shew of fat cattle here today, Rotherham. Beef 8/6 to 9s, mutton 8 to 8 1½, wool quite flat, now talk of 12d per pound. Corn no lower, London market on Monday better for wheat but no higher, 74s per quarter highest. I fancy we shall hardly sell our wool now. John I believe I forgot to tell you how much Mr Bates is benefitted by altering his lime kiln, viz. making it wider at bottom than used as they do here. He sent a mason here to look at the kilns. They make them almost as wide at top as bottom, by which means they keep a greater strength of fire lower down, which makes them do with less coals and cart loads of lime per day. And by means of a little angular hillock built at the bottom called a *saddle*, which throws the lime to the eyes they draw as well as if they were narrow bottomed like yours. I would wish you to look at those in this part when over, and send a workman some time. It is very important. Yours in haste

Geo Culley

Morpeth I hear was not so good for cattle yesterday, but was for sheep and lambs.

134.   *Matthew and George Culley to John Welch*

Wark 11th July 1802

John Welch

I have sometimes thought of the hedge you made above Lime Kiln Bank inclosing a part of the land next the moors, a thing I wished for, only we were afraid since my brother was over that it may be too near the road. Although I believe it may be measured to the quicks it should have been 20 feet from the middle of the road. Now my brother and I think you had better keep fair with Mr Wilson the surveyor, and pay him rather more for the scraping of the road than you may think he ought to have than have any way his anger. Thomas Bowes and he are only midling men, at least I do not like Bowes. We had better keep in with them a little than have the hedge thrown down, and keep fair with them. When men get into power they sometimes make an ill use of it at times, better not to kick too much against the pricks. Also your master G. is seemingly fond of leading the scrapings, when you have leisure, to the bare places especially on to tillage. I cannot tell when we may get to Denton but I believe it should be soon. If you can get slates perhaps we should put some straw on the High House where the wet gets in, as the people will be very uncomfortable and may catch cold this wet weather which we hope you have as well as we. We have much rain now, so as to make the land flourish greatly. Corn, grass and turnips &c grow exceedingly well, so as to promise abundance (floods, moderate ones in the Tweed). The rain now heavy but not to run much, in the lanes only and the burns, but it's very like a rainy season, has rained heavy yesterday with several heavy showers today (Sunday).

I am sorry to hear by Mr and Mrs Peacock that George Dent's son is so very weak. His father will be sore distressed as well as all the family for him poor man. There is nothing but submission to our lot, it will come sooner or later we are to leave this world. I hope he has been a decent young man. There is a great collection of clouds on Cheviot very like rain. Let us hear how your wool markets go, how the prices are with you. 25*s* for 24 lb. here is the general run for such as ours. They say cotton wool is so very plentiful; and low priced enables the manufacturers to have more profit by cottons than by wool. Mr Pearse of Ingleton, a busy fellow, has threatened to indite your road to Darlington on account of the narrowness of the lane. Better perhaps to put him off by good humor and promises than harshness. I mean you may tell him we intend to alter those roads in a future day when we get a little time on hand, as well as we have altered and done to the road from Houghton le Side and endeavor to pat him by fair words. Such men as these are not pleasant at all. I hope these rains will recover the pastures and your meadows. We have a very growing time. Turnip land was well wrought I suppose everywhere, and there is a great prospect of a good crop. Beef has my brother says advanced at Shields from 48 to 52 and Sunderland yet higher. Stock of all kinds very high at Stagshaw Bank fair. Will you not want some stock for wintering? We will have more I think to send you as we have only a small cast of oxen this season. I wish you would let us know soon how many head and of what sort you may want, that we may try to buy some. We have a good quantity of sheep and a good stock of lambs. They were a good crop this year but sell very high. Mr Ralph Compton has 16 to 18 steers &c which we are to look at. Indeed we saw them once but he asked 18£ a piece (guineas), by farr too much, were well pastured are in good condition, are likely to get fatt and heavy. Those we bought of him last year improved and got more weight than those we bought from Beal from Mr Ned Gregson. Mr Buston's are I apprehend the best bred steers that are sold in your country. If they sell them this year, may ask them about their steers, suppose they are fatt. Yet I would like to buy them in preference to others, fatt or lean if healthy I think if we want them and I suppose we will. If we are to do our old house this season I wish that the slates could be got in proper time against the time when we come over, as there are many which I want to have done if I am spared to do them. But we should have wood got to lay the floors in the mill, and that will take many logs for deals for that purpose. Mr Peacock and I, on considering the slate business over, believe the neglect lays with the man at Staindrop. Therefore give Mr Peacock the dimentions of the house at Denton and he will write to Mr Rawes, that will be going to the fountainhead. From

Mattw Culley

PS We mow the laid spots in our watered land or patches with 1 or 2 men, and leave the weak spots to mend when they will. As slates are not likely to be had we think of leting the house alone this season. Therefore you should patch the High House where needfull with thatch.

[*G.C.*] Well John I had wrote to you as well as my brother so have put the 2 half sheets together. My brother wishes you to sow 2 pounds of hop clover per acre still amongst your other seeds, and I believe it will take this very fine growing weather, if you can get any hop clover seeds.

135.   *George and Matthew Culley to John Welch*

Eastfield 15th July 1802

Well John

I sent you another long letter of my brother's and mine today, and answered yours which I received from James Glass yesterday. Kitty Mason came to Wooler today, and at the publick table began to say that John Forster had bought a cheap bargain of sheep for you of *Rawling*. I asked if they were not fairly bought. He said he knew *nought to* a contrary. I told him 2 Sunderland butchers had refused them and that I apprehended you and John Forster had every right to buy them. 'Damn says he, I find you know all about it better than me'. He forgets how his old father jobbed, and have not you an equal right? However I must observe that in our situation the eyes of the world are *upon us*, and we cannot act too *circumspectly*. Perhaps to have kept these sheep a few weeks might have been as prudent before selling, because the serious part of the world will say 'They *grasp* at everything, nothing *escapes* them'. But it is after all perhaps the best to do what is *right*, and not mind the *envious* part of mankind. For surely it is perfectly fair what you have done? Only jobbing is what I would not wish to encourage by any means, particularly in our present situation. We should be *graziers* and *farmers*, not *jobbers*. *Friday July 16th*. This is rather a heavy looking morning and glass sinking these 2 nights betokens more rain. I am just going to Holy Island where our mistress is, and Matty is going to Mr Spours tup shew. Matty and the shepherd have to draw the last few ewes this morning, which finishes that work. We reckon that you and us will have of segs, wethers, ewes and gimmers about 1600 to winter feed. So I think we shall not have occasion for buying many sheep, wintering cattle for turnips will be fewer than usual, about 60 here, and we shall probably buy a few, Mr Ralph Compton's or others. G.C.

[**M.C.**] Wark 18th. We received your last letter. My brother sent me yours of the 15th yesterday. I think you sold your cattle and sheep well considering so full markets. We have just grass to serve our stock well and no more. Thornington Haugh is an excellent pasture with much stock of cattle in it, as we run the river in the drought over some of it. We had a fine rain again yesterday and in the night, so may say have plenty at present so that corn and grass grow apace except the West Broomy Know which was much injured before the rains came. The barley very rickety and short, also full of weeds. The turnips should have been eat on the land, might have fused the soil and made it bear the drought better. Such light soils will not bear hard work in a droughty season. But on the whole our crops now promises well especially the drilled wheats on Park Close, part of Lambs Know, and part of the 5th short side that had seed turnips last season. Turnips promise well, so do the sweeds. We never had them seemingly so good, are on the west side of the Mid West townfield next Shed Law. I think of going with Bessy to Newcastle on Monday 26th and so to Denton for a few days. Mr Ralph Compton and I nearly agreed for his steers and queys, a cow and a bull to cut, 14 in all. Bid 18£, he stands at guineas, says we are to have them. My brother is to see them and conclude the bargain. Are better than last year's ones in as much that these had full (nearly) turnips last year in winter and those last year had none. Although very high I think it wrong to miss them as they have more growth than many that we can buy. You can judge of that as you had them and Beal steers through your hands this spring. Beal steers 20£ per at Ninian fair, Ralph Compton's under 18 about Lammas, I hope may pay some money as they are thriving on a midling pasture. However I do not know where to mend the matter. With respects to all friends

Mattw Culley

PS. My brother says you cultivate the land well. I am very glad to hear it, as cultivation lays at the root of all farming. That's the foundation of all crops. Money if a man has it is ill spared where it is necessary to be laid out. We are now mowing of watered land, rather too soon as it's in a growing state and cuts a fair crop, but we will now leave after cutting Lambsknow end to low meadow (as we were afraid of being caught by a flood), and now mow our clovers for Thornington, Downey's Piece and Bryans Fold for Harry. We have in some of the watered land as much as they can turn off when the land is dry, and was well watered and laid in in time. Indeed is [*sic*: it] was winter feed, the weather cold and then droughty, which prevented a very great crop but it turns off very well. I go this afternoon with Ralph Compton part of the way to Alnwick to pay Mr Grey a compliment and Mr Beaumont.[62] There is no opposition in Northumberland. Matty went to Mr Spours's to his tup shew and meets us. I know not whether my brother (who is at the Island) goes or not.

136. *George Culley and Matthew Culley jr to John Welch*

Eastfield 28th July 1802

Well John Welch

I shall again address this letter to you, as I fancy my brother will very likely be gone from Denton before this reach you, as the cart will not be with you before Saturday evening, or earlier I hope too *on recollection* because the tup should be sent to Raby that night if time will permit, and you must send some other person to Raby than John Story, who has a wish to go to Denton at this time, and then he can see his friends at Denton. And as John knows nothing of the road to Raby you had better send any of your own men that knows the road. But be sure he get the letter for Mr Sherlock which John Story has with him. I know not how you are circumstanced now John, but I confess we begin to think we have got rain enough at present. But God Almighty's will be done. Only we are a murmuring crew like the *Israelites* of old. This is the most serious rain we have had yet, I wish a flood don't be the consequence. I know not at what time it began, but it was raining fast when I got up at 5 o'clock, and now 7 and I think it increases, and it is from the south east, a rainy hole. However we have less reason than many people to complain, and we can take rain every day, and we got it in abundance every day lately, and yet we have had no floods to call floods except in the Tweed, and these

---

62 A parliamentary election was held in July 1802. The MPs returned unopposed for Northumberland, at Alnwick on 19 July, were the Hon. Charles Grey, 1764–1845, MP 1786–1807, 2nd Earl Grey 1807 (*Complete Peerage*, vol. 6) and Thomas Richard Beaumont of Hexham Abbey 1758–1829, MP 1795–1818 (*Burke's Peerage*, Allendale). The Culleys had the vote as freeholders of Akeld.

trifling. But people from all quarters begin to complain of the wheat being mildewed. I hope it will not be very bad, because there is no laid corn in this country, at least that I have seen yet. However it retards the harvest very much indeed. We have not the least appearance of harvest here that I can see, although we often begin at the end of July or early in August. Not many years but there are new oats at St James's fair[63] which is tomorrow week. However I am morrally certain that wheat will advance in consequence of this serious weather. Not but it's wise to sell in part I think, although there is much hazard of the new samples being only midling now, come when they will, and I think we shall not have any wheat harvest in this part before the middle of September. And should the weather continue soft which is too probable, it will not be soon fit to use or grind. We bought Ralph Compton's cattle yesterday 14 at 18. 10s very dear indeed I think. He would not sell us the best quey neither. He is to keep them untill the beginning of September. John Forster bought the lame red ox yesterday at 34£ but he would not buy the 2 bad cattle, so we send them to him to make the best he can of them. They go tomorrow morning for Shields. James Glass is just come to go with this. John I think you sold your last lot of cattle well. Was it to Shillito? Were they the few that went from hence the last? John I must intreat you to pay great attention to your breeding sheep this wet and dangerous time for the rot. Should it come warm weather after as must be the case if we ever have any warmth this summer, you may depend upon it that your very bare eaten pastures will put up a *tatty grass*, extremely dangerous for sheep. You may better continue wethers there than ewes or gimmers, because the wethers can always be sold in case they are tainted. And however extraordinary it is very well known that sheep in a state of *rottenness* will feed quicker than if ever *so sound* to a certain time. But it is very dangerous to trust them long after the winter commences. Old ewes to a certainty will feed astonishingly for a certain time after getting a taint, especially about Michaelmas time. This day is not so rainy as it was, now near to 11 o'clock, but still looks like more wet. I am yours

Geo Culley

[**M.C. jr**] John, Lord Darlington's tup is an old sheep, busted C on the near hip and a small spot of keel on the near rib. Mr Eales's tup is

---

63 Kelso, 5 August, on a green opposite the town at the confluence of the Tweed and the Teviot, important for horses and cattle: Douglas, *General View of the Agriculture of the Counties of Roxburgh and Selkirk*, p. 207.

an old sheep and busted on the near hip and no keel, and the sheep for your own use is a dinmond and busted on the far hip and no keel. You must send Lord Darlington's tup to Raby immediately on his arrival at Denton and keep Mr Eales's tup until he send you word for him.

137.   *George Culley to John Welch*

Eastfield 2nd August 1802

Well John Welch

Being at Holy Island when your last letter and my brother's came to hand, and Matty neglecting to send it to Berwick on the Saturday, and only coming home today, I had no chance to answer it. However I am glad that Matty immediately answered it to his uncle at Newcastle. I think you made a very good end of your sheep, and shews that they have been exceedingly cheap bought. I am glad to hear that your corn markets are better. They are better here also. I believe that very few clips of wool are unsold in this district, and it seems a very uncertain trade. However we must now await the results. The hay time has been very difficult indeed, and is some spoiled. We have had some tollerable days so as to get a good deal piked, and the weather appears to be more settled than it was. You were certainly right in taking your sheep from the unsafe grounds, I mean the keeping sheep ewes, lambs, gimmers &c &c. You are right in fallowing that bad moor, it may grow a piece of red wheat but nothing else. I should not wonder at the market being better at Skipton the next fortnight. The ewes would not possibly have any tallow in them. They were quite green and had got very few turnips indeed. If they had not been very good growers it could not be supposed that they could have got so much meat upon them.

John if you come to Newcastle Lammas fair I think that you had better bring us a 100£ with you, as we are likely to be a little pinched for cash. Either Matty or I will be there. Pray John did you ever get the balance for the butter firkins? It is more than time that it was paid. And have you ever received the money of George Nicholson for the few sheep that Matty Culley trusted him? It would do us good if we had it. I do assure you that we have been very much pinched as all the fat stock went to you. I must know of you what cash you think you can raise, I mean with the money in Mr Peacock's hands, before the end of October. You know that we have a most *serious* sum to raise by that time, and much depends upon you as all the fat stock has gone to you. Pray do you wish that we should buy you James Grey's ewes this autumn, because I am certain that we can have no occasion for them ourselves. However if you think that you can make any advantage of them I am willing to buy them for you. But if not James Grey should know and not be kept in any suspence now ewes are selling so well. *Tuesday 3rd August* is a wet morning as usual and very likely for a wet day as the wind is south east. The weather now really begins to be serious. Corn must take hurt I think from such repeated wetts, and the harvest will be very late indeed. Only there is one thing favorable to corn, there is little of it lodged in this district, so that it the sooner gets dry again. As to hay it is in a most miserable state indeed, and the season is now so late that people wish to have it out of the way before the corn harvest, which must take place soon. Mr Andrew Davison was here yesterday paying his rent, when he said he should cut some oats in a fortnight, and Mrs Atcheson of Yeavering it is said has some early oats now ripe. A thunder storm attended with a fall of hail and pieces of ice has damaged some pease at Wark very much, George Humble says. Indeed he brought some of the pods here, which are much hurt. Their turnips are also sore cut. The turnips may recover, but the pease will not I am affraid.

Wednesday 4th August. A roke or mist into the very doors. Lammas rokes have come early this year. Indeed it has been nothing but rain and mists lately I think. It rained the whole day yesterday from morning until about 7 o'clock in the evening when it ceased, and looked as if we should have a fine day today. And what is most extraordinary is the mercury being high and not altering at all. The wind appears to be north east this morning 5 o'clock. A good deal of corn begins to lodge now, and no wonder, such heavy rains and mists. It must be astonishing if corn don't now take hurt, and it is more than probable that the new grain will be a worse sample than the old and less gifty. It is really most dangerous weather for old ewes and all sheep. My brother was extremely right in making you eat the Mount Huly alias *Holy*. If you have not much hay you will have turnips and straw, I hope enough. If you could by any means ease or rather free a clover field altogether for a while, the seeds would recover their growing time. But when a field of seeds once is got so bare as yours were, they can not recover without ease. We cannot keep ours down now at any rate. They are much too big for sheep. You certainly deserve commendation for plowing the bad *lings* or weathered moor, it could do nothing in the way it was. John I have this moment been summoned for a juryman at Newcastle Assizes which is on the 14th the judges come. But as the fair will be on Thursday the 12th I will certainly attend the fair, where I would wish you to send me the 200£ as I see we shall be short of cash, having many rents to pay and Ralph Compton's cattle. Now it happens badly for you because Skipton market is on the 16th and Newcastle fair on the 12th. But if you can only reach Chester on the 11th in the evening (Wednesday) you can be at Newcastle Cowhill time enough on the Thursday morning to meet me. However if inconvenient to you, you can send me the 200£ by any of your neighbors who are coming. I intend to go from Morpeth on the fair morning myself, as I shall be so many nights away from home, and I shall have further to go from Morpeth on Thursday to the Cowhill than you will from Chester. But if I should happen to be at Newcastle on the Wednesday evening before the fair, and you should also reach Newcastle that night, you will find me either at Turners or Greaveson the Crown and Thistle[64] where Sunderland did live and where we used always to put up. I don't like Turners, it is so far out of the main part of the town, so I mean to be at Greaveson if they can put me up.

We received a letter this morning from your master from Newcastle dated Monday the 2nd inst., and my sister and they are gone up to Mr Bates's Brunton, and will be home soon as he says he wants to see me about an estate near Leeds which you recomend, and I have no objection but the cash. And I can't see it and be at London to buy it and go to Newcastle Assizes at the same time. My brother says oats were 7s per new boll at Durham on Saturday and wheat 10s per bushell. No wonder that wheat advances, as the corn has a very dark complexion this weather I assure you in this country, and turnips grow slow with too much wet. The high price of hay at Newcastle (8£ per ton I am told) will advance both oats and turnips. Indeed turnips are bid 6 guineas per acre for it is said already. Be sure you pay particular attention John to your holding or keeping ewes &c because this is a very dangerous time for the rot I think. Wednesday evening. We have just received yours and my brother's long letter by John Story, who is arrived safe. When I see my brother we will consider about this Gipton estate which is very suitable in many respects, but the money will be a little aukward for us, and being sold in Chancery is sometimes cheap bought and sometimes dear. But it is a long way to go upon an uncertainty, and as I am obliged to attend the Assizes at Newcastle I can't possibly get. My brother will not be fond of going, and in my opinion Mr Bailey is the properest person to go and valued [*sic*], and so go to London, if he can have that leisure. However we

---

64 Probably Charles Turner, the Queen's Head, Pilgrim St; the Crown and Thistle, Groat Market, Newcastle upon Tyne: *Newcastle and Gateshead Directory for 1795.*

must determine on that when my brother and I meet. Corn must advance this weather, but nevertheless I think you are wrong not to sell your old barley at 40s per quarter, but you must be better able to judge than I am. But I hope you will deal with him when you meet again. I am glad that you have removed your keeping sheep into sound keeping. If my brother and I should think of Mr Bailey's going to value Gipton you will have to go with him to it, or give him a line to James Nicholson may do equally as well I fancy, because I shall be disappointed more ways than one if I do not see you at Newcastle fair. I am in haste yours

Geo Culley

PS Perhaps you will write again before I see you at Newcastle if you can do. You are right in giving Mr Peacock what cash you have to spare.

138.   *Matthew Culley to John Welch*

Wark 9th August 1802

John Welch

As you meet my brother at Newcastle, I send this by him. He will inform you best why we cannot bid for Gipton, as we will have no money in our hands we believe then, and it may be wrong for us to embarrass ourselves now when age comes on, although it may be a desirable thing to purchase. My brother seems to think that we should sell many of our shearing sheep wethers. In such case I would advise you to sell at your sheep fairs or at home, Mr Wood's and Longknow wethers also, as our own low country wethers will likely pay more money for keeping on than worse bred ones. But your master and you consider this matter over. We also should in my opinion send you some of our forwardest oxen or queys to feed, as you will probably have them to sell as soon as times offer. These are my thoughts but may be considered off, only I fear our cattle may hardly be so forward as they were last season at this time, only our pastures are now good and I hope yours may get good. Pray if your pastures don't mend, sell some of the stock off so as to let them get up. I cannot think of having stock on bare pasture if it can be avoided by any reasonable means. Rather sell some to loss than injure the whole. I hope you have got the bridge finished to your mind, passable I mean. It may be completed in the filling up when time suits, only the main or dangerous should be put out of danger. I hope you will begin to mow the parts that are grown fitt, and leave the growing parts to mend a little. Ours has mended to be a fair crop in general now, but the floods have sanded some of it a little, but not materially hurt. Saturday being a nice day we got 2 fields piked and into coyle. This Monday a grey morning, hope the weather may take up and mend the crop. Wishing you good weather I am &c

Mattw Culley

PS. Let us know how Robert Dent recovers. Should you not get the plumb tree sawn in pieces so as to dry and be lighter to carry? They may measure the spining wheel and judge of what thickness to plank it. The spinning wheels are made very light at Hexam. Whether the wood should be left at Bird and Bush or at Sunderland I know not. The corn market is near Bird and Bush for wheat.

[*M.C. jr*] Tell John we have sold our wool.

[*M.C.*] PS. Dolly should get the bed down and wash it with the coarsest soft sope (generally called black sope) I mean the courtains. The wood should be dressed with coal spirit, especially on or about the joints and mortoises or any suspicious place in the wood. The bed bottom should have the jacks[65] drawn, and brushed over with black sope. The pinions[66] should also be dressed with coal spirit. These things should be done now, the sooner the better. Jack Dickinson will assist Dolly to take down the bed, and when dry put it up again. The bed stead will long smell of the coal spirit if you do not set the wood to the door for the air to take of the smell. I wish not to set up the bed before we come over if you can do without it. Since writing the above my sister Culley has persuaded me that I might possibly bring the *bug* from some other place to Denton where I had slept on my road, perhaps at Durham.

My brother's pastures at East and Westfield are exceeding good this very growing weather. Turnips promise well but the wheat does not fill well except the red wheat, especially the close eared red wheat. The late peas are blossoming and growing away so fast that they will not load or fill, early peas will be better as they have a chance to fill. Beans may be a good where drilled. Oats much mended, may be a good crop, some will probably cutt in 8 or 10 days. Barley much mended.

139.   *George Culley and Matthew Culley jr to John Welch*

Eastfield 18th August (Wednesday) 1802

Well John

Luckily not much business at the assizes so I got home this evening. I was upon the first jury on Tuesday morning early and we

---

65 Contrivances for lifting weights, by force acting from below.
66 Small cog wheels.

tried 5 causes and then they excused us from any more attendance. So I came to Morpeth last night, saw a very good market indeed this morning and then came home sweat to death. I don't remember ever being hotter in my life. A vast of thunder and rain to the south east all the way, but I luckily escaped the rain. Good sheep were worth more than 8*d* sink, but very few good ones shewn, and it is no wonder when Jere, Green, Hodge, Paul &c &c are all out buying all before them. I will persuade my brother to allow us to sell all or a part of ours. I saw 2 niceish Scots 37 stone sold at 16£ 10*s* per today at Morpeth. *Thursday 19th.* Mr Pease (Edward) came last night and looked at part of our wool today but we did not bargain. He bid 25*s* and my brother was not willing to sell under 26*s*. At going away last night he said to my brother as he was going over the boat at Wark that he would give 25/6 provided we would let him have the Denton wool along with the Northumberland. However my brother would not consent and so the matter stands. Be so good as give us your opinion of this business. The Wark crops and pastures all look well. They have got a good piece of hay together, and talk of shearing tomorrow or next day a few early oats. But I don't think it will be long before they have a general harvest. Corn has come wonderfully on since I went south, and there was little rain yesterday. *Friday 20.* A very promising morning.

[**M.C. jr**] To make maggot water take an ounce of sublimate of mercury and dissolve it in 9 gills of cold soft or rain water, that is let it stand until it is dissolved. When used, pour it on through a quill the same as you use tobacco liquor, only keep it near the bottom of the wool, and do not lay near so much of it on as you would of tobacco liquor. If you wish it stronger, mix one half ounce of sublimate in 3 gills of cold water, only we think the first quota strong enough. It will be fit to use 2 hours after mixing although not entirely dissolved.

[**G.C.**] 2 o'clock. Matthew and I have been at Wark and Thornington with my brother where everything is looking well. We are putting tup hogs &c upon the rape, both at Wark and Thornington. Both places very good indeed. John we have considered that we will let Mr Pease have the wool both ours and yours at 25/6 in case he will stand his *bode*[67] upon his return from his journey which will be in 8 or 10 days from this date, and you must therefore wait upon him on his return, and write us word what he says. Perhaps it may be right in you to call on Mrs Pease the first time you are at

---

67 Offer of a price.

Darlington, to know when she expects Edward home. The time of payment was not fixed or talked about, but you must *contend* for payment in full on the last day of October in Darlington notes or a bill on London at 30 days, as we have the cash to pay in London then. However if you cannot manage that after every effort you must insist on the best part being paid then and the balance at Candlemas or 1st of February the next year, or as much earlier as you can get it fixed. John I must entreat you to use *little ploughs* and not *double mould board plows* at the first plowing of your turnips. Little ploughs without mould boards are undoubtedly the best, because they loosen the earth so well without lying the earth too near the turnips, which is very bad for the turnips. A turnip is never too much *exposed*. The more the root of a turnip is *exposed* the better it *thrives*, while *cabbages, beans, brocola* &c are all better for being earthed up, viz. you can't lie the earth too near them. But turnips are of a very different constitution, for as long as the tap root of a turnip is safe or uncut you may depend upon it that the more the root is exposed the better. You may frequently see them by the side of a grip or cut in the land, hanging by the tap with their tops below the root and yet thriving to admiration, and much better than those which are carefully earthed up. Nevertheless the earth or soil cannot be too much loosened all about them, and that can't be done by the double mould board plough which leaves the ground hard and stiff all about their roots, whilst the little plow without any *mould board* can work quite near the turnips and loosen the earth to the very taps, which makes them thrive as well again. This Saturday 20th [*sic*: 21st] is a squally windy dry like morning. Many early oats cut in this quarter, but we are unbegun yet. May the winds keep moderate and a good harvest take place is the sincere wish of

Geo Culley

I will keep this open until I see how Berwick market is. Ralph got 30s per boll for pease at Kelso yesterday and 53s for wheat, 18s for oats. PS. Perhaps it may be best for you to prevail on Mr Edward Pease to write to us about the wool after you have said to him what you think right. But be sure you write at same time what passed between you. *Berwick 4 o'clock.* A heavy market. Wheat worth no more than 50s, but some men from Glasgow buying beans, willing to give 5 per bushell or better, say the harvest will be very late. Nothing fit to cut of 6 weeks. In haste

Geo Culley

PS. Write soon. John Nisbet has sold his dinmonds at 55 or 56s.

140.   *George Culley, Matthew Culley jr and Matthew Culley to John Welch*

Sunday 22nd August 1802 in the evening

Well John

I sent you a letter from Berwick yesterday. This has been a fine harvest like day, only a little showery in the morning before church time. I think if this week hold good most of the Poland oats will be cut and some potato oats also. Mr Nisbet has sold his dinmonds 200 at 56*s* and 100 ewes at 48, the latter only midling ones he says. Jere Clayton was his customer. Jere is returned home. Had I been in the way I would have shewed him a larger lot. A great many sheep are sold, both ewes and dinmonds. *Monday 23 August.* A quiet morning, perhaps rather frosty as the wind is rather north easterly, glass steady a little above changeable. This has turned out a wet morning since 9 o'clock, and spoiled us a famous hay leading day. *Tuesday 24th August.* Your master and Matty went to the fair at Whittingham. This is a frosty morning, which is sure to bring more rain, but we ought to be thankfull that it is no worse. By a letter yesterday from Mr William Thompson he says they have fine harvest weather about London. He also tells me that Gipton was sold for 20,100£ and the piece of wood 1,070£. What your valuation and Mr Peacock's were I know not, but I thought it was not to be sold for less than about 50£ per acre. But it has gone further. Well I wish the people luck that has got it. You know I said it would hardly be sold at less than 18,000£, and I think you said 16,000£ would buy it. I could only judge by what you said and situation, which must be most advantageous. The two Matthews intend buying a few steers if they are light ones. Very little hay got home yet in this country, and harvest will be upon us I see now. The potato oats are whitening very fast, the Poland mostly cut and the early sown barley, and autumnal wheat is all coming on apace. Pease will be a bad crop. Two men were at Berwick from Glasgow all the way, wanting to buy beans and pease, but could not meet with a quantity at Berwick. It is said that they would have given 30*s* per boll and upwards for them. This has been a very fine day untill between 5 and 6 o'clock in the evening when it came on very rainy. However we have stacked a good piece of hay in tolerable order.

Wednesday 25th August. is a fine fresh morning, likely for showers. It had only been a shower last night. The glass is low, but I am more affraid of wind than rain at this season always. Matthew did not come home from the fair last night. He is probably gone to Prendwick today to buy a few steers of Elly Reed. He talked of going before he went from home. We began to shear today but were put of by the time

you go out at Denton. However we are for being at it again this afternoon which looks well now. But we can't go forward yet. We are
leading hay at the Westfield from bog also. *2 o'clock.* Matty is just
returned from Whittingham which was very good for both fat and
lean, especially the latter. They bought nothing, but Matty bought 2
steers 2 years old of Mr Reed Prendwick at 24£ the 2 and sold them
Wark worst bull at 30 guineas. They received of Stephen Thompson
58s/6d for George Nicholson's ewe, so George will justly owe 1/6
which of a common bargain and ready money we should not have
expected. But considering the good bargain and the credit he has had,
I think he ought to pay the 1/6 but shall leave that to your consideration. A fine afternoon but rather more wind than we wished for, but
hope it would not do much harm except the *top pickle* of the early oats
as it is here called. We cut a few more very good Poland oats this
afternoon and will cut most of them tomorrow I hope if fair. People
are beginning to cut potato oats also, and I think ours will cut next
week. Matty is gone this evening to Wark for Dunse fair tomorrow. A
great many Yorkshire jobbers, Matty says, went from Whittingham.
Thomas Scott and Mr Armstrong also. Matty says some nice queys
were sold for rather more than 8*s* per stone. These are strange prices
for this season of the year, but things are now as dear in summer as
spring. Bob Wright will leave this at Darlington on Wednesday
morning as he goes with the 2 tups to Maltonside, and will be with
you all Sunday on his return. *Thursday 26 August.* A fine harvest like
morning with rising glass and wind moderate although it is windy
enough like. I have this moment received Mr Peacock's letter with the
very accurate account and valuation put upon Gipton by you and
him, which shews that you have taken great pains for little profit. But
never mind that. Pains taking and industry always meet with their
reward at the last, sooner or later, provided they are well and
honestly intended. How many 1000 miles have I travelled in my time,
and many of them not only expensive but made little or no return.
Still I was increasing my knowledge which can't be got or acquired
without pains taking and industry. However it has pleased the
Almighty to bless our endeavors at the last beyond our most
sanguine expectations surely! It would have been very unwise to
have bought it either over dear or beyond your judgement, particularly when we had all the money to borrow. Something will turn up
trumps when we have money to spare perhaps. At any rate I am very
well pleased. I had much rather purchase Manfield if it could be come
at.

[*M.C. jr*] We hear that the corn markets at Carlisle are exceedingly high, consequently Barnard Castle should be your best market now, they sell wheat at 72 or 74 our boll, and other grain in proportion. I am just returned from Dunse fair, which has been very high, full as high as Whittingham. I bought 9 lean steers, 4 three year olds at 3£ per, and 3 two year olds and 2 year olds at 12£ 4 which I think much cheaper than I could buy them at Whittingham. But they are not so well bred although far from bad ones. We want them for drawing.

Mattw Culley junior.

[*G.C.*] *Friday 27.* A fine hard like morning, wind in the west and likely to *blow*. People complain of early oats yesterday having nearly the seed out in hilly places. We had a very heavy sudden shower yesterday again about 2 o'clock made all swim. Then the wind took it off the corn and people fell to work again. London corn markets keep steady although they have had fine weather I am told, and plentifull crops cried up. It may be so, and I wish good crops but it is thought that wheat will not be the best sampling in the world in this country. Too much red stuff in the ears, and such a darkness upon many crops of wheat. Our potato oats are gotten sadly down in places with these very heavy rains. However they feed the turnips which are very good, and pastures. *Saturday 28 August.* A very fine harvest like morning, but very much like wind indeed. But perhaps I am more alarmed with winds than most other people in harvest, as I once at Fenton had a most serious loss by wind when we were not so well able to bear it. We have a large quantity of very nice potato oats just ready nearly for the sickle, our Poland oats mostly cut. We are going to begin the potato oats this morning. Matty and I are going to value Grindon tithe today. I am sorry to tell you that George Humble is very poorly, confined to his bed which is inconvenient at this time. But God's will be done, I hope he will be better again. Ralph Brown sold for no more than 52*s* wheat yesterday at Kelso, oats 16*s* common ones. But their best meal markets were not so good, and new oats are got to market. Nothing likely to be very high but old pease and beans. They are worth now 32*s* 33 our boll. We let Mr John Barber of Doddington 9 tups yesterday at 84£, a fine trade to be all to one man. It is very remarkable that I served his father for the hire of a tup 2 years ago. The old boy lives at Bulmer and this young man succeeded to Tommy Barber at Doddington farm.

Sunday 29 is as promising a morning for good harvest weather as ever I saw, and we cut a very nice crop of potato oats indeed in the Castle hill yesterday, and I fancy we shall cut all our potato oats this week. Those are in all the exposed places, nearly about the half of the seed oat. But I hope we have plenty without. Matty and I were valuing the Grindon tithe yesterday, and I am sorry to say that Willy Brown's winter wheat is sadly injured by the mildew although red wheat. But *remember now* that it is not our old kind that we have had so many years, but the kind we got a few years ago from Mr Brodie in Scotland which grows a short straw and close short *ear*, which I blame for being so much mildewed, because the *ear* being so close set makes the wet hang so long upon it that occasions the disease. Now our old kind is long strawed and long eared and the head is open. But that same openness allows the sun and air to dry it so much sooner as in my opinion to prevent the *mildewing*. However in the same field the short ear is the most mildewed I ever saw, and the other has escaped except a chance grain here and there, which is of consequence compared. Where Willy expected 5 or 6 bolls per acre he will have scarcely 3. His oats, potato particularly, are very good. Indeed all his oats as good as can be expected. Barley tolerable for Grindon. But wheat winter sown and of the short eared kind very bad indeed, pease also bad although drilled. But the land is very foul, and wherever the land is foul or full of quickens I lie it down as an invariable rule never to sow *pease* because they encourage quickens. If you must sow a crop, plow too [*sic*] or 3 times and sow barley, or else common or Angus oats. Ralph sold wheat at both Kelso and Berwick for no more than 52, and I don't think more will now be got. Oats are down too. The Scotch market for meal is dropped. It is said that wheat is dear at Carlisle, but it had no effect upon Kelso although we expected it would. William Brown's rape is very good on the dry parts of his far hill, and his turnips will be good upon the whole compared. They are mending on all strong land this dryer time. Bob will leave this letter and one for Mr Peacock at Darlington as he passes south. Let me hear from you some time. We expected a letter from you as Mr Peacock says in his that you would write in a few days. I am yours

Geo Culley

PS Ralph Brown says the Berwickers were fonder of buying corn yesterday but at little advance. He sold some to Clunie and Home, and they bid him 24*s* per boll for barley, which is more by 3 or 4 shillings than has been of some months, which makes me think that accounts from the south respecting the harvest may not be favorable. But this is all conjecture of me, only if wheat be mildewed as much in the south as here, it will be a very defective crop. The barley may be wanted for Scotland distilleries. Mr Sayle says barley is late in Norfolk but a great crop. Many oats bad in their neighborhood. Write to us as soon as you see Mr Pease because I have had more applications about the wool from a Mr Meggison of Leeds.

[**M.C.**] Well John my brother George says your watered land was much improved. I am glad of it. Am glad to say George Humble is better I hope as he has rode out today with Andrew Bolton who was also poorly this day or 2. But I hope this very fine weather will bring them about again. *Sunday at Eastfield.* A most extra good day indeed. Yesterday was a fine day and the first dry day of a long time. We are going to be very busy with harvest there if dry tomorrow. Some red wheat then oats. Dry Tweed oats much injured by a shower of hail. Hope to cut on till all be done. We hope to hear from you whether Messrs Pease is to have the wool or not. Yours &c

Matthw Culley

141. *George and Matthew Culley to John Welch*

Eastfield 30th August 1802

Well John Welch

I sent you a full letter along with Bob Wright and his two tups this morning, which he will leave at Darlington as he passes through, along with one for Mr Peacock. This has the appearance of another fine morning, and we are willing to flatter ourselves with settled weather as the moon changed so well on Saturday last. A good many early oats are stooked, and the first wheat I have seen shorn was yesterday as we were going to church. However winter wheat is coming on now apace, especially as the weather has been clear and bright for some days. *Tuesday 31st, and last day of August* is another very promising morning. Yesterday proved as nice a harvest day as ever was, and we cut as pretty a crop of potato oats as I would ever wish to see cut. Not too ripe, but in a proper state to reap I think. Nevertheless many of the oats shelled out with shearing. Indeed the day was so sunny that the husk or cover opened prodigiously. We had not quite 30 shearers, yet they cut above 12 acres out of 16. Indeed there are not many of them down, and those lie fair enough for shearing. I trust we shall now have a fit of nice weather, at least there is every appearance of it at present. Matty went last night to Mr Nisbet's and they go together today to let the tithe corn of his estate and ours at the Grange. We have the tithe this way-going crop year, and Mr Nisbet has the tithes of his until the leases expire. Although we are rather weak of force this year, I think we will be able to keep up with our ripe corn. We shall cut most of our potato oats this week, and perhaps mow our pieces of barley as they come ready, and the wheat being spring sown will not be ready yet a while. We are in such a situation for meat as I have not often known us. The rape is coming to flower, and very luxuriant much of it, and our clover fogs grow so fast that we cannot keep up with both. Indeed at this place for want of water we cannot put cattle upon some of our fogs, and we have all our cast ewes upon our rape except 120 that Mr Brodie has to draw 20 from, and we are not willing to put our shearings on rape as we are inclined to sell them, we think it may rush or scour them. We have all our tups and part wether lambs on the rape too. But it grows so fast this nice warm weather. However we will continue to overtake I think too.

Wednesday 1 September. A heavy morning and the glass a little dropped. However it does not seem as though it would rain either. I am just going to Grindon to try to take our corn tithes there. Matty gone a partridge shooting. *Thursday 2 September 1802.* What I feared came to pass in the night or rather about 2 o'clock this morning, it was a very heavy rain. But now a fair morning but everything wet and like to be more rain. I could not take our tithes of Mr Ball for the first time, but we must take them although I think he is more out of the way than I ever knew him. It is a very midling and defective crop of wheat, the oats pretty good especially the potato ones. Upon the road to Grindon I received your letter of Sunday and Monday last 29th and 30th insts. [*sic*]. I am happy that you approve of watering. I do believe it to be a most valuable piece of husbandry where water can be lead. You must have had a pretty lucky time for your hay. I wish you may not have got it too quick, but watered hay comes quickly, I rejoice that your servant Anthony manages so well in selling your corn. It is certainly a great relief to you but does not take you so much from home. I would like a few bushells of either your own red wheat or any other that you approve off by way of a change. Now John respecting the wool. If this reach you before you see Neddy Pease, it is my wish for you to offer him ours *only* at 25/6 to be paid for by Luke fair in bank notes or good and short London bills at not more than 30 days, or at any rate the greatest part of the money then. Not less than *500£* at any rate, and if he will not give you the price or *better*, or at least as much as Head bid you, or you have reason to think he will give, do not give it, I mean *your own*. Indeed I think you will use Head badly if you sell it before you hear from him again. This is my opinion, and I am going to Wark when I will write you my brother's sentiments, and send you this letter already directly per post, so as you will have time plenty to see it before you see Mr Pease. Mr P. does not come this way home from Glasgow, but by Cumberland and Westmoreland and Barnard Castle. We have some other chances to sell our wool. A Mr Meggison of Leeds, who buys Mr Vardy's of Fenton, has desired Mr Vardy to offer us the lowest price of our wool. Therefore it would seem as though there were some little stir about it at present notwithstanding the run for cotton, and as you say some cold countries must have woolen clothes. But you are wrong in thinking our wool not so fine as yours. The ewe wool excepted, it should be finer than yours. I am inclined to think that that [*sic*] they want it to spin upon their own mill, but may be mistaken. John Mr Pease says that there are now several cotton looms and weavers in Darlington, and did not use to be one.[68] I hope you have bargained with Mr Colpitts or his men for your tithes of corn. If not I trust you will do it, at least it is my wish and my brother's I am certain. I am very happy to hear that your Yorkshire acquaintances give you such a good account of the fat markets. That is the country which we must look to in my opinion rather than Shields, and it would seem as though the Yorkshire jobbers and turnip feeders such as Jere Clayton &c &c must expect very good spring prices, or they would not venture to give such prices for stock as they do in this district. I say long may it continue! I am highly provoked at the behavior of Tom Glass. Not to count sheep, especially where there were so few, is really inexcusable, and if he don't behave better and properly, part with him and we will soon get you a better or I am mistaken. I cannot go into a field but I must reckon not only the sheep and cattle but even geese and turkies &c &c. So much for good habits. How does he know but a sheep may be run into a corner owing to *maggots*, beating upon the *head* or any *other part* with flies, or lying *aukward*? What a stupid dog! It is most abominable behavior. Had I been present I should have been outrageously cross. Even your men are blameable. Better *twice* or even *ten* times told than not at all. It is not to be wondered but your fat markets are good, especially fat sheep, because the demand of late into Yorkshire has been such as totally to preclude any good sheep almost going to Morpeth market. I am of opinion that the day I was at Morpeth on my way from Newcastle Assizes, that good fat sheep were sold at 8 ½ per lb. sink at least. This has the appearance of a dirty day altogether, as the hills are thickening in very fast, and without *winds*, which are all *our hopes* and

---

68 The recent opening of cotton manufacture, in addition to wool and linen, in Darlington by a Mr John Morrell was noted by E.W. Brayley and J. Britton, *The Beauties of England: An Original Delineation, Topographical, Historical and Descriptive of Each County*, London 1803, vol. 5, p. 86, and quoted in W. Hylton Longstaffe, *The History and Antiquities of the Parish of Darlington in the Bishoprick*, Darlington and London 1854. The enterprise did not last many years: no cotton manufacture is listed for Darlington in *Pigot's Directory* of 1820.

all *our fears*. We shall have dirty bad weather, and if so which is too probable old wheat will still be valuable, and more especially if the new be a bad or midling sample which is very doubtful from so much mildew upon the growing crop. I am yours

Geo Culley

[**M.C.**] Thursday Wark. A wetish morning now fair. I am sorry to say George Humble is dangerously ill. We sent him in the chaise to Kelso, Dr Douglas has been here, is at a loss what to say, only says he is very ill. Got a part physick, head shaved and blistered, poultices laid to his soles, sometimes raves, is in general sensible, lies very quiet. But his tongue faulters much. Our white wheat the worst we ever saw it I fear, red better but some colour not good, oats good especially pottatoe, what we have cut. We are very busy, Andrew Bolton very attentive but not clever by being ill, gives me little time to say what I wish to say. Hay mostly got, unless we mow another clover field. We are in general well enough except for G.H. who I fear much. In haste yours &c

Mattw Culley

PS. You don't say how your pastures are, and how your stock does.

[**G.C.**] NB. My brother approves of what I have said about the wool John, therefore you must not neglect to write to me as soon as you have seen Mr Edward Pease saying whether he will take our wool without yours and at 25*s*/6*d* per stone of 24 pounds, with payment at Lukesmas in good money or short bills as I have said above, and 500£ of the money at least at Lukesmas. I say you must write directly as we have other applications about the wool. Yours

Geo Cully

PS. I fear the worst for George Humble

142. *George Culley to John Welch*

[Eastfield] Saturday 4th September [1802][69]

A very [. . .] much rain in the night and is very thick and rainy-like with [. . .] east wind. Very discouraging harvest weather. Ralph says Kelso market was no better yesterday. Crops of wheat less mildewed up the Tweed. But I have been a farmer now 50 years and I never knew corn good in such a wet year, nor bad in a dry year. I think a piece old wheat &c will be well sold still. *Sunday 5th September*. We have been attending the remains of poor George Humble to his grave

---

69 The top of the sheet is torn off. The letter may have been started on 3 September.

at Carham.[70] My brother has lost a most valuable servant, and his poor infants an affectionate parent. I think the whole town attended the funeral, and I have seldom seen a man in his station more lamented. May we all be as fit to follow when called upon. He died on Friday evening. He had not constitution to stand a severe illness, and by all accounts it was a very severe one. Dr Douglas who was twice there during his short illness, and he once up at the doctor in my brother's chaise, thought badly of him from the first, and called it a termination of an *old disease* which had been of a *long standing*. To be sure he was a very tender man. *Monday 6th September* is a hard dry like morning, and has been no rain in the night for the first time for many nights. Winds north west. *September 7th.* A decent frosty morning, yesterday a charming day. Glass fell at night and we thought it would be rain, but this morning the mercury is rising again. We fed Thornington shearers yesterday but I think we can do with our ripe corn ourselves now. Going to lead a few oats if it keep fair. *8th September.* A very frosty morning. Yesterday we had from 11 o'clock very heavy showers of rain and hail which prevented leading.

Wednesday evening. Bob just comed well. Mr John Taylor has been here and hired a tup which is to be sent to Denton in a little time to be met there by Mr T. Be so kind as write directly about wool. I will send this away tomorrow morning when you will learn the fate of poor George Humble. What a blessing is health. Do take care of yourself, and may we all be thankfull, truly thankfull, for that most valuable blessing. We are entirely up to our shearing as our wheat is not ripe and winter sown very indifferent. I shall not wonder at old wheat being well sold until Christmas. Indeed you are getting famous prices for it at present. You might have spoke to Head, as Mr Pease will hardly take our wool I think. If we could sell potato oats at *3s* per bushell I would like to sell a quantity, but I don't think that our merchants will give more than 15 or 16s per boll. By all means deal with Mr Colpitts people for your tithes. We shall bargain with Mr Bell also. I am glad to hear that your stock are doing so well, you are right to put them into small lots. You will get stubbles by and bye. I don't know whether you had not better begin with your rape (*rape*), because after eating once over early it will come again. By all means buy a lot of Kyloes, 4£ or 4 guineas is famous pay and they help the dunghill. When your harvest is over you must take a turn over here,

---

70 He was aged 41. DUL, BT, Carham.

and fix on what cattle you would like to have from us. We have a very pleasing lot of spayed queys, and some open ones for you. Mr Compton's cattle please my brother well. I am glad Glass *feels*. By all means try Mr Buston's steers. Please tell Mr Peacock that I have received his parcel and that I will go tomorrow to Wark to get it executed and return it again directly. We cannot possibly keep our dinmonds through winter for want of turnips without buying, which I think wrong for us to do now. But we can keep them until Wooler fair.[71]

143.   *George and Matthew Culley and Matthew Culley jr to John Welch*

Eastfield 9th September 1802

Well John

We sent a letter per post this morning for you, and hope to hear from you soon again about the wool story. I am morally certain we are going to have a gale of wind if not worse, as the glass is comed down so rapidly yesterday and last night, and the little detatched clouds (now 5 o'clock morning) are skimming along the sky which always portends wind. 11 o'clock and continues a very fine day. I have been up at Wark this morning getting a deed executed by [my][72] brother and self which Mr Peacock sent per coach, and my brother has received a letter from Mr Harrison of Walworth from Scarborough requesting my brother to order you to discharge anybody, particularly of the *lower* or *poaching* kind from shooting in Denton grounds, especially in that part nearest to Walworth, and my brother and [I][73] both earnestly desire you to prevent everyone, especially the lower class from shooting, especially in those parts adjoining to Walworth estate. To people of consequence speak civilly, telling them in a civil and quiet manner that you have your master's orders to prevent any person from shooting upon Denton property. To the other classes you may also be civil at the first, because I think lenient measure always the best. But if they should persist tell them that you will be under the disagreeable necessity of lodging information against them, as it is your duty to execute your master's orders. My brother says he gave Mr *John Colling* leave to shoot, therefore he desires, if he should come upon the premises, to tell him that we wish him not to shoot

---

71 The corner of the sheet, with the signature, is torn.
72 A word seems to have been omitted.
73 A word seems to have been omitted.

anywhere near to the Walworth property. Mr Harrison does it to save his pheasants, and at any rate as he has in general behaved very well to us it is only fair to indulge him in this request. This is what my brother desired I would write to you, but I have requested my brother to write to Mr Harrison that he will give him a *deputation* to shoot upon Denton with authority to keep others off the estate, which will take the weight and blame of you. Nevertheless until that is done I would advise and recomend to you to do as I have said initially. *Friday 10th September.* Another windy morning but I hope we shall get some more oats led. We got several stacks up yesterday and in nice condition. I was luckily mistaken about the wind yesterday morning. We had a fine dry day with plenty of wind and not too much. However I will be bound for it that there has been a vast of rain westward, and we had several showers towards the evening and this day looks showery. But our men are out binding and it now looks well but showery-like. However as long as they are dry at the heart a little damp without is of little consequence. About 7 o'clock we had a hurricane of wind and rain and hail. It stopped us from leading near an hour and a half. Afterwards the wind became violent, and I am affraid has done a good deal of damage. It now blows hard 5 o'clock afternoon but is much abated to what it was. However we have had a good many oats in nice order, and have nothing to blame ourselves for as we had no ripe corn and have sent all our young girls to help William Brown to shear. We had Mr John Taylor of Salton the other day. He hired a tup, which will be at your house in about a fortnight, where Mr Taylor's meet us. I name this, that in case Mr Peacock or you have anything to return per cart, the cart will be at Denton if all be well on Saturday the 25th at night. Wheat advanced at London last Monday 2s per quarter. I am entirely of your opinion that fat sheep will not drop in price until the severity of winter may hurry more to market than the consumption may be equal to. But then our sheep should either be sold before winter or go to you before that time, as the cold weather and short days will all be against their driving. In that case you had better sell Mr Wood's shearings before winter if you can meet with good customers. But at any rate you will see our sheep and cattle when you come over, which you must do before winter, say about Wooler fair 17th October, and you and us can then consider what is best to be done. Indeed it is very important for you to come over, as the regulation of our wintering stock is of great consequence.

Saturday morning 11th September is a very fine harvest- like hard morning as can be, with the glass once more rising, but the violent wind yesterday I find has done a great deal of harm amongst corn. By a letter this morning from Thornington Harry says that their corn has suffered much, and all other places I do suppose with the current of the wind, especially such exposed places as Thornington which are better employed by much in stock than corn in my opinion. To be sure it is not so well to be without a few turnips, and these cannot be had without the plough, but perhaps the less is plowed in such situations the better after liming over well. Ralph sold between 40 and 50 bolls wheat (red) yesterday at Kelso for 54 and 55s. We are going to sell some oats at 16s if a better price can't be got. *Sunday 12 September.* Ralph sold 30 bolls new red wheat from Wark yesterday at Berwick at 49s per, oats 14 to 15 new was the common price I was told. Great complaints of the damage done by the wind on Friday. I fancy we suffered very much at Grindon and Thornington. I do believe that there is very little old corn left in this country. I was told yesterday that corn bears a good price abroad. *Monday 13 September.* A clear calm frost. Our men out binding oats. The harvest moon very favorable to nights and mornings work. We have suffered most at Thornington and Grindon with the wind. We shall get the most of our early oats today. The weather very promising now for harvest, but I do suspect that these early and severe frosts will injure the late moorside corn. Mr Brodie just gone from hence with our Matty to Wark to draw his 20 ewes he bought. We had young Wood, whose father bought so many soldiers' horses in the war time. He wanted a few dinmonds for turnips, but when I asked him 3£ per for a part of ours he was quite alarmed, went away and bid nothing. He bid Frank Peacock 52s per for his, Frank would not take the price, and I don't know that he has bought any in this quarter. Mr Ralph Sanderson of Swinoe was along with him. By Mr William Brodie's account they sell wheat as high as you do, at at least 2£ their 4 bushell boll. These Brodies are wonderfully industrious people. *Tuesday 14th September.* Rather soft. There has been a little shower or 2 but hope to lead in an hour. More like wind than rain. Mr Myrtle took 5 tups yesterday. *12 o'clock.* As I supposed it has been very windy but no more than did good here. We shall get all our early oats in today and we are beginning to cut spring wheat this afternoon, which is a very pretty crop but thin, and well filled. *Wednesday 15th September* appears as though it would be a fine harvest day as the wind is west and the glass advancing, and has been no rain in the night. We have not got all the oats out of Blue Bell yet. The potato oats are a wonderfull crop. 14 large stacks have come out of it and Crane Lee, and there are about 3 or 4 pieces of ours to shear yet. But our wheat is neither bulky nor good in quality I am affraid.

Thursday 16th September is a very fine morning, glass high, wind westerly and has been a slight shower in the night but not to do any harm. I was up at Wark. Harvest going briskly on, about 34 stacks in the yard. Wheat much of it mildewed and very small in the sample and lean. I called on my way home at Franky Peacock's, who has been poorly these 2 months, was better yesterday, owing he thinks to a blister put on his head by order of Dr Douglas Kelso, who he had consulted a few days ago. He has thrown up a good deal of blood at different times lately. He is a stout young man about 40 and may by proper care and management get better. But he and everyone should apply for advice at an *earlier* period in case of danger. Well John we have received yours of 14th this morning for which I thank you. Wheat is up 3*s* per quarter more last Monday I see by the London paper, and I am told that wheat is high abroad. If that is the case, joined to the mildew amongst that grain will be a means of its keeping a high price. You have certainly taken your corn tithes well. I should think if your crop be tolerably good the tithe should be worth nearly double. However it is the interest and duty of J. Hodgson and you to keep *whippt*. Upon my word you sell your wheat and oats famously. It is a capital price 31/3*d* for new oats equal to our 19*s*/6. I would wish to sell ours at even less money a good deal. I hope you are right in not selling your wool and ours to Mr Pease. Matty is going to Wooler and will likely see Mr Vardy and get him to write to Mr Meggison about ours. You were certainly right in selling your few ewes and tups. I hope the markets are not going to drop in Yorkshire. If they do it will dampen us all. John for want of a place left for your wafer, I can't make out what price you were for taking of old Head. Mr Myrtle says they had strong ice that morning you speak off on Monday last. The account you give of Mr Peacock's tup is very pleasing. There can be no doubt but ours are as good mutton sheep and as quick feeders &c as any I ever saw, and I should be sorry to give up that disposition for feeding for form and wool &c &c. It is certainly still more creditable to have got such weight and fat upon Thornton land. I still say that we can breed nothing that I know of that will pay so well as our own kind of sheep. I am glad that you spoke to Mr Buston about steers. I hope you have sold all your feeding ewes that you got from us last.

[*M.C.*] Wark. John, my brother and Matty are resolved to sell 300 shearings out of 419, to keep 50 of the best and 60 of the shotts. My view was to have you sell yours and take ours, such part at least as you could take, they are fat and good, only not so big as we have had them, yet possibly nearly as heavy as ever. Shillito would have liked them or some such persons I think. If fatt ones are best, at least you should have had 100 of them for your own feeding, since I got my brother to put growthy big tups to your in form ewes. The lambs are larger and better for wethers. We now mean to try Jerry if keen. But you should buy Buston's steers, got by a bull of Mr Colling's possably, as you must have and keep cattle for making dung for your turnip land. We have a dozen queys which are to come to you some now or later, and some of our fattest oxen or steers as you may judge best for your feeding off early or later in the spring. But never keep a bad thriver instead of a good one. The sooner bad ones are sold the better, experience shews. Ralph Compton's are 4 queys and 8 steers, as fatt or full of beef than ours are. May have those you approve most off. If you can lay any of them loose in a house they thrive best I believe. The freedom of a little moving keeps them more healthy and not so numb of their legs, and you get all the dung. Try different ways. Your queys did well at the High House courtain, and it's right to follow a safe game where you have it convenient. If fogs are reasonable they improve cattle much. About the Aucklands fresh and fresh improve most unless you have other stock to follow them, which saves your pastures at home. This is only advising you, not commands. George Humble is gone, and I fear for his family. He was a most valuable man and I know not who to put in his place. Bessy's letter, what Matt said to her and her own feeling, brought tears from my eyes. Dear children, their natural fine feelings. He had few enemies, many friends. Jenny is a woman of high mind. She has promised me to bestow his effects on the children, about 100£ she to have the interest for her life, then to the survivors. She says 5£ a year will be little to bring them up with, but she had not the way of saving money for a future day. I wish to be a friend to the infants. He was too born down, and lost his speech, that he never mentioned anything off his children &c, but he was an honest fellow. I never saw such a concourse yesterday follow any person to the grave (at Carham). It was on the Sunday, a holiday. 41 years old nearly, a young man, had a weakly constitution, had hurt himself by overworking when young. Ralph Brown and William Short, H. Rutherford most likely to serve me here. He is trusty but rather slow, very steady.

Mattw Culley.

[*G.C.*] *Monday September 20th Eastfield again.* John my brother has this moment 8 o'clock sent back this letter, so I will send it to you either today or tomorrow morning, and I have begun another which I will send by Bob on Thursday who comes to you with a tup for John Taylor of Salton, who will meet at Denton or send soon after. He will also have a tup lamb for old Mr Pullen of Carlton near Appleby York-shire, who will send for him I suppose, as Mr Askew who sends the lamb for Mr Pullen said he would write to Mr Pullen. Be so kind as tell Mr Peacock that I received his letter of 15th inst. He tells me that he has in Mr Thompson's and his own hands 4250£, which with 1000£ or nearly in Mr Nisbet's hand and what little more assistance we and you can make by Lukesmas or even Martinmas will assist to pay the 6000£ due for the Grange estate. But don't you by any means pinch yourself for cash, as you should always keep enough in hand to not only pay your way but to buy stock at any opportunity that may offer. We can always borrow the money of a bank for a time, and can soon pay it off I trust. Mr Peacock writes that he had then Wednesday last, got in 1200 stooks of famous wheat. Is not that a vast for the parson?

[**M.C. jr**] You should inform Mr Peacock that we have a cart coming north on Monday next from you if he or you have anything to send. Mr Vardy has wrote to Mr Megginson about our wool, therefore it is engaged untill we have his answer. But you can sell your own wool if you chuse or wait until you hear from us again just as you like.

M. Culley

144.   *George Culley to John Welch*

Eastfield Monday 20th September 1802

Well John

When I received your letter the other day (Thursday morning it was) I sent one for you up to Wark as there was room for my brother to add anything he chose. But he was here to dinner yesterday, and neglected to bring the letter nor had not sent it to post. However I suppose he will have sent it away this morning. The glass sunk much on Friday yet the weather was most promising. However yesterday it cast up like thunder and in the evening came on in hasty showers, but no thunder heard here although I suspect it was thunder somewhere. And I am affraid it has rained all night as it rains much this morning. And being so exceedingly close and warm is dangerous for corn growing, only all corn was got very dry when it came on. Ralph sold wheat at Kelso at 56, at Berwick 52/6, but half small or out dressings dressed once again. But old wheat will become valuable now I should suppose. Our winter or autumn wheats are in general through all this country very light in the ear and ill filled, although bulky in the straw, particularly bad in Berwickshire I am told, while our spring wheats appear good and well filled. I have not often known ours better. Even our sand hole weak field to the north of this place is a beautifull crop, and it is very weak sandy land and had a great crop of turnips nearly all taken off. Oats are a great crop, potatos in particular, but all seem to be small in the green. I have not heard or seen much of barley yet, although we have led a good piece of ours and it seems well filled. Last Friday morning I went to Wark with Ralph in his way to Kelso and handled the shearing wethers. I find them not so good as some other years although healthfull and thriving. But they rushed sore at Wark last spring and were very much checked, many died. However if we can't sell them now we have good clover stubbles and early turnips to make them very good by Martinmas or Christmas in case you could sell your own in part or all, to take ours or most of ours, as we can't possibly keep them through winter unless we buy turnips which I do not like. I name this again that you may weigh the circumstances in your mind and advise us what is best to be done. You will and may handle them when you come over. For I am quite in earnest for you to come after your harvest, because you should see our stock especially the cattle, that you may chuse what cattle you think will suit you best. I hope we have some spayed queys &c which will suit you well.

Tuesday 21st September. A decent looking morning 5 o'clock. My brother is here and he and Matty are going to the Grange to set out a *march* or boundary fence between us and Mr Gilburn's estate. Mr Gilburn is in the country and this day is fixed upon. My brother desires me to say that you may send your few apples back, or what you can spare, by the cart if you have any and are fit. Also he says if the 2 pears are still to the fore that Bob Wright may bring them to Bessy Culley at Newcastle, Mrs Smith's boarding school. Yesterday was our first day for letting tups, and we seldom do much upon the first day, indeed we had not much company. However we let one to Mr Gregson junior of Low Lyne. The young man was here himself, a very quiet well behaved young man, does not know much but seems open to conviction and very willing to be informed. He had his shepherd with him and a riding servant. We also let one to Mr Nisbet of Ancroft. Mr Wood went from us to Mr Gregson's all the Wednesday night was a week, and bought 300 dinmonds of Mr Gregson at 54s per, cast out 22, these never tasted turnips except a part bought of Jemmy Robson Lowick. But they are bred from our tups. Old Gregson and son were highly pleased with the sort of sheep. They hire of us almost every year, have got from 5 guineas to 14 now. David Green has bought a large lot of black-faced sheep at Falkirk. I believe he has somebody that buys them for him at Fort William, and then meets him with them at Falkirk. The great buyers of Kyloes &c,

Bushwhistle, Waddylove, Armstrong &c &c &c entered into a league with one another and never offered to buy until the 2nd day at 11 o'clock, when it is said they brought the Scots down for the grazing cattle 32£ per head. However they brought them down considerably I do suppose. It is said that there never was more poor Scots immediately from the Highlands, nor never so few fat or grazed cattle which are depastured in the low countries. As that is the case, feeding forward cattle of all kinds will be sold high from the scarceness I do imagine. *Wednesday 22 September* 5 o'clock appears to be a promising harvest day. Indeed the weather is excellent, most promising to be good. We still find our spring wheat very good and we now shall soon be done shearing. Not more than 3 or 4 days work, but will not hardly be all ripe this week. Let Mr Peacock know that we have a cart coming north on Monday first, as he may have some things to send. Mr Peacock seems to think that corn will sink in price. We think not so in this country, and if my information be good, that corn is dear abroad, we shall not have it low at home I think. Mr Peacock writes that on your giving him a survey he will try to purchase the land of Mr Croft his landlord which we wish for. Therefore I am certain that you will not neglect to give him, Mr P., a correct survey and plan of it. Perhaps you had better ask Mr Peacock with our compliments to assist you, or otherwise get some person conversant with measuring and planning land to do it for you by your direction and assistance, as it is of very great importance to be correct in a matter of that sort and it will be a very great accommodation to us. Mr Peacock in a letter to me yesterday mentioned the excellence of the tup he had killed of ours at Darlington, says that he was become nearly incapable of doing his business. I have no doubt that our sheep in general are possessed of as much *inclination* to make fat, and when fat as *valuable* and *beautifull* mutton as any whatever. It would have been very wise to have had him killed on the *exhibition* day at *Darlington* when the agricultural meeting is held,[74] but I suppose Mr Peacock had not adverted to that, nor conceived that a tup's carcase could have been so beautifully fat. But that *stamps the merit*, to have a tup to die as beautifull as a weather in general. I was told by Tomy Batters of Eatal

---

74 The Agricultural Society for the County of Durham was founded in 1783. It held two meetings a year at Durham and two at Darlington, and gave a number of prizes. Rev.Thomas Peacock was a subscriber; Matthew Culley was not. *Annals of Agriculture*, 42(1804), pp. 117–23; Bailey, *General View, Durham*, pp. 366–7.

where Mr David Green reports to, that he, David, had bought 5000 black-faced or Forest sheep at Falkirk.[75]

145.   *George and Matthew Culley to John Welch*

Eastfield 24th September 1802

Well John

I hope you will receive a long letter by Bob Wright, who set of for Denton yesterday. We have surprising roky thick misty morning. The corn keeps wet until 10 o'clock and the stooked corn keeps raw and damp. Very little been led this week. We are leading a piece of wheat this day, begun at 11 o'clock. We have not above 1½ day's shearing of white corn, but part of it will not be ripe until towards the end of next week. I am affraid this misty weather may end in rain, but the glass is very high. I was at Wark this morning, they are not shearing and have very little white corn uncut. Mr Barf is in the country buying a few sheep, Mrs Barf with him, and speaks highly of your calling upon her at Wakefield. I think I told you that Mr Wood bought 300 dinmonds of Mr Gregson out of 322 at 54*s* per, and 8 score of some person who I have forgot. They say there are few ewes left unsold but plenty of shearings yet. Tell Mr Peacock that we sent the deeds to Mr Scruton today per Charlotte coach[76] after being executed by my brother and self in presence of 2 witnesses. My brother only came from the Island last night and I went up this morning early. *Saturday 25th September 5 o'clock.* The mist does not seem to be so bad as it has been for 3 mornings past, but it is coming on I am affraid as we have no little to shear. I have advised to put of until the afternoon before they begin. They may not cut all the ripe but will come very near it in the afternoon, and then we shall not have more than a good day's shearing left, besides the few beans. Ralph sold 15 bolls of wheat only yesterday at Kelso, could have sold more, but thinks that old wheat is falling very scarce and will be wanted. He got 58*s* for old red and 50 for new red.

Sunday 26 September 6 o'clock. Not much mist yet, glass uncommonly high, and every appearance of drought. But what makes me think more of its being inclinable to dry weather is the account I had the other day from Mr Sayle. He dates from his farm in Norfolk and says that from the Trentside to Titchwell where his farm is, there is *no*

---

75 The signature is lacking.

76 The Royal Charlotte coach ran daily between Edinburgh and London. Its route through Northumberland was via Coldstream and Wooler. The daily Mail coach went by Berwick.

grass. He compares it to *a road.* The lean stock dropped much in price and the turnips in Norfolk *exceedingly bad* for want of rain. He says many *dead* and *all bad.* Harvest done, the best ever known. Barley a great crop, but he thinks *wheat* will be found *very deficient* upon his farm, very badly *mildewed.* And you may depend upon it that the wheat in general will be found a very *defective* crop. In some districts it is good I believe, but if in general it be bad and midling, it will make it a defective crop. Ralph sold new at 49 and 50, famous prices in my opinion still, and old at 56s to be delivered at Tom Dun's mill. But the merchants are rather shy and nice in buying, and no wonder as London market is dropping and new wheat is shewn soft yet, which must be the case this close thick warm weather. I do advise John it is dangerous weather to lead much corn in. Indeed we have led very little this last week, none before Friday and yesterday, and did nothing until 10 o'clock both days. But perhaps you had not such rain this day week (Sunday) at night. Our stooks were so wet that we were affraid that it would require opening out, and the bands on the north side keep wet and damp until today. However I now think we are going to get very dry weather, which will make lean cattle drop. I wish you had not gone to Brough Hill, because if it should become droughty further north than Trentside it will drop lean stock a good deal, at any rate it must affect them, and indeed I expect sheep will not be near so dear at Ninian's fair tomorrow as they have been bought lately. Jere is not coming down. Barf is here and is buying dinmonds and highland weathers but is very shy. I don't think he will handle ours, and if he does won't buy at our prices. Tom Alder offered him his yesterday at 56, great strong sheep and of famous land. Many good dinmonds to sell. But I am not affraid of fat dropping at any rate, and I never knew keeping so plentifull in this country except the last autumn. Indeed they are much alike. Many people mowing 2nd crop of red clover and much better than the first, and every appearance of being well won. Turnips pretty good in general and on strong lands much mended lately. Some very dry lands and early sown shew a tendency to *mildew.* You may depend upon it that the sellers have had a bad time at *Woolpit.*[77] I don't know whether you now should not buy some turnips before the account is known, because if turnips fail in the south, and Norfolk is certainly so

---

77 The fair at Woolpit in Suffolk in September was one of the most important fairs for livestock for East Anglia and the east Midlands, with large numbers of cattle for fattening for the London market.

which is the greatest turnip country in England, they will become more valuable in the north, because less stock will go south now. Indeed I expect that very few more lean cattle will go south now for a time. Still I see no reason why fat may not be as scarce and dear in the spring as was ever known, because if Norfolk can't supply the Smithfield consumption some other parts must. Consequently the Lincolnshire sheep and cattle will have to travell south in the spring. Mr Sayle writes that corn advanced last week at Lynn. If so it will soon advance at London.

Tuesday 28 September. Yesterday was St Ninian fair and we had one of those great and sudden changes that has and will take place in both politics and country affairs now and then. People were disposed to think that there were no sheep left to come to the fair, particularly ewes. But behold that there were a greater shew of all kinds than has been known of many years, which occasioned a total stagnation. Indeed the drought in the south, with the very uncommon prices given all along, had disposed the buyers to be very cool. The report of Weighton fair on the edge of the Wolds, and York market being bad, all conspired to depress their spirits. I am unable to say what the drop might be in the pound, but in fact there was nothing done compared. As to the numbers it should be considered now that the Dishley blood has spread into all or most of the southern parts of Scotland and up all the riversides. There were sheep, decent country ewes, from within 2 ½ miles of Hawick! and from many parts all down Teviot and the Tweedside &c &c. I don't think there was much drop in fat and very forward cattle notwithstanding the bad times which the jobbers have experienced at Woolpit &c &c. However we have missed our chance of selling our dinmonds. We should have sold them to Mr Wood. Therefore the next consideration is, what is the best to be done. I am very unwilling to buy any turnips, and the misfortune is that this stop will enhance the value of turnips prodigiously. It is very lucky that we are full of meat, clover fogs &c &c, and this is the most promising weather ever known, and in my judgement it is as likely to hold a good while as ever was known. I can fancy that the drought in the south is now extending north, and if so it will be most fortunate because we have moisture enough and it will make our clover fogs and stubbles and pastures go far and do the goods immense service. And I hope it will enable you to get over by and bye, when we can *consult, contrive,* and arrange matters in the best way we can. It will to a certainty bring in the latter harvest famously. We have had a tryal with Mr Barf for our wool here, but we could not bring him up. He offered to buy a part only at the 25/6. This being the case might it not still be right to sell both this and yours to Mr Pease, provided he will stand his word and yours unengaged. It does not quite square with my brother's ideas, but I think it always unpolitick to refuse a times price for any commodity when offered by a good man, and certainly Mr Pease has offered a times price and such a price as we never knew before. And when things are so high they may like the sheep come down. We have missed it in the sheep, why should we do the same with the wool? My brother says it is of no consequence. I deny it, a man should always be governed by times or he should give over trade. We can well afford the wool and the sheep both at the money, and when people are not content with fair profits what are they to be called? The answer is *plain, obvious* and *short*. They must be *covetous, greedy,* or at best unreasonable.

Geo Culley

[**M.C.**] Well John I own myself like you too fond of high prices, but we have missed selling our wedders. If you have not sold your weathers you should to some butchers or who you can, for you should sell and not buy as things have and are likely to go down very much. William Glass unkle to your herd, has been with Armstrong at Woolpit, where they had the worst fair ever stood, no one to ask the price, especially of the Irish cattle. The Yorkshire cattle are returning home to Yorkshire owing to the drought. Such is the change and a great one it is. Armstrong &c must stop their cattle from Scotland &c as the meat is gone in the south. I never saw so full a fair at Ninian, and so little done at it. Mr Deverel (whose letter my brother has just received this morning) sayth the drought is so severe that the turnips must be eat off directly he says, or be burnt off. The harvest truly good, our white corn cut, beans ready. I fear the drought if it continues may in time injure our turnips and grass, although at present they are doing well and grow fast. We are seldom injured by dry weather so far north at this season of the year. If you have not bought Buston's steers before you get this letter, put them by for the present as there is an appearance of stock falling much here, but we are the last pin that moves. Yours

Mattw Culley

[**G.C.**] John you see what my brother has said above of Deverell's letter.Every post brings fresh accounts of the severity of the drought in the south. It is certain that it will produce a very *great change* in the markets, such a change as we cannot pretend to calculate upon. We have most assuredly made a grand mistake in not selling our wethers, but it can't now be helped. As to the wool, Deverell says that it is advancing in their country. He has sold to his old customers at 31s per todd of 28 pounds, which is about 13*d* and a farthing per lb. or 26/6 our stone. So I think we had better not sell ours at the present but wait a while. So you need say nothing to Mr Pease at present but wait a little. However you may do as you will with your own. *Wednesday night.* Bob is just arrived safe and well. I am glad that you are so well forward with your harvest, but you need not to fear any rain as the drought is evidently coming north in my opinion. However it will not do either you or us so much harm as the people in the south, where it has stood so long. I am rejoiced that your crop turns out so well. Your tithe will not be too dear. The Peases are good people, but all Quakers are attentive to their interest. You are right in putting your wethers forward, but I am sadly afraid that Skipton and other fat markets are likely to sink in price now. But however it may be right to try a less number than a hundred at once at first to Skipton. I also think you will have worse markets for your Kyloes and fat of all kinds at present. But I may be mistaken, and I wish I may. You were right in offering a good reward to find out the sheep thief, but they were bad to make out. We will be glad to see you when you can come north, and to stay as long as you can, at least until we can properly arrange matters well, as this is likely to be such an overturn as we have not often met with. But we will know more about it soon, as I have wrote to all my correspondents, north and south. However I believe you will be right to be selling as soon as possible, because I am persuaded that things will be worse soon until the spring, when I am persuaded that fat will be as dear as ever, and if the Norfolk people can't supply London some place must, and that will fall on Lincolnshire and Trentside, which will leave Wakefield and Skipton for the north to supply. That strikes me, and perhaps if you can buy a few turnips on proper land it may be wise, or to lead of on to pasture land may do. I offer these hints for your consideration. It may be that Yorkshire is the likeliest place to buy in. However I beg that you will now buy no steers of Mr Buston or any one, except lean ones for straw, because the more beasts you can take from us the better able we shall be to winter the rest and our sheep or what we can of them. But there is no doubt but we can keep our dinmonds until Christmas easy. Nor would I wish you to buy any Kyloes at present because you may depend upon it that the last will be the best bought. I will tell you more in my next how Kyloes are selling at Falkirk and the north country fairs. We don't wish to purchase any more at present except it was very convenient or very cheap at any rate. As I propose this to go to the post in the morning will conclude yours in haste

Geo Culley

146.   *George and Matthew Culley to John Welch*

Eastfield Thursday 30th September 1802

Well John

You must not blame me if I don't give you full information of everything concerning agricultural business. You will rather blame me for stuffing your head with too much of it. I sent you a very full letter this morning and now I immediately begin another. I can't expect you to do the same to me, because I have more leisure than you. But I would still recomend you to come as near to this plan as possible, because matters occur to you and can better be transmitted to paper every day than resting upon the mind. The mercury has dropped a good little piece last night but is still very high. I did not advert to your fair, but you can't by any means be from it if well. But I would have you avoid for the present buying any steers &c except very cheap, and those to be kept upon straw only. Nothing will pay so much for wintering as our own sheep, I am certain, and now we are fast with them. Let us try to make the best end we can of them at any time in this autumn, at Christmas or any other opportunity through the winter. Your master and I have made a remark ever since we had stock to sell, that it is the wisest way to *sell* when customers are in the humour to *buy* of you. But we forget to put in practice in our old days, what we have long approved of as a right and good method. However I would wish to recomend it to you and all young men as an excellent maxim that seldom is wrong. We most certainly missed a good opportunity very lately, and I must further observe to you that when things are at very high prices it is generally wise and prudent to sell. And we often find that some unexpected and unlooked for *event* takes place and gives a sudden *check* to markets, like what happened the other day. 'Better to me sell than to me keep' says the old proverb, and I have heard it said of old Tom Robson of Ellerton (who made as much as any in his day) that he always made it a rule to sell whenever he had any goods at a market, and I believe it is right 9 times out of 10. His reason was that if he selled cheap he could buy cheap.

Friday 1st October proved a nice harvest day, a strong wind but not so as to do damage, but good. We sent our young lasses to Thornington and the Wark folk went there to[o]. I fancy tomorrow will nearly finish Harry. His own are going to the Longknow today, where they are also far on. *Saturday 2nd October 1802.* Ralph and I are both going to Berwick to pick up what cash we can to pay our great rent at Chillingham on Monday, and I believe we shall do very well if we do but get what we expect, and I hope there is not much danger. Ralph got nearly 150£ at Kelso yesterday and sold a good little piece of old wheat at 56 and 57s. They have thrashed some white wheat at Wark in good condition, but yields very ill with much small in it, and I understand that they have a good deal worse than it. Yet London and most other inland markets come down. It is likely that wheat is a better crop in the south than in this country. We have some early turnips which are mildewing a good deal, which I wonder at very much as we have had no want of rain. Whether the thick misty mornings followed with uncommon warm, or rather hot days, can have done it I don't know. Thursday week was so hot that Mr Nisbet had either 7 or 9 women were quite delirious with drinking too much water it was thought and the heat together. I believe they all got well again with bleeding and medicine. A young man called Whitehead, a wool stapler from Bradford, is here and has seen our wool, wants credit which would be very unsure in us indeed to give. I told him that he must take and pay, and if he can't pay all to leave the remainder of the wool until he can raise more money. I asked him 26s but he made no answer. He is going to Berwick with us. It looks as though there was some little stir at present however in the wool trade. Will you be able after Yarm fair to assist Mr Peacock with any more cash? Perhaps if you have to buy any quantity of Kyloes you can't. Your harvest wages too will be heavy. However you will have no steers to buy except for straw for us, for I think according to present appearances it would be worse than mad for either you or us to buy cattle to feed, because as I told you before we will spare you as many feeding cattle as you like when you get over. And now we have all our sheep in hand and likely to be so, will take all the turnips we have and more, I am affraid, and I am very unwilling to buy. In short I never liked buying turnips in my life, and it becomes more irksome to me at this time of day. Besides if our turnips turn many of them mildewed they will be obliged to be eat early enough. I don't know whether it may not be prudent to put of buying Kyloes until Newcastle fair, as things will be comed to some kind of level by that period, and I see that cattle are dipping in price in all parts. If we can believe newspapers at Carlisle vast quantities of Irish cattle make their appearance and can't be sold, and if they go south they must return again or be sold for half nothing. Indeed Midllam Moor[78] may be a very proper place for ought I know, but you are a better judge than me of these things. Only one thing let me again repeat to you, that you are never to *straiten* yourself for *money* at any rate, because a

---

78 Middleham Moor, near Leyburn.

grazier or any man to go to a fair &c to buy without money enough, will often be at a loss, and it does not take a very little money to buy cattle with at this time of day. My brother recommends to you to try Auckland for fogs for the cattle you are to have from hence, or any other place that occurs to you. It is a nice time to drive cattle or sheep just now, and if the weather once break, which it frequently does about Newcastle Luke fair, it gets to very bad driving for goods and punishes them much. My brother also advises for you to be selling your Kyloes, which I see by your last letter you were going to do. I only wish that you have not put it too late, but if so it can't be helped. As long as people consider and then act for the best, or to the best of their knowledge, I can never blame them. It is only folks who never think, or never think as they ought, that get so far wrong. After a man has considered and *reconsidered ever so often*, he may still act wrong, but he will be more apt to go or *act wrong* if he never considers or thinks *at all*. If consideration is not a virtue it is always an attendant upon virtuous action. The inconsiderate man does not weigh or think upon consequences, but rushes headlong as his passions prompt him until ruin and devastation come upon him unawares. Now there is a piece of very good moral philosophy for you.

Sunday 3rd October. Very like wind as glass has comed down and it blows strong, but I hope it will not be much. Yesterday proved a little hazy and wet, not so good a leading day. We had a bad market at Berwick, I dare say very little done. Mr Smith wanted to buy lower of us and we would not submit. I could make nothing of my wool man. He bid me 25s and we parted. *Monday 4th October.* Glass very low indeed, and black and very windy like, both rain and wind from the west. Has blown very strong all night but no rain I think, at least not much. We have 3 little pieces of spring sown wheat still uncut which we must take today though I think, although not very ripe some of it. Your master of Wark and self just going to Chillingham to pay our May half year's rent.

[**M.C.**] John I think at least let Kyloes alone till Newcastle fair, sell fatt ones if you can, be doing well with sheep alone. M.C. I like Buston's steers but am afraid you cannot buy worth your money. We have fatt ones as many as we can winter. We want straw cattle &c. They are best which are in best condition. We have 10 or 12 nice queys for you of our own breed and Ralph Compton's. Yesterday we were at Chillingham paying our rents. Brother George went for Alnwick. I hope he will write you an account of that fair. Young Kettle came from Jedburgh fair (which is said to have been tollerably good). As a proof he came back after the fair and bought 20 steers of Barber which were shewn at Ninian. Barber said he sold them at 20£ per, he asked 20 guineas apiece for them. Kettle is concerned with Wood, and they are finding sheep this winter together. We have had some high winds these 2 days, must do harm, but the most of the corn is cut. Harry had a few oats to cut, and some little at Longknow, not much turnips, the early ones and on hot soils rather begin to mildew. Our early ones begin to run so we are giving the wethers a few every night. Our lambs on the stubble clovers are now led. The clovers are very good. Yesterday and last night some cold rains with like frost as the mercury riseth much. It is now cold and frosty. I know not how to advise on buying, but only if you can buy Mr Buston's steers worth the money for wintering, as straw cattle with a swill full of turnips once a day so as to be fitt for grasing next summer. I hope we can buy a few steers here by and by for straw and working. You will have to meet Matty Culley at Newcastle to assist in buying some colts or mares to keep up our drafts, although horses pay a tax. The number I know not as yet that may be wanted. I know not when I or we shall get to Denton as I have more to do than usual, by the want of George. Andrew Bolton and William Riddle &c do very well upon the whole with my help, only it keeps me confined more than I wish, and I know not where to get a good overseer. Gib Taylor is fixed with H. Morton. But I will wait a while. Do you know of one likely to serve me?[79]

147.   *George Culley to John Welch*

Easttfield 8th October 1802 (Friday)

Well John

Yours of the 3rd came to hand in due course when I was absent at Alnwick fair and Sessions, and only got home this morning, and as our people had in my absence sent you a letter which I had wrote and forgot to take with me to Alnwick, I will answer this of yours and tell you about Alnwick fair, which was a very full and heavy fair. Not one buyer from the south except a Mr Oliver from your country who buys incalves and queys, a tall man. Some steers for straw might have been bought lower than of many months, but I hope you will be able to buy better bred steers at Yarm and Darlington fairs. Fat stock were sold from 6/6 to 7s per stone sink, and many were sold to Shields, Newcastle and Morpeth butchers. None that I knew from Sunderland

---

79 The signature is lacking.

but Lincoln and Kilvington, a man who often buys from *Jack* now Mr Robinson of Tuggal, and Ralph Sanderson. I believe he is a relation of Ralph Sanderson. They bought a good few. Rawling was there but bought none. But why John Forster was not with him I know not. I have a very great notion that Forster poor fellow is falling weak, because he puts payments long, and he seldom used to miss this Alnwick fair. We have an offer of 25/6 through Mr Vardy for our wool which we think of accepting, as I do not like to keep a commodity at such a price. You need not name it to Mr *Peases* without they speak to you, and then say it is sold you believe into Yorkshire without saying what at. Well John if you have had such winds as we had your moor oats would be well smashed. Ours was all reaped at every place. Winds were violent on both Monday and Wednesday last. It has uncovered several stacks again. You may depend upon barley being much lower this than the last year. Perhaps all grain will be lower, but barley is the best crop. Oats I think will not be lower, on account of the high price of hay at London. They are a little up this week, and wheat same as last week. I do assure you that you sold your Kyloes well. I had no idea that you would have sold them so dear. It is great encouragement to your buying more Kyloes, and you have certainly bought that score of heifers cheap at Brough Hill, and I think you ought to buy some lean ones at Newcastle or where you can best meet with them, as those you had last winter have done so well. And I would have you buy as many steers as you will for us as we are very strong in straw. Never mind what ages, 1, 2, 3 or 4 years old, provided they are well or decently bred and only fit for straw. And I am quite against you buying fat or fattish ones, either Mr Buston's or others, except you can buy them to pay at straw which is very unlikely. I dislike wintering cattle at those high prices prodigiously on turnips except you can buy a few half fat ones for yourself to eat straw and a few, or rather *half* a bellyfull of turnips once a day by way of preparing them for grazing in the spring. It is certainly necessary to make straw into dung, highly necessary, and the question is how to do that in the best and properest manner. Where people breed, this is done perhaps best by their young stock, but you that don't breed should consider whether Kyloes are not fitter for you than country cattle, and of that I shall certainly leave you to your own judgement because I candidly acknowledge that I believe you know better than I do. Well but what I am still leading to is that sheep to a certainty pay more for turnips than cattle, and improve the ground so much more than beasts. But still on account of straw you must have both or part cattle. But I also must observe that on account of getting certain crops of corn I highly approve of the new method of eating only the half of and the other half on, or better perhaps 2 thirds led of and a third eat on. But that must depend on the state that the ground is in as to richness or poverty. There can be no doubt but turnips will go much the furthest all led of, and perhaps feed as well or better, but I do approve of eating *part on,* if ever so little with sheep, and by leaving *intervals, spaces, lanes* or *alleys* through the field the sheep have bare ground to lie upon, which they not only like but they don't *foul* or *dirty* their meat so much.

The Sunderland butchers would have bought plenty of large cattle at Alnwick, and cheap. Woolpit was a most horrid market owing to the terrible drought in the southern counties. I will be bound for it that your highland heifers will pay you well. As to the Irish cattle I am told that everyone who have tryed them are tired of them, but the right Scots are a valuable kind of cattle. You say that you have not the least idea of the dry weather injuring us here. It certainly has been of vast benefit to our harvest, but I do assure you that our early turnips, and many other folks, are turning mildewed and white and yellow, and stopped growing, and many running into flower already. And what is worse, and shews the eagerness of our winter feeders, they are giving 7 or 8£ per acre for turnips. I know not what may be the end of all this, but sheep bought at *even* the *lowest* price this autumn will have a bad chance to pay decently with these prices of turnips. I do assure you that I should not like the speculation. But as we have our sheep and part turnips, we must do as well as we can. But I have no notion of buying turnips at these prices. Your Manchester butcher may be very right and must know much better than I can pretend to do, but my idea is that if one was certain of turnips standing sound all the winter (which I greatly doubt, from their present white look) I say if we could be sure of their support through the winter, that there is a great probabillity of stock being higher than ever in the spring and early part of summer. Because if Norfolk and Suffolk be unable to supply Smithfield as usual they must have a supply from other places, which must fall upon Lincolnshire and the Trentside. Consequently those not coming to Wakefield &c will at last have an effect upon our northern parts. However we are often mistaken after all our probable conjectures. But it is certainly wise and right to look forward and make the best conjectures we can. I think you are right in refraining from going to Skipton until further on. I assure you that I should feel very comfortable if we had been so lucky as to have sold our sheep to Mr Wood. I think it right to cut your beans. We have cut or rather *pulled* ours, and Wark folk have got the last of theirs in today.

They have a full yard and all their hemels[80] full, and a good few stacks out of the yard. When shall we see you over John? Not until Yarm be over at any rate I suppose. Perhaps you will come, and return to or by Newcastle fair. Had we had you at Wooler fair (which can't be) we might have been encouraged to buy a few dinmonds at Wooler fair as I am persuaded they will be bought low there on account of the high price of turnips, which will discourage the buyers in this country from purchasing any. I will not seal this until I see Berwick market and now late. I will conclude yours sincerely

Geo Culley

Ralph says new wheat and oats were both lower today at Kelso. He still sells old wheat well but we are nearly done. I find that wheat yields badly in Bamboroughshire and all along the coast, and is only a midling sample. Oats and barley good. You still neglect to leave a space unwrote on for the wafer, and I could not make out a word or 2 in your last owing to the wafer. It is best to draw a round or square place with your pen in the middle of the last page before you write so far down, as I have done here. *Saturday 9th October 1802 5 o'clock.* A fine fresh morning but a very low glass with a strong wind. Yet it has not been much rain lately, only slight showers from the west which the wind soon takes up. But as the Tweed comed down pretty strong yesterday afternoon it is not improbable but we shall get rain soon, and I do assure you that moderate rains will do a great deal of good to turnips and fallows in particular. Turnips are blown to death and are all turned yellow or white almost, and in my opinion the *fallows* at present are far *too dry* to sow. The corn cannot vegetate, at least in many rough cloddy fallows. Yet if you don't get your moors sown in the beginning of *October* or before the wet come on, you may not get them sown at all and they will not do for spring wheat like your infield lands. So I shall leave you to do as appears to you most proper. You must be a much better judge than we can at this distance. If the land be worked small and any moisture in it, there can be no fear of its vegetating. I am sorry to say that Franky Peacock is I am affraid in a very bad way. He continues to throw up great quantities of blood I am told every now and then. If he has not made a *will* he ought to do it, and it is rather an unpleasant matter to recomend. Yet duty superseeds everything in these cases, and if he don't make a disposition of his effects his natural child will get nothing, and he has brought her up in an indulgent way, perhaps too much so poor thing.

---

80 Sheds.

Matty is expecting Mr Thomas Charge and Mr Cradock to look at our cast ewes. Tommy wants a few. If they should stay 2 or 3 days I may likely write a few lines to send by Mr T. Charge. They are to come in the coach today. Remember that sudden falls in the weather glass in general denote wind rather than rain. When the mercury sinks slowly for 2 or 3 days depend on rain then, and often a good deal. A little drop denotes showers or a change to fresh from frost &c &c. You may depend on it that turnips will be higher here than they have been of many years. 8£ and 8£ 10s per acre are aukward prices. If you would meet with a few burnt land turnips, pretty good to lead of would do well, and the burnt land ones are the best *worth buying* always. They *feed best*, are *soundest* and *fullest of juices*, and *stand* the *winter* frosts &c best. I would recomend to you to be picking up stots or steers for straw whenever they fall in your way, and of any age. By all means cut your beans for fear of frost, which spoils them sore, and rain must come and when it comes it will probably be a wet time.

Berwick 3 o'clock Saturday. We have done no business today and yet the market is not worse than the last Saturday. But we had determined to take no less for new wheat than 48s, and Ralph was only bid 46, but believes the man, it was Mr Walker, would give him 47s. He refused 17/6 for a few potato oats to a miller, but the carriage was far of or he would have taken it. It has rained gently all day since about one o'clock, but I don't think that the weather will turn quite rainy yet. I still think corn will not be badly sold although Mr Hay will not give more than 12s per new boll for new wheat. Watson of the mill was wanting wheat today but did not hear what he would give.

Geo Culley

148.   *George Culley to John Welch*

Eastfield 9th October 1802, Saturday

Well John Welch

I this day put a letter into the post at Berwick for you, which I hope will reach you tomorrow evening Monday at Darlington. Ralph after I had left the market to go to my dinner, sold his 50 bolls of Wark wheat at 47s per boll, which perhaps was better than running the hazard of a worse market the next Saturday. People are in various opinions respecting the corn markets. I am clearly of opinion myself that the wheat crop will turn out defective, but then it is said that there is and is to be a great importation from abroad. But we shall know more about that by and bye. Ralph sold a few oats (potato) to a miller at 18s per boll. Barley price is still unfixed. Many people think that it will be low. They were bidding 15 and 16s per boll today for it only. John we think that you had better bring over your books with you with your accounts made up to that time as nearly as possible, because it will not want much of Martinmas then our time of settling, therefore can't make much odds and it will save a good deal of trouble. This day has promised wet mostly but not heavy. However the glass is going up again and the wind is northerly, and not likely to be much rain yet. The little we have had will do more good than harm. Messrs Thomas Charge and Cradock are not arrived yet. We have a little spring sown wheat to lead yet, and nothing else except the beans. Very little corn to lead in general in this district. It is said that the winds on Monday and Wednesday have done much damage upon the edges of the hills of Lammermuir &c. I heard a man say that in one field of good barley of near 8 bolls per acre, had not one boll per acre left in it. *Sunday 10 October.* A very pretty drying day with quite a cover of snow upon Cheviot which is likely to lie all day as it is now near 5 o'clock after noon. Messrs Charge and Cradock arrived last night after I was gone upstairs to bed. They stay until after Wooler fair. *Monday 11th October.* A very fine morning indeed, either for harvest or sowing. Glass has got up again to above changeable although there is not much frost this morning. We had an ox died upon clover on Saturday, but he was sore clyered. The cattle had been upon our eaten fog of clover, and this they were put on was very rich indeed, and we had ordered them not to be put on until the middle of the day when the clover was quite dry and clear of damp, which I dare say was litterally done but the rain had come on soon after and although William Brown junior rode up directly (it was Westfield) and took them of and no appearance he says at the time, this one only began to swell to a degree that occasioned his death. But I hope the loss will not be serious as he was bled directly. *Tuesday 12th October.* A very wet morning and like to be much wet I think, as the glass has sunk much last night. Indeed it shewed for rain all yesterday and I shall not be surprised at all if the weather be going to break. Nothing but a gale of wind to prevent it. But we ought to be very thankfull as we got all well in yesterday except our few beans. Indeed we only had some odds and ends of spring wheat to take in. They have led all their beans at Wark. I went yesterday with Mr T. Charge and he drew 40 ewes out of more than 300, but we have not bargained yet. He wants to draw the 20 best out of the 40 again but hold these without more cash than 5£ 5s per head. We had, or rather Matthew had a decent tup day yesterday to hire late in the season. Messrs Charge, Cradock and self did not get home until dinner time. Matty has let 10 I believe to different parts from 5 to 25 guineas. I hope will make this as good season as the last before we be done, and if we can make 1000£ it is a grand trade.[81] To be sure there is a draw back this year but it is well it is no worse. We will get the south country tupped for I hope. *Wednesday 13th October 1802.* After a compleat wet day yesterday of small hazy rain from the north east &c, here is a decent like morning with a rising glass, as though the weather was not going to break yet. However I thought it was yesterday. But we have all reaped, stacked and covered except our few beans. *Thursday 14th October* had all the appearance again of fair settled weather, suitable both for the bean harvest and wheat seed. Messrs Charge and Cradock leave today, and I think our old friend Mr Thomas Charge has not behaved quite so well about these ewes as I should have expected. He has at last bid us 160 guineas for the 40, which is 4 guineas each. But it would be extremely unfair in us to our other friends and customers to take less of Mr C. by one guinea each of them. We propose shewing a part of our hill ewes at Wooler fair. Ralph Brown is going to Grindon today to make of those we shew as well as he can.

Friday 15th October 1802 is another charming morning, and I hope we shall harvest our beans well. Mr Charge is gone and not bought the ewes. I don't think he has behaved as he ought and as I should have expected. I received a letter from the man we wrote to about the wool, inclosing a draft for 450£ which is certainly a confirmation of that bargain. I intend to submit this 450£ bill to Mr William Thompson in a day or 2 as it will answer as far as it goes to make up the money for the last payment for the Grange estate. So you may sell your piece when you think right. I have much reason to think that it is upon the advance, and I am pretty sure that there is little or none left in this district. Wheat advanced in London last Monday 2s per quarter, and I dispose to think that wheat will be found a defective crop. As well as beans and pease, barley and oats rather plentifull crops especially barley, which has a chance of being sold the lowest of any grain in proportion, Oats cannot get very low so long as hay keeps at so high a price in London. Matty has sold Mr Cradock one of our little 2 years old gone queys, open ones which we intended for you yesterday at 20 guineas not quite 50 stone. She is in calf, and could not be got fat in time before calving. She is to go over with the first cattle you are to get from hence. *Saturday 16th October* is still a

---

81 In the year ending 22 November 1802 the Culleys made £698 6s 9d from letting and selling tups. The previous year they had made £1080 0s 6d. See pp. xxxix–xli.

promising day. We got our beans all tied and set up yesterday. If you get the fine rain we had last Tuesday it would put your fallows in nice order for plowing and sowing. Many of the fallows upon the strong lands in Bamboroughshire &c were become so dry that they durst not sow wheat upon them until that rain. It also benefited the turnips much. Nevertheless many of the early sown turnips are so mildewed as to be past redemption and the turnips coming into flower. Ralph sold new wheat at Kelso yesterday at 50 our boll, which is a famous price. *Sunday 17th October.* Another charming morning although the glass is comed down to changeable, but it may be turning fresh or going to be a gale of wind. However I hope you have got your wet moorland fallows sown. John you were in luck at the Brough Hill fair. They have had the dearest fair ever known the other day at Falkirk, dearer than last year by much, and no fair has been like it this year anywhere. It is really a mysterious business that stock should be dearer this moment in the north part than the south part of the Island, a matter that perhaps scarcely ever happened before. But it is to be accounted for in two ways. First the drought in the south has hastened people to market with what stock could be turned into money, and at the same time prevented them and others from buying any more stock, while the north part had been much drained before and no drought there, and the matter has happened chiefly with the fat and forwardest cattle which are really scarce in Scotland it is said, and again the Highlanders are bad to fight upon their own *middings.* I should not be surprised still if you buy lean Kyloes at Newcastle well enough, because the south country people will not be eager to buy them so very dear I think. Indeed one thing puzzles me a good deal, that Tommy Armstrong who is one of the most steady and cautious men of all the great jobbers and a well-informed man, was there and bought a good many at these exorbitant prices. It looks as though fat stock was not over and above plentifull still in the south. However a little time must develop this business now. I had heard of this at Berwick, and when I got home I met with Henderson, a great drover and jobber from the Scotch border. He had a tup of us last season and another this, and was at Falkirk Tryst himself and bought a good many. He and a brother who lives at *Shap* in Westmoreland had 20 score Kyloes at Brough Hill and sold very few. Mr Henderson wants a few ewes but we can't prevail on him to buy a part of 8 score of the Longknow ewes which we are going to shew at Wooler fair. We think that if we can sell them anywhere from 2£ 2s to 2£ 5s they will not pay to keep all winter, and they never get fat before the beginning of summer. However I would not be fond of taking less than 2£ 2s neither. Good fat in Scotland is worth 10 to10/6 per stone, which is better than 8s our stone. It is the wrong end of the Island for things to be dear, even fat, but it will have an effect. It is said that Joe Mills of Glanton Pike bought . . .[82] score Irish cattle at Brough Hill fair. Did you see him there? He is a strange man for doing business. However Henderson says he bought them pretty well but might have done better. I propose sending this and one for Mr Peacock tomorrow about the middle of the day from Wooler fair by Mr Thomas Charge, who with Mr Cradock propose going per coach to Newcastle tomorrow night and will leave them at the post house the next day as they pass Darlington.

Wooler 18th October 1802. Well John this has proved full as good a market as was expected. We have sold the Longknow ewes 160 at 44s per head. Cattle are sold pretty readily at not much reduced prices. We bid for a few steers but could not buy to do good. But I hope you will buy better and lower at Yarm and Darlington. I hope we shall see you here soon, and hear from you in the mean time. It is still good weather but mercury much come down. I am in haste yours

Geo Culley

PS. I hope Mr Charge will deliver this to you at Yarm fair as he talks of going there.

149.   *George Culley to John Welch*

Eastfield 19th October 1802

Well John Welch

I hope Mr Thomas Charge will deliver a long letter which I sent by him yesterday into your own hands at Yarm fair tomorrow the 2nd fair day, as he told me he was determined to be there if well. The dinmonds stuck in hand a good deal yesterday at Wooler fair, but I understand that most of the ewes were sold and at better prices than was expected. We never looked for above 45s for our Longknowers, and we got 44s. Indeed Ralph Compton only got 45/6 for his, and refused 50 long since, but he was bid too much. Indeed at the time he was bid 50, people thought they (ewes) were not to be had for any money. I do believe that there are very few ewes unsold in this country. A shabby shew of cattle, and all sold that were worth buying, but not at quite such high prices as before St Ninian. But they were sold for plenty of money. We could not meet with any steers fit for straw and yoking. Ralph sold the little horse bought the last year of

---

82 A space for the figure is left blank.

George Orwin at 20£ and sold for the same now. Matty also sold a mare, sister to the one I ride and the brown one you sold, but I did not hear the price. I believe little money, but she was useless to us and often fell lame. It is very remarkable that the mercury has kept falling for several days and is now rather lower, and still no appearance of rain. Selling these ewes has made us very clever for cash, and it is very well because we would have had to borrow of a bank as Matty is determined to buy some horses at Newcastle for the draft, and some colts, and probably some straw cattle if you have met with none. *Wednesday 20th October* is a decent morning for the season, yesterday was a blustering showery day but it turned better out than was expected considering that the mercury came down to rain. It must have fallen a deal of rain in the west I think, to have sunk the glass so much. We had some customers in the tup way yesterday, who took about a 100£ worth. We are now very likely to exceed 1000£ which I did not look for. 500£ is certainly a very fair sum in that way but 1000£ is better surely. But I am well content with either, and so long as a person is in the farming line it is certainly right to make all one can with honest endeavour, and when a man turns gentleman let him do so with all my heart. But I am always for having people to act up to what they profess. So long as a man farms, I conceive it to be his duty to cultivate his ground as well as possible and to raise such produce as will make the most money, and the same in regard to his stock &c &c. Otherwise he neither benefits himself, his *family*, nor the *community* so much as he ought to do.

Thursday 21st October. A windy morning but does not appear to have been any rain in the night although like showers now. We had begun to lead our beans yesterday about 10 o'clock but did not get a fourth in until very stormy showers of rain and sleet put us of. It fell white upon the hills. Indeed the air is much altered of late from almost warm to cold. But it is the season when we are to expect cold and winterish weather. We have had so fine an autumn as ever I knew, and except the early turnips stopping and turning white and mildewed, I never knew clovers and pastures fresher. We are still cutting clover for our horses. Young Wood and Castle, a younger one than the squinting lad that buys cattle, came here in the afternoon of yesterday and bought 2 of the ewes which Tommy Charge ought to have had, so that with the 10 that Mr Sitwell has got, we have quitted 31 of the 40 ewes which T.C. bid us 4 guineas per head for. These 31 come to within a few pounds of what Tomy bid for the whole, and we have 11 ewes left, because I offered Mr Charge 1 into each score. But my old friend T.C. did not behave as I would have wished. Had he said that the ewes did not suit, all would have been well, because he knew that we don't sell below 58s to breed from per head when picked. Messrs Wood and Castle had seen and drawn the ewes at Grindon on Tuesday on their road to David Lee's, whose 6 score shearings they bought, and then went over the Tweed and bought 200 of George Logan's at Ednam, so that these boys have done a good deal. I think they said that they have bought about 1700 dinmonds. They complain most of old Mr Gregson's. I understand that they are very lame in the foot rot, and they were dear bought, Wood says they are 100£ over dear. I am of opinion that Mr Barber's are as good and fat sheep as any they have bought. Wood said that they would buy ours any time if we would drop them a line at Christmas or Candlemas &c. But we could not absolutely promise, because I always hope that you will be taking most of ours some time between [. . .]83 and Christmas, probably if you sell your own. But we must consider over these matters when you come over. Mr Wood paid me 100 guineas directly so that we are now very strong of cash. My brother and Matty will take a good deal to Newcastle fair I dare say, and I am thinking that it may be very well if a part can be paid to Mr Peacock to make up his last sum to remit to London. It will be better than borrowing of a bank. But that must be considered of at Newcastle or here, because if many cattle, *steers* and *Kyloes* are bought they will take a vast of cash, and cattle we should have of some sort, both you and us, to eat and tread down our large quantities of straw. I don't know of a large lot of shearings now but Mrs Grey's 300 of Kimberstone. I do believe that the low country sheep are *nearly sold*, ewes in particular, and many shearings are also sold although the Yorkshire jobbers did very little at Wooler fair (I don't call Wood &c jobbers). It was the ewe and lamb men from the Tyneside and the south of the county that made the fair so good for ewes.

Friday 22nd October. A fresh fine morning but still showery-like, has been some rain in the night, and yesterday was very wet and stormy all the forenoon with a most violent tempestuous wind from the south west. I begin to wish for our beans in, but I don't think the weather has the appearance of being very rainy still. *Saturday 23rd.* A very fine dry windy morning indeed. I hope we shall now get our beans well in. Ralph Brown sold some old wheat yesterday at Kelso at 56s per boll all red, and he sold 11 pork pigs at 7/9d per stone. The Berwickers would only promise 7/6 the last Saturday although they

---

83 A word seems to have been omitted.

can certainly much better offer to give a price than the people of Kelso, but they are not willing to work for so small a profit I suppose. However it is a capital price, and only shews that butchers meat is not still over plentifull at London. *Sunday 24th October*. A thick morning. Your mistress and I went down to Ancroft on Saturday afternoon and came home last night again. Berwick market keeps steady for wheat, at least no worse, oats better, worth 16s and 17s, I think we shall thrash a few. Barley not started still. It looks as though oats would be as dear sold as barley in our markets. Ralph sold some very indifferent red wheat from Wark at 45s, many samples in he says but all sold, and people begin to think that wheat and oats will be well and better sold soon. *Monday 25th October*. A very thick morning. John you have not been so kind as write us an account of Yarm fair, but I can suppose you would be very busy with your lime accounts &c &c. However we have heard both from Mr Peacock and Mr T. Sayle that it was the dearest fair that has been this year. What will be done for steers &c to eat our straw? Matthew and you must try to buy us Kyloes if you can meet with nothing else at Newcastle, but I dare say everything will be high at Newcastle. However although we are in want I am better pleased that things are looking up than down. Thomas Sayle also says that the markets at York, Wakefield &c &c were better for sheep considerably, but not much for cattle. I am always glad when sheep keep up. It arises from a selfish motive I acknowledge, because sheep are our *sheet anchor*. The Yorkshire jobbers will now repent that they did not buy any sheep scarcely at Wooler fair. Indeed if they had bought it would have been a capital fair. As it was, most of things were sold, especially ewes. I should not wonder now if the weather is going to break, as the atmosphere is turned very thick and the glass comes down gradually. Indeed we have had a blessed season, and we may now get a very wet one. But His will be done who gives us such weather as He pleases, says

Geo Culley

PS. We have got the beans in, in nice condition. This thick dark weather may go away with wind but it now rains. I have wrote to Mr Peacock in a letter about money matters today, to tell you to come north at all events from Newcastle fair and to be there as early as possible to help Matty to buy horses, colts and cattle. *Tuesday 26th October*. A most pleasant morning indeed, seems like more good weather still. We let a good few tups yesterday and have some low priced customers still to come next Monday. Indeed we have only a very midling choice left now. We have got all the spayed queys down here now to be ready for you to look at and take away. I hope that several of them will make nice ones and please you. As Matty sets of for Newcastle today along with Grieve Smith, I shall conclude and send it by him G.C. When you return from this place after the fair it will be very well if you and us together can muster one hundred pounds or more for Mr Peacock to send to London. It will make less to be borrowed of the bank, and the less the better and the sooner that is paid of the better which we shall be obliged to borrow of a bank.

150. *George Culley to John Welch and Matthew Culley*

Eastfield Wednesday 27th October 1802

Well John

I sent a pretty midling full letter yesterday by Matty, who with Grieve Smith, George Ripon, Tomy Batters &c set of in the coach for Newcastle fair. This is as fine a fresh windy morning as ever was in October. Stock upon turnips will get a good start, however, and all the wheat fallows have a good chance to be sown, and all the last bought sheep have a nice time in driving to Yorkshire, Cumberland and Scotland &c, and it would be a fine time for our cattle to go to you. What a charming thing is dry weather in general to the husbandman! Let us be gratefully thankfull to that beneficent Being who is so kind to us. John I have this moment in my road to Grindon got yours of the 24th inst., am glad you bought Mr Buston's steers, and only have to lament that you did not [buy][84] 4 times 6 steers. I will [be] bound for it that none will be bought so cheaply by 2£ per at Newcastle. 7 ½ *d* per lb. is plenty for Cleaveland old rotten ewes and shearings. Cheese high indeed! But never mind how high cattle and sheep are, because as we buy so we are likely to sell. I am glad that you had so good a receipt for lime, but I blame you for pinching yourself for cash. You should never be bare of cash, because it prevents you buying when you might. Very likely you could have bought more steers at Yarm if you had had more money. You have pinched yourself to give to Mr Peacock to send to London, which I am sorry for although it arises from a good principle. But a grazier should never be without money at command, and I consider you as a grazier who should always have a command of money to buy when the opportunity offers. Had you bought more at Brough Hill it would have been good, as it was perhaps the best fair to buy Scots &c at that has happened this autumn. I think you give Christopher Charge long credit, but it may be very right if he is a good man. I am glad that you get such good

---

84 Words seem to be missing here and in the next line.

prices for the hogs still. If your weather is as good as ours, you will have a nice harvest for your beans. I hope John Forster will pay Matty at Newcastle for the cattle he owes us. By your account the Shields butchers must have a decent time. Whenever offal sells well the butchers do well. We beat you sadly for pork, which I never knew before. John you ought to make some little of pigs as you are both a pretty large corn farmer and a miller. Do you heed me yet? You are perfectly right in my opinion to eat of your bad turnips to sow red wheat on. I think Tomy Charge right in refreshing the farms. A plowman is bad to meet with such as one can recommend, so all is well. Kitty Mason and Kitty Wright are two very different men. By all means buy what steers you can, even after Martinmas, and of any age ever, calves or yearlings. I don't know John what the spring may turn out in regard to fat. By a letter this day from Mr Vipan they have no turnips in Norfolk worth anything. Cole and rape also bad. From whence then is London to be supplied? I mean in the spring. We are very apt to be mistaken in our conjectures, but I can't [but][85] be of thinking that fat stock has a chance of being dearer in the spring and even summer than ever was yet known. But it will do very little for us because our turnips cannot last us through winter although we have scarce started with any yet, and this uncommon fine weather will make meat go far indeed. Mr Vipan says they had neither meat nor water until lately they have got water. He says the wheat is a fair crop, barley and bean remarkably good. But in the Fens wheat is all mildewed, and oats a bad crop. Wheat is on the advance with them. He has bought us 2 quarters of red wheat for seed, which is to be sent to Newcastle to Mr Darnel. John my mistress desires that you will not neglect to bring over the account of everything relating to George Thompson when you come, both board, cloaths, schooling &c &c.

Thursday 28 October. What a charming day! William and Ralph Brown and self have been up at Longknow and Thornington, and I may say that I never saw the hill country look so well nor never viewed it in so fine a day. A clever little stack of good corn at the Longknow, and the sheep and young cattle all looking very well indeed. Thornington also has a very good appearance. William Brown has fixed upon a very nice lot of steers for his yoking in the spring. I wish Mr Buston's may be any better. John you must not forget to gather a quantity of haws. We have them very scarce here, and William Air has all the country employed gathering. Even in this

---

85 A word seems to have been omitted.

vicinity we can hardly get our own saved from his people! *Friday 29th* has proved a saving day although it has drove very thick clouds all day from the south east and had been a little rain in the night. I have been at Wark where everything seems doing very well. The wether hogs are beginning to eat turnips nicely and look very well. The ewes are taking the tup cleverly also. *Saturday 30th October 1802* is a very thick misty-like morning, dark as midnight at 6 o'clock, glass is steady but low, and dropt a little yesterday. Ralph sold oats at 24s the Kelso boll yesterday. It is very near 20 our boll. A wonderfull price, also red wheat at 52s for seed. Old midling red from Grindon at 56. The millers fond of buying oats. I understand that they are a bad crop in the moorish parts of Scotland, and oat meal is advancing. Some little fairs which have been at Lauder, Greenlaw &c lately, all kinds of their lots of cattle were the dearest ever known. Jack Smith of Wark gave 12£ for a little half bred Scotch quey coming 3 in calf. John I hope you removed your ewes and gimmers out of your moors in time to preserve them from that fatalest of all diseases in sheep, the *rot*. I saw Mr Grey of Heaton's[86] shepherd just now as I was returning from Grindon today, who told me that every ewe upon Heaton was rotten as a pear, that he had told his master some time since but he would not hear him speak. However the other day he prevailed upon his master to kill one of the best and one of the worst, when they both proved completely rotten, so that they have put them to turnips. Now I have never known rotten ewes in our country get very fat on turnips by Christmas, nay we have bought ewes knowingly rotten since your master and I were farmers, but at a *rotten* price, and turnipped them untill or near Christmas and made handsomely by them. But they must not be kept too long, and especially into bad weather. But I am affraid those of Mr Grey's are too poor to do any good on turnips (which is the only meat we know that will do rotten sheep good, but a little oats in trough and hay will still be better), because Nicky Douglas the herd says that the 2 which was killed upon tryal only weighed 9 and 10 lb. per quarter, which shews that they must have been poor indeed. But he is a starver of sheep at all times. We often keep our ewes very poorly here, but then this is very sound land and the ewes too fat. However if you have any suspicion of yours, I would advise you by all means to keep them well and give them both hay and corn in winter. We had at Denton to do so at different times, but

---

86 Probably Edward Grey, uncle of John Grey of Milfield. *History of Northumberland*, vol. 11, p. 247.

Denton was a dangerous place for the rot, although a famous farm take it altogether, otherwise we never would have done what we have. It has showed for rain these 2 days and I think it has comed at last, at least it rains heavily and seems set in for rain all round. I just got home about 2 o'clock before it came on. But I have frequently known the weather break just at or a day or two after Luke fair. A little rain will do us good, as our land is too dry and the turnips want it much, although it is a strange thing to wish for rain in the end of October. I wish we had our Newcastle folks home now. *Sunday 31st October.* A very fine day after the rain. Matty and William Brown and the 5 horses all arrived safe but very sore wet with the rain indeed. John [you][87] should try to buy from 10 to 20 more cattle for us of any age.

Dear Brother. I hope this will reach Denton before you leave it, and you will then see that all is well with us thank God. We can keep John Welch's cattle (meant for him) until he has brought us some more lean ones for straw well enough. And John must write when they are to meet by the road, or one of our Glasses go all the way and then bring the straw cattle back. It will do us good to eat some of our mildewed turnips with them. But this fine rain last night will do us much good here both to turnips and pastures &c. But John must come over himself before the fat queys &c come away from this country, because he should and *must* see them to chuse those only that suit him. This is a charming day indeed after the rain. Matty is going by the Roads to pay my sister[88] Nelly's quarterly payment 52£ 10s, and I go to Mr Nisbet's on Tuesday evening to be at Berwick on the Wednesday the High Market to remit 1000£ (tell Mr Peacock) to Mr Thompson London, which I wrote to Mr Peacock about before, and he will I am sure take care to remit the balance or rather 300£ to enable his brother William Thompson to pay for the Grange and then send us the deeds &c. I am in haste yours affectionately

Geo Culley

PS. Ralph sold more oats yesterday to millers, *potatoes* at 19s/6d per boll. We sent your half letter to my sister today from the meeting to Wark.

---

87 A word seems to have been omitted.
88 James Culley's widow, Margaret Pickering, and their daughter Eleanor now lived at Etal Rhodes.

151.   *George Culley to John Welch and Matthew Culley*

Eastfield 31st October 1802

Well John

I sent a letter to you per post today, that it might reach you while my brother is with you, that he may know all about us, and my letters are a kind of journal of our proceedings. I did not know until the above letter was sealed that you are in want of fat cattle to tie up. Now that makes it the more necessary that you should come and chuse, because it would be a pity to tie up altogether the nice queys which we had intended to send you, as they would make uncommon nice ones by having a little grass in the spring, being almost all spayed heifers that should not be sold I think before the months of May or June. Indeed you might tie them up at nights and let them into a curtain or fold by day, which would make you full as much dung as being close tied up, and keep the hair better upon them to turn out in the spring, and you have plenty of straw to bed them with in the courtains. We could give you heavy cattle (oxen) to tie up, but they are only fit for Shields while I conceive those heifers to be fit for the south country markets or anywhere, and I do suppose that there may be a great demand to the southern markets in the spring or in May and June. But if you prefer or it be thought right for you to have other cattle we can keep them here to be in a fold at nights and out by day. Indeed we put our untied cattle in both in the middle of the day and nights. By this means they make nearly as much dung as being tied up, and fits them much better for a little grazing in the spring. But we can frequently sell heavy Shields bullocks as well as you, while you can sell nice cutters for 1s per stone more than us often. However I thought it right to suggest these hints to you at all events, and then do whatever is most proper. This has turned out a very quiet day. The colt you bought is only 2 years old, you had not looked in his mouth. *Monday 1st November.* The last month closed, and this has begun with more the appearance of winter than any weather than we have experienced this any time. Yesterday was the most promising morning, but the wind whipped about from the west to the north east where it still remains, driving thick dark clouds before it, but no rain since yesterday about 4 o'clock after the wind turned northabout. Nevertheless the hills are covered with these dark clouds and the air is humid. *Tuesday 2 November.* A very decent-like morning. We let Harry Morton ½ a dozen tups yesterday at 5 guineas each, fearing Jemmy Brodie did not put up his appearance. Mr Hogarth sent down a cart with one of the tups he had which did not answer, and got another, also one for his nephew at Pinless I believe they call the place, so that we have done wonders now.

Thursday 4th November. I went to Ancroft on Tuesday evening with an intention to go to Berwick High Market on the next morning, but it had rained from the north east most of the night, and was so very wet a forenoon that we took dinner at Ancroft and went to Berwick in the afternoon, when Mr Nisbet paid me £940 16s which I made out 1000£ and remitted directly to London to Mr Thompson in a draft at 30 days. The High Market was very high for cattle of all kinds notwithstanding the wet day. Little done in the corn way. Ralph had received a good deal of money both at Berwick and Kelso, and I drew for John Masters 60£ at 20 days, so that we are very strong in money now and will begin to pay our bills. I gave my sister at Wark some more money the other day, and spoke to Mr Thomas Forster junior to attend your quarterly meeting at Wark for the poor rates[89] on Tuesday next, which he promised me he would. Either Matty or I will also attend if the wool weighing don't come in the way. The oat market has dropped a good deal in Scotland, so Ralph sold few at Kelso last Friday. James Brodie and son came only this morning. Matty let the young man 3 tups at 12 guineas, but could not deal with his father. Jemmy grows worse and worse. His son says he will come no more with him, because he will neither take any himself nor let him take. I fancy we are done for this season, and so we may seem as we never let so many before by nearly a half a score, and never let for so much money one year excepted, when we took in ewes for a good deal of money. John Mills is buying up all the dinmonds that are left I think. He has given Mr Watson of the Mill 3£ per for his if we are rightly informed, but I dare not offer him ours if he would even give me 3£ per. Turnips are selling for 4 to 6£ per acre, which is considerably lower than was talked of. *Friday 5th November*. A very wet morning indeed. If my brother be at Denton now he will have nice sport, but I hope he will beware of catching cold. Water, water, plenty of water. Well, it was to be expected after such hours of dry weather and the finest harvest ever known. We are now putting all to turnips, cattle and sheep, and not a good time for hogs learning. However all the weather hogs at Wark, and I fancy they are about all there, have

---

89 Under the existing Elizabethan Poor Law parishes were responsible for the relief of their own poor. Poor rates, assessed on the income of occupiers of property in the parish, were raised by the vestry and administered by the overseer with the assistance of the churchwardens. The overseer was appointed by the justices.

learned to eat turnips well. William Brown's turnips are become the best we have again this year. Indeed Wark turnips are very good. Harry's are the worst, and ours where mildewed are not good, a strange growth upon the ground as I ever remember seeing it at almost Martinmas, and our dry grounds are absolutely still growing They were wasted before with the drought. The wheat has dropped again at London this week but nothing else, and now thrashing time is coming on it will be seen whether markets will come much down. I have no idea of much drop in anything except barley, which is confessedly the most abundant crop.

Saturday 6th November. Showery morning but not much, and some slight showers at different times today. Matty and I have been at Wark and Thornington, all well, a good deal of water on Thornington haugh. I saw water there and Andrew Tate attending at Wark watering. They have sown no wheat at Wark yet and the land is now too damp to sow. But a dry day or 2 would make it fit to sow. Andrew had tied up the cows which I dare say is right, but I think he had tied up the oxen rather sooner than needs. However we will let them be, only we advised the turning out of Cuddy Harvey's steers and a spayed quey, a nice one which you had ordered to be cast, and a light-weighted beast or 2. The Wark wether hogs are doing very well indeed. They are upon rape once eat at Jock's Rigg one part, and the other upon the rape and bad ruta baga in Park Hill, but are also well master of eating turnips. I wish we had the ewe hogs as well learned. The wethers are also doing very well, but we advised a change of part of them, that lying wet below the Jock's Rigg turnips. They are plowing the pease and beans land in the haughs. Harry has plenty of water to thrash with now, as well as for watering, his stock all looking well also. Last night we received a letter from you directed to my sister at Wark, so took it up to Wark this morning, and my sister was so kind as give it at home with me, so I shall now answer it and send away this answer tonight if I can so that you may get it on Monday at Darlington. But first I must tell you that we met young Willy Smith and young Ned going to the funeral of their Jamaican uncle's only child which died of the hooping cough on Wednesday evening last. 'Sic transit gloria mundi' is here litterally applied. Poor man I believe he was wound up in the child, and I fancy there can hardly be a more afflicted set than he and the two aunts Miss Moffatts. Well sir, my sister and Nelly and Jane are all well. We breakfasted with them this morning. I wish I could say so of my Nelly, but she is very indifferent at times and sometimes finely. But God's will be done. My sister at Wark is glad to hear of her son and Bessy being well, and that you keep well, and we are all also glad, and that the Cleasby folk are well. You should have gone and condoled with poor Tommy Charge. However it is not quite so bad as losing a sweetheart. But I don't like Kitty Mason *caking*, if he did *pledge* himself to take it. I would not wish a Culley to act so. But we will say no more until we hear both sides of the question. I should suppose six months will be the time for paying in the mortgage of Balston, when I trust we will be able to land a good *thrust*. By all means don't pinch the port if you want the *bubly jocks*, and by all means take care of yourself, and if you get a wet foot be sure you get dry stockings on. I am sorry that John still has most of his beans out. Tom Smith and many others have theirs out here. However beans are a thing that will bear a good deal if kept on foot. If we can't get any more [. . .]90 we must do as well as we can and be kind to those we [. . .]. I am glad to hear that John's turnips are so good. We are not to expect his wether sheep to be so good as ours. I do believe ours are strangely mended, but no sheep will mend more in the grazing time. But tell John that we will keep ours as long as we can, but it is impossible for us to keep them through winter. We have above 1600 hogs, which are more than we ever had. However they are good company. In the mind I am in we will send John a dozen pigs, I think they will travel along with the cattle. But I sincerely wish John to come and look over the cattle and pick out which he likes. He certainly must be the best judge. But we will send the pigs and keep the cattle until John can come. I am glad that Mr Peacock has remitted the money. You will see that we sent away 1000£ last Wednesday. Milliary fevers91 are also in this part. Mr Henderson of Ancroft has had 4 children in it but all recovered. We are in no hurry about the cattle going from hence to Denton, and you have no occasion to hurry yourself so long as you take care of yourself. I think barley will be more than 3*s* per bushell but perhaps the cheapest of any grain. William Brown and Matty got well home and the horse all well. Yours in haste affectionately

Geo Culley

PS. If John wants money we could spare him a 100£ or 2 by putting of the payment of our lime.

PS. Has John sold or likely to sell his wool? Not that I think there is any fear of its running in.

---

90 There is a small blot on the page here, affecting two lines of text.
91 A disease characterised by a rash resembling measles.

**152 George Culley, Eleanor Culley (Wark) and Matthew Culley jr
to Matthew Culley and John Welch**

Eastfield Sunday 7th November 1802

Well Mr Matt and Master John

It matters not, for I suppose you will both see it if both be at
Denton. This is a very fine morning, the most so of any we have had
since the commencement of these last rains, and as the glass has kept
ascending gradually during the time of the rains one may still flatter
themselves with the hopes of a few decent days, so that the beans
which are still out may be harvested, especially as the wind is now
westerly and west winds are the only and best chance for drying
beans. I need not recomend the making of windows or hollow air
holes through the bean hemmels, because you are both up to that,
and indeed in late seasons no time should be lost nor no contrivances
or exertions left short. 2 pieces of wood set up in the manner they are
done for teazles &c is perhaps best, and so build the beans around it
or them, or still better to be erected long ways like a low *roofed* house,
because *oblong* stacks are the best at all times perhaps. However in
late harvests they are undoubtedly so. I had forgot to say that I sent
off a long letter last night to the post office after seeing my brother's
directed to Wark. If more cattle can be got at a reasonable rate it may
be very well to buy, but I have no idea of buying things beyond par.
We are not in want of cattle. I find none to *draw* or fill up the ox
draughts, and to buy more too dear is absurdity, but still greater
absurdity to buy for feeding, because sheep always pay better for
turnipping than neat cattle, and we never had so many sheep I am
sure, and as turnips are not plentifull is it not wiser and better to give
what we have to our *own valuable sheep* than buy feeding cattle. The
matter is just here, we could take a few more cattle to consume and
turn the straw into dung, but had we not better be more liberal to
those we have than buy others over dear. Then let us rest contented
and thankfull that we are decently of for neat cattle, and abound with
the still more *valuable animals, sheep,* and those of our own kind which
will pay more for what they eat than any other I ever saw. And we
have plenty of *pigs* too thank God, which are at this time a *most valuable article.* And to keep you in good humour, if you are *good lads* we
will send you a *score* instead of a *dozen.* We had a consultation last
night and it was directed, after many learned arguments and consid-
erations pro and con, to send you of a lot in the beginning of this
week. We only want the approbation of William Brown and Harry
Rutherford, two of the privy council, who were not in the council
chamber last night. We had once determined not to send the pigs

until the cattle went, as we thought that a boy along with a man would do for both. But then we considered that the pigs would be wanted immediately for the bean stubbles, mill offals &c &c. But I hope after this Master John, that you will need no more pigs sent so far, but breed your own. It is absurd and ridiculous in the highest degree indeed not to breed a good many pigs upon a farm so extensive as you have at Denton now, and a mill besides in your own possession or occupation.

I have this morning received a most sensible and intelligent letter from Mr Proctor of Glammis dated the 5th inst. The contents are really important and are as follows. He says 'near Montrose (which by the bye is a very fine part of Scotland) last Wednesday I saw a number of fields of corn unladen, and in the counties of Aberdeen and Bamff I am informed by a servant just returned from the north, that there is a great deal of corn still to be cut down. The weather of late has been dreadfully wet here and our crops in general are very balking.[92] But the wheat nowhere clear of the mildew and the grain the poorest I ever saw. Nowhere half a crop except in the Carse of Gowry where I am told it is very good. But the farmers there are at no time *disposed* to *confess*, although it be otherwise. This puts me in mind of *Kinnear* your old friend who was not a good man, and if the rest be like him, they are not good or liberal people. One would think them a kind of people so illiberal as to look with contempt upon the rest of farmers. Much of our barley will not malt, being frosted and very inferior in quality. Nevertheless some fields very good. Our oats do not meal well. Mine only give 14 pecks to the boll when the last year they gave 18 pecks. The prices of grain are not yet fixed. Wheat has been sold at Dundee as high as 35s down to 30s the 4 Winchester bushells, which is very high considering *quality* and softness. 16s has been mentioned for barley of the best quality, and 15s for oats, and the boll of barley and oats are same as yours. But as yet very little grain has been thrashed. Our lean cattle pass all comprehension, and are advancing every day. Without joking they are in some instances sold higher than fat ones, not only one year olds but even calves are bought to market. Our very best fat sells for 10s per stone Amsterdam weight (this is 8s English). The black-faced sheep about 12 to 14 lb. per quarter have been sold as high as 1 guinea and a half per head to put to turnips, and I think it impossible that they can pay much'. I think we may infer from this very intelligent letter of Mr Proctor's

---

92 Off the straight line (Scots).

that neither lean nor fat stock can be sold low for some months, except the importation of live stock should affect the prices. But I have no idea that they will come from the Continent in any quantity. However it is a fact that about 40 were landed the last week at Holy Island. They were drove in there by stress of weather, being intended for Shields. I only heard of this the other day, but will know more about it on Thursday our high market at Wooler. But report says that many *thousands* are coming. I would have gone to the Island to have seen and learned more about them, but have not had leisure. They are said to be very pretty ones and in great condition. They are probably from the banks of the Elbe which is a fine country I suppose although cold. Holstein is at any rate famous for horses, witness Mr Stephenson's famous stallion.

Sunday noon. I find I am not weather wise, because I really *flattered* myself that it would be a fine day, but the wind these two last mornings, though west in the early part of the day, immediately on the day getting up whipped about again to the north and north east, and one heavy shower of hail after another have taken place which have covered not only Cheviot but some of the *nether* hills quite white. Alack a day for John's beans and many others now. But beans will bear a great deal if bound and kept on foot. But George Archbold has many pease still out which will have a bad chance. We were none of us at Berwick yesterday as the High Market was only on Wednesday last, but I hear that wheat was much down. However I would be sorry to sell wheat cowardly yet, for although the great breadth of corn country must be allowed to be in the south, still if a want is in the north of *England* and all *Scotland* it will be felt by and bye, especially if the wheat crop is not an average one, or rather below an average in the greatest part of the Island besides. Oats were well sold and I think will be, and barley was 18 to 20s our boll. Your amiable daughter is here and desires me to say with her love, that she wrote to you a few days ago but supposes that you had not received her letter when you wrote. James Laidler goes to Mr Gibson's today, and I desired him to enquire if you were at Brunton, because if you had I thought you would be glad of company home although an old servant. I had not received yours when I saw James. *Monday 8 November.* Last night came Mr Meggison of Leeds, the person who with Mr Charlton has bought our wool. So I am going with him to Thornington first to begin to weigh the wool. Matty is obliged to stay at home to wait on George Purvis about hiring a tup. He is not very well of a cold neither, and wool weighing is cold work this very wet season. I hear it rain while I am writing, which will put us to difficulties in packing of it.

But we must do the best way we can. I have got up at 5 o'clock to finish this letter as it is to be sent to post today, to prepare John Welch to meet about 20 shots (pigs) which we propose setting of for Denton tomorrow (Tuesday morning). They are to go no further than Mr Robson's of Humbleton on Tuesday evening, and where to afterwards I cannot say, because James Glass, who I suppose will go with them, must have discretionary power to lodge at nights wherever he finds it most convenient. But I should imagine as they are clever active young swine that they will travel for 12 to 16 &c miles a day, and if John could set Tom Glass to meet them or any other, it would set Jemmy the sooner back to take the oxen south, and we have no other so proper to drive as James Glass. However this must be left to John Welch's and your consideration and situation. I think this will reach you before you leave Denton, as it will go from Berwick tomorrow morning and will be at Darlington that afternoon, and you will either get it at the northern neat alias beast fair at Darlington or it will be forwarded to Piercebridge tomorrow night, where I am affraid that it may lie long enough. However it must take its chance, we can do no more. Nelly is still very poorly, I know not what to think about her poor thing. But God's will be done. Matty has only a slight cold I hope. Wishing you better weather your affectionate brother

Geo Culley

PS. I had almost forgot to say that the pigs must and will be sent the Corbridge road, so you and John Welch must advert to that and be sure you send to meet them, which I hope you will.

Wark November 8th

[*E.C.*] Dear Father

As you have not dated your letter, we cannot exactly tell what time you think of being at Brunton. However if all be well my Mother intends setting of for Brunton on Thursday first.

E. Culley

[*G.C.*] Dear Brother

We have received a letter from you today the 8th November, but yours has no date. We are at Thornington weighing Harry's wool, and he will set of for Denton according to your request tomorrow and we will send this letter by him. I am glad that John has bought cattle at Midlam Moor.

[*E.C.*] We are all well at Wark. E.C.

[*M.C. jr*] Tuesday morning 9th November. Harry is just setting off. I have given him 300£ for you to give Mr Peacock to pay of the money he borrowed to send to London, as we could spare it here and it is better to pay of the bank when we can. You should keep Harry as short a time at Denton as possible, as he comes very ill prepared and Martinmas is drawing nigh when all our accounts are to settle. The lad is just setting off with the pigs 21. Yours affectionately

Mattw Culley

Harry comes on horseback. You can send the Galloway by the pig driver, or if the pig driver bring the cattle that John Welch bought, Harry can bring the Galloway in his hand, which will be the cheapest way.

Matthew Culley junior

153.   *George Culley to John Welch and Matthew Culley*

Eastfield 10th November 1802

Well John

One of the Glasses set of yesterday with a drove of pigs for you, so I hope we shall hear some more on that head from Mr Matthew, but you will contrive to breed your own pigs in future. Harry Rutherford also set forward for you, but I have been engaged weighing wool these 2 days as Matty has been ill of a cold and this is the 3rd day. 'We make a long harvest of a little piece of corn'. Old Starforth of Durham would have done it all on a day easy. However this is a nice civil lad, he never touches the bundle, we weigh it ourselves. This seems a milder morning although it has been a severe frost in the night. But I think it is going to be a change as the glass is sinking and the wind blows from the west this morning. Our wool weighs better this year than the last so far a good deal. If you had as much rain as we had the last Wednesday and as severe a frost after it, your beans will be frosted I doubt. Mr Sayle writes that fat cattle (he heard) were bought at 7s per stone at Wakefield annual fair last Wednesday, and sheep not more than 8d per lb. sink the best. It is very extraordinary but very true, that both fat and lean stock are dearer in the north part of the Island than the south at this period. I want much for you to come over to fix what cattle you will take. We are quite in suspence on that account, and cannot tell what cattle to tie up or how to arrange matters. We have everything upon turnips now *old Martinmas*, but not before. Many of our wethers have been upon our young clover still and doing very well. I do believe. I hope the 300£ that we sent per Harry will lend you a hand to pay the bankers, you and Mr Peacock. Now I think you could not buy 30 cattle at Midlam Moor without borrowing at the bank, and I do not like to be under obligation to bankers if it can be avoided. But it can't always be avoided and they are necessary *evils*. But I had rather lodge money with them than borrow. But do not you suppose that by what I am saying that I would blame you for borrowing at a bank. By no means! I would rather have you borrow of a bank for a while than be in want of money, because to pinch yourself for money is the foolishest thing that can be in your situation. But I would likewise advise you not to give long credits. The seldomer the better, depend upon it, *pay as well and give as little credit* as possible is one of my rules in life, and I have always benefitted by it. And you as a servant can *press* people for payment much better than masters can. Not that I would wish you to be severe. Things of this sort must be modified according to circumstances. It would be as wrong to press a *good, honest* well *disposed* man for money as it would be *absurd* to give a known *rascal* credit. I only suggest these hints for your consideration. I know that your own good sense will apply them properly. All young men are better for being advised, and you are a young man to your master and I. If I had had any idea of Thomas Paul turning out so basely I certainly would not have given him so much credit, both to our loss and disgrace. I am affraid our giving him so much credit might lead others to do it. So cautious ought one to be. But I really thought far too well of him, and it only shews that short credits are best, because we are often deceived.

Thursday morning 11 November is a frosty morning and coming on of small snow I think. We finished our wool weighing yesterday at Grindon, which wool did not weigh so well as expected. Very pretty and clean, but the highlands hog wool was very light. The whole of our wool weighed 723 stone and amounts to 921£ 16s 6d, and the balance 471£ is to be remitted to us in good bills by the 1st of March. Wooler High Market was not so throng as I have seen it, although there were several lots of sheep and some cattle, with 8 cart loads of pigs selling at the most enormous prices I ever saw. Mr Joseph Mills extremely busy. He bought John Bell's dinmonds yesterday. In short he is litterally buying all the sheep and cattle, especially the former, that he can lie his hands upon. He is a most astonishing man. I will be bound for it he would buy ours, and I would have no objection provided he would pay ready money. But I shall not ask him, for fear. He thinks he can buy turnips at a low price, indeed according to present appearances there are not fat stock in the country to consume the turnips. But turnips are more apt to fall short in the spring than sheep. Most of the turnips about Crookham are sold at 6£ per acre, but Mr Thomas Reed's and many others are unsold, particularly westward. Poor George Reed of Marden yesterday had his arm broke and sadly smashed by the thrashing machine. We have not learned the particulars yet, but it is to be lamented that so many unfortunate accidents happen by these usefull engines. *Friday 12th* is a fine morning and not much frost. Mr Jobson and others were leading beans yesterday morning, but the wet afterwards put them of. This is the day of payment for the Grange estate at London, but by a letter the other day from Mr Thompson he will be well able to meet them if Mr Peacock has sent what he promised, of which I entertain no doubt, but it was not then comed to hand. I paid several of the Wark poor and Mrs Alice Smith's and John Lowry's interests, and will pay Mrs Bell's on Saturday. *Friday 12 o'clock.* After a shower it turned out a fine day, wind easterly. I have this moment received yours from Darlington dated the 6th at the beginning. It is not appear [*sic*] that Harry had arrived before you set the letter off, but now I recollect that he went on horseback and would not be at Darlington. I will answer this letter immediately and send it away, and perhaps it may reach you before you leave Denton. But Matty is not at home nor Ralph, the latter at Kelso. I intended to have gone with him as far as Wark this morning, but the wet morning hindered me, which is as well as this letter of yours is come and I can get to Wark tomorrow. I am glad that you have got cattle enough now I hope. Mr Eales's would have paid for his tups. Does he say nothing about them? We will set of 16 cattle for you, but you will find by the letter by Harry that one of the Glasses is going with the pigs. But we will send somebody with them. I hope John is well Kyloed now, and probably worth his buying. John should have an assistant. But when I pressed him on that head he said he expected to get Harry again, a man I was not fond of his having after insulting Dolly Welch so basely. However I wish he had one. I believe Laidler would have answered very well for John, but he now goes to Mr Gibson. It would appear that stock of all kinds is scarce. The cattle now at Holy Island from the Elb, they will not sell below 8s per stone we are told. Something like Scotch cattle but good enough beef, because Stephen Thompson killed one that was not thriving. They were sadly knocked about and bruised I understand. I would have sent Matty to have seen them, but after I heard how high they held them I prevented him. However more may come, and no one can tell as you say what turn things may take. We shall have no war now I think, at least not soon, but a great armament by sea and land must be kept up, which will help the consumption. I have often heard that Midlam Moor was a famous place for well fogged steers and queys, but if as Ben Sayle says they were sold at Wakefield at 7s per they would turn the penny badly. I hope the 300£ will help which Harry I trust would get safe with to Denton. We can put of our lime bills a while. It was well done to intercept those steers from Raby and

Langleydale &c &c. I shall be glad if you do remember to bring the account of stock from John, as it often stops my getting that done.

I thank you for your good advice to my daughter and your good moral reflections. Nelly is again better, and should attend to her health, but young people and old ones think very different. Crozier[93] has lived and will die a *brute* I am affraid. I am glad that John has got in a part of his beans, I think Colpitts must be a confessed animal! I am in haste yours affectionately

Geo Culley

154. *George Culley and Matthew Culley jr to John Welch*
Eastfield Friday 12th November 1802

Well John Welch

I shall now address myself to you, as I suppose that before this reach you my brother will be gone. I sent a letter away today in answer to one I received from my brother this day also. I addressed that to him as I hope it may reach Denton before he leaves. *Saturday 13.* Ralph sold 100 bolls oats yesterday to a merchant at 18*s* our boll, *pottato oats*, nevertheless a famous price I think. Wheat much down. He bought 37 bolls of particular nice plump red wheat (and white) of Ned Dixon Norham at 45*s* per boll. Barley same price as oats. He could do no business at Kelso. The Scotch market for oatmeal is quite down again at the present. *Sunday 14th November.* We received another letter this morning from my brother. I am affraid the one I sent away on Friday will not reach you before he and Harry left Denton. No matter, tomorrow morning we will send Jemmy Shillito's man away with 16 of the forwardest oxen or steers we have and the blue Kyloes also. I am affraid that you will not find the oxen so far advanced in beef as you could wish, but we can't help it. Some of Ralph Compton's steers are full fatter, but we thought them such nice as will lie out, so full of hair and never have been housed, that we thought we would lie them out here and send them to you in the spring. If we have done wrong it is not designedly nor without consideration. But John I must put you in mind to omit no opportunity of selling either cattle or sheep when an opportunity of good market offers, because you may depend on it that we have far more fat stock than we can possibly winter here, be the winter ever so favorable. We have 2600 sheep that necessarily must have turnips. That is fat *sheep, tups* and *hogs*, besides all our holding stock, and

---

93 Crosier Surtees. See No. 226.

nearly 60 fat cattle and I know not how many cows. However we have plenty of meat at present, indeed a part of our wethers only went on turnips the other day. Old Hallow fair at *Calton*, Edinburgh, was still very dear for lean Kyloes, but the forward things were not so ready sale. A great number of cattle shown too. Mr William Pickering being at the Roads, he and Matty Culley, James Darling and some of the Smiths are gone down today to Holy Island to see the *German* cattle. Your master writes that butter firkins have got an advance, but does not say what nor present prices. Will thank you to say the prices, and whether we should sell now or try longer. I am glad that you have got your beans in at last. Perhaps you could have done without the pigs now. But my brother made such a rout about them the last year that we durst not but send this, and to have sent only 10 or 12 in my opinion was not worth a man's while going all the way with them. I find that my brother now wishes we had sent a sow also in pig, but we could not divine that and he did not order us. But surely you will in future contrive to keep a sufficiency of pigs. Too many become a nuisance, but they certainly have their use in a reasonable degree, and where you have a mill to grind any refuse barley &c it is not prudent to be without some. However if we have sent too many you can always sell a few of them. It is said here that pork has dropped 6s per stone, but I think it will be up again. I hope you have sent us an account of your stock, dead and alive, by my brother. The mule can be considered about. Mr Cradock's quey will also come in this drove, which with the 2 Kyloes will make 19 to the score. You must write or send word somehow to Mr Cradock that the quey he bought of my son is arrived, and I suppose he will pay you for her the first time he sees you, 20£ or 21£, I know not whether is the price, but he will give you what is right.

[*M.C. jr*] Mr Cradock's quey is to be 20 guineas and 5s for driving her to Denton. He will send for her and pay you for her. You will easily know her, as she is the only open quey, she is a thick fat red on white quey, mostly red, and the fattest of the lot. I am just returned from seeing the Dutch cattle. They are really ripe runts, or as like them as possible, mostly black and white or mouse colour and white. There are 36 of them come for Morpeth market on Wednesday first, and then to the horse fair if not sold at Morpeth. They are none of them fat, and seem of a bad sort, have never started on turnips in their lives.

[*G.C.*] *Monday 15th November.* Now John we have this day sent away for you 14 of the best oxen we could pick, a large white spayed quey which we think would not do well to lie out, and George

Humble's blue cow which makes the 16 you wrote for, 2 small Kyloes which my brother wrote for us to send you makes 18, and the quey in calf for Mr Cradock, which makes 19 in all. Wishing them a good journey I am yours in haste

Geo Culley

PS. John I will send this to the post by William Brown who is here from Grindon.

PS .We send these cattle by Jemmy Shillito's man until he meet Willy Glass who will exchange droves, Glass returning to Denton and the other will bring the steers for straw here. The weather is mild and growing, but now with slight of showers which prevents us from getting our beans and pease land sown with wheat. I think the wind is got rather more south and is fresher, as the snow is wasting of the skirts of Cheviot where there was a very great cover of snow as I have often seen early on. Some evil-disposed rascals have destroyed Ralph Anderson's 3 ploughs last night in that field next to our General's Close, cut the beams through with a very sharp axe and broke the mettle mould boards to pieces, and wrote on one of the pieces to take care of the stackyard as it was in danger. These are dangerous villains, however I would like to make them out and punish them. G.C.

155.   *Matthew and George Culley to John Welch*

Wark 19th November 1802

John Welch

We got well home on the Wednesday from Halton, left Mr Bates well and the Misses Hoppers. I staid at Black Hadley all night. They pull many turnips off their wet ground, cut of the tops, which they give to the stock, lay up in houses where they keep well for several weeks and are not hurt by frosts as in the fields in a hard frost, so that the cattle eat them well. I think it's a good way, as then the land may be plowed and sown with wheat when a fit of good weather comes. This is a soft morning and no wheat sown at Wark although much plowed ready for sowing. I hope you would get the pigs safe, but the sending you a score of pigs was not thought off. We have 1 which we think of sendin[g] in a tup cart. Our old sows are unwieldy so as not fit to be sent, we send her to Grindon to an excellent boar if not breeding them of too great a size. We have 3 litters of nice ones, intend keeping some sows from them for you and selves. One of the pigs looked ill. I thought you should if wanted get some troughs from Gatherley Moor quarry. The cattle I met on your side Thrunton planting, they had come too far for the second day (the steers got to Eastfield yesterday, a little footsore I fear, the weather so wett would shear their feet). My brother who came on Thursday says they sent a few nice oxen, 2 Kyloes and George Humble's quey, a grey one injured the least by the fire off any of the poor animals by fire, took off Jinny's hand at prime cost (a quey of her first calf) 15£ big with calf when George got her of Jemmy Bruce of Kelso, is seemingly a decent quey. He bought one in lieu of her to milk, far short in milking of her. Our cattle &c should always lodge at Cambo as Mr and Mrs Armstrong use us as well as any on the road, have hay and grass, charged 4d a piece, 6d at Glanton for the steers. 12 pigs were a new to send you, but you will have to grind them some oats or barley meal if wanted for them. There is a great want of a fold at the straw house and stable door for pigs and cattle, at least a wall on the west side, and a gate way for the carts going to High House. Should a pig or hen house be built on the wall next the Unthank north side within the fold, the hen house above or both single houses? These things to be thought off as well as many others. If Mr Colling's sow eat chickens &c, should when her place can be supplyed be disposed off. We should contrive to send you sows in pigg as they are usefull when not too many are kept.

(Friday) It continues to wet. Ralph Brown has not come for Kelso. I think we will cut off all the turnip tops which our housed fatt cattle get. They are worst to cart in wet weather but they dirty sore in wet days, so a quantity should perhaps be cut in dry weather and led when convenient for the cattle. Rain may then fall on the heaps and wash off some of the soil so as to clean them a little. I fear you will have more work cut out than you can accomplish this season. Do those things which appear to you most needfull, only the waterway I prefer where proper. The ditch from Lime Kiln Bank well should [be][94] scoured out or well dragged, the willow opened and that part planted south of the race, the new barn in Long Burn and the hedge up the cow pasture and well scours, some stops in the burn, and the spare water used in the tenement on the grass land in floods, and on Sundays from Saturday evening to Monday morning, the swallow under Hall bridge well taken out when dry and filled with earth. You must shovel off the stones down to the quarry, then you'll be where the water gets on. Quicks are now 20s per 1000, Mr Dixon of Hawick[95]

---

94 A word seems to have been omitted.

95 Messrs Dixon of Hawick did a large business in shrubs, fruit trees and forest trees in southern Scotland and the north of England: Douglas, *General View of the Agriculture in the Counties of Roxburgh and Selkirk*, pp. 227–8. They also had a nursery near Perth: Culley, *Travel Journals*, pp. 92–3.

said their selling price. Gather plenty of haws, some acorns, ash keys, ask Mr Harrrison's (send Bob) leave to gather larch &c cones of the older trees above 10 years old, Scotch fir cones and their seeds should be sown in drills in your orchard or Plain Tree Garth, only they take more kindly on rich land than poorer, yet are they more likely to thrive and grow on rich soil. I wish my brother would send you the mule as a common hack, as it would save your mare and be of great use I suppose to you. It is master of your weight. I was at Thornington, the watered land looking well, the cattle graising where no water has yet been on in the Hind's bog, wonderfull fresh where the land has been limed, the land is very fresh. My daughter wishes to have Winstone's grey Galloway. If the mule be sent to you it can be sent by the man who brings the mule to you. Mr Forster has oats out upon Downham. The weather has been far more mirey here than with you I suppose. I am with respect &c

Mattw Culley

PS. Do gather haws &c.

[**G.C**] *Eastfield 22nd November 1802*. My brother sent me this long letter yesterday when I had wrote one to you. However as we can't send both I will give you the heads of mine and will send this after we hear how this day's market proved with you. The cattle from Germany are gone to be sold at Newcastle store fair this day. They began to try to sell them at Belford by the scummer but bought them in again as they did not like the prices. A son of Jim Parker's of Sunderland has the management of them. But I find by Saturday's Newcastle paper that many more are come over to Sunderland, Shields &c &c. What effects they may have upon the cattle markets is not easy to calculate at present, but I should not wonder if they have some little effect at the first. Mr Joe Mills is buying all before him, cattle and sheep. I met 8 capital steers he had bought of Mr Alder of Horncliff on Saturday as I went to Berwick. I am told 27 or 28£ per, a lot of ewes 55s per which did not appear to me to be above 18 lb. per quarter. Since his buying those 500 Irish cattle at Brough Hill he must have laid out many 1000 pounds. Mr Spours tells me that skins such as our shearings are worth near 6s per, and that skins are going to be as dear as last year. Have you heard of this advance John? And it is a fact that Mr Barf wrote to William Spours to buy him all the wool he could and ours also, but ours was gone. However I would have you to be upon your guard with your piece of wool and to be in no hurry until you inform yourself. As the idea of war is over for the present wool will advance, and pork and butter drop in my opinion, and pork is dropped a little here but tallow is high, 7s per stone, and if skins get to last year's price it will make the offal of old ewes and shearings valuable. Our Brankstones from Wooler have bought Mr Bell's skinnery &c at Haughton, they are very rich people. The 15 steers got well here on Wednesday night, one a little lame and several footsore, but this wet season is bad for cattle travelling. I would recommend to you to put the cattle we have sent you into a grass field for a few days to recover their feet before you tie them up. We have done so with the steers, and you would be surprised how they have come about. To tie up heavy cattle with sore feet would be very wrong. You know we are under the necessity of putting soft dung a little under their forefeet during the winter frequently or they would be apt to fall lame. Dunse fair was exceedingly dear. It is said that Mr Mills sold some of his Irish cattle there and others, about 40 in all, very well. You see how the tables are turned. Instead of sending south we now send them north for better markets. John you ought to have sent the account of your stock, your turnips &c by my brother. It will stop me much from getting that account done, and this is the very best day for taking it in. I have got all but yours again. Be so kind as send it per post as soon as you can after receiving this. I am affraid you would not recollect to send it by William Glass. We did not take in the 19 cattle sent to you, nor the pigs, but will take in these steers here if we find you have not. At Berwick wheat very flat 2£ 2s about the best price and down to 1£ 1s. You can form no idea how bad we have some mildewed wheat. Barley 19s to 1£ 1s, oats 14 to 16s. The millers are the best customers for barley, as they have a pretty brisk demand for milled or shelled barley just now. The maltsters and merchants are very cool about barley. We had a prodigious wet day on Friday, indeed the weather has been very wet lately. I hope at all events after your markets for fat are over about Christmas that you will make a pass over. I want very much for you to see our fat stock. I should wish you [. . .][96] think of no one for an assistant to you. You should have one surely. I believe that Laidler would have answered very well for you, but my brother had a dislike to him. He is now hind to a Mr Gibson near Stagshaw Bank. That rascal Jolly is now in Ireland and has been swindling money from Kerr Richardson a correspondent of mine. Calls himself my nephew, a pretty relation truly! It hurts me very much nevertheless that this villain should have it in his power to raise contributions upon the public by pretending to be my relation and brought up with me &c &c, and I never saw the outcast that I know of.

---

96 The paper is torn here.

Tuesday 23rd half past six o'clock. It is very dark, and not one star to see, but the wind blows from the south and the east I think, and it does not appear to have been any rain last night. I hope we shall have the pleasure of a letter from you in a day or 2 with an account of Martin market &c, and an account of your stock &c. Ralph sold William Marshall Kelso 4 pigs yesterday at nearly 8*s* per stone we thought. *Wednesday 24th*. This is surprising *wet warm* weather for the season surely. Were it April or May how acceptable it would be, but God's will be done. It is a bad start for turnip sheep, especially those upon turnips. Ours are many of them on stubble, and although dry land they are dirtier than one would wish. It is also particularly bad for rotten or tainted ewes &c. However it is much better than frost and then fresh changeable weather. But we shall have a change soon I think, as I perceive the wind is got north again or north east, and winter must be on soon. In my opinion turnips have improved since the rain, especially those on soily[97] land and not too early sown, and grass grows on dry land. *Thursday 25th*. Wet all day. Much disappointed John at not hearing from you how Martinmas day was, and we want an account of your stock. No account of William Glass yet. *Friday 26*. Still rain and more rain. Good sheep well sold at Morpeth, a bad and thin show of stock at the Newcastle stores I am told. *Sunday 28*. We are a good deal disappointed at not hearing from you John about Martinmas day, and no account of William Glass. We have put of sending this still because we so fully expected to hear from you which we would answer in this. What is the matter with Mr Davison of Greenfield near Croft that he is selling all off? It is a slight frost this morning and appears to have been a fair night for the first time of a long time. Wheat markets have got very low, oats stand best, barley is flat. Ralph sold beans and pease (Blandings) at Kelso at 35*s* our boll, which is almost the average price of wheat now at Berwick.

Sunday 2 o'clock. I am happy that we have at last received Nelly's Galloway and your letter. I am glad that you approve of cattle sent. I wished them better but we sent best we had. It is a mistake of your country folks to alter Martinmas day or any other fair in my opinion. It is certain to breed confusion. I wish butchers meat may be as well sold as you suppose. Certainly offals are very high. I think you may get your price for wool provided we have not war. I am glad that you are to lead a part turnips of from Mr Harrison, certainly your burnt land ones will keep the best of any. It is impossible to sow bean land

---

97 Deep, heavy.

this very wet season. This was a sweet-like morning but now rain again and like rain. But be plowing your bean land by all means to be ready. It will be better for being busted. Pork has dropped 6s per stone here. Is Robert Usher safe to trust? Your price is good for the Kyloes. I think little of the Hambro[98] cattle, only the quantity will have an effect, and some cargoes are come in at Holy Island. We will be glad to see you when you come. John we think that you must have more than 189 wethers unless you have sold more than we have heard of, as you got 170 from Mr Wood and 160 of our own hill hogs which makes 330! You will consider that and write directly. I send this today to save post if possible. I am in haste yours

Geo Culley

156.   *George Culley to John Welch*

Eastfield Sunday 28 November 1802

Well John

I have just sent of a letter in answer to yours by William Glass. This was as fine a morning as could be, but began to rain again about half past one o'clock and still continues now 4 o'clock. It is a strange time for sheep upon turnips. I desire that you will lead turnips of to your sheep at Walworth if you have a right. If not you must give our cast sheep to Mr Harrison and tell him that we desire that he will either alow you to lead of or to take them of the turnips altogether, as they will be ruined this wet weather in eating turnips upon the land, and Walworth is not very dry land. When you give such prices you ought to have leave to lead off or eat on either. I am far from blaming you for buying the turnips, but really to eat upon turnips this strange season is ruinous altogether, especially as the land is not dry of itself. But no land is dry at present. After all John I would not advise you to be over stiff with your wool. One often suffers by being too stiff in selling. There is reason in all things. You must know more about Robert Usher than I do, but I cannot help cautioning you against selling in trust. 40£ or 50£ lost takes a good deal of making up. I hope they will tire of buying so many cattle from the Continent, but they may hurt a little in the mean time. Mr Mills is no rule to anybody. He has very seldom been right in his speculations. We are thinking of selling a few sheep at Morpeth soon, but as you advise to wait still, we shall do so. But I understand that good sheep are very scarce and well sold at Morpeth at this time. I fancy you have also forgot to include your

---

98 Hamburg.

rape. We always put rape, turnips and potatoes in altogether under one head in all our farms, and it is very much if you have not missed some wethers, I hope 100. *Monday 29.* This has proved a better day than I expected, but by a fresh cover of snow upon Cheviot and the glasses going up, with a north east wind, I fancy we are going to get frosty weather. My brother is just gone home. By a letter from Mr Peacock today I am sorry to find that Mrs Peacock is again danger-ously ill. It is a pity but you had told him of Glass's being [at]⁹⁹ your house, because the letter might just as well have come by him, because as it contained an inclosed account it is charged *treble* postage, although we need only pay double postage neither. Indeed it would always be right of you to acquaint Mr Peacock when you have a conveyance of that sort, as he frequently has to write to us. *Tuesday 30th.* I met my brother today at Longknow to set out someone to plow about the herd's house and to make an exchange of commons between Mr Morton and other land near to Longknow house, but it proved a very wet day indeed. *Wednesday 1st December.* A hoar frost which makes the sheep eat their turnips with avidity. *Thursday December 2.* A very hard frost all day. *Friday 2nd* [3rd] Fresh again but a very cold wind from east south east, dry all day but very gloomy and dark, and has been rain in the night. It seems that the Morpeth markets are very good for good stock, especially good sheep. We are half tempted to be trying a few or to sell them at home to Mr Mills, for ready money though I assure you. He was at Wooler yesterday as keen as ever, and Clem Stephenson and William Spours and the whole of them. But shall wait until you come over now as you desire it, and I hope you will not be long now of coming. I believe there are plenty of moderate stuff shewn at Morpeth, both sheep and cattle, but not many good ones, which makes the better chance for good ones. At the same time I really believe that there are not any great number of fat stock laid on in this country this autumn, because although turnips were not supposed to be plentifull here at all, still a great many remain unsold. One thing indeed is that they did ask great prices, 8£ per acre, but are now comed to 6£, and if the winter should not prove severe they will come more down yet. *Saturday 4th December.* A very wet morning, has rained all night but cleared by about 10 o'clock when Ralph Brown set of for Berwick. Corn of all kinds low at Berwick. *Sunday* a very hard frost.

Monday 6th a fine west wind and likely to be a gentle thaw. John

---

99 A word seems to have been omitted.

you must enquire if Lord Darlington and Chris Eales's tups are done, and if so fetch them home to Denton. On reading over your letter again I find that I was mistaken about Mr Harrison's turnips, and that you have a right to lead them of. John you behave not well in not saying what rape you had, and if you were wrong about your wethers or not. It prevents me from getting our annual account of stock done. *Tuesday 7th*. A good day. *Wednesday 8th*. A particular fine day. We have been sowing a piece of wheat in the Watery Close with the machine. I like this month the worst of any winter months to sow in, but we ought to finish the dry part of that field. They got to sow at Wark. *Thursday 9th December 1802*. John as I have waited so long for an answer from you to my last, I am determined to send this by post today. I hope you are coming yourself or you would have wrote sooner, because by not knowing how many acres of rape you have, and my hoping that you have forgot a part of your wethers, has prevented me from getting the annual account of stock done. So I must beg that as soon as you receive this that you will immediately write me an exact account of your wethers and acres of sheep [*sic*: ?rape], except you be coming which I would fain hope is the case. But then I think you would have dropped us a line saying when we might expect you. By a letter this moment from Mr Vipan from Norfolk he surprises me by saying that they are still forced to drive cattle several miles to water, and fetching it for the house uses many miles in carts. They have had so little rain yet. Turnips are much mended, will be usefull for store stock he says, but will not do for fat cattle. Wheat 7s, oats 2/6, barley 3, beans 4, and says these are the top prices. He also adds that they have had a wonderfull advance in both lean and fat stock since Woolpit fair, and that he now expects things to be very dear, beef 9s and upwards, mutton 9d per pound sinking offal in both, and they are expecting them higher, and he says at the rate that they now buy lean stock they must be higher or they will lose by grazing. This another very fine morning, which makes me hope that dryer weather is coming. Pray write directly, which will great[ly] oblige your friend and well wisher

Geo Culley

PS. It is surprising that they have had so little rain in the south east parts of this Island, and we so very much.

157.   *George and Matthew Culley to John Welch*

Eastfield 10th December 1802

John

I waited until yesterday (Thursday) morning in full expectation of a letter from you, but being disappointed I then sent the one I had in hand away per post. However this afternoon brought me yours of 5th inst. which was Sunday last, but I suppose you would send it from Darlington on the Monday as you have been so kind as give me an account of that day's market. But if it was put into the post at Darlington last Monday it ought at any rate to have been at Berwick on Tuesday evening and with us on Wednesday morning. Now how it happened to come only today, 2 days later, you may be able to explain, but I cannot. It has no later date than Sunday (5th), but it could not come before Monday or you could not have given an account of the market. I thank you for an explanation of your *wethers*, but you have still forgot to name the quantity of rape. This is *all* for want of examination. You are much cleverer than I am, but you certainly have much to do, and I can excuse your mistake about the rape but not the *wethers*, because that was Tom Glass's business not yours. He is the person that ought to give you a proper account of the stock, especially the sheep, and to miss 100 in less than 300 shows him *stupid* to the last degree. However it was the better way to mistake surely. Your slip of paper cost double postage, and it is of no consequence to me, but I am sure you did not expect it or you would not have done it. Poor unfortunate Mr Wood is said to be a dying now for certain, as nothing rests upon his stomach. I am glad you have got your bean land plowed. It is then ready for a season when it pleases God to send one. I am also glad that your cattle are doing so well, and Mr Wood's sheep at Walworth. But I wish sore for you over to consider what is to be done with our wethers, as we cannot possibly keep them long. But we will put off until you come now I think. And I believe as much as you do, that markets will not be worse and may be better. But then we must either be selling a part or buy meat, which last I am not fond of. Or here are plenty of famous turnips to buy at present. If I was as young and active as I have been, I would have no great objection to buy a few turnips for our wethers, as they are growing apace I do believe, but I think it wrong to hazard so much. You and us have 750 wethers besides 300 ewes feeding. Now it is too many to *game* at I think. If you could have quitted Mr Wood's, then we could have sent you 200 in their stead, and then 200 more after that, and either market the ewes or send a part to you, or write to Mr Lincoln to come and buy a lot or your friend George &c &c. However

I hope you will not now be long of coming, when we must consider and reconsider the best way we can. Young Wood said that on our writing to him he would come and buy our wethers at any time. You are certainly right in giving your tied up cattle room, those foreign cattle will have some little effect. Pork has dropped considerably here just as it is getting up again at London. One could hope it would advance here also. My brother would make us send you pigs whether we would or not, and once he gets a thing into his head it must be done or no peace in Israel. We are to send you a sow in pig also when the tup carts go to Denton. We would gladly have had him allow you to buy one of Mr Wright Cleasby or someone else, but we could not prevail. Davison of Oxensfield must have some I believe in his hand. Mr Peacock wrote me word that he and Mr Harrison had parted. I am sorry for Mrs Peacock and I think he will have to regret her loss much. In eating turnips 5 drills are too few to take away 7 or 8 are better taken, leave as few as you like, because a cart cannot turn well upon less than 7 or 8 drills. I like rails much better than nets, indeed nets are the worst and most expensive things that are used in my opinion, seldom last more than one year. You will see our rails when you come. We think them the best solution. Did those south jobbers buy anything? Being there is nothing without buying. L. Gibson had gone out of curiosity. I am told that nice cutting were sold at Morpeth last Wednesday for 8s per stone sink, midling 7/6, sheep not so good as the week before, 8d per lb. sink about the best price. But there are such heaps of trash go to Morpeth at this season, and I again repeat that I am very much disposed to think that sheep will be well sold and very well sold by and bye. But the misfortune is that we can't do every-thing, and some things take place that we don't expect or foresee, *naked frosts* to destroy turnips, or even deep or sudden falls of *snow* &c &c. Better and wiser to be doing a part. Besides, few people are fond of breaking up at this season, but a naked severe frost would compell many to sell, and perhaps glut the market and so sink prices 1d per lb. or more at once, which take some time to advance again. Such things have been and may be again. At the same time I am convinced, as far as probabillity can convince one, that fat stock will or have a chance to be dearer in the months of April, May and June than we have yet known, Nevertheless strange checks take place, and unforeseen events. Who would have expected live cattle to have been imported? It was never allowed before, and if things go any higher Government will be importing them themselves or offer high premiums. No one can tell or foresee what may take place, and we can well afford to sell at these prices which is always prudent to do, at least in part.

Saturday 11 has proved a very fine drying day, but cold and like frost but seasonable. *Sunday 12.* A very fine morning indeed, a slight frost but temperate and pleasant, I dare say many people will be sowing wheat upon early eat of turnips, bean stubbles &c &c. Not but I consider this as the deadest time for vegetation of any others in the year. However vegetation has never ceased this fine autumn, as we have had no frost to call frost yet. But the danger is that frosts, perhaps naked frosts, may take place, which stops all vegetation, and then the grass is apt to founder altogether. But when the spring sets in every day gets longer, and the sun has every day more potency. I always think that if the grain only germinates, even within the ground, it will then bear a good deal. I well remember that in the autumn 1778 we sowed wheat which never appeared above ground until February, but it was in our kind soft soily loams at Fenton. It became a wonderfull crop. Ralph Brown sold a few cartloads of potato oats to Clunie and Home yesterday at 18*s*, which is full more than the price of the best barley and more than oats are worth at Kelso. But I understand they are buying these in speculation, as they expect a good many will be wanted in seed time to be shipped to different parts of England &c where they are still unknown. For to a certainty they are the most valuable oat I ever saw, inasmuch as they are early, yet stook like late oats, they are the nicer than most early oats except the red Scotch oat. They make more meal than any other oat I ever knew, and they are short and beautifull without tails. John you write the word oats thus *ots*, which should be wrote oats. You will excuse me this, for I wish you to spell well. Wheat is still very flat and likely so to be for a time, and as their crops are so good and abundant in the south it may be below you the summer or even all the year. It is from 1 to 2£, but mostly from 36 to 40*s* per boll. *Monday 13th* has proved a nice drying day. The land has got into wonderfull order. They begin to sow wheat today at Wark for the first time in nearly the middle of December. Ralph Brown was at Wark selling a cow that [my][100] brother bought of Andrew Bolton to William Marshall. She has lost the use of her limbs. William Marshall bid him 25£ for a cow bought of Ralph Compton along with the lot of cattle which we got of him. I think a fair price, at least she will pay us well at that price, but Ralph would not take it. Several people were sowing wheat today. *Tuesday 14th* a most promising morning, although a little frosty it is

---

100 A word seems to have been omitted.

temperate and not cold. How long this very good weather may continue I know not, but I know that we ought to be thankfull just so long as it pleases the Almighty to give it to us. It is now as good as ever it was bad, and I think one may perceive fat (outlaid) stock to improve almost every day, Besides, although turnips are not large nor a great crop, yet they seem to go far, and not eat so fast as one might have expected. But people were longer of starting with turnips this year than usual I think, and I would gladly hope that the fresh tops they have put, and the growing state they are in, will or may make them preserve the better from rotting, particularly if this fine season should continue a while longer. One thing which made people longer of putting to turnips was the great quantity of grass in the autumn, and the open growing although wet season. Indeed many people have their ewe hogs still at grass. What a pleasure it is to see the poor sheep eat their turnips so clean and comfortably now [compared][101] to what they did a while ago when the weather was so wet and the meat so uncommonly dirty. *Wednesday 15th* a most pleasant morning. Not the least frost, although a thin cover of snow has fallen on Cheviot the last night, which certainly denotes frost in the air. G.C.

[*M.C.*] Well John. Marshall bid 26£ 5s for Ralph Compton's cow, 66 to 70 stone, 2 years old and calf in her. We should sell her. The cow Marshall bought was clyered very sore. We are busy sowing wheat, the land easy to harrow, but I only let them give Willow Bay one single and then a cross or stout single, as I suppose it a bad way to harrow land to tread it much when of a strong nature. We should sell some stock, ewes especially, as they (I believe) gather no weight or little after Christmas, and not found out for run too great risks. Moderate profitts are best I believe. We have lost nearly as much by too much speculation as moderate speculation. Why be over anxious for very high prices, man's judgement is fallible. We have one harrow, rather 3 harrows with a long swing tree cover a ridge, the horses go in the furrows. It requires strong horses to draw it, soft land should not be poached and made so as to prevent the roots shooting fresh. New piping land is better from being fixed by treading. I am badly off for a forman, Andrew Bolton does all he can and is honest, but is only fit to follow, or be under such a man as George Humble was. George was clever, very clever, and so just and honest, but you knew him well. I have got a deed ready and have little doubt Jenny will sign it, that

---

101 A word seems to have been omitted.

will secure 100£ to the children (if I stand good) and her 5£ a year for her life. Your obliged

Mattw Culley

[**G.C.**] *Saturday 18th* was frosty in the morning but soon changed to fresh, and in the evening fell a good deal of rain. Ralph sold oats still at 18*s*, potato ones, but every other grain exceedingly flat indeed. Barley scarcely so high as oats, and wheat from 20 to 36*s*, very nice up to 40*s*. At Morpeth the last Wednesday I understand that some nice little West Country cows were sold at 8/6 per sink, many at 8*s*. It is said that the shipping butchers will give 8s for good shipping oxen. I believe sheep were more plentifull and not so saleable, except very good ones which sold as well as ever. *Sunday 19th*. A charming day, a fine fresh wind from the west. Monday 20th. A most delightfull morning, still a fine fresh wind from the west. It will be getting people on a-sowing wheat again. It begins now to put the winter of a good deal, and the poor sheep get their backs dry and have not to wade so much among moisture. Wednesday 22nd a frosty day, but we have had some very good weather lately, only it now looks very like settled frost.

Thursday 23rd. Well John we have received yours of Tuesday last the 21st duly this morning, which should always be the case and which also proves that your last letter had been neglected somehow by the post people, because I have no doubt but you put it into the post yourself. We propose the carts for the tups to set of from hence this day week (Thursday) to be at your place on Saturday Jan 1st, to rest all Sunday and go to Thirsk on Monday where we expect to meet Mr Taylor's &c tups, as we have received a letter from Mr John Taylor to that purpose. We would wish you to send one of your men to Thirsk with fresh horses and let our horses rest, as ours will be sore loaden back here again and our mare will work at anything for you until the other return. As you can hardly be very busy at this time we think it will be the best to do thus, and as our own man is a stranger yours will likely know the road to Thirsk better. They meet at Mr Casses. Mr John Taylor has unfortunately said *Monday 4th*, but Monday will only be the 3rd of January. However we have little doubt but he means Monday at any rate, and to prevent any mistake we have wrote to Mr Taylor again today. But at any rate if they are not there on the Monday night, your own man must wait while Tuesday. We also expect Mr Buckley's man at the same time at Thirsk to meet his tup. We have received a letter from Mr Buckley's man today, and shall write to him again this day to meet us at Thirsk. We think that you had better send us 3 bushells of rape seed, as both Wark and

Grindon people propose sowing a good deal of rape, and I am certain that you will endeavor to have all the other things as well as rape seed ready by Tuesday night for our man to bring with him, viz. iron axles, brooms and besoms &c &c, and be sure you ask Mr Peacock if he has anything to send. We are rejoiced to hear that poor Mrs P. is again likely to get better. I hope by this time that you have dealt with them for your old and new barley. Remember Darlington is an easy delivery compared, and too hard selling is always wrong. I would give a good deal to cure you of that complaint, but your own good sense will cure you in time, a man that means well will leave of all bad methods in time. We always put in all the turnips and rape we grow, whether eat at the time or not, as well as potatoes. We will be very glad to see you when you can come. You must chuse your time, only we wish to be selling some of our fat sheep, and we believe our markets are full as good as yours yet, although I think as London market is mended much that Wakefield must follow, and we don't wish to begin to sell until we see you, that we may have a thorough consideration. As you have got your wheat mostly sown I hope you will be able to stay two or 3 days longer with us when you come. All the early sown turnips have run in all countries, and more especially if the seed be of a bad kind. I am certain that ours except on above have grown a great deal all along. I am glad that you have sold your wool. I always like to hear of your selling. It is a capital price. You are certainly right in consuming barley straw at this season, and you may assure yourself that no straw is so proper for turnip cattle as barley straw. Remember I say so! All markets are full of bad fat stock always at this season, and that makes good fat stock sell so much better in proportion. But the bad should begin to fall short soon I hope. The reason why I had not sent you this letter was that we were expecting you every day. I was quite against sending you pigs, but my brother had fought about it for more than a year, It was right to sell your weak Kyloes. I am sorry for poor Mally Wade. I think the Shields butchers will want shipping cattle much soon if the season keep moderate. My brother has got only part wheat sown I believe. I think this by much the worst and most dangerous month in the year to sow wheat in, still it is right according to circumstances. I hope you will not forget to send your saddle and the saw. I am in haste yours

Geo Culley

### 158.	*George Culley to John Welch*

Eastfield 24th December 1802

Well John

I sent you from Wooler yesterday a full letter in answer to one we received from you some time put in the post at Darlington on the Tuesday by your shepherd, I think you said. Now it is plain that there had been another letter sooner here, respecting the letter next before the one received yesterday, because that put in at the Monday by yourself did not reach our post until the Tuesday morning whereas we received the one put in by T. Glass yesterday (Thursday). I generally leave your letters open until I get to Wark, that I may give you any information I receive there respecting Morpeth &c &c. But we had stupidly and inconsiderately sealed it before I went from home, which prevented me saying that large shipping cattle were much requested by the Shields butchers, and at about 8s per stone, the little West Country cows 8 to 8s/3 and 8/6. I had this from Mr Mills, who says fat cattle are very scarce in this country, and will be high sold. By the bye, he says that the reason of these little West Country cows selling so well is that their hides are sold so uncommonly well at this time, about 6s per stone, and many of their hides are from 7 to 9 stone, while the cow may not weigh above 45 stone. Sheep were not near so well sold, such a deal of trash still in the market. He says one would be surprised where they all come from. Best sheep hardly 8 ½ per lb. sink, and although that best mutton sells readily enough at 7d at Newcastle the other is sold at 5d down to 4d and even lower. But I also received a letter yesterday at Wooler from Mr Wood of Hillington near Easingwood, who wanted to bought [*sic*] our wethers so much when he bought some of Mr Gregson's &c before Ninian fair, and he now wishes very much to come and buy them as they have turnips so very plentifull he says. Now a penny for a counselor. I am myself perfectly inclined to invite him over and sell him a pretty good lot of both wethers and ewes, if he will buy the latter, but I would give rather more than 1d for your advice at this moment. I shall see my brother today and consult with him what is best to be done, but as I have all along said to you, I am certain that we can't possibly keep all our sheep and few cattle without buying some turnips, which I am unwilling to do, because I suppose he will give us 8 ½ per lb. sink. At least I will take no less to my knowledge, and at that rate they will give 3£ per. I mean with very little or no short of the wethers, the ewes at 56s or better. Now these are capital prices, and though more may be had afterwards in all probabillity. For it is not certain, yet so many things happen that we do not or cannot foresee, and we shall be the better able to maintain our *1700 hogs* &c &c that it strikes me as highly wise to give him them at the above offer. We cannot send them to you without some expence, and then such sudden and unaccountable alterations take place. Only recollect what a sudden *depression* has taken place in the price of pork, and how do we know what may happen to mutton or beef. To be sure they can bring no more cattle at this frosty season from Hamburg, but they may come from France &c into the south of England, as our ports are open to these importations at this moment from any country, and our high prices may tempt speculators to do what we cannot at present foresee. However if mutton should advance to 1*s* per lb. I shall not repent, because I think it wise and prudent for us to sell as we are circumstanced. You may try yours on as long as you think proper. I have no kind of objection.

Monday 27th December. We have had a most severe frost for 3 or 4 days, nevertheless the glass has kept coming down (after being very high) for 3 days, that I expect either snow or rain, and this morning it has fallen a little, very thin cover of snow, and is given over now and has again the appearance of frost. However if much snow should fall it will be very aukward for us as we have to send for tups into your country, to the Firth of Forth and to Hawick, and also into Bamboroughshire, all this week. It very unfortunate that people at a distance are generally so long of giving intelligence about these tups coming home, and the season and roads have been and are now very good, provided the snow in great quantity don't come. John there is a young man whose name is *Eales* who came to Coldstream some months ago and practises as a butcher. He is from your country and has a brother who I am told you are acquainted with. Now my reason for naming this *Eales* to you is whether he is a man of character, as he has expressed a desire to deal with us. We have determined to send you this per post, that you may either come over or write your sentiments what we should ask per lb. for our sheep to Mr Wood. I am in haste yours

Geo Culley

159. *George Culley to John Welch*

Eastfield 28th December 1802

Well John

At my brother's request I sent you a short letter yesterday, which I wish I had sense to have sent you earlier, because it is more than probable that Mr Wood will be here before we can have your answer. Ralph Brown and I are going up to Wark this morning to handle the wethers and form some value upon them according to 8 ½ per lb. The weather is now extremely frosty, and likely to be so. Jemmy Armstrong is gone north to the Queensferry today, William Dodds goes into Bamboroughshire tomorrow, Wark folk send up to Hawick on Thursday, and the same day our Jemmy Rutherford goes for your place and I will send this by him. My principal reason for writing now is to say that it appears to me foolish to send your own tup north again. If he is good enough he will do for another year. If not, kill him or get him killed in summer, and we can send you another against the next season. I think it nonsense to *fag* him back and forward. The only difficulty is you keeping him, but you can keep him with wethers or wether hogs. However if you rather chuse to send him home you may, and I leave you your choice, certainly after making the above remarks. Well wishing you and yours many happy returns of the season I am in haste yours truly

Geo Culley

PS. Wednesday 29. A very severe frost indeed and no snow still. Ralph and I went to Wark yesterday and handled the weathers, which we found very good, very much improved since putting to turnips. We think 250 will be 22 lb. per quarter, and 100 best at 25 or 26 lb. per quarter. This weather begins to alarm me for the turnips! There are some rhubard plants breed for you from Wark, besides a sow &c &c by James Rutherford and his cart. I think the rhubard plants will be best put into the fore garden. There are 2 scuflers[102] from Mr Bailey at Wooler for Mr Thomas Charge, but I am affraid Jemmy will be so loaden as not to be able to carry them both if either. The corn winnowing machine and *sow*, and *tup* fills up much beside many &cs. I really think John you had better not send your tup back here, as the lad will be so much loaden. I am in haste yours

Geo Culley

You may [put][103] Mr Charge's letter into the post, and send him word about his scuflers that he may send for them.

160. *George Culley to John Welch*

Eastfield 31st December 1802

Well John

Before this reach you I hope you will have received our man James Rutherford and all his live and dead stock. Not a little I assure you, and you may depend upon it that *Peter* the mule is not the worst

---

102 Implement for stirring the surface of the ground, horse-hoe.
103 A word seems to have been omitted.

although a mischievous dog for getting wrong &c. I should not have troubled you so soon again after writing per cart, but thought that it is right to let you know that I received another letter yesterday from Mr Robert Wood junior of Hillington near York 'that he will not be here before the 15th or 17th of next month and *next year*'. So I wrote to him that we would wait till then of him and no longer. Now as I suppose you will be coming soon you will have time enough to be here and give us your opinion of our sheep before Mr Wood and his friend come. Or if you would rather, you can contrive matters so as to be here perhaps at the very time. But do you do that as seemeth best to yourself, only I would wish you to be here at any rate a little before the 15th that we may have the benefit of your *advice*. This late storm of about a week in length has gone away charmingly and without much rain. I think yesterday was one of the finest winter days ever seen, and not over time. However this is a very fine dry fresh morning 6 o'clock, and our two young ones and their cousin Nelly of the Roads going to Berwick a-shopping I suppose. As I was not at Wooler yesterday I have not heard at all how Morpeth was on Wednesday last. William Marshall of Kelso had promised Ralph Brown to return this way from Morpeth, where he had gone to help to sell some sheep of a Mr Scott of Martin, but he did not call. The week before was wonderfull for cattle, some say as much as 9s were given for some particular nice casten. John I think it would be right in you to come by a Morpeth market, as you could then perhaps form some idea how things are sold there. However you are to do everything of that sort as best suits your circumstances and situation at the time. But I hope you will write again before you come north, and apprize us when you will be here. I want much to consult you about Messrs Lincoln and your George Nicholson about buying a few ewes of us. I am in real haste yours truly

Geo Culley

1803

161.   *Matthew Culley jr and George Culley to John Welch*

Eastfield 20th January 1803

Well John

We have this moment received your letter of the 18th inst., but rather wish we had seen yourself as this is the deadest season of the year and the best time you could come, and if it comes fresh we rather think that you will be disposing to sow the wheat and that will stop you. We think you are wrong in not selling your pork pigs at 8*s* per stone. We would be very thankfull to take 7*s* here and cannot get it. What chance has pork to rise if there be so much corn in at Liverpool and they sell it so low? We think you had much better sell at a times price, you cannot keep everything. We sold George Nicholson 65 ewes at 3£ per, 5£ to be returned. They went away last Saturday. I wish we had all our sheep as well sold, they are more than 8 ¾ per lb. It happened just as you expected, the drop in markets has stopt Wood &c, indeed we rather suspected they would not come. We had the young man Eals from Coldstream yesterday handling 70 of our wethers, but he could not buy things as he says he cannot give more than 8*d* per lb. sink, and we are hardly willing to take much less than 9*d*. He handled our fat cattle at Wark and tried to buy 6 of the best (or fattest). He would have given us [. . .]¹ for the 6 but we thought it too little to pick 6 out of 12 [. . .] we bid 8*s* per stone so we parted. My father is going to Wooler today to see if he can bring Clem Stephenson home with him to try to buy some of the wethers, but I fear the time is past for the present. If so we may blame Wood for it, as Clem was very keen last Thursday and wanted much to come home with my father, but we were then engaged. We have no snow here but a very severe bare frost and high winds from the south east which will hurt the turnips we fear.

M.C. jr.

[*G.C.*] Clem Stephenson be not at Wooler. I am determined to begin to market our shearings and ewes at Morpeth directly, and continue marketing until all sold. But your letter giving an account of the drop everywhere so suddenly may alter our minds again. Indeed I am rather in a *dilemma* what to do. I am affraid of the weather for turnips, otherwise I never knew so many to sell in this country at this

---

1   The figure, and a word in the next line, have been obscured by the wafer.

season. But I would rather not buy. However necessity has no law, and we must do the best we can under existing circumstances at any rate. I should be glad to see you over. Come you must, and the longer you [. . .]² it, the nearer seed time and the more difficulties will present themselves, until you will be putting of coming altogether. Pray write at any rate again if you do not come the next Wednesday. I may still be at Morpeth next week but not certain until I see Clem S. or not today. I am in haste

Geo Culley

PS. You don't say whether you have returned Mr Pickering's Galloway or not. Corn keeps very flat indeed with us, oats excepted, and they are on the drop. Ralph sold 120 bolls mostly potato on Saturday at 17*s* and 17/6 per boll. Barley is not worth so much, and some decent common oats sold at 13*s* per. Watson of Warren Mill we hear has got a contract to supply Government for all our sailors with oatmeal. If so it may help oats a little. Wheat the highest 36 and 38*s*, we have sold none a good while, indeed we have sold nothing but oats lately. You were lucky after all in selling your old and new barley. I hope you will learn wisdom by experience. I know you did it all for the best, and to serve your master's interest, but it is always wrong to be too stiff. I have sometimes been called a dear seller, but I always thought myself a *times seller*. My brother used often to say 'I had a soft spot in me' and it may have been so, but I adhere to the old proverb 'Better to me sell than me keep'. We can't *always hit the nail* exactly on the head, but it is wise always to be selling in part when times are fairly good, and if we had not waited on Wood &c I believe we had sold a good many sheep to Clem, and that shews the absurdity of engaging any stock. It is best to keep ourselves free I do believe. Wooler Thursday 2 o'clock. I cannot hear one word from Morpeth market. It snows all day, but will not lie on the low ground I think. Mr Turnbull, that married one of the Miss Howeys and was an attorney at Berwick, is missing and supposed to be drowned or lost. Since Saturday no account of him. Yours in haste

Geo Culley

---

2   The sheet is torn here.

162. *George Culley to John Welch*

Eastfield 24th January 1803

John Welch

I am this moment favored with yours of 22nd inst., and my brother being here had an opportunity of seeing your letter. We have sent 60 sheep to Morpeth, 20 good wethers, 20 small ditto, and 20 ewes which will be a fair sample. I shall go up, and we have fixed on Harry Rutherford to go also, that he may be learned by degrees to sell sheep, as we are badly of for some persons to sell as I am growing too old, and although Matthew is not unwilling, yet we cannot expect him to do as his father and uncle have done, and so long as my brother and I are concerned we are sure to have a good few sheep to sell in the spring of the year, and particularly the next year as we have such a famous lot of hogs, and we shall have more turnips the next year here than common. We shall more than probable winter our wethers and ewes both. If the market at Morpeth is anything tolerable we will keep going on, and as you propose being at Morpeth that day I shall not go up. Therefore I beg that you will not neglect being there to assist Harry. But and if you find that you cannot conveniently be at Morpeth on Wednesday come a week, I must beg that you will write, that I may be there in your stead, because Harry can't be expected to be able to sell properly until he have a few weeks practice. I am glad that cattle still keep up, and I have no doubt but all stock, both fat and lean, will be high if not higher sold than ever. But I am very unfond of buying turnips, and am satisfied with the present prices, and except they drop again we shall keep going on. What I never knew before, these 35 years that we have been north, fat cattle have been dearer at Morpeth than Darlington until this time, when Morpeth has exceeded Darlington 6*d* per stone at least, and fat cattle are so dear in Scotland too! Still it is right for us to be doing, as we have so many more stock than we can maintain. We shall be glad to sell George Nicholson a few more ewes or wethers, but I am very ill pleased at George for driving the ewes such long journeys as to leave some upon the road, although we sent them only to Milfield one day, and the next to the Haugh Head. But George's man would take them to Powburn, and no wonder that he tired such fat ewes. Had he gone 20 miles at the last it would have been wise compared to going 20 at the start. I am only affraid now John that you will not get north at all, because if good dry weather come you cannot come for sowing wheat. However from this coarse snowy day (but it don't lie) it looks like very bad weather. I would like to have J. Hodgson's farm into our own hands if he engages with founderys and manufactories. We have had no snow to lie here at all, although it has laid upon the hills all around this good while. Saturday seemed quite fresh, and people began to plow in the afternoon. Sunday rather frosty, last night fresh, and this day a good deal of snow but does not lie in this low country. I should like very well if you could sell 100 or all Mr Wood's wethers, as we could supply you with more. But why not sell your housed oxen now if cattle sell so well, because we could send some queys to lie in a fold, which would make you as much dung as those inlaid cattle and pay more for keeping. And we must send them to you as we are running fast through our turnips as you will see when you come. Indeed I begin to be very uneasy about wanting turnips soon, and my brother makes me more so as he says they will be out of turnips at Wark very soon, and it will be a pity to starve our nice hogs.

Morpeth Wednesday 11 o'clock. Well John we have not had a good market, nor I can't call it a bad one, so we will shew 50 or 60 next week, and as Harry Rutherford will come by himself you must not neglect to come along on the Tuesday evening to assist him. Pray don't neglect on any account. If you can't come, write, and I will come up myself again as Harry can't do alone yet. I am in haste Yours

Geo Culley

PS. We sold our little wethers at 9d per lb. sink, but the large ones and ewes for less.

PS. I wish you write at all events.

163. *George Culley to John Welch*

Eastfield Wednesday 9th February 1803

Well John

I think it best to begin a letter and have it upon the stocks, as I can then keep adding to it every day as it suits me. On Monday we propose sending you those queys, they will be at Mr Robson's on Monday night, Glanton Tuesday night, Horsley Wednesday, some way between Morpeth and Newcastle on Thursday, say Bull Head 6 Miles Bridge, Chester or near to it on Friday, and about Kirk Merrington Saturday. But I hope you will send a man to meet them as you promised, and in that case he will meet them between Newcastle and Morpeth probably, as I will send this letter away on Saturday which you will probably get on Monday or perhaps Sunday evening. I am the more earnest for you to send a man because we are so very badly off at present for drivers owing to Wark shepherds being ill, and on the Monday following again, which will be the 18th inst., we intend setting of 200 or more wether hogs to you by the Rothbury road, which you must also send to meet for the reason before given,

or we shall not be able to get the fat sheep to market, and I am determined to keep selling let things be as they will, because I am
convinced that there will be a serious failure in turnips. For which
reason I would recommend to you to be taking Mr Wood's sheep to
Skipton and to be quitting your *inlaid* cattle as soon as possible also,
and if you are so lucky as to prevail on a reputable butcher to come
and buy the Wark cattle it will be a fortunate event I assure you. For I
was at Wark yesterday, when I was sorry to see their turnips going
like *dike water*. Thank God your master is no worse, better I hope, and
agreed to Harry going to Wark but seems to prefer George Winter's
going to Thornington. As I do not know the stages for the hogs the
Rothbury road I can't name them, but I hope you will not neglect to
send a man to meet them in good time. I asked Bob Tate also
yesterday if he thought the great red ox would travel, and he says
there is no danger if he only has time given him, which we will take
care to do because we will send George Achison with the 2 oxen all
the way by Newcastle away so that you need not send to meet him
and I will answer for it that George will not hurry them, and he shall
have discretionary powers. We are not determined yet when we may
send them, but if the red ox keep well we will not be long. This day
has the appearance of fresh at present. I pray God continue it, as it
will be of wonderfull service to the country in many respects. We
have several ewes lambed and several twins.

Thursday 10th. We were sorry to hear by Harry this morning that
you had only a bad market. However we are determined to persevere
and have been chusing near 70 wethers besides some shepherds
sheep for next week, when either Matty or I will attend Harry at
Morpeth. This day continues rather soft and very mild, but nevertheless the glass is very high and keeps ascending, which is a certain
proof of frost *in the air*, and indeed the snow never moves in the east
upon even the lower hills. I do conceive that markets for fat will
continue to fall, and would advise you in the strongest terms to get to
Skipton with some of yours. *Friday 11th.* I have been up at Wark and
am happy to say that my brother is considerably better. This day the
wind has got to west, and it has much more the appearance of fresh,
but the snow does not offer to break at all from the high hills and the
glass keeps very high. Were it not for the glass I should think it was
real fresh. However the snow is gone in the low grounds, and
ploughs are going everywhere. Nay they have sown and harrowed a
good deal at Wark upon Broomy Know. I wish you good weather.
Yours truly

Geo Culley

Saturday 12 a wonderfull fine dry morning and quite fresh, snow gone of all the lower hills but I can't see Cheviot. People will all get to sow wheat on Monday if this blessed weather continues. I hope it will also help the fat markets by sending the ships to sea. This Mr Mills poor man now begins to be talked on more than ever. He owes everyone almost, and bills are returned non accepted. He bought Tom Smith's cattle of Berryhill in the autumn, and sold them almost immediately at Dunse fair for a small profit. But he has not paid Thomas yet, and Mr William Smith junior told me yesterday that they had drawn upon him for the money but the bill was returned not accepted. Mr Watson of Warren Mill got a number of his Irish cattle to answer for his wethers bought some time back, which are to be valued by two indifferent people chose by the parties. I thought it was right to let you know this, but I would wish you to be cautious of speaking except to those you can confide in, and then in a cautious manner. It is a delicate matter to speak off, although it is very proper for people to know that they may be aware of such customers. I have been affraid all along that he would be asking me to buy our stock, but I would rather meet with ten bad markets than *sweat myself* with selling to such dangerous customers. I have had enough of that and wish for no more. The less trust the better depend upon it. No man I would venture to say is more disposed to pay than I am, and I don't deny that I like also to be paid well. I am yours

Geo Culley

PS. I am also told that George Tate has secured some cattle of Mr M.'s at meat in Scotland. What difficulties and unpleasantnesses does this foolish man run himself into. I should first sell my estate directly, and then determine to live upon what is left, were it only 50£ per annum.

164. *George Culley and Matthew Culley jr to John Welch*

Eastfield 12th February 1803

Well John

I sent a letter to you this morning to Berwick to go by the post, which I hope you will receive safe. And having received a letter this moment from Mr Sayle wherein he says that fat cattle were from 9 to 10s per stone sink, at both Rotherham and Wakefield, and sheep (the best) 9d per lb. sink, I am rather affraid that this intelligence may not reach you before Holdgate and Shillito are with you. But I am determined to hazard it however. At the same time I would not wish to prevent you selling your fat stock by any means, but still recommend you to sell both cattle and sheep, as we can soon send you more as we are hard fast here and must be selling. Therefore don't consider this as a letter to prevent, but to recommend you to sell, only it is certainly right to put you upon your guard so that you may be enabled to get as good a price as you possibly can, but still to sell by all means. However if you don't sell, by all means take a part of your stock to Skipton the next market.

Mr Sayle fears the turnips being damaged by the frost very much, and should that be the case the markets will fill again over full I am affraid. Clem Stephenson bought Mr Smith's wethers of Hay Farm, 7 score at 3£ 3s it is said, but I should scarcely think so much as your market was so heavy at Morpeth. This has been a charming day indeed, and has all the appearance of being good weather I hope, which will have as good an effect on fat and lean markets as anything. However the corn market continues very heavy at Berwick. Ralph could sell nothing today at all. I am in haste 10 o'clock Saturday evening yours

Geo Culley

[*M.C. jr*] These high prices for cattle begun at Lincoln for the London market, and so they may travel north.

165.   *George Culley to John Welch*

Eastfield 14th February 1803

Well John

I am beginning again. I am affraid you will not receive my last short letter at Darlington today. Mr Ratcliff came to Wark last night, and Ralph and Matty went up this morning, but he would only give 7/6 per stone so they parted. But I am not affraid but we will get 8s or 8/6 or better per stone soon, by the account of Mr Sayle's which I wrote to you, and John Batters told me yesterday that they had a letter lately from David Green that he was coming out soon to buy fat cattle, and desired that they would be looking out for some for him. I told him that we would sell him 12 good cows at Wark. I am thinking that Northallerton land fair, which is today, will spoil your fortnightly day at Darlington. This is a nice day, and we are very busy sowing wheat. *15th February*. Matty goes to Morpeth. Very windy and showery but pretty fresh, and snow coming of the hills. We hear that there was a good deal of snow to the south last week. We had none here. Mr Smith has sold 7 score wethers to Clem Stephenson at 3£ 3 per, cast only 15 a good price. I hear that Mr Mills had some cattle at Northallerton fair yesterday. *Wednesday 16th*. A most dreadfully cold morning and has been a little snow, and surely will be more on fall, it is so very bitter and *ill natured*, and the mercury is very low. *Thursday*

17th. Well John we are sadly disappointed and mortifyed at not hearing from you. I had no doubt but we would have had a letter this morning at the farthest. You must best know why you have not wrote according to your promise, but we cannot account for it. Matty, who was along with Harry Rutherford at Morpeth yesterday, met with a better market than the week before when you were there. I think they sold the little sheep at fully 9*d* per lb. sink. I had a letter this morning from Mr Barf of Wakefield saying that sheep were fully 9*d*, but does not say that cattle were quite so high as Mr Sayle said, which I wrote to you in a second letter. He says about 9*s* was the best price. By a letter also received this morning from Mr Proctor of Glammis, he says that fat cattle are sold at 11*s* the Amsterdam weight, which is nearly 9*s* English. Now when they are so well sold both north and south we shall certainly come in for a share by and bye. Mr Barf says 'that if he was to give an opinion, he does not think that fat will be down again of some months'. We have very rude windy weather, sometimes snow and frost, and sometimes rain. *Friday 18th February* is very frosty again, with a very cold strong westerly wind. However these west winds will have one good effect by forwarding the shipping to and from London in the coal trade. We sent to the post again this morning John, but no letter. I only wish all may be well with you, or your letter may be lying somewhere by the road. But what can be the matter, I can't help being a little uneasy, because after having been here a week and so many important matters to write about, something very extraordinary must have happened that we have not received a letter from you. However we will hope that the letter has miscarried, but that can hardly happen unless some abominable neglect takes place at some of your post offices.

I met with little Tommy Bates yesterday at Wooler, who wished to buy our remaining wethers. He came this morning but never bid anything. He wanted to have bought the queys with you, but I refused him an offer. I have been at Wark and found my brother pretty well thank God. As you will not write to us I am determined to write to you, and shall send this tomorrow.

Berwick Saturday 19th. Everything flat here and very little doing in the corn way indeed of any kind. Little Tommy put up his appearance but said nothing to me. John I am very anxious to hear from you. I never was so much so. Not a word after so many days fills me with alarm. I am certain that if you were only sensible how much pain you have given us you would be much concerned. If alive do write 3 lines per return of post. Your much disappointed friend

Geo Culley

166.   *George and Matthew Culley to John Welch*

Eastfield 21st February 1803

Well John

This is the 5th letter (I sent one yesterday Saturday) for you since you left us. However I shall begin this with more composure than the last, and endeavor to answer your last in some parts more fully and more cooly than I did yesterday, and am in hopes that you will not again offend so egregiously as you have lately. You did not only keep my mind in suspence, but all this family and the Wark family. I dare say that you may expect a rebuke, and very justly, from my brother. And if Glass only get back to meet poor sackless George Hutchison before he reach Morpeth it will do well enough, otherwise poor George will not know what to do. This wonderfull little animal Thomas Bates, who is in my opinion possessed of more requisites to constitute a jobber that I have known combined in one man, has connected himself most wisely with the very respectable Mr *Moorhouse*, generally called *Morris* of Skipton, who you must have seen. Mr Moorhouse is to sell at Skipton and Tomy to buy in the north. Tomy is possessed of the subtility of the *serpent* and the *apparent* innocence of the *dove*. He has a wonderfull share of *persuasive* language. He is perfectly *sober*, able to undergo any *fatigues*, rides *catch weight*[3] and at the rate of 14 miles in the hour if necessary, has the full command of *money*, situation in regard to land *convenient*, and above all not over *scrupulous*, which last requisite I would wish him and all men to be divested of. But God knows I am affraid that too many of them will take advantage when in their power. But I would wish to guard you and all my friends against it, and instead of taking an advantage if in your power, to avoid it by all means and act with that honesty and *liberallity* which constitutes the *good man* and the *Christian*. 'For what signifies if a man gain the whole world and lose his own soul'.[4] I am no *Methodist* John, but I would like to recommend and also follow good *methods*. No doubt of it, if Mr Wood's wethers weigh 22 lb. per quarter they should be worth more than 3£ per head, but I am affraid that your cattle may not have done well with you that went from us, as I think that they should by this have been worth more than 30£ per head. To be sure they might not be the heaviest of ours, but I am certain that we thought them the fattest when they went from us. The time of taking from you you do not mention, but it

---

3   Without weighing (horse-racing).
4   Mark 8.36: 'For what shall it profit a man, if he shall gain the whole world, and lose his own soul?'

would have been a considerable bargain altogether. Neither do you name the number of Mr Wood's wethers which they offered to purchase. However I hope you will have a decent market at Skipton and the *moneys* home with you, and get these *bankers* paid of in the first place. I think it is as well that you did not send the 2 oxen to Wakefield as things are, and things will be better understood by the first Monday of March when you can shew them at Darlington with much less expence at any rate. And if not sold, you can still take them to Wakefield. You were lucky in having snow when we had none. However turnips have saved wonderfully here. They have sown most of their spring wheat at Wark and we are busy here, but there is time plenty this 3 weeks provided the land is only *fresh, clean* and in good *humour*. But old land does not do to sow late, nor strong land. Your land will bear up better for being plowed a while beforehand. Joseph Hodgson's mother would be old I think. I believe Mr Mills has paid of many people lately, indeed they were become exceedingly *clamorous*. Laird Burn was arrested on Saturday by Mr White of Hexham I am told. Poor animal, he must go to be a soldier or some such thing.

[**M.C.**] What have you done with Mr Colpitts about your corn tythe John Welch? I think you should try to get Bob Dippie to table himself with somebody in the town. Suppose you give him 25 or towards 30£ a year. M Culley. And on thinking things over you should have wrote a short letter no matter how short, as we were in suspence till your letter came, and my brother has so many apprehensions about him. Indeed we wished much to hear how our Matt is from you, as we were told he was not looking well. The postage for 2 letters when one might do, is of no consequence, in future be so good as write on half a sheet or a half filled letter, as easing the mind is of great consequence to us who grow old. I know not when I shall get over, as servant hiring comes on now and Cornhill hiring (and comes soon)[5] is on Monday, the 1st Monday in March. We have agreed to take Harry Rutherford to Wark, and George Winter goes to Thornington to manage watering the land and going with the women &c, and William Mabel to manage plows &c and to assist each other with advice. As Harry cannot get here (Jenny being in the house) to live, I shall have less benefit from him, and cannot get so well from home to see you which I much wish for. Yours &c

Mattw Culley

---

5   These words in parentheses should perhaps come after 'now'.

[*G.C.*] Eastfield Monday. You must know that little Tomy was at Morpeth, and on Thursday last he dined at the ordinary with us at Wooler, when he asked me if I would sell him our sheep. I told him we had so many at Eastfield marked out for market, that the rest were at Wark. He said he would be at Eastfield by ½ past 8 in the morning. Being obliged to be at Wark I was gone, and he and Matthew could not bargain. He went to Kelso and came to Berwick on Saturday, where I happened to be. He then asked me if I was agreeable to sell him the sheep. I told him they were then on their road to Morpeth. He said that made no difference. I sent for Ralph and we considered to abate 1*s* per head of what Matty had asked, and stand to it. He would have 5£ again, and when we came into the room they all set on me to split, as there was only 15*d* per head between us. I told him I would neither split nor *rive*.[6] Mr Nisbet said '*Bates* if you mean to buy these sheep you may as well do it, for I am certain George will do no otherwise'. He said he would buy them no dearer, so no more was said until I was on horseback when he came in my will, and then Mr Nisbet abused him well, told him that he was no better than Jere Clayton. It stopped us late, and I had to give him a line for Glass. The sheep were to be delivered at Morpeth, and the 19 ewes stay until he has more sheep going, at 6*d* per week. I received in full for them. I could almost have wished that you had taken George Nicholson's money and paid the bankers. T. Bates is to look at the rest of the wethers if unsold on his return about the 2nd of March, but we are at liberty to sell when we like. I could wish George Nicholson to have these 50 best if they suit him, because they are uncommon sheep and should die well. George Hutchinson came home last night. *This is Tuesday 22nd.* Glass met him at Whittingham, and we shall set George of with the 2 large oxen on Monday next at the farthest if all be well. My brother and sister at Wark are not pleased with my dealing with T. Bates, but as he and I never differed, why should I refuse a fair price and ready money? This is a hard frost with a white rind. We had an uncommon fine day yesterday and sowed a good piece of wheat. *Wednesday 23.* A charming day indeed, quite fresh. Wheat sowing going on at Grindon and everywhere. The poor grey mare is recovered, at least recovering. You must know the lad riding one day past the wood we bought at Ford, she boggled at a piece straw, and jumped against a forked stick and tore her thigh most shockingly. We all thought she would die, and old Robin Tate had bad hopes.

---

6 Neither split the difference nor give back anything.

However with great attention, warmth and plenty of bark and port wine she is now in a fair way of recovery. This is the 9th day. Had it been her blind side poor thing one would not have been so much surprised, but it is on her seeing side. It is a dreadfull wound indeed, but she was in great health and spirits. However I hope she will still be able to bring a foal or 2.

Friday 24th [25th] *February.* A very fine morning, indeed since the change of the moon we had remarkably fine weather, and sowing of wheat goes on much, beans also I am told in some places. I find that Morpeth was only a slow market on Wednesday, mutton 8*d* to 8 ½, beef 7 to 7/6, not much at 7/9*d*. If this continue to be the case we shall have to send you the whole or most of our fat cattle, provided your markets continue to be better than ours for cattle. As to wether sheep we shall be done on Wednesday next most probably, as we have marked most of what we have for Morpeth, except the 50 best at Wark and a few segs. There will be 100 or better of very nice gimmers which we propose sending to you. It seems Ratcliff was out again this week, but have not heard what he bought. But he had called one day at Fenton but Mr Vardy was from home. Corn markets continue extremely dull indeed. Well John we received your very pleasant letter from post this morning after I had wrote the above, but I have been at Wark since, and unfortunately forgot your letter. However I will endeavor to answer it as well as I can. It would appear by your beast market that Mr Sayle has overrated the prices. However you have proved your judgement respecting the sheep. I must say that I consider you not only possessed of good sound judgement but an able salesman. Your favorable account has stopped us from sending any more sheep to Morpeth, although we had 70 marked and ready to set of, and would have gone today had you not wrote. Morpeth was a worse sheep than beast market, although neither good. If you don't contradict it, we will send you what sheep we have left, wethers I mean, first about 150 including the 50 best ones at Wark, which will drive by giving them time, and I have no idea of George Nicholson buying them, Be so good as write about this as soon as you can after receiving this. The 2 large oxen will set of on Monday to you, but I am rather affraid of the red ox, not but he is sound of his feet but does not fill himself. However you will soon see if he don't mend upon meal, and if he does not I would recomend to you to quit him as soon as you can. The other ox is perhaps as wonderfull a shewing animal as ever was, as we have no objection to you keeping him another year if you approve, or at least as long as you think right. There are 100 nice gimmers for you, which we shall also send to you by and bye, and we have no objection to your clipping them provided you think nobody will be tempted to keep them to tup. That must be left to you, and I think if sold at Skipton there is no great fear. We will not send the 100 hogs so soon in case you take the wethers &c, and that you will be sure to acquaint us of. Little Tomy is a surprising animal however. He not only *chirps* but flies I think. He had gone to consult with Mr Morris alias Moorhouse at Skipton I suppose. Mr Maynard has sold his wool famously indeed. Is old Head not going to take up yours yet? Matty will go to Morpeth at all events to meet George Nicholson on Wednesday. My brother thinks they can keep Wark cattle a good little while yet, but I am affraid that you would not like to take them. Housed cattle drive in so much. I hope you will now get out of debt to the banks. But John there is one thing which strikes me forcibly. The first Monday in March will be on the 7th and the Skipton market on the 8th March, which will be very awkward for you indeed. How will you manage? And you should shew your large cattle (inlaid) at Darlington by right. However you will consider it well over, and do the best way you can. You are well mounted, and can go as far after Darlington market is over as you can, and so into Skipton the next morning. It will be very severe upon you certainly, but I should not be at all surprised if Skipton be a better market on account of Darlington fair. I know not whether I have answered the whole or not of your letter, but I hope I have answered the most important parts, and I hope we shall receive a fuller letter from you in a day or 2 as you have promised. However I am so pleased at your letter that I could almost forgive you that, and will now conclude Yours

Geo Culley

We think we have as many hogs of our own this year as we can cleverly maintain, otherwise Mr Wood's would be as proper to buy as any, and they have done wonderfully with you after all, considering how they were pinched in the drought with you. I should be affraid of their proving badly for tallow from being so pinched in summer, but they were exceedingly well wintered. *Saturday 26* is a very blowing showery morning. It began to be blustering and showery from south west when I came from Wark yesterday about 3 o'clock, and it rained a good deal afterwards, and Surtees, who was sowing one of north fields called Planting Close, a wet field. However it does not seem so bad this morning 6 o'clock. Berwick. A bad day indeed. It has rained all the way and I repent of coming, but had rather a particular piece of business. The corn market and sheep extremely heavy indeed. Poor Laird Burn is seized upon and his sale is to be on Monday next. Let me hear from you again as soon as convenient. G.C.

167.   *George Culley to John Welch*

Eastfield Sunday 27th February 1803

Well John

I sent you yesterday from Berwick a pretty well filled letter which I hope you will receive today or tomorrow. We had a very coarse wet stormy day yesterday, this morning the hills were white and this forenoon we have had several stormy showers of snow from north west. The *wheat* and *pease* and *bean* seed is over for the present. Ralph sold 110 bolls of potato oats yesterday to George Richardson, merchant in Berwick, at 18s. Barley is not saleable, 13s can scarcely be got for it. Many people are breaking it and boiling it for their horses. Laird Burn's sale of everything, even furniture, is tomorrow, and the farm is advertized to be let, to enter to immediately. Monday a decent freshish day, sowing of wheat again. Brother and sister were here from Wark, have got most of our servants hired again. *Tuesday 1st March.* A charming morning indeed. Matthew just going to Morpeth to meet George Nicholson, but no sheep up this week from us. I have just been up at Wark where all is well. Ralph and I went to handle the 50 good wethers and put a value on them in case George Nicholson should come home with our Matty tomorrow. They are the fattest and best lot of wethers I think we ever had, and I do wish George to buy them because they are particular, and he shall not have them over dear. *Wednesday 2nd March.* A very wet morning and rained most of yesterday. Last night just as we were going to tea came little Tomy Bates and young Moorhouse, a genteel youth indeed. Handled the 70 sheep which we had stopped going to Morpeth about 22 lb. per quarter overhead. I asked 3£ 3, he bid 3£ and we parted. They say they will call again. Met the 2 oxen and wanted much to buy them. I said they were not to be sold at present. I also said they were too large for Skipton. He answered no, and said that Mr Moorhouse had one customer that would take 2 or 4 oxen ever so large, at any time. Seemed to intimate that they were fonder of cattle than sheep. He had been at Wakefield I found, but could get nothing more from him than it was a thinner market than Skipton. He tells me that Shillito &c and Jere's son have stock every week at Skipton. I had no idea of that. I could not learn that Mr Bates had bought anything, had been at Robert Thompson who has 2 fine Fifeshire beasts, has kept them one year and means to keep them a year longer. I thought nothing remarkable except being so much alike as not to be easily distinguished. They had seen cattle at Harry Howey's also, and been at Mr Jobson's and Mr Vardy's and Mr Smith's, and were going to young William Smith Shedlaw, so to Mr Oliphant's &c &c. &c

Thursday 3rd is a very frosty cold morning. It rained all yesterday forenoon and began to snow about noon, then fell frosty and the hills all covered with snow down to Ralph Forster, but it melted as it fell in this low country. Matty did not come from Morpeth last night, but will I hope meet us at Wooler hiring for hinds &c today. I am affraid George Nicholson is not coming. Tomy Bates told me that Wood sold a part of Mr Gregson's sheep at Skipton that day you were there at 5*s*/6, only 2*s* per for wintering. He must lose a deal of money at that rate. John, George Nicholson has behaved badly, neither to meet Matty at Morpeth nor send the money, nor write. However I have wrote to him that I shall draw on him tomorrow at Berwick at 10 days, as we have need of the cash. Friday morning, have received yours of 3rd this moment, and am glad you have sold Hodgson the rest of your wethers. We have sold little Tomy the 70 at 3£ 3, 5£ returned, and got the money. Mr Smith had a lot of cattle, would have suited you but Clem bought them. The 2 oxen are gone, and I heard they were well. The gimmers set of on Monday for you. We will consider about giving the Wark cattle meal. This goes by your master and mistress, who are going to see Matty in the measles. The bankers all take 5 per cent, it is the principal part of their profit. You are right John to pay all clear of, and then you may put into the bank or give to Mr Peacock as you and my brother think right. As I am affraid of missing my brother and sister as they pass Blue Bell, yours in real haste

Geo Culley

168. *George Culley to John Welch*

Eastfield 4th March 1803

John

I had not time to answer by my brother your kind letter received this day, I had so little time before they, my brother and sister, passed. We had much rain here at the time you had, but not near what you had by your account. You have really made a capital conclusion with Mr Wood's wethers, but the best of it is that it shews your judgement *sound*. We have not now above 60 wethers left, except the segs which are backward, and Tomy Bates bid us 3£ 5*s* for 47 at Wark and said if we could not make more at Morpeth that he would take them at that price. Now if George Nicholson does not come here on Sunday first, as I still hope he may, we will shew these 60 sheep at Morpeth on

Wednesday come a week, and if not sold to our mind there, will send them forward to Denton. What do you think? Indeed we are pretty determined to do so, and told T. Bates we would, and he says he will be at Morpeth on every Wednesday. I told you in my letter today that we would consider about giving the Wark inlaid cattle barley meal. Perhaps my brother will write about it. Indeed it is only giving them meal at the last, and turnips for water. The gimmers will set of on Monday and will be very nice to drive to Skipton or sell at home as is best thought of. I believe you will find them good in general. I am not fond of buying you inlaid cattle. If you want we can perhaps afterwards get some outlaid ones. John I think you should write to George Nicholson. Indeed if he come here on Sunday I will let you know. I hope it is only his ignorance of business that made him neglect. Poor Matthew had [an][7] unpleasant and as it happened unnecessary journey to Morpeth to meet him, and no [. . .][8] thing to do besides. You are certainly right to pay all your debts, and then I hope you will begin to *deposit* as you properly say, and I wish you much to do so. My brother scolds me for being anxious about the money which we have to raise against May and June. But if somebody don't think I don't know what may be the result. By all means sow barley rather than risk wheat too late in your situation, where you can sow barley decently. However my brother can advise you well. I think seeds do the best in or amongst spring wheat, but perhaps your barley does not lodge like ours. My brother and you can now also settle about Bob Dippie. If you could drill your barley it would be a very great advantage to your seeds, even suppose you cannot hoe the ground.

John as my brother and sister may stop a day or 2 at Houghton if Matty be poorly, I will send this per post tomorrow, and you will get it either on Sunday evening or on the Monday at Darlington, and I hope you will not neglect to write on or as soon after Monday as you can. If this should reach you before my brother and sister get to Denton, you must know that they are gone this morning for Houghton to see Matty, who we heard by Mr Bailey at Wooler yesterday was in the measles. Mr Bailey's sons had wrote so to their mother, and I think it very odd that Mr Raws has not wrote to my brother about Matty. They go from Houghton to Denton. Ralph has sold William Marshall of Kelso a burnt cow and bad steer today 7/6 per stone sink. This is the last of the poor burnt cows. T. Bates says he

---

7   A word seems to have been omitted.
8   There is a blot on the page here.

will give 60£ for Mr Compton's cow. Marshall bid 50£ for her a long time ago, and I understand she is very much improved since. I will finish this, and send it from Berwick tomorrow, and am with every good wish yours

Geo Culley

Saturday 5th. Frosty, but a change is taking place I think as the wind has got south east. The weather has been exceedingly changeable since the new year came in. I think I told you Ralph sold 110 bolls potato oats last Saturday for 18s per boll, a fair price still. Berwick Saturday near 2 o'clock. Nothing selling but potato oats from 16 to 17/6, but I hope Ralph Brown will sell for 18s. Barley bought at 12 and 13s by the brewers they say. I think they [are]⁹ fools who sell it I would eat it and sell oats. Wheat no alteration. T. Bates not appeared yet. This day looks like onfall again. Yours in haste

G.C.

169. *George Culley and Matthew Culley jr to John Welch*

Eastfield 6th March 1803

Well John

For fear my brother and sister should not reach you so soon as a letter, I sent you away one from Berwick yesterday, which I hope you will have recollection to send [for]¹⁰ to Piercebridge tonight, because I generally send you one on Saturday and it is sure to be at Piercebridge on Sunday. Ralph Brown contrived to sell 150 bolls more of potato oats yesterday at 18s, although they begin to be a little nice about them now, as these are for seed to send to different parts of England &c coastwise I think. I told you that the maltsters are buying barley as low as 12s per boll. I think I have not known it so low of many years. Wheat much the same but very flat, and very little done. However we hope there will be an alteration for the better soon, as the old corn laws take place some time this month, when an exportation is expected.¹¹ Little Tomy came into Berwick just as we were coming away. I only just saw him, had bought no more but had many

---

9   A word seems to have been omitted.

10 A word seems to have been omitted.

11 George Culley's information was wrong. Although the prices of wheat, barley and oats dropped steadily during 1802, and continued to fall in 1803 and early 1804, the average price of wheat in the early months of 1803 was still higher than the level below which export was allowed under the Corn Law of 1791. The Acts forbidding export of grain and permitting imports duty free were due to expire on 31 Jan. 1803, but on 29 Dec. 1802 they were renewed until January

people to speak to whose sheep he had handled, is to be at Darlington on Monday and probably Skipton on Tuesday. Do you intend to clip the hill sheep? I am affraid they will clip light. John you must not neglect to speak for scythes in time. John have Messrs Graves and Peacock ever paid you for the large bargain of barley yet which they bought of you? *Monday 7th* is a blustering ill-natured morning, neither frost nor fresh in the low lands, but plenty of snow lying close upon the mountains. We sent to Morpeth this morning in hopes of meeting with a letter from my brother, but none yet. We shall send every morning until we receive one. *Tuesday 8th.* We were very happy in receiving this morning my brother's letter from Houghton containing a very favorable account of my nephew Matty. We directly sent the letter to Wark. This has been a fine drying day, several people sowing wheat again. Ralph Brown has sold 38 hill ewes today to the Berwick butchers at 47s per head, 16 lb. per quarter fully 9d per lb. or nearly however a famous price, considering that they are the shots neath of those sold at Wooler fair. Ralph says that they are very handsome mutton. Nothing like things that are well bred, they will turn out. William Glass set of for Denton today with 120 gimmers, the greatest part of which I hope are very handsome mutton. I would send this letter away to you per post, but that I expect a letter every day from you giving us an account of the March fair at Darlington last Monday. But we shall not get it now until Thursday morning, as Wednesday is no post day. As George Nicholson has neither comed here nor wrote to us, I think it would be right in you to write to him John, to let him know that we shall shew those capital wethers at Morpeth tomorrow week, when if we can't sell them to our minds I am determined that they shall go forward to you. You may assure yourself that they shall be particularly carefully drove. We intend to set them off on Thursday first. My brother would tell you that I wrote to George Nicholson by him, about my drawing on him at ten days for the £190. I don't wish to offend George, but we really could not do well without the money or not least the greatest part of it. *Wednesday 9th.* Frosty, but a decent day. I wrote to George Nicholson last night to acquaint him that we proposed to send those famous wethers to Morpeth on this day week, which we wished him to buy at home here, and if not sold there to go forward directly to Denton. Eales sent

---

1804, and then again renewed until March 1805. Barnes, *History of the English Corn Laws*, pp. 85, 88.

12 housed cattle south the other day, 5 of the last of Mr Marchbanks and 7 from Fenwick Compton.

[**M.C. jr**] We will thank you to tell your master to buy us [. . .][12] as he returns through Durham, as what we can buy here will not keep. If you could meet Will Glass with the 120 gimmers and set him back, it would be right, as they have no herd at Thornington at present but Jack, and the ewes begin to lamb next week and he knows nothing about that. You should set a man away immediately on receiving this, as he will be within a day or two's joining of you, and Jim goes away with those 61 wethers for Morpeth, and if not sold there will come forward to Denton with them.

[**G.C.**] *Thursday morning 10 March.* Received your and brother's kind letters in same sheet, and although your fair has been what you did not expect, yet let us be thankfull. I can't say but I would advise you to sell your hill wethers. Can you in reason now expect to sell for more than 9*d* per lb., and though they might pay well for clipping, will it not pay as well to keep our hogs better? Depend upon it that it is as broad as it is long. We have often told you that we have a larger number of hogs than we ever had, and very nice ones indeed, and is it not wise to maintain these well than pinch them for the sake of making 5*s* per head more of 100 wethers? And how do we know that 5*s* per will be got, although it is very probable, yet not certain. Times are very singular, and may alter even for the worse. Don't forget one thing, viz. whenever things are very high the consumption is less by a deal, and it is always wise to play a safe game. It was for the sake of our valuable hogs that I have been pushing the sale of our fat stock here all along, and I rejoice that we have got so well through, and I am certain our hogs will clip more wool and pay as much for their meat as any fat sheep whatever, and I am still affraid that we shall be hard run. However we will keep of sending the hogs as long as we can, and give meal to fat cattle &c &c. I hope George Nicholson will meet Matty at Morpeth on Tuesday evening as I have wrote to him, and you talked to him, and Mattew will give him something fair. But it is impossible that his ewes should have weig[h]ed so ill if he had not drove them too hastily to Painshaw. But butchers will never grow wiser. Our wethers go today for Morpeth 6 days for 40 miles, but we are determined to shew them fresh if we can. You never mention poor Matthew, but I hope he is doing well. John the gimmers will sell any day you chuse as well as the hill wethers. Therefore no hardship, and

---

12 Three or four words have been crossed out in a different ink, and are illegible.

at 9*d* per lb. too. You are right in selling the inlaid cattle, or take them to Wakefield and quit them at all events, and stick to your outlaid queys &c. I am affraid Beany's pride won't let him buy your oxen at 50 guineas. I much approve of your letting Mr Peacock have the money so long as he will take it. We had a charming day yesterday and this is a wet cold rain from the north. I have seldom known the weather more changeable ever since the new year. I had no fear of George Hutchinson and his 2 oxen. He is made on purpose for such a job. Matty was at Wark and saw his cousins well the other day. Ralph was there yesterday, but they were walked out, and I hope to see them tomorrow. I don't know if I told you that Marshall has got the steer and is to have the cow next week, and a fat wether, and one is to be killed by the Halls along with Mr Jobson's cow. Now I think that I have answered all that is important and will send this away from Wooler. Yours in haste

Geo Culley

Wooler 1 o'clock. Morpeth a good market I hear for both cattle and sheep, but I have heard no particulars yet. But I am vexed about the son of your Will Glass. He has tired some of the gimmers and turned again when he met James Glass, and sent him with gimmers to Denton, and Will should have gone and James to Morpeth as James best knows all about Morpeth, and the wethers this week are so very critical a lot to drive. I am affraid of trusting Will at any rate with them, and know not what to do. Had George Hutchison been at home he would have done. G. Culley. John do send to meet the gimmers I beg, as Harry has no herd and ewes are lambing. Brother you should know exactly when Mr Bates is to pay his mortgage, the exact day.

170. *George Culley to John Welch and Matthew Culley*

Friday Eastfield 11th March 1803

Well John

I still address you because I am uncertain of my brother's stay. We have got some snow and frost this morning. It now begins rather to impede the bean seed time as well as the spring wheat &c. Indeed people would now like to begin with the oat seed, but whatever has been done any way yet has all been done by catches. However we have known seed time deferred until after Durham fair[13] by severe weather, and a good crop ensue for all that. One good thing is, in this country the plowing is so forward and the plowed land so *tendered*

---

13 31 March.

and *mellowed* with the frosts and changeable weather, that it will not only harrow very readily but has a much better chance for a good crop. It used to be a foolish custom in the country where you are, formerly to plow none until Candlemas was turned. But I hope you know better than to pursue or advocate such *antiquated* and *absurd* customs, for as I have observed before, early plowing not only *faclilitates* business but benefits the ensuing crop, a matter we know by long experience, the best of all tests. And as far as my observations have reached, I have always found that good plain *common* sense, founded upon *experience*, has succeeded better in the agricultural line than great *schemers* or *projectors*, or perhaps any other line. Projectors benefit mankind but frequently ruin themselves and families. I think it was yesterday week, no I believe it was from Berwick the last Saturday that I sent you a letter which neither my brother nor you take any notice of in yours of Tuesday last, which we answered yesterday. I like to have you say that you received such and such letters, but it is possible that it might be lying at Piercebridge and not comed to your hands, or you would certainly have noticed it, because it contained a bargain with T. Bates &c &c. It may have escaped you, but I am happy in saying that you are upon the whole very attentive to dates and taking notice of receipts of letters &c &c, which in business are more important matter than many people will allow, or trouble themselves to think important. Mr Sayle often forgets to date, and sometimes forgets to put his *signature*, but luckily I could swear to his hand were it necessary. I do not know *Jobson* of Shields who you sold 2 beasts to. I am disposed to think that your Mr Maynard has as much pride as profit. Oil cake at 12£ 12s per ton is very expensive feeding indeed. To give barley meal to cattle, particularly when at these low prices, I think prudent and even wise. I have heard Mr Maynard boast of giving his milk cows *cake*, when I observed the troughs in the field where the cows were. The grey bull has done Mr Charge an immensity of good, perhaps more than any other person ever received in that way by a cross. His cattle were strong big usefull beasts, and the cross hit to a hair. Yes it puts me in mind of Togstone. Mr Matty and Ralph Fenwick recollect how much they were benefitted by Charles Colling, but Togstone most, and those were large big-boned cattle, still larger than John Charge's. When I saw John Charge's yellow steer last July I thought I never saw a sweeter in my life. Morpeth market was very good on Wednesday, particularly for cattle, and I have little doubt but fat cattle will be well sold yet. I am inclined to think, and information confirms it, that usefull fat cattle are not plentifull. The Scotch folks were *wisely* tempted to sell last autumn from the uncommon prices. I understand cattle were readily sold at 8s at Morpeth. Were it not for this severe changeable weather fat stock, especially cattle, would sell well, and as barley is low it is wise I repeat, to give it and keep on a while. You may rest satisfied that George Nicholson's ewes weighed all their weight when they left us, but if butchers will continue foolish I cannot help it, and it is hard that we are to be punished for his folly. Did not his man take the ewes from Haugh Head to I know not where, when our man was to stay there all the night? What do you think of double stages at the beginning of a journey? But it is impossible to convince ignorant butchers, and when we call to mind that those ewes were *uncommonly fat*, were to be *kept on*, not *killed directly*, the absurdity of double stages strikes me with double force.

You confess your sins Mr Matt, and that blunts anger, but you should not have missed sending your letter from Houghton. Crisp was always a cunning man, and a proud unpleasant fellow to deal with. *Friday afternoon.* I have been up at Wark and saw your daughter, who would have had me staid dinner, but I came round by Thornington. I went to Wark with an intention to go to Kelso as it is a great day there, along with Ralph, but the day proved so extremely stormy that I altered my plan and came by Harry's home, all well thank God everywhere. James Phillip continues better. *Saturday 12th March.* A very severe frost indeed, weather quite in the extreme. But God's will be done. Ralph could do no business at all yesterday at Kelso. Marshall had one of the fat wethers, not the largest but very fat, gained credit and sold him very well. Ralph also got little money, indeed he did not expect any, but a Mr Lees from Lauder for 5 bolls wheat which the man sent. John I find by a letter from Mr Sayle received this day that fat stock was worse sold at Wakefield last Wednesday, mutton 8 ½ to scarcely 9d sink. So you may consider yourself well of that sold so well at home. Many left unsold. Beasts, not so many spared, but lower was his expression. However I hope that will not prevent you sending your cattle to Wakefield the next market day, for I find that we are going to be scarce of turnips here upon every farm we have, notwithstanding the great quantity we have sold and sent away to you. Indeed this weather is so very severe that grass cannot grow, and turnips go very fast. Mr Sayle says turnips are very plentifull still in Yorkshire. By a letter from Ireland I find that fat stock is high, and lean also, and horses, but corn very flat. Mr Sayle says that wheat advances with them. This will assist your markets. That compleat rascal Jolly has been raising contributions all through the north of Ireland. However he luckily only got a solitary guinea from my correspondent Kerr Richardson. Mr Richardson, as I wished him, sent an advertisement to Dublin, but unfortunately the mail it went by was robbed and he does not say that he sent it afterwards. Brother, will you be so kind as call upon Mr Anthony Surtees[14] our cousin and attorney at law in Westgate Street as you return through Newcastle, and you will probably receive Simon Smith's dividend from him. As we have heard nothing from him since, I requested him to pay the money into the bank of Surtees and Burden, and write to me about it. But be sure that you pay him his bill at the same time. Ralph Brown has sold 120 more bolls of potato oats today at 18*s*, and it is understood that this *war story* will advance grain of all kinds. But pray God grant us a continuance of peace, let the corn sell as it will. However it is wise methinks to sell oats at these prices at all events. The weather has turned quite fresh this afternoon, if it only continues, but the wind turned more about to the west, which makes me affraid that it may not stand. *Sunday 13th* is as fine a fresh morning as yesterday was frosty in the extreme, and a nice moderate breeze from the west south west. However in the afternoon of yesterday we sowed a good piece of wheat which I hope may do very well. Your people got Jock's Ridge sown a few days back, very well done. Harry was there that day and advised the measures.

Monday 14th March is a very fine fresh morning. The wheat seed will go on again I hope. Nelly Culley of the Roads dined here yesterday on her road to Wark to stay a day or 2 with her cousins. Andrew Bolton came here yesterday on his road from the meeting, and I desired that he would come every Sunday until your return. Matty is going to Wark this afternoon to cut some calves, and I intend him to take this letter with him to Morpeth tomorrow and send it thence after market per post. We have got an account of 3 bags of red at 90*s*, which is 19*s* cheaper or less price per cwt. than what we get from Haddington. The white is 112 and hop 56. It was not got to Berwick on Saturday. Mr Thompson speaks well of the red, and badly of white and yellow. We have got no rye grass seed yet but are to fix with Mr Air tomorrow. They are going to be very high as I expected. Dear Brother I know you incline to the best or at least part of the best hogs going from hence to Denton, to which I ought to have no objection, only allow me to observe that I see we shall be distressed for turnips and the hill hogs will (I think) do in a week or 2s time in John

---

14 Of Purvis and Surtees, attorneys at law, Newcastle upon Tyne: *Newcastle and Gateshead Directory for 1795; Directory for the Year 1801.*

Welch's moors to run them thin in parcels. You will recollect how well our large Bishoprick hogs used to do in your moors formerly. Indeed I don't see what John Welch can eat his moors with so well and so advantageously as with these hogs. However I sincerely wish you and John to give this a dispassionate and impartial consideration. The hill hogs never were so good in my opinion as at this time, and are doing exceedingly well indeed, but in a week or 2 their teeth will become loose and then they will lose condition. A very bad backcast,[15] especially for wether hogs. William Brown is so far run with his turnips that 200 ewe hogs must come from his direction. Now those ewe hogs might do at Harry's if they are furnished a little, provided John Welch would in a week or 10 days take the hill hogs or 100 of them. This need not prevent 100 of the best hogs going to John afterwards. There are 180 of the hill wether hogs or better, and sweet fellows they are. I am not clear but if well managed now but they may come near to your best. However I hope you will turn this over in your minds and then write as soon as you can.

We have this moment received yours of 9th inst. We agree to what you say about *declining* breeding sheep at Denton, either all or in part. In that case John can feed off his ewes or most of them in the autumn, but I don't see that John need to eat either Hall field or Fitts. Could he not find sound fields without those, and dry? What is your *Low Quarry field* alias near *Galla field* doing &c &c? However he must know better than I can pretend to do. I suppose it more convenient for Tom Glass. I think your Bishop Auckland Quaker is very high with his clover seed. I have not seen Mr Thompson's red seed yet, but by his account I have no doubt of its goodness and 91*s* is a very low price compared. But I see that clovers are dropped much at London. How far the war story may affect again time will shew. It is impossible that John's hill wethers should be good. They were so hurt last spring and then starved in John's upland last dry summer &c &c. That makes me so desirous that John should take the hill hogs this year before their teeth punish them, and they are so much *superior* this spring, at least they appear so to me. And I am fully convinced that they will do abundantly better in John's moors or even in a common than upon turnips. I see all our hogs are beginning to peel the turnips now, and you can see them turn the corners of their mouths to them, which is a certain sign that their foreteeth are loose. John will recollect that it takes a week or more in going. If we only have his consent we will

---

15 A reverse, setback.

still keep them as long as they are doing any good on turnips, and longer is madness! By sending the hill hogs early we can be the later in sending the best hogs. You may rest assured that Matty will sell the wethers at Morpeth if possible. John has not answered any part of our letter, but I hope he will say more in next. You told me about Jemmy Thompson, and I can tell you that Jemmy Armstrong is gone north to buy cattle. I hope John will deal with Holdgate and Shillito, if not he is right in taking them to market. I am glad that you name poor Matty in this letter. I thought not to have sent this until Wednesday by Matty from Morpeth, but I am now determined to send it away today, and Matty will write from Morpeth if he don't sell, and have leisure. I am in haste, yours affectionately

Geo Culley

171.   *George Culley and Matthew Culley jr to Matthew Culley*

Eastfield 14th March 1803

Dear Brother

I will address you this time as you don't talk of coming north yet. We sent you a letter today per post, directed to John Welch, and I am now beginning another to you and will answer yours more fully now. You talk about the Quaker of Bishop Auckland sending some samples of clover seeds to John of Dutch or French, but it is all Dutch to me as I cannot understand it at all. However if the red is 112*s* and the white 144*s* as I suppose, they are both sadly too dear. Clover seeds are coming down in price every day, and we can buy capital red of William Air at 5 guineas and Mr Thompson's is only 91*s* and I suppose a good sample by Mr Thompson's account. I never heard of Thomas Richardson being dead before. I requested you in my last to call on Anthony Surtees for Simon Smith's money, and I now desire you will at same time pay Luke Kidd his account. I am agreeable to breed fewer or even no sheep at all at Denton. In that case his shepherd ought to work, or he will have very little to do indeed, in the winter especially. However he may attend turnip cattle, but I am certain John will not keep him idle if he will but work. But herds are like old soldiers, 'sweer[16] to yoke'. I am also very sensible that John Welch can and does sell fat stock better than we can, and I know him both able and willing, and the more practice the more perfect he will become. But in that case we shall more than ever need a constant driver, and these Glasses begin to sicken of the trade. It is best for Jere

---

16 Difficult.

to tell John Nesbit himself, and Mr Wood may also complain to Mr Gregson of Low Lin, and what better will either be for complaining. Rather let Mrs Grey of Kimmerstone &c complain of not selling over winter. I will be bound for it that those that sold in the autumn acted the wisest part. You nor John have never noticed a letter which I wrote and sent from Berwick Saturday was a week, to apprize John of your coming to Denton. I could wish to hear from you respecting the hill hogs. Indeed we shall be under the necessity of sending them to Denton without leave soon I am certain, because William Brown was here today, and says he is nearly out of turnips and is going to send 200 gimmer hogs to Harry at Thornington, and Harry, who was also here, says he does not know what he will make of them. However we must take Harry's 11 outlaid steers and do as well as we can. It would be very well if you could prevail on Mrs Boldron late James Robson's widow, to assent to the selling of their Escomb land. I am glad Mr L. Wilkinson has been so much her friend. I am glad John is forward with plowing. His hedging will be got done. This is a blessed day indeed, we sowing wheat and other people beans and pease, but we can't get ours sown yet but will as soon as we can. Matty just returned from being at Wark. James Phillip mending and going about but very silly. They are obliged to put part ewes and lambs upon the seeds, indeed we shall be sadly whipped for meat everywhere, and no wonder, we all had fewer turnips at the beginning than usual, and more hogs *thank God* than common. However we have done wonderfully upon the whole, and if this fine weather stand we will still struggle hard. *Tuesday 15th*. A very fine fresh morning, not unlike showers, glass comed down a good deal. Yesterday we had the pleasure of seeing Cheviot not without snow but sufficiently broke, and quite clear on the top which is more than it has been for many months. I do believe we never had a real fresh until within these few days. I hear the coots in the pond this morning. I will send this by Matty to Morpeth, so that he may finish it and send you an account from there at any rate. He is just going to set of for Weetwood to meet Mr Nisbet for breakfast and have his company all the way to Morpeth. I am yours in haste

Geo Culley

[*M.C. jr*] Morpeth Wednesday 10 o'clock. I have just [set][17] William Glass off [with] the 48 best wethers for Denton, as I could not sell them here, never was bid money for them. T. Bates would have

---

17 Two words seem to have been omitted from this line.

bought them at 3. 14 or 15, but that I would not take, sold 11 at 3. 1.6, about 21 ½ per quarter. A very loaden market and many butchers, but will be a great many both beast and sheep unsold. This rumor of war and a hot press at Newcastle has caused all the sailors to run away from the ships, and all the ships that are ready for sea are laid up again and likely will be for some time, which causes the check in markets, but those that can keep over until times turn again will likely be well paid for it. In haste, yours

Mattw Culley

172.   *George Culley and Matthew Culley jr to Matthew Culley*
Eastfield 20th March 1803

Dear Brother

I will venture to address you in this, as I think you will be scarcely left Denton before it reach you, as I propose sending it by Matty, who with his friends propose being at Darlington next Friday on their road south. I was at Wark yesterday morning all well. Nelly of the Roads there. I saw James Phillip for the first time, he looks wonderfully well but his feet swell, which I understand is common after fever. Beans nearly done, begin oats tomorrow. We have put of beans two days on account of some of our land for oats being likely to get too dry, but begin with beans tomorrow. Harry also has sown no oats yet, being busy with beans and rape. Watered land looking well at Thornington. I would gladly hope that we shall be able to put of sending the hill hogs a few weeks as they still eat turnips so well, they are surprisingly good I think. The other best hogs both at Wark and Thornington begin to peel their turnips and turn one side of their mouths, so I gave orders to have their teeth examined and put those that have bad mouths to grass. Jemmy Glass was not got home yesterday. I am affraid the gimmers have drove badly. However not many of the ewes are lambed yet and the weather charming, so they have done very well. Several twins at Thornington. Harry has got all his quicks run. Last night Matty and Ralph brought me your long and kind letter begun on Monday last 14 inst. and concluded on Thursday 17th, the principal heads of which Matty tells me he answered in a letter at Berwick which we had wrote to send to you. However I shall also add some observations in this. I am glad the watered land at Denton looks so well, much in my opinion is to be done by water there. I am glad John is sending away the oxen and sheep to market. Perhaps this may be better than the last. It sometimes happens so. However I now don't see any probabillity of much if any advance. Indeed we can hardly expect sheep to exceed 9*d* per lb. It is a strange price, but I should look for cattle being higher, but it does not look like it. If it do take place I think it will be May and June. As you are so kind as mention money for our rents, I think it will be right if you can bring 2 or 300£ to assist in paying Mr Orde, and the tithe rent which is next Tuesday, and I never got James Bell's letter about it until yesterday and we are not provided for it, but I don't mind much as we can pay Mr Adams at the Alnwick May 12th fair. Lord Tankerville I am provided for, and I was to take care of it rather than the tithe rent. But Mr Orde wants 300£. If you had comed home by Morpeth might have paid him 300£ as that is the exact balance. But probably you will come by Mr Bates's, and if so don't forget to ascertain the time his mortgage is to be paid us. You rejoice me much about Mr Peacock being about to purchase the Escomb land and to pay in the 500£ and settle all matters. We buy it no truly, but as you say we may sell him the Hutchison piece afterwards. The 500£ will be very acceptable to me, and I am very willing to accept his note for the remainder. I am glad you have got the wool weighed and paid for, 6 lb. per fleece is as much as I should expect from hill hogs or dinmonds. I think they will weigh better this year, at least they look full of wool and full of health. I have long thought that nothing is so good for mankind as to be *usefully* employed. I am persuaded that we then fulfill the end we were created for, and does that duty which I trust will be found acceptable to our gracious Creator at the last day. We have secured Mr Air's rye grass seed. I hope you will allow John to sow hop or yellow clover seed this spring. I will be bound for it that that was a great loss to John's seeds. You certainly should do your house well when you begin with it. I am quite rejoiced at John letting a kiln so well and wisely at Lime Kiln Bank, and should approve much if he would let your lime kiln also, only he must take care in that case to guard against their wasting too much of your good soil. All that land may be made as good and perhaps better than before, by preserving the surface soil when boring part, and then spreading that properly at proper time upon the worked parts when levelled. If John could let the kiln it would leave him much more at liberty to pursue his much more important business. True Mr Matt I believe our cattle would be as good, *perhaps* better, at 4 years old than when drawn at 7. We are glad that Matty Culley is so well after the measles. I hope the Baileys, Hogarths &c &c will all do well poor fellows, and live to be usefull members of society.

Sunday afternoon. Andrew Bolton has been here, all well at Wark, and we have given Andrew your letter for Nelly and wrote to her to send a line over to Mr or Mrs Hogarth to let them know what her brother Matty says about their children. This is a fine day but very frosty. Andrew says that they finished sowing beans last night, and will begin with oats tomorrow. *Monday 21st March* is a very calm frosty morning with a rind as white as a slight cover of snow, but quite mild and is likely to prove a very warm day. *Tuesday 22nd.* A decent morning and midling fresh, yesterday turned out a very ill-natured day with a violent wind from south east and by east. However we got begun to sow our beans and will soon be done. I am going to meet Harry at Wark this morning to look over the farm with him. You talk of our going sometimes, but we have seldom been 2 days without some of us being there. Now after Matthew leaves us I cannot possibly get so often, but I think you will be returning soon. I intend to take my dinner with the tithe stewards at Mrs Barisford's, notwithstanding that I cannot pay them. *Wednesday 23.* A sweet mild morning, with mist rising in the hollows. We are going to sow seeds here and at every other farm I expect, as we got some rye grass seed from Mr Air yesterday and the clover was comed before. The London clover 19s per cwt. cheaper than our Haddington friend in price, and much superior in quality. Thompson for ever I warrant him. I wonder John Welch and Mr P. does not apply to Mr Thompson. Well sir, I must acknowledge that you have been wonderfully attentive in writing at every opportunity. Yesterday came William Glass when I was with the tithe gentlemen at Cornhill, and brought yours of 18th inst. after we had given both Will and Jemmy up, thought that they had gone to be soldiers in their warlike line. However I am glad it is no worse. But these gimmers must have been somehow mismanaged at the first. The young boy Glass with the assistance Harry lent him, has managed the lambing ewes cleverly. To be sure we are all much more obliged to the Almighty for sending such a fine lambing season, seed time &c &c. May we all be truly thankfull for His mercies. 'They are more than can be numbered'.[18] We are all rejoiced at Matty's being got to Denton, and although weak he will soon come about I trust, in the nature of the Culleys. Your daughter who I saw yesterday wrote immediately over to Mrs Hogarth, who and her young daughter is doing very well I hear.

I met Harry at Wark yesterday morning and was over the whole farm, everything looking delightfull. How pleasant and suitable is dry good weather to the husbandman, Beans all done, and oats are spreading upon the ground everywhere. I almost wish to be young

---

18 Psalm 40.5.

again, it is such fine seed time. I am sorry that John's oxen had been neglected by Kirtley, but I was certain something had been wrong because we sent him the best and forwardest oxen to a certainty. These unfortunate Hutchinsons, like too many others, have been their own enemies. I have no idea of John doing without a herd. I only meant that the herd would have rather too easy a time of it if he was not taught to put his hand to labor, or job at any rate. We shall not send John any hogs now of a while, because the seeds at Wark and Harry and watered land is all coming on nicely. Harry has put his toothless hogs on seeds, and we gave orders to do the same at Wark yesterday. I really think that the hill hogs are doing better than ours now, they keep their teeth longer, and have been wonderfully well managed this year so far. I certainly congratulate Mr Peacock on purchasing Escomb and rejoice to hear it, but I had no conception that he would do it. He must be obliged to that little *black man* who attends *stock exchange*. I fancy Mr P. would get part money on the death of Mrs P's other brother. I am told that Griffith the Durham attorney[19] is wonderfully clever. Mr P. is also an *investigator* or *sorter*. I don't perceive James Phillip to be reduced in flesh. He was walking amongst ewes and lambs yesterday, walks badly but does not look shrunk. We have caused each of your people to plow the *Pepper Hill Knows*, and I hope you will be home (although I don't wish by any means to hurry you) before the east townfield be plowed, because Harry and I think it should not be all plowed but a part should be left to that side of it next the shepherd's house and a temporary fence made. However we rather wish for your advice about it. But it will not be plowed of a fortnight or nearly from this. I thought that old *fox Head* would have paid when the wool was weighed. You must bring 300£ at least I think to pay Mr Ord &c &c. Wark clovers are very pretty. We viewed them all yesterday, and will have all the toothless hogs on them directly, and the rest ere long. But I again repeat that the hill hogs are the best, not the largest but more blooming in the wool than yours, and I should suppose fatter. They have never met with one check yet. I wish them well at Denton, and if they *luck* on will be as good as Mr Wood's. I hope better. I hope you will not be come from Denton before this reach you. I have already told you that Matty will leave it at Darlington as he passes on Friday next, when I hope you will get it that evening from Piercebridge, or next day to a certainty. And I hope you will not come away before Sunday and will always

---

19 John and Thomas Griffith, attorneys, Durham: *Pigot's Directory*, 1820.

recollect to send to Piercebridge before you leave Denton surely for letters. And I am sure you have no need as far as I know to hurry yourselves, as everything out of doors at Wark appears to me to go on as well as can be. Within doors I can be no judge, but I am sure you have a *nice, healthfull, plump, well disposed* and I have no doubt *attentive housekeeper.*

There was not a large company at the tithe rent dinner yesterday, I thought, but great plenty of everything. I was much hurt to see Mr Wrightson's hand shake so much poor man. When I first came down stairs this morning I could just perceive some *mist* rising in the hollows, but now ½ past 7 o'clock it is quite thick to the very doors. Ralph Brown has been the healthiest and in better spirits than we have ever known him, until within these 3 or 4 days he has fallen quite low. I have seldom known him so bad, and I am the more surprised because we always thought the weather has a great effect on him, but he now began in this very nice weather. But I trust he will come about again by and bye. *Thursday 24.* A fine growing morning, indeed I never saw a finer season than the present for both seed and lambing and grass growing. I leave this open for Matty to add anything that may occur to him on his road to Darlington, and am with every good wish yours truly

Geo Culley

[**M.C. jr**] Darlington Friday pm 4 o'clock. Dear friend, I am so far on my way to town, should be at Northallerton or Boroughbridge tonight. I left all pretty well at home. There was a very slow market at Morpeth on Wednesday, and a drop in both cattle and sheep, a good many set up as many people are unwilling to submit to take the prices going. The country looks better here than at Eastfield. I fear this will not reach you before you leave Denton. If it does not it is no matter, as John is just come in. I must conclude with respects and duty, yours in haste

Mattw Culley

173. *George Culley to John Welch*

Eastfield 25th March 1803

Well John

I shall address this to you as my brother is sure to be left Denton before this reaches you. I sent a long letter for Matty to leave at Darlington post house this day as they pass. He and Mr Jobson set of for Morpeth yesterday afternoon. I was sorry that Mr Nisbet does no[t] go. He was countermanded and I think Matty was disappointed. But I hope he will do well enough poor fellow, only Mr

Nisbet is such famous company and has such inexhaustible spirits. John what wonderfull fine weather it is. We are very busy indeed. We only got beans done yesterday, have a little piece spring wheat still to hazzard in Purdey's Holes, and are sowing seeds and oats. A nice calm time to sow seeds. Baily Hair has taken Ralph Laidler's Flodden farm for his brother. Ralph has drank himself stupid, and James Laidler is pursuing the same course. John I think Will Glass behaved well in getting those 48 sheep up. I dare say the long drive would hurt them much, but I thought them a grand lot. I know not who you will get to buy them. It is very aukward to drive them north again to Painshaw, and I fancy George is in the pot.[20] But I must blame bad driving entirely for his ewes weighing so badly, because the rest weighed so well at Berwick. But goods overdrove is the unwisest thing that can be. However if George had met Matty at Morpeth we intended him 5£ again at least, and I am certain that those wethers would have suited him well. And I have no wish that any honest man should lose by us, if they will but speak the truth, the *whole truth* as I am of opinion George will. *Saturday 26* is a most delightful morning indeed after several peals of thunder yesterday. I was up at Wark, and ordered all the wether hogs upon grass directly. The thunder was north west and went into Scotland. Our folks heard but little of it here. *Sunday 27* is another pleasant-like morning, although frosty, will I hope prove a warm day as there is little wind. I sent to the post this morning, in great hopes of hearing either from my brother or you, how you succeeded at Wakefield and Skipton, but no letter which has disappointed me much. However it leads me to think that my brother and sister must be on their road home, or you would some of you have wrote. *Monday 28* is a fair fresh morning with a west breeze quite moderate, and every way like a continuance of this blessed weather. May we be truly thankfull for these mercies. We have sent to the post every day in hopes of a letter to give us an account how you came on at Wakefield &c, but as no letter arrives I suppose my brother and sister may be upon their road home. George Ascough is in the country trying to buy a few fat cattle, but good fat ones are not plentifull I am told, either here or in the Merse. People sold all their older steers last autumn and are feeding *stots*. The lad is returned from post, no letter, so you have either forgot our wish to hear or my brother must be on the road. But a few lines would have *satisfied* if not gratified us. *Tuesday 29*. A very fine morning, but no Mr nor Mrs

---

20 In difficulties.

Culley nor any letter. We were paying our heavy Chillingham rents yesterday at Haugh Head. They are altering the castle.[21] There are 2 drivers here for the cattle bought of us and Mr Compton, and his sheep for Ratcliff. I was sorry to hear it said yesterday that Wakefield was a bad market last week, and that accounts for your silence on that occasion. But we must know it sooner or later, therefore I think it wrong of you not to inform us. However neither you nor I can make markets, we must take them as they happen and ought to be thankfull for things as they are. We that have much to *sell* must meet with *them*, both good and bad. I hope you quitted and got your money, and then never fret about the times, because I am sure you would do your best.

Tuesday 29 March continued. We have this day received the last letter from your master from Denton I imagine for a while, as he says my sister, he and Matthew were to come from Denton on Sunday morning. But I can't quite understand when they will be home, as they come by Brunton. I am glad that Mr Thomas Bates came to see them at Denton, and I congratulate you John on being either so lucky or so clever as to get through your cattle at Wakefield, and your sheep I conceive were well sold at Skipton. I do assure you I have not been better pleased a good while than by Ralph selling Ratcliff our *Wark oxen*. By a letter from Mr Sayle received this day he says 2000 sheep were turned out unsold at Wakefield. Mr Sayle says Matthew and Mr Jobson called at Wentbridge on Saturday last, but he was at Pontefract, the laidies went to Thomas in Norfolk, but they left their names and we heard from Matty at Darlington on Tuesday by a letter to our Nelly from Miss Bessy Greathead. How well you sell wheat by what we do. We can't get above 6s and you get 9s. Admitting that yours is much better, 3s per bushell is a strange difference. Your master talks of our sending you some Angus oats, but we have no good ones at present, we must try to get some better sent you. John let me advise you to sell your pigs. If this warm weather should continue much longer you will hardly get them sold, and corn is too high priced at present with you to consume with pigs. But this day has turned out very cold and raw and like showers. I am really glad that Mr Peacock has bought Escomb estate, pray give my respects to him and tell him that I wish him much luck and enjoyment, and I am glad that you have lent him so much money. What with the cash you have

---

21 A number of rooms in Chillingham Castle were remodelled in 1803: Pevsner, *Buildings of England: Northumberland*, p. 229.

put into Mr P.'s hands and the 500£ he is to pay us in May, what you will be able to raise more by that period, and what we shall raise with what we have in Berwick bank will I trust go far towards what we may want without borrowing any quantity however. Mr Peacock really is a wonderfully clever man. Mr Peacock and you must make your calculations with *accuracy* as I know you can, what money you can raise for Mr Bates by 27 of May, and then I can better tell how to *arrange* our money matters here. John I can't say that I would wish you to clip these large fat wethers. They will be far too fat for warm or hot weather. It would be well if you could prevail on George Nicholson to buy them after all and give him time to kill them, and money also if necessary. May tell him that we will return him 5£ on account of ewes, indeed Matty intended that at least if George had been at Morpeth the week Matty went on purpose to meet him. The rest of your sheep you may clip if you think right now, because I think we shall not now be under the necessity of sending the hill hogs so early, so long as they go before the weather turns very warm. When I see my brother we will then consider whether it is better to send a part of our best hogs now or not until the autumn. We have a dozen or 14 beasts, Ralph Compton's steers &c &c which will be to send to you by and bye, but we can keep them a good piece yet, I would fain hope until you have grass for them. We have got all our clover &c seed sown here except the land for barley, and also at Wark. It has been the finest nice calm time for sowing than I almost ever remember. I still think the fat markets will be good in May. Mr Sayle says that the turnips in Norfolk are *all rotted*. That must hurry the fat goods from that country to market, and although that fills the markets now, may make a want after. It is a bad job that your herd should fall to the millitia![22] I know not what you will do or how you will manage, but a herd you can't do without at any rate. 17*s* per new boll is certainly a famous price for white pease. They are worth growing at that price. *Wednesday 30th* is another sweet morning. It turned mild in the evening of yesterday, and this is a promising day. I am going up to Wark to see how they are going on.

Ralph and I have been at Wark. Wether hogs all on seeds very good and must do well, but I think they have lost ground of late. This has turned out a very blustering day, and cold, not unlike March weather

---

22 The militia was raised for home defence by ballot among men between 18 and 50. Paid substitution was allowed. The Militia Act of 1802 provided for raising 51,500 men.

neither. They have got a vast of lambs at Wark. Seed going on nicely everywhere, but it will [be][23] thought too dry for some countries. *Thursday 31st.* Yesterday was cold and windy, this morning windy but not so cold, very droughty though. My brother got home last night, wrote from Morpeth and left letter at the Blue Bell. He also wrote to you by Mr William Spours, who was going to Durham fair. I hope you will be there, or ought I am certain. However Mr S. will put it into post if you are not at Durham. A good market day at Morpeth for both cattle and sheep. I do think things are not plentifull here, cattle especially. I hope you will still prevail on George Nicholson to buy your few wethers. *Friday 1st April* has been a fine day. I was at Wark to see your master and mistress after their return. My brother says the seeds at Wark are 10 days before yours, but my brother forgets that he has been a week from Denton, which makes great odds at this time. They desire to know if your little boy has got the measles. My brother is going to sow some early pease as you do, by plowing the land twice and then drilling with a double mouldboard plow. We drill ours all with the wheat drill by Mr Bailey's directions, and answer well. I wonder you do not use yours so. My brother thinks a substitute may be got at Newcastle for Tom Glass at 10 or 12£, at least they are now hiring them at that price. The influenza is bad at Newcastle and Alnwick. Horses we hear were dear at Durham fair, many gone past for Edinburgh. You are to give my brother's old coat and waistcoat to Jonathan Cape, or make them for Robin, only the coat has been turned. These are my brother's own words. *Saturday 2nd April* is a very fine morning, but the wind has fallen low again into the south east and the east. However I think we have had some of our coldest weather lately from the west. Corn is got quite low with us again. Ralph could sell nothing yesterday at Kelso but poor 10 bolls at 40*s* per boll. I dare say we shall not need to send you any hogs now until you ask for them without you be very long of asking indeed, and it is my inclination to send you the hill hogs, as we are likely to have a large portion of turnips this autumn. Perhaps we had better keep our best hogs here. However I am willing to be advised for the best. But if the hill hogs come to you, I would advise you to assort them by putting the best into your best hay meat and the smaller ones into the moor or worst meat. Not but they will all make fat, only some of them are much larger than others. But all our hogs are fallen of lately owing to our being too long of giving them

---

23 A word seems to have been omitted.

grass, and we might (as the weather has proved) have put them a week or 10 days sooner to grass, which would have made a deal of difference between getting and falling off. My brother says you intend to clip those few best wethers, and I have no objection but that their mutton will be too fat for warm weather. George Ascough bought nothing I am told at all. As I intend Ralph to take this to Berwick to put into post I will conclude yours sincerely

Geo Culley

Grindon [...]²⁴ *o'clock*. As I have no letter from Matty, which I rather expected to meet with as we came here this morning, I will seal this, only I will leave it open until my brother come here if Ralph can wait so long, as he may chuse to add a little. If not I will seal and send it away by Ralph. I think my brother is not like to come. G.C.

174.   *Matthew Culley to John Welch*

Morpeth Wednesday 29 March 1803

Well John

Here is an excellent market for both sheep and cattle, from 8 to 9*d* per lb., beef 8*s*. Mr T. Blackett here, bought highland sheep at 52*s*, Boyd's sheep 52/6. They cost Clem Stephenson above 50, Mr Spours sold some of Ned Smith's or William Smith's at 3£ 11s, and Blackett said 21 to 22 per quarter, liker 21. A shabby show of cattle as could be, hardly fit for Shields. Mr Scott, banker at Newcastle, had hired a militia substitute at 10£, says he had several offers, the night before. I wish Glass had been there to hire one. The bill and notes were accepted at Surtees's bank thankfully, I backed the bills and paid our rent to Mr Bennet for Mr Ord. Forgot to ask if you would be at Durham fair, horses will be dear likely, many buyers as we came through Durham, were standing at all the roads into Durham to catch them as they came in. You might get a cart one in lieu of the mare I am to have from you. If I can, I will send this by Thomas Blacket. Yours &c

Mattw Culley

PS. I hope John is better. All the folk in Newcastle ill of a cold (influenza), not fatal but very common indeed. Write from Durham fair. The wethers would have been sold here today readily I think, as here are plenty of butchers from Placey used to buy your fatt ewes, says he sells good mutton at 8*d* per lb. I forget his name.

---

24 There is a blot on the page here.

175.   *George and Matthew Culley to John Welch*

Eastfield 3 April 1803

Well John

We sent you a full letter yesterday and this is still the very finest morning that we have yet had. It is most astonishing weather indeed. Pray God make us truly thankfull for all His mercies. Ralph could sell no corn yesterday, was only bid 37s per boll for wheat, and offered it at 38s. *4 April.* Frosty with a white rime but calm, mild and has the appearance of being a very fine day. We are all made happy with a letter from our Matthew from London this morning, but not a word about Smithfield. Indeed he had no time to see it when he wrote last. Grass forwarder. John if you want any Sweedish turnip seed alias ruta baga we can furnish you with it pretty good, only it will have to come by waggon. Had we known it could have gone south by Matty in the chaise he went in. I am clearly of opinion that they are better for horses than potatoes, and to a certainty injure the land less. I believe you should sow 2 lb. per acre, and we think that they are fond of a dampish soil, but the fresher the better, and are for sheep feed beyond anything of the vegetable kind I ever knew. By improper mowing and meat they are become impaired, but I hope we will be able to put them in right again by strict attention, which is required (happily for us) in most things in this world. And so nice our tup hogs become that they will *pick out* all the *red ones* and leave the white ones.

Geo Culley

[**M.C.**] John I believe we must do the house according to Richardson's plan this summer, but now we will come over when it is ready to begin with, only things should be got ready so as we may loose no time when we come. I wish you to get the level of the cellar and then bring up the level to near the back courtain door, then when the kitchen wall is down the drain can be carried up all the way to the church yard wall or nearly I think. If Richardson approve the back kitchen loft may be supported if necessary, but this Wass &c may consider off as Bright Wass should draw a plan, but I will write more fully in a little time. Why should we do more at the house than need when a good house is intended to be built above the town? We mean to take the roof lower and have the kitchen fully to join the passage, which passage is to go straight through, then a passage past the foot of the staircase to the pantry and milk house. Your mistress prefers a convenient house to a genteel one. Yours &c

Mattw Culley

I wish for Purdy Starforth, who was our carpenter before we left

Denton, to make the windows and roof the house &c, as I know he is a compleat workman. Mattw Culley

[**G.C.**] *Tuesday 5 April* is cool and rather windy but good weather. We are still plowing Lee and sowing potato oats, but will be done this week I think. Many people are fallen out of love with potato oats, but we have no reason to quarrel with them yet. We had the pleasure of hearing from Matthew yesterday, safe arrived at London. *Wednesday 6 April* is a sweet morning after a cheerless day. I am going around by Thornington to Wark, and I was at Grindon yesterday, seeds doing wonders everywhere this admirable weather. We are turning our ewes and taking out the barren ones to put on good meat and get them forward, and sell our hogs live and wethers are now on grass. I don't believe that any sheep *enjoy* or do *equal good* upon turnips whenever they are running to top. Indeed I have long been of that opinion, and am more and more confirmed in it. Nay they will do better on very midling grass, and they are unanimous desirous to be at grass. But cattle on the contrary never do so well on turnips as in the above state untill they are *yellow* and were podding. Long experience convinces me of this, and I would account for it by natural reasoning were it necessary. But experience is the best of all tests or proofs for the farmer. But I have known clever men err much in that way, the Mr Taylors in particular. I have known sheep that were fat [?reduced][25] to *bones* by punishing them on turnips in the growing state, and these money-making men at that time led the country or leading. So much for bad examples. But remember John that all knowledge is derived from experience, but people are often led wrong for want of making proper observations, and deductions from experience. *Deduction* is too hard a word, *considering* the matter well over, or *viewing* it *on all sides* until you are satisfied that you have found the reason why. Mr Taylors made a *great fortune*, but had they the world to begin now, I am almost persuaded that they would have *lost* one. However I will continue to say that they would not have made half. I don't make these remarks out of *envy* or *malice*, God forbid, but because it came across my mind at the time and may be of use to you. They were no more to be compared to Robert Colling, nor nothing at all, but they could not discriminate, nothing was *good* that was not *great* in their eyes, and good land and long keep made all up, and totally *deceived them*. It is the same with many Lincolnshire men at this moment, at least it was so when I was last in that *fertile country*.

---

25 A word seems to have been omitted.

The goodness of the land *blinds them*. But we now have cattle and sheep that make fatter on *midling fare* than Mr Taylors or Lincolnshire men ever showed any from their *good land* and *long keeping*. I wish you saw a cow bred by Charles Colling and feeding now by Mr Compton. She is a *phenomenon*. I never saw her equal I think, nor would any person believe what she has done in a few months.

Thursday 7 April. Windy and droughty but a good morning. We have got no newspaper this morning so can't tell how London markets are this week. A man called Thornton who I believe buys for Lunn has been buying several young turnip-fed young steerish tied up cattle of Tom Tulip &c. They must be for grass. I handled Tom's one day along with Eales and I thought them uncommonly green and raw. Let us know when we are to send you any hogs, recollect that they will be to send at twice, and will be 8 or10 days upon the road each time and during which grass is growing. We certainly after the new year commenced had a succession of very changeable weather, severe frosty winds and thaws, with sleet and showers of snow which have laid 48 hours scarcely. But since this fine weather began I know not that I recollect a spring so good, spring wheat and oats have come up the quickest, and with the strongest blade I ever remember. It is a shower of rain this moment. But I am always affraid of rain in this month because it is so apt to bring cold weather and fall snow on the mountains. *Good Friday 8th* is a real good Friday. I never was out in a sweeter morning after a succession of warm mild showers all yesterday. I never knew so fine a spring, everything is alive and all Nature is rejoicing. Shall not man then return thanks and praise to that beneficent Being who so liberally supplies all our wants. I understand that Morpeth was a full market on Wednesday last, but that cattle were readily sold at 8s per sink, which however extraordinary has been near about the price for a number of weeks or rather months. For excepting a little while that fat cattle were sold at 8/6 at Morpeth when you were barely 8s at Darlington, they have kept pretty steady at 8s, sometimes a quicker, sometimes a slower market, but the variation in price trifling. But it is a good price still and a price at which we can afford them at very well. Sheep from 8d to 8 ¾ rather too many, consequently not so readily sold, but I fancy all were got through, and I begin to think that things will fall scarcer here away soon. In short I know of very few lots of fat cattle and no great quantity of sheep. I know not what Jemmy Thompson of Boag End is doing, but I have not heard of his being at Morpeth yet. Perhaps he intends to clip his. Our paper did not come to hand yesterday but I saw one at Wooler, wherein I was sorry to perceive that corn in general is in a sinking state in London, oats excepted which keep price the best. Wheat lower and markets full of it in the south, and we shall be under the necessity of selling a quantity as we are so full, and unfortunately a good deal of it is not good in quality by far, especially Grindon wheat this year. I now wish Ralph had sold the last Saturday. However we must *grin* and abide it as the lad did when his mother got him behind the door. *Saturday 9th April 1803* is a very fine morning although a little frosty after some very fine mild showers yesterday and in the evening. Whether it may continue or not cannot be known, but to a certainty there never was a greater growing time at the beginning of April before. *Easter Sunday 10th.* A fine morning. No letter from Matthew. We sold 200 bolls wheat yesterday at Berwick at no more than 37s, but we were so full that we were obliged to sell.

Monday 11th. A dry but good morning. No letter from Matthew which disappoints me much. I hope we shall hear from you now as soon as this Easter Monday is over. But I can tell you that the first market will be bad and that I have this moment a letter from Mr Barf of Wakefield being shockingly bad last Wednesday, sheep only 8d per lb. sink, and butchers meat comed down 1d per lb. in all their markets, and the general opinion is that things will be lower. Corn flat and sinking, wool much down, especially hog wool which is not now marketable. I know Barf will try to sink our spirits if possible before clip respecting wool, but I am affraid this war, which I doubt will take place, will sink everything, especially wool at the present. However we have been very lucky in selling so much of our fat stock, and your piece of wool, and I am not much affraid still but you will sell your queys well enough. At least I am certain that things are falling scarce in these Borders, and at any rate you must now be strongest in keeping, and grass growing apace every day so you need not hurry in selling for some time. They will always gain *weight*, and I should not be surprised if markets still take a turn, and we can do for money, I think, very well. We certainly should have taken little Tomy's price for those few best sheep, but as it was done for the best never mind, and they have gained an abundance of credit I assure you. They say the like were never seen at Morpeth. We missed in not writing to Tom Humble. He has said to somebody since that he would have bought them, and that he has sold very little mutton below 8d per lb. in joints all this spring. He was the boy for us, and I would fain have wrote to him at the time that I wrote to George Nicholson, but Matty thought it unnecessary. Well they must now take their chance, but I am quite of an opinion that they will be best sold to the northward now, and all sheep I do believe, as I am clear that fat, especially sheep, are falling scarce in this district and we have now such a flush of grass upon our seeds in this country that people will be in no great hurry with their fat stock to market. Indeed this fine warm dry spring, and such vast quantities of seeds in a flush, will make fat early and heavy lamb. But still they can't live on lamb altogether, and it will be very surprising to me if the war don't still advance fat cattle. At least I never knew it otherwise in my time, only the idea of war came upon us like a gun shot and has thrown [. . .][26] a damp on all kinds of things. But I hope we will recover again from this panick and then things will go again. G.C. Eastfield

[**M.C.**] John Monday, my brother gone to see Mr Bailey at Chillingham to sign the papers or deeds for Mr Peacock for Escomb. We now have grass and turnips plenty. I fear you mist selling your swine, we can keep our small ones as we have pottatoes, but you must sow Swedish turnips rather than too many pottatoes, as they are more valuable for sheep and cattle or horses. Our early grasses are fit for cattle, so good. We never heard when Matty went to school. I think I may likely be with you at the end of May, as I shall likely be with Mr Bates on or about the 23 May, when I wish you to have the drain brought up for the cellar, and stones got for walling. Winter was to have bricks burnt by June, but he should not be hurried as good ones are the word. Yours &c

Matthw Culley

[**G.C.**] *Wednesday 13th*. Misty to the door, but mild and calm although frosty. We got a letter yesterday from Matty. He and Mr Jobson would leave London on Monday last, and are now I fancy at Mr Boswell's at Piddleton Dorsetshire. Smithfield he says is very flat. I fancy that this war story affects everything more or less. If fat sinks in value, lean must also drop. *Thursday 14th*. No letter John which mistifies me much not to hear of Easter Monday. Would you but write half a dozen lines I would thank you. I am sure it would be bad for sheep. 500 set up at Morpeth yesterday it is said. But cattle sold at about 8*s* and not too many, nor will not be I think, except sheep hurt them. I begin to think now that there will be too many sheep. This sunny weather will feed and swell the lambs so fast, they have nothing to do but sleep. Mr Mills has sold Armstrong all his fresh oxen at last. I am much disposed to think that good fat cattle are falling thin in your country. I will not send this until I hear from you,

---

26 There is a blot on the page here.

if it be a month to come. Yours. *Friday morning continued*. John Welch I am unable to say how much I am *hurt* at not receiving a letter from you neither yesterday nor today respecting Easter Monday. The lad who I sent to post every morning to receive your letter is returned and no letter from you. In my anger on the other side I said I would not send this until I heard from you if a month elapsed, but you not doing your duty is no reason why I should not do mine. I will therefore send this tomorrow at all events. You may be ill. But in that case, if not insensible, might get Bob Dippie or rather Mr Peacock to write. However I may conjecture a thousand things until knowing what can be your reason for delaying writing. Good God, would you only write 3 or 4 lines it would be most agreeable and satisfactory. But – no I will say no more than that I am made happy by receiving a letter from my son about an unfortunate youth John Proctor who was once with me, a young Scotchman, who has left his father abruptly. Too good a father for him! How happy am I and I pray God to make me truly thankfull for a son that is dutifull and attentive, and I know nothing so pleasant as dutifullness and attention *in us all*. Your well wisher

Geo Culley

Grindon Saturday 16th. I received your letter by the road and thank you, but would have thanked you more if it had come last Wednesday or Thursday as it should have done. Nothing so absurd as waiting for our Matty, who is not coming home these 3 weeks, and then it is great odds that he will not come by Denton. Indeed I don't see how he can, as Mr Jobson and he will chaise all the way. They are now at Mr Boswell's. We had a letter yesterday from Matty from Egham in Surrey all well. I am much rejoiced that you have sold your pigs. I still think fat cattle may be better and not worse sold. Sheep I think differently of, because fat lambs and hot weather will hurt them, and indeed that may hurt cattle also. Only queys and Kyloes will sell best in hot weather. If you can sell your 2 big oxen decently do so, if not do what seemeth best to you. We can do for money, which you will hear from Mr Peacock is to be paid on 23 May. We will keep hill hogs altogether if you chuse. Only you should recollect that if they do come in their wool the hot weather will be much against them and they must come at twice. Your few fat wethers you must just do the best you can with, but they must be clipped now. I am happy that the rest are all sold. I would advise you to sell wheat as I do not think it will be any better. Do you think Mrs Barf is with child or only fat? You will see what my brother says about the house in this letter, and had you wrote sooner you would have got this sooner. You make a bad excuse of waiting for our Matty, but a bad excuse is better than none they say. I begin to fear you are incurable in writing to us so often as you ought. However I cannot say more to you, any reasonable man would see the benefit and propriety of letters passing often.

Geo Culley

176.   *George and Matthew Culley to John Welch*

Eastfield 17th April 1803

Well John

I was sorry to say so much to you in my last letter, but really you should not neglect to write directly after Easter and Whitsum Monday &c &c. God sake, what is a sheet of paper and 6*d* postage to be compared to the benefits received by communication in letters. I received a letter from Mr Peacock this morning in answer to one I wrote to him, about Sir Ralph Milbank's mortgage &c,[27] which Mr P. had advised you properly about. The money you see is now to be paid by Mr Bates on 23 next month. Consequently you or Mr Peacock will have to meet Mr Thomas Bates at Durham, most probably on that day, but my brother proposes being along with Mr Bates at that time, and he will *certainly*, or I *ought*, one of the two, to name the place of meeting before that time. The 23 is the day to a certainty, but when Mr Bates wrote that to my brother the other day he had neglected to say where the meeting is to be. But most likely Durham I should think. John you say '2£ 2*s* per head for 11 pigs', which I suppose is a mistake and should be 3£ 3*s* I conceive, as those pigs were worth nearly 2£ 2*s* when they left us, especially at or near 8s per stone. The money George Nicholson makes of the offal of the sheep bought at Darlington is most surprising. What a pity he did not receive my letter in time. I always calculate you sell wheat at 2*s* per bushel or 6*s* our boll above our price on an average, but at this time your price exceeds ours 15*s* our boll. Our best price is 37 and yours per our boll is 52*s* or 52/6. Indeed I never knew corn flatter or less done in that way at Berwick. I am glad my nephew Matthew was got stout again, and I am happy to hear that your folks at Denton are healthy. The influenza has been fatal to many old and weakly people about Berwick and most large towns. You ought to attend Durham fair if possible, and all publick markets near you. There is always something to be *learned,*

---

27 The piece of land at Denton which the Culleys bought from Sir Ralph Milbanke in 1798 had been mortgaged. The mortgagee, Lady Headley, was claiming possession for non-payment. See Nos. 184ff.

heard or *seen*. Whenever you wish for the hill grazing hogs we will send you them, on account of warm weather perhaps they should not be too late. However the weather is altered colder here since yesterday about 3 o'clock. Sir H. Vane[28] is like all folks that don't know what they would be at. You could not expect to be otherwise with Mr Colpitts about tithes. I am very glad that you are likely to have a good crop of lambs. Take care of them, as you are to breed no more at present. Your Quaker's seed appeared remarkably good, but it needed to be no better than we got from Mr Thompson London. As we tried it, all grew or nearly so. *Monday 18* is windy but fresh, and as the mercury has been sinking for many days, it is likely we shall be getting showers which will do much good if it don't turn frosty and cold after, which is often the case at this season. Eastfield Monday.

[**M.C.**] John I wrote this day to Mr Bates desiring him to write to Mr Peacock to let him know when and where the money was to be paid, but Mr Bates will let Mr Peacock know off the day and place, the day is on 23rd May, I suppose Durham will be the place of payment. In my letter to Thomas B. I said we would be if all was well at Brunton by the 20th of May. We shall likely be at Durham and so to Denton, where we hope, according to my letters to you, the drain up to the coal house end will be forwarded, and we will build up the back parts of the house to the churchyard and cover it in. I hope we can do all this without much inconvenience to you and your family, especially if the weather be dry. Mr Wass will have wood I hope, and Purdy Starforth will manage the roof and staircase. John Dickinson will have enough to do in your other work, he charged enormously for his brother, and so Robin Dickinson thought 6*d* per day too much. I mean to take off the roof and put the garret all in one according to Richardson's draught of the house, but with Wass's windows. Overend will be able to get slates ready for the whole roof according to the plan, by widening the north wall. He is taken 18 inches so as to leave a door between the kitchen and coal house. Morpeth was a slow

---

28 Sir Henry Vane, 1771–1813, 2nd baronet of Long Newton and Wynyard, took the name of Tempest on being named heir to the estates of his maternal uncle John Tempest of Old Durham. His daughter Frances Anne married Charles, 3rd Marquess of Londonderry: *Complete Baronetage*, vol. 5. He was a founder member of a society set up in 1803 in Co. Durham, the members of which undertook to conduct experiments on their farms and inspect each others' results. Matthew Culley and the Colling brothers were also members. *Farmer's Magazine*, 4(1803), pp. 283–6; *Annals of Agriculture*, 41(1804), pp. 1–20; Bailey, *General View, Durham*, pp. 269–70.

market for fatt stock on Wednesday last, very slow. I know not for want of Richardson's plan whether it was 18 in. or 2 ft 6 in. Richardson knows, but I wish to follow Richardson's plan of the width of the house we think. Winter &c will have bricks as soon as wanted I suppose. I wrote to you saying you should get the drain forwarded, and the conduit walled and filled up to the back courtain door, also to get stones of dry quality for the house walls, again I get over. I hope you got my letter to that purpose. We should have our plan fixed and the wood for the back room, joists, window frames and doors &c for the back of the house ready, and above all the roof and slates ready in a little time after we begin to build, which I suppose may be the last week of May. Yours &c

Matthw Culley

[*G.C.*] *Tuesday 19th* is a very cold morning indeed, has been a shower in the night, and Cheviot has a strong cover of snow, which is not at all uncommon at this season but is a very unpleasant change after the extremely hot weather we had only the last week.

To talk of the weather it is but a folly,
While it snows on the hills, it rains in the valley.

Mr Ratcliff called this day at the Westfield and looked at 4 oxen we have there, Ralph asked 28£ per, Ratcliff bid him set them at 26£ so they parted. He had been up at Redding and bought a large lot of Mr Nisbet, says he can buy things very low to what they did a while ago. I do suppose that things are very flat at present in most parts of England, and you may hug yourself and rejoice that you got quit of your large oxen at so fair a price at Wakefield. I do assure you I feel extremely pleased at having sold so much stock so easily as we had luck to do. But still either lean stock must drop a good deal or fat can not come much below 8s per stone. However we can't have forgot Alnwick fair last year, how low fat stock fell, and it may be the case this year again. Nevertheless things advanced again in the autumn higher than ever, so very fluctuating are times nowadays. However I would not run away from good fat stock in a very cowardly manner yet. Ratcliff says he bought 2 pigs which weighed 14 stone 3 lb. each last Wednesday at Morpeth for 5£ 10s the two, No wonder that he should get rich with such bargains, and you may be thankfull that you have sold your pigs even at 2£ 2s per, although I hope you had set it wrong down in your hurry. One should always read over what they wrote of possible, because all mankind are liable to mistake. I often think upon an observation of Ralph Brown's. He says that one should always be content with at or near 9d per lb. for sheep and 8s per stone for cattle sink. As he says, what would people have? Cannot we afford them at those prices? This has turned out a most tempestuous day indeed. I think I never recollect a more stormy day, with constant showers one after another of hail or snow. What a sudden and terrible change from extreme heat to extreme cold. I really never felt a colder day than this I think. Ratcliff talks of 7s or 7/6d per stone at most, says that they have reduced beef from 56 to 52, but that the Sunderland people did first drop the beef and that he, Binney and Moffitt and Swan kept the price up to 56 until very lately, and some time after all the others had dropped to 52s per cwt., and after all I cannot help thinking that fat cattle are not plentifull in this district, only he, Ratcliff, has bought Mr Tom Nisbet's large lot, is going to buy Mrs Grey's of Kimberstone, and being now full handed is determined to buy low or not at all, and who can blame the boy? In fact he has the ball literally at his foot, and he will take care to play it well you may be assured. Indeed he is the only butcher that has come this spring into this corner. Moffitt and Swan, Binney, White, Stephenson nor Forster &c &c never come further than Morpeth, and very right too I should think. Nevertheless it is right of Ratcliff to come, because he kills more I understand than almost all the rest, and cannot depend on Morpeth at all for his immense consumption. *Wednesday 20 April.* As cold as Christmas, but much worse to bear after such nice weather as we have had, the snow lying close on *Cheviot, Newton Tor* and many of the lower hills, and has laid all night. Yet the frost is not so severe as last May by a deal, when we had that bad weather, and the wind is westerly, not so violent as yesterday but extremely cold. Glass very low but now rising a little. John as I can forsee no probabillity of corn advancing, I would advise you to keep selling as much as will serve to pay your way at any rate, and that will keep your money whole which is in the bank, and I think nothing of borrowing a few 100 pounds of the bank to make up what you have 2000£, which is the sum wanted on 23 of next month for Mr Bates. But Mr Peacock will manage that matter for you and us, he is more conversant with money matters and more clever in business of that kind than most people I know off. Indeed he wrote me word the other day that he would transact that business without giving us any trouble. Not but your master seems to be over at same time, and so he should as he seems to wish to make considerable alterations about the house you live in, which may be right but buildings are things I don't much understand, nor do I think it prudent to do much at the present. However he either does or ought to know best.

Saturday 23rd April. John I have been so ill of the influenza since Thursday that I have been keeping my bed most of the time.

However thank God I am much better this morning but far from well. However I trust to be soon well. Your mistress is ill, began after me, but has not hitherto been so severely held. William Brown (young Willy) began first, and had it slightly, Bella our maid servant has been worst of any of us, and I suppose it will go through the village.

This day much milder. Otherwise we have had a week of very severe weather, violent frosty winds, and snow upon all the mountains, and you should think ofhaving some hogs soon I do think, as it now be May before they reach you, and you will likely be for 2 droves. In your answer to this I hope you will let us know. Nothing at Wooler on Thursday, I have got no account how the Morpeth market was. I am yours sincerely

Geo Culley

PS. My hand shakes so that I can scarcely write so as you will make it out.

177.　*George Culley to John Welch*

Eastfield 28th April 1803

John Welch

I was in bed and so poorly this morning when your kind letter arrived, that I was glad to employ Nelly poor thing, who I am affraid is beginning with this unpleasant disorder. Few people in this village have escaped it, but I thank God we seem all to be in a way upwards now. Your mistress and myself have been confined now a week, were the worst but are much better. The weather is so bitter cold that we dare not stir out. I hope you will receive the short letter sent today in due course, and then the ruta baga which we have sent per Howey's waggon. The coach is so very expensive, and the end of May is soon enough to sow the first. We sow again about 8 of June and a 3rd time about 24th June. The first is generally largest, but the last the richest and will keep until July if you chuse pulled, and after taking the roots laid up between a rail and an old back of a hedge, or rails on both sides, with straw on the top. It requires air below, or it hurts and ferments both. You may either take the tops of or not, but the roots and dirt must be off. If you had a machine for cutting them with, how they would have grown your 2 oxen mixed with their corn or meal, or any other particular cattle at a future day. Has Mr Colling not a cutting machine? If so any of your clever ingenious smiths could easily make one by it. Our smiths at the Blue Bell the White, I dare say have made some dozen after seeing ours. The only danger is that workmen will make improvements, which unfortunately generally turn out quite different. John I can't understand how you make out upwards of 200 fat sheep. I thought you had sold several at Skipton. John you don't acknowledge your mistake about the 2£ 2s pigs, you may as well confess now as afterwards, we shall know. Don't complain of your wheat market my lad, yours are famous prices upon my word. To be sure we beat you in oats. Mr Barf's payment is great news indeed. Never fear Mr Peacock, he knows where to get money, I hope he will buy our price next. He is and will prove himself a clever fellow. I should be very glad if your Skipton or Bradford friend would come and buy your 2 oxen and large sheep. The drop for fat sheep especially has been so great at Morpeth lately that no butchers are likely to go to Darlington from Shields &c &c. I now begin to fear the fat markets mending, I wish they may only stand where they are. However I still am of opinion that fat cattle are not plentifull in this district. Alnwick fair will decide the matter I think. *Saturday 30th.* Well John we had the happiness of getting Matthew safe last night, and Miss Greathead, they brought me your letter of 27, also your mistress and I are thank God both getting better although slowly. All the young people have got easy through it, but it is much severer upon old folks. I am glad that Matty got a part of a day to look over your farm, and I am happy to say that he gives you great commendation, says that he has seen few farms so correctly managed. This gives me very great pleasure John I assure you. I am glad that you have sent 8 queys away to Skipton. The money will be usefull. Not that you need to sell on that account, as we can always raise the money on a bank, but the less is to raise the better certainly. It shows that trade must be very flat indeed when Crisp does not buy such valuable cattle as Lord Darlington's. John I beg your pardon about the pigs. I really was in hopes that you might be mistaken. We have these 8 outlaid steers of Ralph Compton's and 2 or 3 more which must come to you as soon as ever you can take them I hope you will secure the Sweedish seed safe. *Sunday 1st May* is a fresher day than we have had of some time although yesterday was a decent day. The glass is comed much down which makes me affraid of a gale of wind. However if it only keep fresh grass will grow, the sun has such power. By a letter from Mr Sayle he says Wakefield was good for cattle and rather better for sheep than had been, so I would fain hope that you would meet with a decent market at Skipton for your queys. Ox beasts 8 to 9s per stone sink, small cattle higher, and what is rather remarkable, Morpeth last Wednesday was one of the dearest markets which had been there of a long time. Nobody knows what may turn up *trumps* still. I still remain of that opinion that fat stock, *cattle* especially, are not plentifull in this district, and shall remain in that way of thinking unless Alnwick fair set my opinion aside. Ralph Brown sold a very decent good cow which we have at the Westfield at *8s/6d* per stone. Ralph was offered 28£ for her again and again, but he chose to back his judgement for once and very right, although butchers generally win on those occasions, and if they do beat him never mind. They can always continue to make a difference in the killing. Be as it will, she is a good weigher or I am mistaken, a deal of lye and a deal of fat, 7 years old only. I hope with Ralph that she will make more than 28£ by weight. If not we will not be disappointed and you shall know the result. I know the butcher beat you with the ox which went to Manchester, George Ascough told me so. I am glad that you are selling corn. I have a bad opinion of the corn markets. Ben Sayle speaks of them in a very disponding way. I am better today than I have been yet, but weaker than I almost ever remember myself by so short an illness. I think my wife is not so well today. Be so kind as let Mr Peacock know that I have signed his Escomb deeds right now I hope, before Ralph and William Brown who signed as witnesses, and that we will send them by my brother and sister when they go south about the 20th I suppose. It will spare Mr Peacock the postage of a letter as I have nothing else to write to him about at present, and tell him that I sincerely wish him to buy the other Escomb estate provided he thinks it will do him good, and that if it was in my power I would lend him a few hundred pounds, but that at present my hands are tied. *Monday 2nd May.* A wet morning, and if it come fresh and warm after, must necessarily do a great deal of service to the country, as the land was rather beginning to want moisture.

Tuesday 3 May. Well John I know not what sort of weather you have at Skipton today, but I do assure you that it is very coarse, blowing and boisterous here, attended with sharp cold showers, but no snow or hail yet. However if you have but a good market you will mind the weather the less. It will be very happy if you can take some more cattle and sheep to Skipton soon, or at any rate to take a few of our cattle and hogs as soon as possible, will be very charitable to us I assure you. I am happy to say I fancy myself better today than I have been any time since I began that illness, and could I only get out either on foot or on horseback I think that I would, please God, soon come about. I think your mistress is worse than I am now, but it has been very severe on both of us. *Wednesday 4th May.* Pretty calm after a sad cold windy day yesterday, the wind a good deal north east this morning, but has fallen some slight showers of rain. Our people almost all gone to Wooler fair except myself and Ralph Brown, who both should have been there, but Ralph is badly this morning and I am sadly affraid he is going to take this unpleasant disorder. Matty has had it very badly, indeed the most of the village has had it one way or the other, and although I am sensibly better daily, yet it is very slow amending. John I hope we shall have a good or pretty good account from you in a day or 2 about your Skipton market, and that you will begin to think and contrive about taking some stock from us. I do believe grass would gladly grow if we only had a fine day or two. *Thursday 5th May.* Yesterday at Wooler fair stock for grazing was very dear sold, several hog buyers also out from Yorkshire, but it's said that they are not fond of giving great prices, but the matter is people will not take bad prices. I am glad to inform you that Jemmy Thompson told Matty Culley that good sheep rough were worth 9*d* or nearly sink per lb., and clipped sheep in proportion, and little difference in cattle. *Friday 6.* A very fine fresh day, and I am happy to say that I am considerably better today indeed, and your mistress is a good deal better. Ralph Brown I hope will not be so severely held, but Matty is very ill still. Well John a lad returned from Grindon with eggs and has brought me your kind and good letter, near 8 o'clock, but however late it is very acceptable. How pleasant are good markets. How pleasant also to have good stock to dispose of at these times. I have prophesyed long that a turn in markets was to take place, and it has at the last but rather longer than I expected. I hope you have got your sheep washed this sweet day, and I approve of your not sending them until the 2nd market. However that you will consider of better than I can advise you. The sooner before hot weather the better, I should think. I am rather affraid of your being too late with your 2 large bullocks. You are right to take your fat sheep all to Skipton even if you should suffer, because you have been so lucky on the whole at Skipton that you can afford to run a little risk, and to wait on your friend George and West is only babling. Ready money is a very pleasant thing always, particularly at this time. We have had no rain of any consequence here. I know the gimmers are small, but they had been mismanaged in driving to you. We will put of the hogs as long as we can for you. I don't know what to say about our fat cattle, they are not very fat but some of them very clever. I am of opinion that there are very few such like nice fat cattle as you would like to stand by at Skipton. I am glad that you have sent away a part of your turnip seed to near Leeds to sell. I will send you this tomorrow, and send yours up to Wark for my brother to see, and am with a wish for good fat markets yours truly

Geo Culley

Saturday 7th May. A very fine bright but rather frosty morning.

However grass will grow such days as this well in my opinion. Ralph not quite well yet. Matty is going to Berwick from where he will send you this per post, and we may likely send you another short letter at or about Alnwick fair.

**178.	*George Culley and Matthew Culley jr to John Welch***

Eastfield 8 May 1803

Well John

We sent a letter to you yesterday per post in answer to the one containing the good account of Skipton market, and this morning has brought me a letter from Mr Sayle wherein he says he was at Wakefield 'where the market was particularly good, cattle from 8/6 to 10s, unclipped sheep rather more than 9*d* sink, and everything sold. Corn very flat, some wheat looking well but much badly, barley looking red as a fox tail'. But barley often looks ill with frosty mornings at this season. At Berwick yesterday next to nothing done. William Pickering called here yesterday and says that Alnwick will be a bad market for fat. It may be so, but I will not believe it until I hear it from good authority. *Monday 9th*. Very dry, but good farming weather for getting quickens out of the turnip land. Your master Matthew is highly pleased with the price of your queys, 29£ 10s overhead for them is a most capital price. That man Oliver from your country was at Wooler fair it seems buying queys and fresh steers. Ralph Compton is resolved to sell his steers 10 and 2 spayed queys. It is a bad time of year for us to buy them, but we must look at them. He wants the money amongst friends, and then he will keep them for us a month or 2. If we can buy them to do good we can manage thereafter. Matty is gone to Harry Howey's sale at White House today, I dare say there will be a great number of people there. *Tuesday 10th May*. Matty attended Mr Harry Howey's sale yesterday at Brandon White House, ewes and single lambs were sold from 55s to 3£ per, twins 3£ 11s, ewe hogs 35s to 44/6d, wether hogs 43 to 45/6, tups hardly sold, but the sheep upon the whole were pretty well sold. Last year's calves were sold up to 9£ per, 2 yrs old 12 to 14£ per, 3 yrs old 16£ 10 to 18£, cows from 13£ 10 to 19£. So much for the sale. Brother was here yesterday and we fixed for 100 hogs to set of for you on the next Monday. Perhaps we may clip the other 100 or whatever you can take. The weather is now very kindly again, but we begin to wish for rain in this cold country. My brother says that the Scotch fat markets are better. It is very curious but very true, that the fat markets generally get all better together or drop altogether. This has proved a most stirring droughty day. *Wednesday 11*. Very cold and droughty, makes vegetation very languid. Matty is just going for Alnwick, at least to Mr Spours overnight. I shall say adieu

Geo Culley

Since signing my name received a letter from Mr Barf, says Wakefield was higher for sheep than any day this year *9d* to 9 ½ *d* sink, clipped in proportion, best cattle *9s*. I rather think his account as likely to be correct as friend Ben Sayle, corn all flat but beans. By a letter also from Mr Stubbins he says corn very flat, but hogs at Lincoln fair sold from *40s* to *55s* per. A letter from Scotland says fat markets much better and likely to continue, so we hope. G.C.

[*M.C. jr*] Alnwick 12 o'clock. Well John here has been a very high fair, very little fat in and then sold at high prices, 7/6 to 8/6 and nothing very nice shown almost. Ratcliff's man has paid me but has bought nothing, says he has plenty to do with and will not buy at these prices yet. Several other Yorkshire men, and here stock very high, also milk cows, a quey sold at 24/6 to calve, and the buyer sold her again directly for 2 guineas profit to a Yorkshire buyer, not a beast left in the street some time since. I have seen John Forster, says he bid you 50£ for the 2 large oxen but you would not take it. I think there is every chance for markets to stand high at present, at the same time I think it wise to be selling at *9s 9d* or near it. Very fining weather with high winds. I am in haste yours &c

Matthw Culley

179.   *George and Matthew Culley to John Welch*

Eastfield 11th May 1803

Matty[29] took a half sheet letter away today for you, to give you an account of Alnwick fair and then send it per post. This has proved another very cold day indeed. Ralph Brown is just come home from Berwick High Market, says Jemmy Armstrong, brother to Mr Armstrong near Easingwold, was there buying small weedy queys at strange high prices. He got about a score and then away to Alnwick. Ralph says no decent work horses of the Scotch kind to be bought under *70£* per pair! What are things to come to? Men servants lowered a good deal, and women a little but not so much as the men. *Thursday 12th* is another very fining droughty day, the wind extremely cold from north west. *Friday 13* is a dry fining morning but not near so cold as yesterday yet. I am glad Matty sent you a line from Alnwick fair yesterday. I understand that it was a very good fair.

---

29 The greeting is lacking.

Ratcliff would not promise to come and buy our 4 inlaid oxen. We were obliged to put them to grass yesterday as our turnips are done, but I hope somebody will buy them by and bye. This severe weather begins to pinch us for grass, and I suppose will you also. But I think you should take our fat cattle before it be long, or your markets drop, which one would hope will not be very soon neither. This has become a very fine pleasant calm evening, thank God after all. *Saturday 14 or St Helen day*. Dry and cold as ever. How long it may please God to continue this fining weather, He only knows. It is such weather as we frequently have at this season, but after the 3 weeks of such uncommon weather in the end of March and beginning of April we put up the worse with this. If you have clipped your sheep they will be starved poor things. *Sunday 15*. Cold as ever. Yesterday a poor market at Berwick. This suspence that the nation is kept in respecting peace or war keeps everything in a state of suspence. Several Yorkshire men out buying hogs, but unwilling to give much above 40s. But people will not submit to take little prices, and why should they? If war come stock will be higher, at least not lower. By a letter from Mr Stubbins he says lamb hogs, as they call them, were sold at Lincoln fair from 40s to 55s per head. Wool still dear, some sold at 35s per todd[30] in Lincolnshire, which is 15d per lb. at 28 lb. to the todd. It is now absolutely snowing! *Monday 16* is a very cold morning but turned rather less cold towards the latter end of the afternoon The 100 wether hogs are to set forward for Denton tomorrow morning, and as soon as my brother gets to Denton I hope he will write word for the 11 fat steers to be sent to you. We are now very much pinched for grass indeed, therefore you had better sell what fat cattle you have at Skipton for what they will bring, than distress us so very much for meat, because it punishes everything else we have.

[*M.C.*] We are to be at Newcastle for the 24, where Mr Peacock is to meet us, and come forward to Denton where I shall stay 4 or 5 weeks with you. Purdy Starforth and the mason may be told this, I mean to employ Purdy in the job. He is a nice house carpenter and an old servant, very clever and trusty. This need not hinder Jack Dickinson from doing your other work unless his pride hurt him. Try to sell your fat stock now when everybody be content to sell now, when times are high enough surely. Yours &c

Matthw Culley

---

30 A weight used in the wool trade, usually 28 lb.

PS. We have been through our pastures, they are hardly fit for fatt or feeding cattle, but hope the weather will change for the better soon, either rain or warm weather or both, but we may be thankfull as our pastures are better than many of our neighbors here. We have selected our queys to put to our bull, 16 for the bull, 11 or 12 for feeding, may do to sell in the spring if they get fatt. By all means try to sell your fatt cattle and sheep. M.C.

[*G.C.*] John Brown told my brother at Kelso the other day that good work horses are now worth 40£ per horse, and midling ones 30£ per.

180.   *George Culley to John Welch*

Eastfield 18th May 1803

John Welch

I inclose you two letters, one for Mr Sherlock, Lord Darlington's agent, the other is for Mr Eales, son to the late Christopher Eales. You must be so kind as put proper directions upon each of them, as I don't know how to direct to Mr Eales, not knowing the name of the place where he lives, and in regard to Mr Sherlock I am not certain that he is living. If dead, you must not send that letter except he has a son that has succeeded to the same employment. If his son has Lord Darlington's agency you may send the letter, or if the old gentleman is living you may send it. But and if the old Mr Sherlock be dead, and his son has not succeeded to him, then don't send the letter for him. I think Mr Sherlock lived in Staindrop, and I should think your best way is to put both the letters into the post at Darlington or rather Piercebridge. I find that I am obliged to be at Newcastle upon Mr Thomas Bates's business next Tuesday along with my brother, and I should have liked very much to have paid you a visit. But as my brother and I can't well both be from home at same time, I must deny myself that pleasure at present. But am with every good wish yours truly

Geo Culley

PS. I am happy to say that this is so far a mild though frosty morning, but changed to very cold after. You must take care to seal the letters before you put them into the post, and direct them properly. *Wooler 19th.* An overloaden market at Morpeth yesterday for sheep. Between 40 and 50 score big sheep clipped sold at 7 ½. It would have been a good market but all is confusion at Shields and Sunderland &c with the embargo on the shipping and pressing &c.[31]

---

31 War on France was declared on 17 May.

Few cattle, but none to buy them from same cause 7/6 to 8s. *22 May.* Still dry but thank God calm. We go for Newcastle today, and I send this and 2 inclosed letters for you to deliver by my brother. Berwick market was better for grain on Saturday a little. Ralph Brown and we all got pretty well now, except Mally Brown who has a grievous cough. We are beginning to be very much distressed for grass in this country. I own that I think you should give us leave to send you over what fat cattle we have, as they are hurting essentially and doing little good here, General's Close burning fast.

181.   *George Culley and Matthew Culley jr to John Welch and Matthew Culley*

Eastfield 25th May 1803

John Welch

We received yours of 20th just as my brother was setting of for Newcastle to meet Mr Bates and Mr Peacock &c, where I hope they would get their money matters properly transacted. We are sorry for your account of John Peacock, and lament for poor George Dent losing his son. David Green came here on the day that my brother went south. He wants fat cattle or sheep, but rather cattle I think by what I could gather. I hope you will meet with a good market at Skipton. I have not heard what D. Green has done, but when I do will let you know. I am not fond of your selling your large oxen by weight, the butchers always contrive to beat us in these cases. I wish you had taken Eales's 80 guineas. I would not buy Ralph Compton's nor any man's cattle at present. Is it not madness to buy in the face of such a drought? We were flattered with hopes of rain today and some little fell, but it has turned quite frosty again. The war will necessarily make laborers and servants scarce, and want of men makes women more valuable. If you could have done without a horse or horses until the autumn we could then spare half a dozen, as our land here, *the infield*, will be all laid down now. John you are exceedingly unwise not to sow your Sweedish turnip seed, never mind how dry the weather is, when sown it is always preparing and will come the sooner when rain comes. We sowed some the last week, indeed we never stop for dryness, but prefer it to sow small seeds in to a certainty. *Thursday 26 May.* Cold and dry, with the wind westerly but not the least appearance of rain. By all means work and sow all the land you can with turnip seed, because this has much the appearance of the year 1762, which was the most severe drought I remember, when our cattle had to subsist upon ash branches which it was my daily employment to cut for them. But we won't play that trick here. You may cut part at Denton, but there are not so many ash trees now as at that time. However my brother and I sowed all the turnips we could, and contrived to sell hay rather than buy, but we were *young, active* and *anxious* at that time, and had been sore put about, which *redoubled* our *exertions*. The turnip caterpillars or black worms attacked the turnips that dreadfull year, to add to other difficulties. We hired some oxen out at so much per week to a Methodist *hypocrite* who starved them, so that we were obliged to take them away. We sold an ox the spring after at 20 guineas, which was the first we ever had sold at that price. When the weather broke it came dreadfull *thunder* and *hail*, which destroyed the corn about Forest in Yorkshire and *ruined* several farmers.

Well John we received a letter this moment from my brother, giving a good account of you. I am glad that you have sold your 2 big oxen so well, and also a few of your little sheep to a Liverpool man. I see by this day's paper from London that wheat is considerably advanced. I wish you a good market at Skipton, and advise selling by all means as the drought is so very severe and likely so to be. I will take our 4 coarse oxen to Morpeth on Wednesday next, where I am obliged to be at any rate as I have to meet a clerk of Mr Liddle's of Hexham there to sign some surrenders &c. Mr Matthew, how is it that you paid £3 12s for the reeds? (kassops as you call them, a *shocking word*). Have they advanced the price of *reeds*, or have you paid over again for those we got last year? Do explain this matter. Either they have advanced their price double or have charged for 2 years. And I can prove that I paid his sister for the last year's ones when I was attending the assizes at Newcastle. But I hope you will clear up this matter in your answer to this. There was to be just three dozen of the reeds for us. Matty is just come from Flodden sale, the dearest ever known, ewes and lambs 3£ per and other things in proportion, only 4 months credit. It is really wonderfull how things sell at these sales, notwithstanding here is as winterish a day as needs to blow, attended with showers of hail. David Green was there at the sale, not buying anything there but soliciting Tom Smith to sell him a few nice fat steers, but Tom will not be tempted. Matty understands that David Green has bought nothing at all yet. I have long prophecyed that a dear time would come, but it has kept long of, and if Bonaparte had not given us a bit of a spur perhaps it had not happened yet, although I think it would have taken place now or now abouts, and should not wonder that fat things become as scarce as was ever known. *Friday*

27th, flitting Friday[32] in this country, is not so cold as other days last past, but dry and frosty. We sent Mr Colling 14 lb. ruta baga per waggon yesterday, which will be at Darlington tomorrow I hope. I also wrote to him per post, and you can tell him on Monday if this letter reach you by that time. I am just going up to Wark along with Ralph Brown in his road to Kelso. We have still decent demands for turnip seed. We have now sold nearly as much as the last year.

6 o'clock just returned from Wark where I dined with my sister and daughters, all well thank God. Harry sowing rape, and like to fall short of seed. Clipping the tup hogs, very midling mutton indeed. If those here are not better we shall make a terrible exhibition. Washing wether hogs. Pastures bad but not quite so bad as George Logan's, his are litterally *bare lees*. Indeed the country begins to have a winterish appearance. This weather has been made on purpose for cleaning foul land. However we must not despair nor despond. The Almighty is always kind, rain may come to save us still, although no appearance at present. By a letter from Mr Deverill says he could have shown Matty the best wethers in Brittain 5 years old. Mr Buckley gone to Ireland but expected back in 14 days. Quick work say I. 'Does not wonder at his tup lambs not being very smart, as he thinks Buckley's sheep less smart but thicker in carcase than many others'. Poor Mr Thacker is dead, but has left 2 clever sons who are fonder of spinning cotton that breaking wheat, Mr Breedon mending very slowly. He is in great spirits about the war, no fear while our fleets are treble his.[33] Beef is advanced from 8/6 to 9/6, never knew beef *scarcer*. You see this is a general complaint. Honeyborn[34] has been bid 10/6*d* per stone for 3 beasts. He thinks wool will be lower, but not much. Stores still very high although the drought begins to have effect in many places. Corn advancing everywhere, but he thinks oats will be very high. He says it is expected that the usefull sheep will take the lead of the gay ones this tup season. I am exceedingly sorry to see your favorite bull calf at Wark got by Fenwick's bull turning clyered. It is a very unpleasant circumstance amongst our cattle that some clyered store, and how we are to get clear of it is the matter. We have a very nice kind of cattle too, which makes it all the more to be lamented. Your last year's calves also make a very poor appearance.

---

32 The day, about Whitsuntide, when farm servants hired by the year usually moved from one farm to another.
33 I.e. Napoleon's.
34 Bakewell's nephew, who continued the farm at Dishley after his uncle's death in 1795.

Ours look a year older that lived upon next to nothing all winter, at least they litterally scarce ever got a whole turnip but all shells and refuse. *Saturday 28.* Ralph sold 25 bolls of your wheat at Kelso, at 45 and 46s, but some sold as high as 50 and 52/6. It is rather unfortunate that our wheat this year is a worse sample than most people, and used to be among the best. I attribute it in part to sowing the same seed too long, especially the white wheat. The red is the best of the two. For that reason John Welch should endeavor to procure us a few bushells of hedgerow wheat, or any *approved new kind*, either red or white, which may come by carriage or tup cart if any are sent south. Ralph received 40£ of Mr Brodie of Ormiston for turnip seed, got the last year, which was a seasonable relief at this time as your people wanted money to pay the servants, Mr Brodie grows his own now. This will be the case with many, and no wonder when prices have continued so high so long. I see your neig[h]bor Mr Hogarth is sowing a large piece of rich-looking seed. Well well, we have had a good time and have no occasion to either complain or give over the trade, so long as we sell some when we do, or even half as much. I can't help thinking that your shepherd James Phillips is an *ignorant* and *negligent* man. After I had spoke to my sister yesterday morning and given her your letter (I am happy to say they are all well and send love and affection) I went to where they were clipping the tup hogs, and 2 of them being uncovered I said What are these Jemmy? They are *sturdied*[35] Sir. 'Why the d – l did you not tell Ralph Brown that he might have told Marshall to have called and bought them?' I *forgot* says stupid James. *Forgot* replyed I, what are your brains for, what have you to disturb your mind that you don't remember to do your duty! But the man is a weak man and does not seem to interest himself as he ought. 'Is easy whether the dog kills the hare or the hare the dog.' The 2 hogs are worth 40s or 42s each, and if killed at Wark won't make near the money. I am not in anger Sir, but dislike to perceive *negligence* and *remissness* in business. I have continued to fill this pretty well with one bauble or another so will send it to you per post by Ralph Brown. It is not quite 5 o'clock yet, and I have had a pretty long walk this morning besides writing from the last date. That I may long be able to rise in the morning and be of some use, is I am sure your wish, and the Almighty surely deserves the gratefull thanks of your affectionate brother

Geo Culley

---

35 Afflicted with sturdy, a brain disease of sheep.

PS. What an amazing difference between the tup hogs wintered at Wark and here. You must really turn over another leaf next winter respecting the tup hogs. It is not the way to eat turnips at home. A steady man with some *understanding* should have the charge of them, and a temporary house erected in the field where the hogs are considered to go. If James Phillip had any feelings he would *blush* up to the ears to see the difference. If to be sold in a market there would be *12s* or *14s* per head *odds* at this moment, and that will pay for extra trouble and expence. Jack Murray has shewed his sheep much fatter this year than common, which he says is owing to being fed on ruta baga without cutting, and says that in a few years more ruta baga will be grown than common turnips. I am very much inclined to think that a period is not very far distant when people in general will feed their wethers and wether hogs with ruta baga cut with *machines* certainly as we do for our tup hogs. And it will pay well. But we don't know what is to be done yet. We think we know a great deal but we really are only beginning to *know*. But Mr Matty, James Phillip lies all the blame upon you about your tup hogs. The quey Marshall killed weighed barely 45 stone. Is it worth while to cover cart sadles all over in the *Yorkshire fashion* with leather at 23 ½d per pound? We paid 17s for leather at Berwick last Saturday and your people will still want more.

[**M.C. jr**] Besides you got all we had here, when the sadles came first to you. David Green has bought Mr Oliphant's sheep of Marlfield, I do not know the price. He has also bought Mrs Grey's wether hogs of Milfield at 40/s, ten score and eight, and only the 8 shots. He bid Ralph Compton 46/- to have 3 guineas again, did not agree.

182.   *George Culley to Matthew Culley and John Welch*

Eastfield 28 May 1803

Well Mr Matt or Mr John

Just as you like, one or both I am to address. We sent you a letter this day by Ralph Brown to Berwick, and neglected to say that Morpeth was a good market. Indeed I did not know, but Ralph had heard so at Kelso, and Matty had learned so by some means and he named it behind the letter after it was sealed. This is a decent day and we are going to wash our few sheep over again at the Shelley (Till) as they are no better for being washed in our dirty pond. We have this moment one o'clock received a letter dated Denton 26th inst. I am glad you called on Mr and Mrs Harison. As you and John desire it, I will not take the 4 oxen to Morpeth, but intend to be there myself to meet the young man. They are 3 out of 4 very plain oxen. Nevertheless I think that the ship butchers must advance their price of beef soon, and then the remaining fat cattle will be better sold. Had I known this sooner I might have gone to Stagshaw fair today and not gone to Morpeth, but now I must go to the latter. I shall have no opportunity to see Mr Bailey before Tuesday at Whitsun Bank, and ten to one but I am gone for Morpeth before Mr Bailey come to the fair. Matty may see Mr Bailey but can't say so much to him as I could. However I shall tell Matthew to talk to Mr Bailey about it. I am disposed to think that Hallowell is a very clever place. I have no doubt but Mr Colling knows it well, and I am sure you can't apply to a properer person. I am sorry to hear that John Peacock is so poorly. We have had no rain. Old Mason bought Mr Nisbet's cattle very cheap. I am glad that Matty Culley is so well, and the little Hogarths. I hope Mr Colling will have sown the ruta baga seed we sent him before this reach you. I wrote to him also. *Whitsunday 29.* Thank God we had a gracious shower began about 6 last night, but when it gave over I know not. It rained in the night, I heard it, but at 2 o'clock I looked out and was quite fair. However I see it has entered between 3 and 4 inches into the new plowed land before you come at dry dust, and will do much good, but as it is fresh we may hope for more, and it looks like showers in the hills this morning. Ralph sold 50 bolls of indifferent red wheat from Grindon at 41*s* per to Clay, some good samples sold at 45*s* per to bakers. Mr Nisbet says some cattle were sold at 9*s* per stone at Morpeth, sheep plenty. David Green has got no cattle I believe, but some hogs and Mr Oliphant's first sheep. One Longstaff a queer fellow from Sunderland came out and bought Mrs Grey's cattle and sheep cheap of Kimmerstone it's said, bid Tomy Vardy for some cattle which Mr V. would not take, then sent the money Mr V. asked by his driver, but Mr V. had altered his mind. James Glass is here just now and represents Thornington to be in a most deplorable way for grass, and says that John told him that we might send away the fat cattle, Mr Compton's 8 steers &c &c whenever we chuse. Now you have not said one word about their being sent in this letter, and it is the more to be wondered at as Glass says the Denton pastures are *very good* while ours are *very bad*. However we certainly will not send them until we get your positive orders, although they certainly will do better at Denton where there is decent meat, than here where we have very little indeed, and they will be 7 or 8 days on the road, and surely the sooner they are at Denton the sooner and fatter they will be to sell as things are circumstanced. I will send this letter per post today, although I did not intend sending it until Wednesday from Morpeth, but I can write again then. Be so kind as say whether we are to send these 4 plain shipping cattle to Denton along with the 8 of Mr Compton's &c &c, as I hope your answer to this will instruct us to send you part cattle at any rate, and it is all one trouble to send 15 as 8, and you will be able to say in your answer how your markets stand, as you will hardly write until Wednesday that John returns from Skipton. I am yours truly

Geo Culley

PS. We have luckily received 50£ in a draft from Mr Bramley for hire of a tup, he lives near Hull. I assure [you][36] it has come very opportunely.

183.   *George Culley to Matthew Culley*

Eastfield Sunday 29th May 1803

Dear Brother

I direct to John as the simplest, but address you as I am rather earnest for our fat cattle being sent from this country, since James Glass told me how much grass John Welch has compared with us. I sent you a letter today, and propose sending this also from Morpeth. I am now informed to a certainty that David Green has bought no fat cattle and no fat sheep except a few of Mr Oliphant's. He bought Mrs Grey's hogs Milfield at 40*s* and John Batters's but don't know price. He bid Tom Smith 30£ for his cattle, Tom offered them at 30 guineas. Thomas Paul has bought Ralph Compton's hogs at 47*s*/6*d*, and Tom Smith has sold his at 48*s* but I know not who to. But *Mick* Hodge &c &c &c are all out, and now after this shower I do expect as dear a fair at Whitsun Bank as ever was, and the dearest for all kinds of live stock. Mr Joe Mole died yesterday morning poor man after a long and tedious and I suppose complicated illness. This is turned out a delightfull warm fresh day indeed after the pleasant shower, thanks to the Almighty. If more showers succeed we may still have a tollerably fruitfull season. *Monday 30th May* is a very fine fresh morning, very like wind from the west and showers. Our Wrights are all for America, old Hannah and all. I have been trying to hire Willy this morning, as he does not go until after they hear from Tomy, who goes along with Lisle and his *Darling family* directly. I bid Willy 7/6 per week, and victuals until the end of harvest, or 2*s*/2*d* per day and no victuals, but he will not touch. They have had a letter from Joe this week, who is doing very well I understand, and has been selling a

---

36 A word seems to have been omitted.

piece of land to Tom Aynsly, who by the bye I don't think is a man of principle. The Wrights, having money amongst them and a brother there before them, may do, besides they are handy men and can work, but I think Lisle a useless animal, can work none and has no constitution. Otherwise I am given to understand that there is not the encouragement now that there was, to go to America. Property is so much advanced unless you go far in, which is the case I suppose with Joe Wright. With the advice of the stewards, who are here today, we have come to the resolution of sending away 100 hogs to Denton tomorrow, in their wool, although washed, because we are so exceedingly distressed for meat at Thornington in particular. The hogs will go from Wark, as the hill hogs are upon the haugh at Thornington, and as it is your wish that none but our own bred hogs were to go to Denton this year, although I think a part of the hill hogs would have been full better for John's moors and high ground. However we have considered, to take out the least of what are at Wark, and take the smallest from these again, that would not drive so well. It is very lucky [. . .]³⁷ contrived to send me back 200£ which I hope to receive on Thursday. We [. . .] because William Brown wanted 100£ or better today and Harry for [. . .]. However I am sorry for Robert Harrison poor fellow, and am agreeable to assist him. But you write in a manner that I do not comprehend you, I have received no letter from him, but one may be on the road or may be at Eastfield by this. However if you are willing I am to assist him with 100£. Matty and Nelly were at Wark yesterday all well, but Mally Sinclair who I suppose has the influenza. Mr Bailey has been exceedingly ill but is better, and was at Whitsun Bank today. Lord Osulston³⁸ is going to raise a troop in this country. I wish our Matthew were quit of Mr Askew's cleverly, and he shall join Lord O. John Forster is comed this moment to buy our 4 oxen he says, but they are at home. Mr Nisbet has 3 horses, I hear of no one on the road. What quey does John Forster owe you for? I hope Mr Colling got his ruta baga, you do not say he has. *Wednesday morning 1st June.* Mild. Here is going to be a very dear market, near as many butchers as sheep and next to no cattle. I wish I had brought our 4 oxen, they would have been sold at 9s. Mr Nisbet sold John Forster his 3 last night at more than 9, a seg, a

---

37 The bottom corner and edge of the paper are torn, affecting three or four lines of text.

38 Courtesy title of Hon. Charles Augustus Bennett 1771–1859, eldest son of the Earl of Tankerville, succeeded his father as 5th earl 1822.

quey and a steer, clean neat beef. I never saw so many butchers nor so few goods.[39]

184. *George Culley to Matthew Culley and John Welch*

Eastfield Sunday 5th June 1803

Dear Brother

We sent a letter from Berwick yesterday, and another just now, because as Ralph sold Ratcliff the Wark cattle (Compton's steers and 11) we thought it was right to acquaint you, because you could then arrange your matters accordingly, as you can now expect no more stock from us of a while except the barren ewes if your sheep or the cattle coming forward, both which we can keep as long as you wish, and you can by this means push your best Kyloes the faster. I forgot to tell you that Lord Osulston has wrote down to Mr Bailey that he wishes to raise a troop, to be exercised once a week, armed and clothed by Government, and not to be called out without an actual *invasion* when they are to be under millitary law. A letter came here yesterday to be sent to Wark, to try what recruits can be got there and thereabouts. They have got 50 at Wooler and Chillingham and 55 at Doddington. What do you think of old *Dodginton*? I could have wished for you at home more on account of this than the tup shew. I dare say my son will join them, as Mr Askew's troop will fall a pieces, as he wrote to our Matthew who called a meeting, when only *two* subscribed. In your absence my sister of Wark was served with an *ejectment*[40] last night by Mr Willoby's clerk, who behaved very delicately on the occasion. My sister and your 3 daughters are all here this moment, and I am happy to say that my sister seems not in the least alarmed, but took the young man in and treated him with *hospitallity*. But I do assure you it is an aukward business, and I could like to know how it is to end Mr Matt. It gave me a good deal of uneasiness at first, but I began to view it in its very worst light, which is that we can only lose it, and when that is done we shall still have more property left by much than ever I thought to have been possessed of. Only what punishes me is that we should be so foolish as to purchase an estate under such circumstances. *Monday 6 June.* The wind has changed from north and east to north west or nearly west. By a letter from Sir J.E.Swinburn he and Dr Fenwick are engaged to breakfast at Turlaws this morning. Matty is gone to Mr Robinson's show but I

---

39 The signature is lacking.
40 See No. 176, n. 27.

think I shall not go. We had invited Sir J.E.S. to breakfast here. We received a very polite letter from Mr George Rennie of Fortafin the other day requesting to send him a cwt. of turnip seed. He sold a few days too early 126 cattle at one stroke at 10/6 per Amsterdam stone, says if he had been one week longer he would have got 6*d* per stone more, which would have been some money in that quantity. So you see the very cleverest can be mistaken. But I think we have hit the nail on the head for once. *Cuddy*[41] for that, steers never drawn or *worked*, and kept so as not to lose their *calf lye, weigh* like *lead*. William Brown has been here and got 80£, Harry 65, besides I settled with them both and paid them their annual wages and interest &c &c. I had settled and paid them before. Pray did you ever make a bargain with Harry for his Wark servitude? He must have some more wages I suppose with you. Harry has never showed any greediness, and should not be the worse used for that. I have much reason to believe that he is a very honest man. I think you should go and call on George Dent if you have not been there already. Give my respects. Matty is gone to Mr Robinson's tup show.

Tuesday 7th. I staid in bed until near 5 o'clock this morning, and the white frost rind was lying still where the sun did not come, yet the wind is westerly. It was exceedingly cold all the fore part of yesterday and some showers of hail fell in places. However the grass is much mended, and I have not often seen oats look so well. I think they have a great chance of being too big. I am going to meet Sir J.E. Swinburn and Dr Fenwick at Cornhill this morning to show them Wark farm. Matty says there were not above half the number people at Mr Robertson's tup shew this year as the last, but Kitty Mason was of the number and Billy Spours. His sheep I find cut a bad figure. Very few let and those at prices below the last season, and if it were not his tenants I fancy he would have done very little indeed at any decent prices. Harriot John junior took the highest, 53 I think. Sir John and Dr Fenwick were gone from Cornhill before I got there this morning although I went from this place by 20 minutes past seven. However I overtook them before they reached Wark and attended them all over the greatest part of the farm. We called at your house, and my sister offered them a glass of white wine. Dr Fenwick seemed much pleased with your watered land, which is looking very well indeed at present.

---

41 The meaning is not clear. Cuddy is usually a donkey, but can mean a sucking lamb or calf. It is also familiar for Cuthbert. Some of the Culleys' cattle had names, e.g. 'Miss Hill', Nos. 43, 50, 53; 'Blue Cap', No. 216.

Sir John was all politeness and civillity, invited me to go to see him, which I shall certainly do at some opportunity. They gave me a worse account of Mr Robertson's show than our Matthew did, and say they will go no more to it. They had not staid dinner but came to Cornhill. I set them as far as Brankstone Moor on their road to the Haugh Head, and came home by the Westfield. We have had several showers of hail today and it has the appearance of more, I dare say fell very heavy in places. We are going to wash our ewes here. We have sown full 30 acres of turnips and ruta baga, but shall sow no more of some days. We have such a surplus quantity of fallow this year makes us begin early with part of our neeps. *Wednesday 8* is a very mild fine morning, rather frosty, wind north west after a very cold day all the day yesterday. Two o'clock a prodigious rain. Ralph and I were going to Wark to draw some segs for next Wednesday's market, but the rain has prevented us, so Ralph will go tomorrow when I am at Wooler. I forgot to tell you that your friend Justice Dixon was at Mr Robertson's tup show and was making very great enquiries for you. Somebody told him that Matty Culley of Eastfield was there, but he took no notice of Matthew. Perhaps if you had been there he would have *paid you* for wheat, turnip seed &c. Pray did you ever write to him about payment? Only one single *brattle* of thunder after the rain was gone. John Welch, we begin now to expect some mowing may come on, but where are the scythes? *Thursday 9th June* is a fine morning rather frosty. The sun glashes or rises bold, but if I am a good prophet we shall have rain before night. The lad is returned from post, and no letter from you, which is not only a disappointment to us but keeps us in suspence about these ejectments and proceedings of Lady Headley. I am rejoiced that Ralph has sold so much wheat these two weeks, because I see that the London market has comed down for all grain, but wheat in particular is dropped 6 to *8d* per quarter. At Wooler I heard that Morpeth was very good for cattle and sheep again. We are going to draw a few segs tonight to take up to the market next week. Our first sold 4 oxen went to the Haugh Head today to meet Ratcliff's driver with some other cattle. Ralph Compton has had 21£ per head for 10 of his 3 year olds and 20£ for the 2 bad ones. I can't say but I am glad of it, as we could not have kept them through the summer and it will do him good. James Pinkerton is the person who bid it, and I still think that it is a strange price! 40 men have subscribed to a paper at Wark, to serve under Lord Osulston, provided they are paid for the day they exercise and are freed from the millitia. I thought it very fair, but Mr Bailey says they will not be accepted on those conditions, but said he was obliged to us for our attention. I received from Mr Bailey the money he had got from Anthony Surtees. Mr B. had quite forgot it and thinks he received something less than 20£ but cannot remember what. So it must rest until we can ascertain the matter. Anthony Surtees ought to have wrote to us by him or per post what he paid Mr B. I suppose he has deducted his own charges, I desired that he might. Your niece my daughter and James Darling were married at Ford church this morning.[42] It looks very like more rain tonight I think. Walter Heron said today that the yeomanry corps was still liable to be balloted to the millitia! and if so I think it will take all the infantry from Collonel St. Paul's Cheviot legion.[43] The weather is so cold that pastures don't mend. Yours at Wark are very indifferent and I understand that the pastures at Thornington are shocking. However the rain has done corn a great deal of good.

Friday 10th June is a very compleat wet day indeed. At what time the rain began I can not say, but it appears very likely to rain for some time. The wind east or a little to the north and the glass has sunk very little, which may be owing to the wind being rather north east. We, that is Ralph, Matty and I were all going to Berwick fair today, but I think I shall not venture now except the day alter for the better very soon. We shall begin to have a decent drenching now I think. Well we have never have had the land sufficiently wet until now, still the weather is so cold that grass does not grow so well as might be expected, after a long dry time rain in plenty coming in the month of June when long days exist and warmth should be expected. Corn is much mended indeed, particularly spring corn, but I think all or most of hay crops upon new land (seeds) will be light. Old land may and possibly will recover much. If Mr Peacock be gone to Lord Mulgrave's[44] I wish you would ask Mr Johnson of Darlington, the

---

42 DUL, BT Ford. The witnesses were the bride's father and brother, and her friend Elizabeth Greathead.

43 Count Horace St Paul c.1729–1812, of Ewart. While a student he fought a duel and had to flee abroad, where he served in the Austrian army. His father gave the Ewart estate to his younger son Robert, from whom Horace, having been pardoned in 1765, bought it back. He rebuilt Ewart Park to his own design, planted trees, and improved the estate. Col. St Paul raised the Royal Cheviot Legion in 1798. In 1801 it had nearly 900 men, with centres at Whittingham, Wooler, Lowick, Belford, Bamburgh, North Sunderland, Beadnell and Alnwick. *History of Northumberland*, vol. 14, pp. 183–7; George G. Butler, *Colonel St Paul of Ewart, Soldier and Diplomat*, London 1911; Pevsner, *Buildings of England: Northumberland*, 2nd edn, p. 274.

44 Henry Phipps 1755–1831, created Baron Mulgrave 1794. His wife was daughter

man of the law, about these *ejectments*, as I think he Mr Johnson was consulted on that purchase. Or you can get a letter sent to Mr Peacock by getting the proper direction to him. Anything of that nature in suspence is *torturing*. Sir John and Dr F. dined with Robert Thompson on Wednesday. Mr Bailey also says R. T is never without company. R.T. showed some nice shearing tups. *Half past 2 o'clock Friday*. Still rain and has rained all day, I think it will be a flood. It comes on quite calm, no wind, but what there is is nearly north. Ralph went to Berwick fair, but it stopped both Matty and I and probably many others. *Friday 7 evening*. Ralph just come from Berwick fair, very few people there, few cattle shown and but a small day. Ralph sold 25 bolls of morelease red wheat to Mr Clay, received 10£ of Mrs Grey of Milfield for 5 bolls seed wheat, got some small orders for turnip seed. It still rains. *9 o'clock at night*. It has rained all day. But as the wind has got about to the north west I think it will be fair soon, indeed it does not rain near so much as it did. We are going to send all the carts we can to lime tomorrow, as the land is so wet. *Saturday 11*. Blows dry from the west. I am just going up to Wark and around by Thornington to see if James Glass be returned from Denton, to take the segs to Morpeth. *One o'clock*. I am just returned from Wark and Thornington, all well, and I hope Thornington very bare pastures will mend by and bye, yours are growing better. But Thornington has a lamentable appearance and we have taken the hill hogs out off the watered land, as I began to think that if the weather turned warm they might not be safe. My sister is much disappointed as well as myself at not hearing from you, as she was affraid you might be alarmed about the ejectment being served upon her. She got Nelly to write to you. This has continued a fine day after our *quantum sufficient* of rain. The Tweed is considerably out, the Bowmont not very much.

Saturday evening 5 o'clock. I have this moment received yours of Sunday 5th inst., which I think should have been here sooner. Indeed we had nobody at the post yesterday and Ralph went and came to Berwick by Etal. It was so very wet a day. We would have clipped the hogs we sent to a certainty, but you can't conceive how badly they were at Thornington so sent them away at all events. It would have been wiser to have sent the barren ewes, and we will still send them if you and John approve, because we don't like to sell them at Morpeth and we have nothing to send you now of cattle kind except such as

---

of Christopher Marling of West Herington, Co. Durham. *Complete Peerage*, vol. 9; *Oxford Dictionary of National Biography*.

are feeding. But shall not send anything more until we have your orders, as we shall now get grass in time. But when land is beat so hard as all our farms were, and such a dry time, and *cold* as prevailed here you know well enough what a deal it takes to bring it about, especially when hard stocked as we have been until lately. It is very extraordinary that R. Colling's ruta baga is not comed to hand, because we sent it immediately, It is lying somewhere I will be bound for it. Bless me, we have letters from Glasgow, Perth &c that all their seed is safe arrived. Why not to Darlington? We will be happy to see Mr Colling's Major Trotter. I have just been handling our tups, and I own they please me more than ever. By a part of the letter wrote by Mr Peacock he advises me to send the ejectment to Mr Scruton to give to Mr Smales,[45] Sir Ralph's attorney. I will think of it. But both your wife and I were served with ejectments. I am very glad you have let poor Bob Harrison have 100£. If we never get it again, never mind. Let us do good while we stay here, or try to do good. I am glad John Welch has got a bull. By the bye we should have put some cows to Carham bull, he is a nice beast and well bred, healthfull and nothing akin to us, and ours are so shockingly addicted to clyer. Never mind taxes John so long as we can keep of Bonaparte! We are able and ought to be willing to pay them. I am rejoiced to hear that grass grows fast with you. I do assure you it is not the case here, although I hope it is coming now. We have been very much pinched indeed and still are. If you will allow us we will send you a few forward cattle, or we can keep them until you give us orders to send them. George Humble's cow, the brown stot and 2 or 3 more are saleable, and I shall be selling them if you don't chuse to have them sent to Denton. You certainly know by this time that we sold Ratcliff Ralph Compton's cattle, and I am very well pleased with price. To be sure if John had agreed to take them we should have bought Ralph Compton's other cattle, but they are now gone and I wish Mr Pinkerton luck off them. I have so [many][46] times said all were well at Wark, but I sometimes think you don't read my letters with attention as I do yours and John's 2 or 3 times over. You talk of writing, I am certain that where you write one line I write 10, and where you write one letter I write 10, but then I write in the morning while you and half the world sleep! and lose half of the best of the day. Well may you long have health and ability to write is the sincere wish of yours in haste

---

45 Francis and Henry Smales, attorneys, Durham: *Pigot's Directory*, 1820.
46 A word seems to have been omitted.

Geo Culley

Matthew at Akeld all day, at least he is not come home now 5 o'clock.

185. *George Culley and Matthew Culley jr to Matthew Culley and John Welch*

Eastfield 11th June 1803

Dear Brother

After receiving yours this evening of the 3 and 5 June (which I think should have reached us earlier) I immediately dispatched a long letter per post to you which I hope will be with you on Monday. I will send this you have sent me to Wark tomorrow, and I thank you for being so very attentive in writing. You will certainly know that we sold Ratcliff our cattle before the last reach you, because I wrote directly as I thought it right to acquaint you, that you and John might arrange matters accordingly, and I am glad that John has plenty of grass for some, although we are still much pinched but in hopes to be pretty well soon. I dare say if you were at home you would be mowing some of your watered land in a week's time, but Harry has too many irons in the fire poor fellow, with a large quantity of turnip land, to mow any yet. He staid until 11 o'clock last night paying the people at Thornington last night. I sincerely wish that these affairs of Sir Ralph Milbanke's were settled and the mortgage paid off to Lady Headley. Mally Sinclair got well and went away. We cannot hire a good man here under 2s per day. I think Tommy Wright need not to have gone to America. Well I hope you will prevent John from keeping any more breeding ewes. I can read your letter nicely on greasy paper. I really understood you, and Matty and Ralph also did, that we were not to take the 4 oxen to Morpeth! but you wrote your last letter or 2 very compressed, and this letter today is as well wrote as anybody can write. What a pity that you can't always write so as to be understood. John Welch does not say what sheep were sold at Darlington and then one knows nothing. Please to tell Mr Peacock that I have wrote this evening to Mr Scruton of Durham and inclosed the ejectment with which I was served, and requested Mr S. to hand it to Mr Smales, Sir R. Milbanke's attorney. And desire Mr Smales will by return of post write me a satisfactory account of the above proceeding. *Sunday 12 June* is a very fine warm mild morning. Everything seems now to be thriving in the vegetation way except the hedges which have never recovered that severe frost and subsequent dirty winds. It is surprising how long James Glass is always of coming back from Denton, and we want him to go with some segs to market, and sheep clipping comes on and hands scarce and as dear as with you. I bid a man 2*s* per day until after harvest from this time and he refused it. *Monday 13th.* A charming morning. A letter from you and John Welch had been sent to Wark, which came by Glass and came here some time yesterday when we were at Mr Pearson Weetwood meeting Mr and Mrs Nisbet. I am very glad you have got into good humor again about our selling the 11 cattle to Ratcliff. I assure you I was pleased, and still am, and I am rejoiced that you abound so with grass at Denton, while we have been shockingly distressed upon all our farms here, not yet recovered, but I hope this good weather will in time recover us. But when land is burnt and eat to the bone it takes much recovering. However you must allow me to say that it was entirely your and John Welch's fault that those cattle were not sent to Denton. We waited and waited and no orders came, and I assure you I was glad to quit them. Now you say these barren ewes are to be sent to Denton, and John in another part of the letter says 'we are not to send them until further orders'. From want of clear and proper dates in your letter I don't know whether your or John's orders are the last. However we will most certainly wait until we have *clear* decisive positive orders to send them. The 11 cattle were sold for something more than 9*s* per stone. Say also, pray, if we are to send you the brown stot, George Humble's quey, and several others which are in a feeding way and going very badly at Wark. I think John ought to have sold John West a few gimmers in my opinion. We generally find that it is the wisest way to sell when butchers are in the humor to buy. We can not tell what may be the consequences when things get to such a very high price. Things come out of odd holes and corners. Clipt hogs will many of them be sold, and when things are so very dear meat goes much further and people in warm weather live half upon vegetables. Nothing like striking when the iron is hot. Geld or barren ewes come on and lambs in abundance, besides the nice hill cattle in a very green state as well as sheep, and things grow strongly fast at this season of the year before flies plague or too hot weather come on. Mr Dinning was upon Mr Rae's farm lately. The lease has 2 years to go I am told.

Yes Sir I admit that John Welch is a very correct driller, good corn farmer, and will in time be a good salesman and grazier I hope. But I would not wish you to interfere with him or check him in his order, because you must permit me to say that for want of *practice* and *attention* you are becoming a very bad judge of the weights of cattle. John certainly fell in with a very midling market at Skipton. That is the fate of war! but we should recollect that he has often fallen in with good markets there. We can't always forsee when the precise time arrives for good markets, although we may have a great perceivance that the time is approaching. Therefore we must hazard now and then. However a good calculation is a good thing, and it is prudent and wise to consider and reconsider things over and over again. It is consideration and prudence that brings well disposed men the nearer to perfection, heightens their understandings, and enables them to transact their business to the satisfaction of their employers and to their own minds, whilst precipitate and hasty proceedings run business into *ruin* and *confusion*. We have been wonderfully fortunate in meeting with good servants, and few farmers have done so much by agency as we have, and perhaps the very unsuspicious and liberal encouragement which we have in general given to our upper servants has contributed in no small degree to both their advantage and ours. If we had been so unfortunate as to have met with dishonest men, it would have been very aukward, but they would have been found out. However how thankfull ought we to be to the Giver of all good that we have been so happy as to meet in general with honest and many of them intelligent men. Surely much of our success in life has been owing to the industry, activity, and energy of the people we have employed as upper servants.[47] I am glad that the fleshed ox did so well. He was a true grower and shewed your judgement in picking him out. That is your fort, and were you in practice you would know weights better. But the cleverest man cannot judge without practice. I am certain that many people were better judges than I ever was, nevertheless whilst in *practice* I could have judged of the weights very near. In short a man of a midling capacity with practice will judge better than a man of strong natural judgement without practice. There is not a truer maxim than 'that practice makes perfect'. You see what bad agents, bad attorneys, and an extravagant wife has brought Sir Ralph Milbanke to! I sincerely wish that matters were got properly settled however, because until that mortgage is paid off we can have no proper title I am affraid. You know the deeds must be with the mortgage. I inclosed the ejectment served upon me to Mr Scruton, and if it is necessary when I am told I will also send him the one served upon my sister. You can ask Mr Peacock if I need send the other. John's high priced bull will surely be sold for prime cost after

---

47 George Culley left small annuities to William Brown and Ralph Brown, noting in his will that he always regarded them as friends rather than servants: DUL, ASC, Durham Probate Records DPR/1/1813/C15.

doing his duty. I believe that all the queys at Wark have been bulled by the red stirk bull got by Fenwick Compton's bull, although from his appearance of being *clyered* (which I certainly named to you) it is not prudent. Why were no cows put to Mr Compton's young bull? I saw him the other day and think very highly of him indeed, and if Mr C. will allow you to have your cows bulled by him at this overturn I think there should be no hesitation in availing yourself of it, so near hand and so valuable a brute

My mistress desires me to thank you, John Welch, for procuring the knives and forks, will be glad of them as soon as convenient, but would not wish to load your master with them. I am glad my brother has bought most of Birkbeck's houses, but no reason is given why the one occupyed as a publick house is not included in the bargain. It is imperfect without that one, nor anything said what they are to cost. I also thank you for the scythes which are not come, but by your account I hope will come in time. But John I beg you will date when and wherever you begin, even in the same paper that my brother has wrote in also, because for want of that I can't tell whether these barren ewes are to go now or not. So pray say in your next, when we are to send them. I should suppose, Mr Matt, that windows with 2 panes sideways and 6 up and down will not pay for more than one?[48] But this is only my own idea.

[*M.C. jr*] Dear Uncle. My father wishes that I should go to Mr Colling's &c's shows, which I should like very much myself. If I go I will likely be at Denton either before or after them, they are to be on Tuesday and Wednesday the 28 and 29 of this month. In your last letter you copy one from Matthew in which he speaks of taking home the bay mare from Denton, and you seem to wish him to be at Denton a few days before he comes north. If it should happen and he be at Denton I shall be very happy in his company home, and I am persuaded it will not be disagreeable to him. If he gets to Denton on the 23rd it will leave him at least 6 or 8 days to be there before I return. I hope this will meet with your approbation, and am dutifully yours

Mathw Culley

---

48 The window tax was levied on the number of windows in a house, not their size. By a provision of 1780, in the case of two or more lights fixed in one frame, if the partition between them was twelve or more inches wide, each side counted as a separate window. The rates were increased several times during the war. Dowell, *History of Taxation and Taxes*, vol. 3, pp. 168–77.

[**G.C.**] This continues a most charming day indeed. I hope we shall get pasture grass now as well as you who live in a warmer climate. But I am pretty certain that new laid meadows and pastures both will not be good. Spring corn is very promising indeed, oats in particular. Now be so kind as write about the barren ewes and gimmers, and the cattle that are feeding. We can keep them a while, but I do assure you we can do very well without their company. Now we are going to weaned our lambs &c &c and cast our work oxen. I am with sincere regard

Yours affectionately Geo Culley

186.   George Culley to Matthew Culley

Eastfield 13th June 1803

Dear Brother

We have this moment sent a letter to you by William Brown to put into the post. *Tuesday 14th.* A very frosty morning between 4 and 5 o'clock, but likely to be a warm good turnip sowing day. Ratcliff's driver had come here last night after we were all in bed, and taken away 9 cattle comed from George Tait and our 11, *cuddy and all,* to the great joy of the whole *Commonwealth* of the Eastfield. We have set him of with them this morning and he is to be at Tom Allison's tonight. Now it would seem as Ratcliff's consumption must be great, because we expected our own cattle were to have staid a week longer. But I do assure you that their *absence* instead of their *civillity* is very agreeable to us. *Past 8 o'clock.* Just returned from Wark, Mrs Culley and her two daughters at Berwick, I suppose my sister is gone to leave Bessy at Mrs Coulter's to attend Mr Bellamy in the musical line. Saw Jane very well, and all well and very busy sowing turnips, pastures freshening, clipping wether hogs, pretty and well layered but not very fat. But you have got 200 of the tops. It is an unpleasant thing to find fault with your servants. I never thought *James Phillips* overloaden with understanding, that we should not blame him for, but ourselves for hiring a man with weak intellect, but I really think since Jemmy had that fever that he has been much stupidder! A dog had chased several hogs in the back green into the river, and after *escaping* him there he had drove them into the lane. Poor innocent creatures, they must have been terribly scared, as they had rushed past the gate post and other places, several were found in the mid closes, and 2 down towards Cornhill in Ralph Compton's close. It was in the grey of the morning, Mr Marjoribanks's fishers say, and a blackish dog. But I don't find that these brutal fishermen offered any assistance to those inoffensive animals. How cruel! All this I don't blame J. Phillips for, certainly not, but I blame him much for not 'pouring wine and oil into their wounds',[49] and if I had not happened to have gone up today by mere chance (as I was not for going until tomorrow) nothing would have been done to these sheep. Indeed he said he was for doing something to them this evening. However I got Robert Tait, and he dressed their wounds and I hope they *will* all recover as none of the bites seem very deep, mostly about and above the houghs, buttocks and twists. We sewed the skin together again where much torn. Some carts were going down the lane at the time, the fishermen told Jemmy. How unfortunate that those carters can't be put of taking dogs with them. James Darling and George Logan had each some ewes worried lately, part were obliged to be killed. Collinson has been watching all day the return of the dog, only saw one mastiff dog but the man pretended that it could not be him as he never left the cart. Few men but would have got something done to the sheep as soon as found, and a doctor to hand. But enough said at once. *Wednesday 15th June* is a very sweet mild fresh morning, wind easterly. I hope the turnip sowing will go on grandly. There were 6 of the hogs bit, but my admiration is that none was killed out. The hogs being new shorn, young and mettled, may have enable[d] them to escape. It is not unlikely that some of the tears might be occasioned by rushing so violently through a dead piece of hedge just close to the lower gate next the lane, and the gates into the turnip field across the lane to the south being open, most of them had taken in there, and probably he might pursue the two which were found in Ralph Compton's fields down the lane, until he had overtaken the carts he belonged to, and not unlikely the brutal carters would enjoy the chase, and instead of hanging the dog, commend him. *8 o'clock.* Matthew just returned from Morpeth and thinks he has had a pretty good market for his segs at 8 ½ per pound sink. Sold the least at 3£ 2s and the best at 3£ 18s, which were the cheapest he thinks. Several beasts in and butchers would not buy because they are all full of cattle and the ships all upwards. Robert Thompson sent his home again all the way to Chillingham Barns, and many people turned out unsold, scarcely any sold except Lilleys of Berwick.

Thursday 16th. We have this morning a little after 4 o'clock received yours of the 8th, 9th and 10th inst. to your daughter Nelly and myself.

---

49 Luke 10.34: 'But a certain Samaritan . . . bound up his wounds, pouring in oil and wine.'

I will send Nelly's part up today and the whole. Also a letter from Mr Scruton and one from Mr Smales, and a very gratefull letter from poor Robert Harrison who we have made happy at present poor fellow, and I sincerely hope and wish that it will be productive of comfort and happiness to him and friends in future. It is nothing to us thank God at present, and I trust that we have not nor will not forget our situation at our beginning of life, nor what struggles we had, not neglect to return thanks to that kind Being who in so wonderfull a manner has blessed our endeavours beyond our most sanguine expectations. I only sent the ejectment which was served on myself to Durham. I can also send the one served on my sister, but by Mr Scruton's letter I don't see that it will be necessary. Mr Scruton says 'I received yours with declaration in ejection at the suit of Lady Headley, and given it to Mr Smales, who in this instance acts as the attorney of Sir Ralph Milbanke, and have given him orders to enter the necessary appearance &c. This is a business in which Sir Ralph is extremely ill used by Lady Headley, it being merely a question as to the regularity of the notice given to her to receive the money, and you may rest confidently assured that no harm can possibly ensue to you. I am &c R. Scruton. PS. As Mr Smales could only give you the same information as I have done, it is probable he may not write to you'. Mr Smales says 'I have this morning received from Mr Scruton a copy of the ejectment served upon you at the instance of Lady Headley to which I shall appear, and do what is necessary to keep you indemnified. By a letter received the other day from my agent in London, I have every reason to expect that the business will soon be brought to a conclusion and am &c Fra: Smales. PS. I shall do what is necessary on behalf of your brother'. Mr Scruton's letter is dated the 15th and Mr Smales's the 14th inst., and both came to our hands today. And I hope you will think with me from the contents that it is unnecessary to send the other ejectment to Durham. However it may be right to ask Mr *Peacock* or Mr *Johnston* the law man at Darlington if Mr Peacock be from home, whether I need send *the one* served upon my sister also to Durham, and I will act accordingly. I am glad that you now approve of our 11 cattle being sold to Ratcliff. I am affraid that we should have *roared* and not him, if we had still kept them in our possession, as it appears now by your Darlington market the last Monday and yesterday's market at Morpeth that things are much flatter. Not that things can come much down for a while, but it makes me recur to that maxim which we well know but sometimes neglect to put in practice, that it is always right to sell when the butchers are disposed to buy. And it [is]⁵⁰ probable that you think now that John Welch had better sold half a score gimmers to John West when he solicited so very hard. Robert Thompson could only be bid 8*s* per stone sink for his fat cattle yesterday, and Mr Lilly, who was the only person who sold, was obliged to submit a great deal. In short, from any experience that I have had in business, which has not been a little, I have always found that to strike when the iron is hot is a good old maxim. Now observe, that our old segs sold at 8 ½ per lb. sink, and only 8*d* you say at Darlington. Consequently our barren ewes would bring much more money at *Morpeth* than *Skipton*. Nevertheless we must not sell many of them at Morpeth except a few plain ones, as Matthew says there are many of them exceedingly handsome and tempting also very fat, and I could wish them away as we can well want them, although pastures are now freshening. Your rains came earlier than ours, and I also suspect that you had warmness also earlier. Do say also whether we are to send you the forwardest feeding cattle as well as barren ewes, or we will keep them longer if you chuse. We can do better with the cattle than sheep as we are now going to wean our lambs, which will make us want all the grass we have fit for sheep meat. John certainly sold his queys decently after all. Little Ben Green's brother showed 10 fat cattle yesterday at Morpeth but could not sell. He had 60 in all I think Matty says, which are now gone south, and will consequently will fill Skipton. David should have bought before, when he came out himself. But as you say he is like a flea in a blanket. All fits and starts. Matty will take this letter up to Wark this evening himself. You talk of lean stock being high at Darlington. I ought to blush at valuing Ralph Compton's steers at 18£ per, and James Pinkerton and Ned Smith have bought them at 25£ per for 10 and 19£ for the 2 shots. To be sure we saw them on a very cold day, and I understand they are wonderfully improved. However I think they can't be cheap. The scythes are comed to hand, thank you or rather John Welch. I am glad that the Peases are giving 1*s* per pound for wool. Our trade will be so protected that I am not much affraid of wool selling. However when 1s and upwards it [is]⁵¹ wrong to speculate to far. I am glad that Mr Colling has received his Sweedish turnip seed. I think our first sown will hoe in 10 days. We are all busy clipping. I am always affraid of the bill trade, it is always dangerous, and Morpeth is the best market in the Island for money.

---

50 A word seems to have been omitted.
51 A word seems to have been omitted.

However it shows that John is cautious, and behoves him so to be. How well grain is sold with you compared with this country. Beans 6/6*d* a strange price. But I do imagine beans and oats will sell on account of the war. I desire you may stay and finish your building and not come home yet without it be your own choice. I am certain I wrote much the same as this to you before. By this you will see in our last letter that Matty has an intention to see Mr Colling's &c tup shows if you approve. You certainly have wrote a great deal, and more than we could reasonably expect, and we are much obliged to you for it. But you might set John to give an account of markets, it would relieve you a little. You really have wrote a great deal, but I beg you may not fatigue or mistime yourself on that account. My rising at 4 o'clock gives me much advantage, but everyone can't do so. When I came down stairs at little past 4 today, perhaps not 4 as our clocks are very fast, our lad was returned from post, but this is an uncommon [. . .]⁵² pity he was so large, the lasses were not up scarcely! Matthew's plantings as you are pleased to call them are looking much better since the rains, but Akeld is too *arid* dry a ground for wood, and we have too little rain in general in this part of the Island for growing wood compared with the west and midland parts. However we must plant part, and they will do by degrees. I believe many of the plants died upon the Sandy Know planted this year. The rain was too long of coming for them also. I think I have now answered the whole and am yours sincerely

Geo Culley

PS. We have been a sorting our shearing tups and we think them decent, much better than the last year, and the old sheep are quite preferable. We have had no rain since Friday last. We had none yesterday and Matty was twice wet on coming from Morpeth, but we have no want of rain at all at present.

187. *George Culley to Matthew Culley and John Welch*

Eastfield 16th June 1803

Dear Brother

I sent you a pretty full letter today from Wooler in answer to yours which we received yesterday and which Matty has taken up to Wark today. Daniel Duncan has given me another 100£ today which makes 200£ now, and I altered the note to this date for 200£ and left room for

---

52 The paper is torn here.

your name above mine. I would [have][53] had them put it in some other hands where they would get 5 per cent, but Bella said they would trust no other people with it, but desired we would not name it as it might affect other customers, and added that they had great industry to save it. *Friday 17* is a blustering and rather cold morning. I did not know that Kettle or Sam dealt in fat cattle. Do they job at Wakefield? Nobody could behave with more propriety to Mr Willoby's clerk than my sister did. I have heard no more about Lord Osulston's corps. Mr Bailey seemed a little cool upon it, I thought the other week. He was not at Wooler yesterday and it appears by yesterday's paper that Government are not very fond of encouraging volunteer corps. They seem to wish for more regular and better disciplined soldiers. *Saturday 18 June* is a fine calm morning after a terrible blowing day. It has blown all the pease and things in our garden out of their senses. Ralph brought home 95£ from Kelso and sold a good piece of about 30 bolls at 46, 47, and 48s per boll, and I am in hopes by drawing upon Ratcliff for past money to be able to pay James Darling the 1000£ I promised him today. I think I told you that I meant to give them 1000£ to start with and then let them feel their own weight. 800£ of this is the money we owe my wife and her sister Mrs Pearson and the other 200£ is Matthew's, but we will explain more fully to you at meeting. *Sunday 19.* After a sweet day yesterday this morning is raw and cold, much like wind and slight showers. I was enabled yesterday to do as I have said above to James Darling, by drawing on Ratcliff for the first bargain at 10 days 124£ and taking all the money out of the bank with interest, and will be able to carry things on still I trust. We have now got through all our heavy payments I know of except N. Culley at the annual settlement in the next month. These taxes will bear heavy, but we ought to pay them chearfully and much more, to keep the *tyrant* and *scourge of mankind* from us. I do admire the *candor* and *straightforwardness* of Mr Addington.[54] He may not be endowed with that stretch of *capacity* that young Pitts and Foxes[55] are possessed of, but what is not *honesty* and *candor* worth. It seems as though we must *stand* or *fall* by ourselves, the Russians[56] appear to

---

53 A word seems to have been omitted.

54 Henry Addington 1757–1844, prime minister 1801–04, Viscount Sidmouth 1805, home secretary 1812–22: *Oxford Dictionary of National Biography.*

55 William Pitt 1759–1806, prime minister 1783–1801, 1804–06; Charles James Fox 1749–1806, Whig politician: *Oxford Dictionary of National Biography.*

56 Russian distrust of Addington hindered the formation of a new understanding, which was not reached until June 1804 after Pitt's return to office. Meanwhile

stand in awe of the brutal governor of France as well as the weaker European powers, and if he should overpower this little Island (which heaven avert) he may then tyrannize over all the world.

Ralph could do nothing yesterday at Berwick, very little done in the corn way at present. On our road to Berwick yesterday we received yours of 16 June and have sent a letter to James Glass to come and take away the barren ewes. I will be able to say afterwards how many there are, but suppose above 100, so that we need to send no other sheep to make them up. Indeed we should keep all the male sheep we have now to consume our turnips in winter, as we have more acres sown and to sow than usual, and our prospect is good. Besides our cast of cattle will not be large. We advert to what you say about sending John Welch the queys (spayed) and smaller cattle, and keep the *ship timber* here, and when casten shall arrange them accordingly, and will try to make the best end we can here of the brown steer (he is now nice beef) and the others. It is possible that they may be as well sold here as with you, driving so far taken into the accounts, and there is no doubt but these barren ewes and gimmers may be sold for more per pound here than with you or even at Skipton. But we conceive it very dangerous to sell them here, consequently Skipton is the properest place to sell them at. Perhaps some of John's Brough Hillers may be fit to be sold along with them at Skipton. I hope you will find a part of them handsome mutton now, and the remainder in a state to make readily fat, and as they are clipt I hope by carefull driving that they may not suffer much by their journey. The gimmers that went to Denton before had been injured by some means to a certainty, and it should be a rule for the same man to drive them all the way in future. I own it is very wrong to drive hogs &c in their wool from hence to Denton &c in warm weather, but when I determined upon sending you those fattest of wether hogs we were distressed for grass above measure, and at this moment we are not very flush of meat. It is not the climate I think, but the land being *beat* and eaten to the root in a cold and severe drought, whilst our rains did not come so early as yours. George Logan has a field at the Bare Lees like the lane still, the poor sheep watch every clover leaf and shoot of grass as it comes, and we were nearly as bad at Thornington on all the hillsides until we sent you those hogs, which enabled Wark to take some of the gimmer hogs and Grindon a part, although

British war policy was purely defensive, and Napoleon was free to develop plans for an invasion.

Grindon had not much to come or go upon. James Pinkerton says that except fat cattle be 9*s* per stone in the autumn Mr Ralph Compton's cattle will not pay. The 10 best are 23£ per and the shots or casts 19£ per, and at these rates what can they pay? The coarse stiff handling quey you speak of, both Matty and Ralph Brown think was one of Ralph Compton's. I am rejoiced that the Fitts &c are such rich nice pasture. You talk 'that John will take any of our smaller cattle which may get fat by Martinmas'. Except those already feeding we have none that have any chance to get fat by Martinmas, as our draught cattle are none of them casten yet except the 4 including Ralph Compton's seg. Consequently what chance have they to get fat by Martinmas? However I hope these good pastures will enable John to make his Kyloes &c good and leave him some roughness or *rowan* as Mr Young calls it, for the winter. My dear man 'why are you to injure your health by going to all the tup shows'? Has anyone advised, much less insisted on your going to all or any of the tup shows? You certainly are a free man yet, and quite at liberty to go or not as is most agreeable to you. You still get rain by your account. We have had none since Friday gone a week, nor are we much in want yet, although our pastures would like a little more very well. John and you must best know whether it is right or not to sell your clipped hogs or part of them. If they can be sold at 8 ½ or 9*d* per lb. like Morpeth, it can't be wrong, and even 8*d* is a nice price although shearings gain fat and weight wonderfully to Michaelmas and after, when I have no doubt, if the turnips prove a good crop again or midling, that sheep will be as dear as ever. However clipped hogs at 50*s* per and wool now pay, and what is unwise perhaps to refuse, and many hogs have been sold at Morpeth at 48 to 52*s* per. A man called Wright, who lives at Prenderguest beyond Berwick, sold a large lot of clipped hogs at Morpeth overhead at 52*s*, a grand price. I believe he got 48*s* last year. I have no doubt but good management such as John Welch's will make a strange difference in Denton estate, especially Sir Ralph Milbanke's part which had been ill farmed for years. But too much tillage I think wrong at Denton as your corn tithes are high and the others reasonable, and after all a few ewes kept in the bare sound parts of Denton will pay more than any other stock. Suppose 50 ewes or even *30*, which few would not need to eat any of the grazing parts? These are my ideas, as you say sometimes, but brothers differ and why may not people give their opinions *freely* so long as they are delivered *temperately*. Extremes of all kinds are wrong, and to breed no sheep at all where you have the wool and lamb tithe taken for the vicar's life is to me more absurd than keeping too many. You can *perceive*, I find, how much better condition the hogs are in which were bred at Denton than those drove from this country, and no wonder, although you have got the very picked ones sent this year. Let them be ever so carefully drove they will and must lose condition and be *checked* for a long time, besides the very great expence attending the driving stock so far. I believe it is a wise way that John proposes, by lying the lime on over winter. This appears to be a good bean year, indeed all spring crops look well in general, but the Friday's wind hurt our beans and pease very much, blowed the pease all topsy turvy which grieves pease to all intents and purposes. I am not so fond of pease as beans, you can't clean the land nearly so well with pease as beans, and pease measure so gifty. Besides the keeping the land *clean* is the great matter and it can't be done well with pease. Your observations on calcarious matters are quite philosophical. We can hardly spare one driver at present (when clipping is so throng) let alone 2, but we shall send no cattle until we hear from you again, because we have more beasts than sheep meat at present, and our cattle to cast will be lean and very unfit to drive at present, and as you say the others may be as well sold here as with you. At any rate we will await your answer to this. I will take this to Wark tomorrow, and after giving you an account of the tup show at Wark[57] will forward it per post. You will see by my last letter that I sent *only* the *ejectment* served on *myself*, and not the *one* served on my *sister* to Mr Scruton. I also *transcribed* you the copies of both Smales's and Mr Scruton's letters. Therefore hope that you will consider with Mr Peacock or Mr Johnson of Darlington whether it is necessary for me to send Mr Scruton the ejectment served on my sister also, and write me word what I am to do. Do observe what I say here I *beg*, because I shall be much hurt if I have done wrong in not sending the ejectment served on my sister also, and if it be necessary for it to be sent I will certainly do so as soon as I receive your answer. I am yours sincerely (with respect to Matthew)

Geo Culley

PS. If your sheep be like to heat and be sore with flies upon their heads there is nothing like caps and capping them early. Tom Glass is not up to it perhaps, although Tom is a decent fellow and best of the Glasses. I suspect they are rather an indolent set, but I am not easily

---

57　The Culleys advertised their tup show, to be held at Wark on 20 June, in the *Newcastle Courant* of 22 May 1803. The tups were to be seen at Eastfield from 19 September to the end of the season. Most of the hiring was done then.

pleased with herds. Few of them without faults. Ours is a dilligent sensible fellow but makes too much to do about religion and is very greedy. We have our first sown ruta baga within a few days of being fit to hoe. We find that sheeps caps made of old *sail cloth is* best. The flies can't strike or bite through them. I spoke to John Forster to get us some. *Monday 20th June.* A very cold morning indeed, I don't like the moon changing in such cold weather. Our cherries are turning red and are only the size of sloes! *Tuesday 21st* is another cold morning but pretty calm so far. An ox has died at Thornington yesterday, a letter came this morning to acquaint us. Drawing oxen in part is very usefull, but we have great losses by it. If one could make men use them carefully it would not be attended with so much loss, but they are often barbarously treated I am affraid. Yesterday was our tup show at Wark, when as usual we did no business at this season, However we refused 50 for half the rode[58] of a tup called Major, from his being used in a shearing at Major Rutherford's. It was Mr William Brodie who bid it and we thought proper to refuse it, whether right or wrong will time discover. He is in my judgement a very capital sheep. Indeed our sheep made an excellent show, and I believe were well thought off. We dined 40 or rather better I believe, and had few except people of business. Mr Robinson and Major Johnson were there, and Mr Nisbet came here all night, but I could not prevail on Mr Bailey to stay all night at either Wark or here. He went home with Robert Thompson. Mr Bailey says he has heard nothing more from Lord Osulston about his soldiering business, imagines it is quite dropped. Indeed you will observe that Government have set their faces against volunteer corps. We received 105£ for 20 ewes of William Brodie and 15£ 15 for the tup that he would have, which neglected John Logan's ewes the year before, but I fancy he means to give me nothing for the other tup. What is to be done about wine think you? Let prices be as they will, in our situation we must have a pipe somewhere before winter. As Mr Darnel has declined will you try Monkhouse and Hopper, or your cousin Surtees[59] as you come through Newcastle? Mr Bailey always says a deal for Monkhouse and Hopper. I would have saved 20£ at least had I had luck to have bought that pipe of Bob Wilson a while ago. William Glass is gone this morning with 138

---

58 The breeding season of a tup.

59 Hopper and Monkhouse, wine merchants, Side, Newcastle upon Tyne; Aubone
    Surtees of Surtees and Johnson, wine merchants, Dean St: *Newcastle and Gates-
    head Directory for 1795.*

barren ewes for Denton by Ingram, Prendwick and Tosson &c. William Wright is ballotted for the supplementary millitia, and I am affraid that we shall have some difficulty in procuring a substitute for him. They are giving 30£ a man for them it is said. Men are so scarce now a war is certain, and unwilling to go.

Wednesday 22 is a very cold blustering morning, I am affraid there will be a bad moon. *3 o'clock.* I have been at Wark and Thornington. The latter place will have no grass this year. The little that was come since the rain is stopped again with this cold windy weather. Sheep all crept under the walls and hedges. Harry will be very late with his turnip seed as well as us. Thornington is the forwardest. My sister at Wark intends to meet you at Mr Bates's, and our Matthew will be at Kitt Mason's show before he come to Denton. We have this moment received a letter from you dated 18th. We shall send no cattle until we hear again from you, because we have casten no cattle yet and are so far back all afar with our turnip seed and they will be casten in low condition, and I find that you don't approve of our sending you the forward cattle that have been feeding some time. In my opinion it depends entirely whether your fat cattle markets or ours are bad. Ours is not near so good as when we sold about well made ones. I do assure that we can spare you a dozen or 14 cattle very well, but by the time we have your answer to this we can better tell what our own casten cattle will be like, as they will be casting a part by that time at any rate. We will send the ejectment served on my sister by our Matthew if we don't forget, and we will not neglect to send a tup for Lord Darlington I think against the time they wish for him. Lord Darlington owes us for some tups. Query whether it may not be right to send in our account by and bye as the old steward is dead and a new one appointed, for fear of any mistakes. And if young Eales is for none this season, would it not be right to let him know that his father when he died was in debt to us for 2 tups. If an exchange should take place respecting the glebe I think you should try if they would take it of the two church fields and Settle field rather than of the Howlet Flat and Gilla field. However you must do as well as you can to have the benefit of such an exchange. It certainly lies very aukward at present. The stripe from the south east corner of churchyard along by the Howlet Flat and to square all the way to the narrow east corner of the glebe next the church field gate as you go to Walworth might be given I think without incommoding you much. I feel much for my poor old friend George Dent. As Skipton market is good for cattle I think John may take some of his Kyloes or queys along with a few sheep, I mean our barren ewes, by and bye. I have no fear of things dropping of a good while more than one day will be better or worse than another. I do not suppose that fat will be low sold all this summer or autumn. I am sorry to find that old John Hope keeps so poorly, but my wife has successfully recovered the influenza. *Thursday 23rd*. A fine mild morning to what we have had. Wooler 10–12 o'clock. I hear that Morpeth was pretty good for sheep about 8 ½ sink, lambs too many and sunk in price. Fat cattle also heavily sold at 8 to near 9s per stone. Now brother Matt, is it not right to adhere to my old maxim, viz. to sell when the butchers are inclined to buy? Only 2 wool buyers come yet. If we had no war, we should have had a score by this time. Should we take 1s per lb. rather than keep? I am inclined to do so, and I will thank you to say what you think is right in this case. It is uncertain, very uncertain, God knows how long we may have war, therefore I think it extremely unwise to keep or speculate in such times. I am yours affectionately

Geo Culley

PS. Mr Howey is to pay Bob Curry 250£ on account of breach of convenant. Matty will explain and he sets of on *Sunday* and will be at Mason's first. But write again a few lines about wool, it is very important for me to have your opinion. Thank you for writing so often.

188.   *George Culley to John Welch and Matthew Culley*

Eastfield 24th June 1803

Well John

I fancy I must now address myself to you again, as my brother talks of leaving you about the end of this month. However whether or not, it matters not, we shall keep on the journals. Yesterday I sent a pretty well filled letter from Wooler, where we had a thinner High Market than I have observed of some time, and only 2 Yorkshire wool buyers and 1 Scotchman out up their appearance, a bad prognostication for the wool selling! Little doing in the corn way except oats, which are scarce and dearer. You will see by yesterday's letter that sheep are still dearer per pound at Morpeth than at Darlington. I find that good lambs were also well enough sold, but many bad ones in which were sold as such, and cattle (fat) not near so well as for some weeks back. *Saturday 25th*. Dry again and mild, I have sweat all night. We are busy taking of our lambs and began to milk the ewes this morning, and we have a great deal of turnips to sow yet, and I think that they will be behind at Wark. Mr Askew, who came home the other day, says he saw no hay in cock until he came to Wooler. They were only just beginning to mow about London. *12 o'clock*. It is one of the best and most promising seasons I ever recollect so far for turnips, and what are come away look so healthy and little eat by insects, our first own ruta baga fit for hoeing, but we can't get to it, we have so much quickening &c. The beans also need a 2nd hoeing much. But the worst of all is our grass lands, meadows very light and pastures all going wrong again. But it is the Almighty's doing, and His will be done. I came through our General's Close just now, and 'it is going like ditch water' with a hot sun and long days. The mercury also uncommonly high and steady, with every appearance of drought. *Sunday 26.* A mild morning, and from the hills being covered with a kind of thickness or mist, one would imagine it a prognostication of rain, was it not that the mercury is higher than I remember to have seen it. However as extremes sometimes produce changes, we will hope that it may by degrees bring rain. We yesterday received 63£ from Col. Graham in a bill, for the hire of the tup 2 years. Matty put 250£ in the bank again as we think we can carry on without it and it is always ready for us in the bank. Ralph Brown sold 30 bolls wheat from Grindon at 42s per. Mr Thompson, Col. Graham's steward, writes that the colonel had wrote to him from Ireland that he was to send the shepherd here to chuse a tup. The shepherd is from Leicestershire. I had forgot to say that Matty was at Bob Thompson's show on Friday when he exhibited some very capital shearings. Mr Bailey thinks one of them the best he ever saw. Your haugh wheat at Wark is going to be a very light crop, and I am convinced that *you err* in sowing it too late after beans. That *old* worn land requires all the time it can get to do its business in, while any weak *fresh* land will grow decent spring wheat (recollect Fenton) good old land can not do it. I think your spring wheat on the haughs as good as can be expected, for it wants youth for spring wheat. You know that compared with your outfield. But Ralph Compton's wheat after turnips, and half eat on, is much, very much worse than your Willow Bush. But I said if the Willow Bush did not produce a decent crop of spring wheat this year it never would, because it was early plowed and well mellowed with frosts, and it is (although not good) as good to the full as could be expected. But it is not the spring wheat that I blame, but the wheat sown or to be sown after beans, because you used to get a very strong crop after beans until these last 2 crops, which makes me suspect that it is from sowing too late those last 2 crops. The Willow Bush wheat exceeds the Mid Close much, and that may be owing to the former being a more smothering crop of beans or beans and pease. To me it is altogether absurd to sow clean pease, because they are only a crop once in 7 years while beans are a crop every year or nearly so, and much better and certainer crop of wheat after beans or beans and part pease perhaps on very old land.

Ralph Brown is quite of opinion that he could have sold oats yesterday at 22s per boll. They are not to be had at any rate, and he thinks and very properly that this must enhance the value of bean and pease. This balloting for 50,000 fresh *Millitia men* or call them what you will, will make a vast scarcity of laboring hands and make substitutes very dear to hire for such as will or can't go. Nevertheless I approve of the measure, or any measure properest to keep Bonaparte from these Islands. Matty may pay the printer for advertising our tup show as he passes back or forward through Newcastle, aa a cart will be lightly laden with only one tup for Lord Darlington the next month. Anything else may be considered of to send and bring back from Denton. *Sunday evening*. All the Culleys, Mrs Culley of Wark and Nelly, sister of Roads and Nelly and my wife, Matty and self were all dining with James Darling and Nelly today. My sister told me today that she proposes setting of for Brunton on the 29th, and reaching Mr Bates's on the 30th. Still like drought. As I send this short scrawl tomorrow morning by our Matthew. I will conclude yours affectionately

Geo Culley

PS. I go for Yetholm Kirk fair in the morning.

189. *George Culley to John Welch and Matthew Culley*

Eastfield Tuesday 28th June 1803

Well John

We are sadly catched again with drought and very likely to continue I think. Yesterday morning between 3 and 4 o'clock our son Matthew set of for the debated tup shows in your country, the morning dry, cold and cloudy but calm. This morning not so cold quite but dark, with appearance of rain, hills covered, but glass very high. At Yetholm (Kirk) fair yesterday a vast number of sheep shown, chiefly hogs from the lowland parts of Lammermoor and Lauderdale, sale heavy but prices high, fresh or fattish and cattle especially, Kyloes readily sold, some went home unsold after being bid full 9s per stone, Mr Oliphant's Eckford. They called them 40 stone each. I was not a judge of these animals but could not think them so much, was bid 18£ per or nearly for them. Very few wool buyers. Mr Barf talks of 21s our stone and says he will buy little. I think we will not sell at that price. This day William Brown brought us your (that is) my brother's letter which has stopped my sister at Wark until Friday. She was for setting of tomorrow, but is altered in her intention and hopes to be at Brunton on Saturday evening. I sent this letter to Wark and had the above answer from Nelly of Wark. I am sorry to tell you that Robert Dickinson is very poorly, and Jack Johnston, but Jack is young and may recover. Harry Rutherford's wife is also poorly. I feel for poor George Dent and family. We shall be glad to see Miss Dent, but poor thing she may never be able to come again. Your reflections on the decline of life are very appropriate, Sir. Good health and common sense are invaluable blessings when properly applyed. I have wrote to Mr Peacock to be here on the 7th or 8th of July if he can, as Matty Culley has to attend Hexham sessions on 14th and will be going about the 12th, and I should wish to have him at home when Mr Peacock is here as he understands the accounts much better than I do, and the money in your and my hands due to him and his sister will want settling which Mr Peacock can best do.

Wednesday 29 June. We foolishly forgot to send the ejectment served on my sister by Matty, but must now send it by the lad Will Glass when he goes with some cattle, which must be as soon as he returns from taking the barren ewes, because we shall be worse than ever for grass if rain don't come soon, and I have no hopes of it except the glass coming down a little from the highest ever known. But it is still at fair and has not moved last night although a very thick mist this morning. I would have been glad to have had your advice about what cattle to send, but as we can't wait for your return we must send the best to our knowledge. As you don't set of from Denton until the 4th July (Monday) I will send this per post tomorrow, that you may see it before you come away. No doubt tithe is a grevious tax, but you are better at Denton than many spots because the clergimen at Gainford have always been better to deal with than laymen, and that is the case 6 times out of 7. I think by taking a strip all the way from the south east corner of churchyard along Howlet Flat and quite through Gilla field as far as end of glebe on that side, and from the north east corner of churchyard next your orchard to the north of the glebe, a large strip all the way down the south of the 2 church fields will be more than 12 acres I should suppose. If not right go beyond the orchard foot. The ox at Thornington had been long ill, had got a *thrush*[60] as they call it, harm by climbing the hills perhaps. We could not do well without some few ox drafts, but you may depend upon it that we have lost a deal of money by working too many oxen.

---

60 A disease of the frog of the hoof, with a fetid discharge.

Whether is an ox *undrawn* and on same keeping *better* at 4, or being *drawn* at 7 years old or even 6 more or less? Every impartial man will say at 4 undrawn, and if so what loss have we sustained! Besides many of our drawn cattle fall amiss from one injury or another, which most probably they would escape if not drawn. But more of this when we meet. Since cattle have advanced in price so much the difference has been more glaring I think. I will thank you, or our Matty if he sees this, to pay printers for advertising our tup show, and settle with Anthony Surtees. That is I have received 20£ of Mr Bailey, but Mr B. had neglected setting down what he received, consequently it will be right to investigate the matter at Mr Surtees's office before it be too long, particularly as Mr Bailey thinks Mr S. took his fee off. This business you are perhaps fitter to transact than your nephew, but do that as seemeth good. I feel very happy in not buying Ralph Compton's steers, we have not meat for our own! You talk of buying steers to work to save ours, and we must plow less or we sell some of them. A considerable part of the ewes went to Denton, all good at an early period, and the rest will follow fast in case William Glass has drove them well; which I don't fear much. If Glasses will not drive we must change them. Drivers must be had so long as we are connected with Denton. I perfectly agree with John Welch that it may be wrong to apply to Lord Darlington or Eales yet awhile. Indeed I know nothing of Eales or his circumstances, but there can be no fear of Lord D's money some time, and although not the best of customers for price he is very steady and pays well at such prices. 12 o'clock. I am just come from Wark, all well except Bob Dickinson and Jack Johnson, both of which Bob Tait thinks badly of. My sister bids me say 'that she will leave Wark on Friday, reach Brunton on Saturday, and meet you at Newcastle on Monday to take you and Matty to Brunton'. After the mist went of between 10 and 11 o'clock it is insufferably warm, but thank God very calm. Corn, spring corn in particular, looks well and turnips grow fast. But winter sown wheat is and will be very thin. Beans are also short and are stopping again I think. Harry has mown some places of laid grass in the watered lands, he thinks to be done sowing turnips next week or nearly. We shall be after that. James Moss sold about 60 of late Mr Wood's clipped hogs at Yetholm at 44s per, a famous price. I will send this from Wooler tomorrow after learning how Morpeth has been if I can. Ratcliff's driver took 23 fat oxen from here this morning, we were determined to have set them of this day or jobbers, our Matty will explain. I am affectionately yours

Geo Culley

PS. If I could advise you I would give up the Tyne Mercury[61] and pay Mitchell the publisher when you are at Newcastle. I think it a very useless paper and the Courant a valuable paper. We have all the news that is in the Tyne Mercury in the Courant 3 days earlier and in the London paper. *Thursday morning 30 June.* A thick mist again. Wooler 12 o'clock. I hear that Morpeth was worse yesterday for both lambs and sheep, chiefly clipped hogs, but cattle are better than they were again. So the tide ebbs and flows. Cattle from 8/6 to 9*s* I am told. A gentleman just by me, a traveller who has been at Inverness lately, says cattle are higher there this year than last. Mr Meggison is here, and is rather desirous of buying our wool at a price, however I am determined to take no less than 24*s* at present. G.C.

190.   *George Culley to John Welch*

Eastfield 30th June 1803

Well John

I think my brother will have left you before this reaches you. I sent a little letter today per post, which I hope will be with you before my brother leaves Denton. We hear of all hands how beautifull your country looks, while we are burning top and bottom. However I am glad that you are so well of. Nothing at all done in the wool way today. A number of buyers now, but they want to beat me down, and we that keep are unwilling to submit very far. I am determined not to take less than 1*s* per pound. If we can reach that I think it might be unwise to speculate. *Friday 1st July.* Cold and droughty *Saturday morning 2nd July.* Thick mist but droughty-like. John I believe I shall send you this short letter from Berwick today to caution you against taking any of Surtees and Burdon's Newcastle or Berwick *bank notes* until they have matters properly arranged again. The two Matthews got here last night and brought the news from Newcastle about the bank being stopped there, and Ralph Brown brought the account of Berwick bank being stopped from Kelso market yesterday.[62] We have

---

61 The *Tyne Mercury* began publication in 1802. It was a livelier paper than the older *Newcastle Courant* and *Newcastle Chronicle*. These both appeared on Saturdays, the *Tyne Mercury* on Tuesdays.

62 In June 1803 there was a commercial panic in the north of England, with a run on the banks. On 30 June Surtees and Burdon of Newcastle and Berwick suspended payment. The notes of three other Newcastle banks were guaranteed at a public meeting, but Surtees and Burdon was omitted from the list. The bank put its affairs into the hands of a committee of management, which announced on 3 November the winding up of the bank. Phillips, *History of Banks*, pp. 385–400.

250£ in the bank but we owe them 200£. But I feel very comfortable because a very little while ago we had 1250£ in the bank. I would you should be very cautious *now* about taking bills of any kind. These are and will be (God knows how long) critical and perilous times, get all the gold you can. Perhaps your wisest way will be to sell as little as you can until we have some idea how things are to go with the banks people, as more banks may fail as well as Surtees's. Berwick 2nd July. Well John here is no kind of business done at all at this place today. If Bonaparte had been upon the coast it could hardly have been worse, all confusion and dismay, and I see by the Newcastle newspapers that the principal tradesmen and gentlemen in and near Newcastle have subscribed to support all the banks in Newcastle, Surtees's excepted. Now that is so exceedingly *pointed*, and shews that they are determined to set their horns against them, and I am persuaded that they will be no longer bankers. Indeed they have none but themselves to blame, by speculating at everything. So you will observe that you may take any other Newcastle bank notes but those of Surtees and Burdon. Let me hear from you as soon as you can, and tell me about wool. None sold here yet. I am in haste yours

Geo Culley

191.   *George Culley and Matthew Culley jr to John Welch*

Eastfield 4th July 1803

Well John

I sent you the shortest letter on Saturday that I have sent you a long time, but I hope you would need no caution about Surtees's notes. I do assure you that there was a very flat and stupid market at Berwick. I don't believe that there was one boll of grain sold. As I shall have an opportunity to send this by Mr Peacock I may as well tell you all we know as not. I do imagine that there will be much disappointment if not loss by these bankers stopping. Thomas Tulip has a good deal lodged in the bank, and all the farmers who have any money. If we had not put the 250£ into the bank on Saturday week we would have been quite clear except 100£ or better of their notes in hand, but we got a bill for near 300£ of them about 3 weeks ago, and I am affraid that it may be returned upon us if it was not *accepted* by their bankers in London before this stop took place. However we must take our chance now. Your mistress of Wark had got as far as Cambo to meet your master at Brunton, but Mr Peacock and my brother came to Cambo when she was there at breakfast, so she returned back with them again. They were here to dinner yesterday, and Mr Peacock, after staying all night, is gone there today. We had a very slight shower this morning between 5 and 6 o'clock, and I hope we shall get more soon as the wind blows fresh and clouds are collecting. I do assure you we are in much need of rain. John I always forget to remind you, but you are so clever and attentive to accounts that I have no fear but that you keep accurate accounts of the alterations in my brother's house at Denton, as it is his own concern. Not only the timber, slates, bricks &c &c but leading all the materials must be attended to,

[*M.C. jr*] and accurate and separate accounts kept, also burning the bricks and the lime used and carters and men employed, and an account kept of the value of the old slates, timber &c that may be used in the West farm which must be paid for at the joint expence. We have just been chusing a tup for you and one for Lord Darlington. The tup for Lord Darlington is the larger of the two and is three shear, the one for you is only two shear, is very lean, consequently looks less than he would have been if fat. He was the largest dinmond we had last year and has a very good fleece, but you can judge whether him or the one you now have will suit you best, and send that one back which you do not like.

[*G.C.*] *Tuesday July 5th*. Lambs were well sold in my opinion at Yetholm lamb fair today, country lambs from 20 to 24s, one particular lot of Mr Walker's of Wooden near Kelso were sold at 27s per, but they were very large and fresh. As long as the Scotch notes lasted there was a quick fair, but when they were short there were no more sold, and I am of opinion that if the Berwick bank had not stopped lambs would have been as dear as last year. Not much done in the wool way yesterday, some little sold at 21s it is said. However our last year's chap would not give me 24s and I am determined to take no less at present. So we parted, and it is not unlikely that we may not sell until a favorable turn takes place. This has been a very droughty day. *Wednesday 6* is a very dry morning. I am glad to tell you that by a letter received yesterday from Mr Smales of Durham, Sir Ralph Milbanke's mortgage money is paid to Lady Headley, so that my mind is now relieved of that anxiety. *Thursday 7*. A cold morning. We expected a letter yesterday but we had no doubt of one today. But behold, we are totally disappointed. It is not kind John, as in particular you promised to write after Monday's market to let us know how you are going on, and in particular about wool, with which we are entirely in the dark. It is really a very great pity that you can't spare time to write a few lines, a very few would give satisfaction. However we must rest contented, what can't be cured must be endured. Mr Peacock says it may be 4 or 5 days before he reach home, but it is no great matter as we have little of consequence to inform you of, but you have much of importance to inform us of, particularly wool as I have said. But I suppose much of that article will remain unsold in this country, at least little is done yet. It is said that Mr Barber of Doddington and some others have sold at 24s our stone or 1s per pound, and I am determined to take that price if offered and no less, and there I fix my resolution at present and my brother is agreeable. The man we dealt with last year will not give that price, so we are at liberty to deal with anyone. Wishing you more inclination to write some time I am yours

Geo Culley

PS. I had forgot to say that we have 750£ to remit to London, but in the course of 2 or 3 months will do, and Mr Peacock seems to think that there will be no difficulty in you and he raising the above sum in the above time when Mr Peacock will remit the money, only I am sure that you will not neglect to talk to and consult Mr Peacock upon this *matter* and acquaint him with the chances you have of raising money. Mr P. talks of owing you for barley, with rents and interest that he will receive very soon to the amount of 300£. And your fat stock and corn must do the rest. The building will also take part money, and we can make no assistance I am affraid as this aukward bank going wrong, and rents coming on, will take all we have I think. I intended to have sent this by Mr Peacock, but as you have not wrote I will send it per post, which I think a mere trifle indeed compared with the benefit of intelligence.

Wooler 12 o'clock. I have just seen William Marshall of Kelso on his return from Morpeth market, says lambs were sold badly, sheep heavily at 8d to 8d odds per pound sink, and cattle 8 to 9s all heavily sold he says, and admitting that I consider it a good market considering the stagnation from Surtees's Bank &c &c. Few sheep except clipped hogs, barren ewes and tups. As Mr Peacock will not be home of some days I shall send this per post, and hope you will favor us with a line or 2 in a little time. G.C.

192. *Matthew Culley to John Welch*

Wark Tuesday 5th July 1803

Well John

We had some thunder and I suppose much rain fell on the Tyneside on Saturday, fell only slightly more near Cambo, where we met your mistress. We left Mr Bates that day, only staid with him the Friday night. We dined at Cambo, so went to Whitingham all night. You should make the herd cap your sheep. Our pastures are pretty good, although we want rain much here, had only some slight showers yesterday and today so as to be off in an hour again totally. Mattys got to Morpeth first night, to Eastfield next day, my son to Wark and his cou[si]n the 3rd day, found all well but Robin Dickinson, who I hope is in a way to mend but is sore shrunk. Jack Johnson is still worse, I fear him much as he gets no strength. They have both been at Kelso dispensary. There are several cattle feeding here and several of them doing well, but not fit to drive this hot season so far as Denton, may be sold or fed here. The wedder sheep are doing very well in all the fields which I have seen here yet. We are likely to have good turnips and a surplus quantity of them both here and at Eastfield, some oxen cast but are larger than you like to have, are looking well although leaner than they should be. The stop of Surtees's and Burdon is likely to be a most serious affair, I wish it may not ruin them all, but the country is most to be pitied, who will suffer ruin by it. What folly to live above your income. As we get rich, then comes pride, ambition and such-like folly. Our Savior preached and taught *humility*, pride belongs to the Devil. Crops of grain seem light on all old land, our spring wheats especially on the East Willow Bush after turnips is very weak. We press our wheat over close. Nature delights in variety. There is nothing like change of crops with good tilling &c &c may go on for years with proper skill and by raising dung in plenty, then lime when wanted (*calcarious matter or earth*). The Almighty (I apprehend) meant the land to grow a crop every year, no bare fallows, drilled crops, and make use of the seasons to plow, hand and horse hoe &c &c, and what may not good land do, even poor soils may do a great deal by proper course of crops, but few white crops only a variety of pulse and green crops laid off to grass. I have not been to Thornington. Paty carried Matty very well he says. Matt fished till 12 o'clock last night, caught a fine dish, the water so low, will not take but at night. There are no salmon, even very scarce at Berwick. I hope you get some showers, as there is some appearance for rain, but we seldom get rain in any quantity in a westerly wind. With respects to you and all neighbours

Mattw Culley

PS. If you see Mr Harison make our compliments to them and say we hope Mr Harison has got free from the cold.

My brother will give you an account of wool and lamb markets.

193. *George Culley to John Welch*

Eastfield Tuesday 8th July 1803

Well John

Disappointed of a letter yesterday, and Mr Peacock not to be home of some days, I sent you a letter from Wooler yesterday. This is a very fine mild morning after a fine rain yesterday which will do corn and turnips much good. Indeed we have between 20 and 30 acres want hoeing much, and we are so busy picking quickings, and so late with our turnips, that we can't possibly get to them until tomorrow when I hope we will soon give them a broadside. *12 o'clock.* Well John yours of 6th inst. is this moment come to hand, and although later than I could wish I thank you for it. At same time I think it very idle of you not to write an account of Darlington Monday market before Wednesday. If you could not find time to write on Monday you might have wrote on Tuesday. Mr Peacock was desirous to hear before he went away, if all was well, and your master was certain that you would write from *Darlington direct*. However John you should certainly have wrote on Tuesday, why defer *one day* on letters of business? Well it cannot be helped this time, nor would I mind it at all if there were any *hopes of amendment in you,* but I really despair of you in the letter way! Would you just write five or 10 lines I would be rejoiced and satisfied. What is the expence of 10s postage to the value of one valuable letter? I am willing to admit that you have a wonderfull deal to do, but a man must be strangely hurried if he cannot spare *two minutes*! and that is quite enough for you that carries the pen of a ready writer. I find that Morpeth beats you for sheep quite, and Yetholm beats Morpeth for lambs. It is wonderfull that people do not geld their male lambs intended for fat, and then if Morpeth won't do they can try Yetholm, which has never been overdone of several years. John Barber of Doddington has sold his twin lambs, ewe and wether these 2 years at Yetholm, and has got better prices than he can reach for his single lambs fat at Morpeth! His gelt twin lambs were sold at 24s per, and ewe lambs (twins) at 21s, what great pay. I hope it may be right, your deferring going to Skipton a fortnight. It is not to be wondered at that Darlington was bad last Monday as it was Stagshaw fair, besides the failure of Surtees's Bank. But I don't see that the failure of Surtees's Bank could at all affect Wakefield or Skipton. We shall keep our cattle another week, but you have no idea how bad we are for grass! My brother had no idea of it until he saw our situation. However there is water enough still in the

Tweed, and they must be content with the dry *windlestraws*,[63] it will make them drive nicely as they will be *light bellied* I assure you as well as *dry skinned*. We can also consider about keeping all our cattle ourselves, if you can supply yourself. I dare say we shall need them, or others, if the turnips thrive, as we never sowed so many before I think, and they are most promising so far. But John why tye all your cattle up, to make manure? Why not keep light queys in an open fold, and give them turnips in *troughs* raised upon feet, or *cribs* in the *corners* of the *fold*? It is the way Mr Nisbet and all our first-rate winter graziers do now. You know that they turn much better out in the spring to grass from an open fold than from being tyed up, and get fat in ½ the time. Better with shades in the folds, but in your warm country they will do without shades pretty well. I am amazed that you should prefer tying up. Give it a proper consideration, and make a tryal for once in part. And if tyed up do the best then stick to it by all means, but I am perswaded you will find the contrary to be the case, without you are not determined not to see it. However I would recomend to you to try it fairly once. Besides, to me it is the greatest absurdity to tye up light cutting up cattle, such as you very properly wish to feed. And we have a lot of very nice queys of our own, that would answer well to lay in open folds and shades, but to tye them up would be the *greatest absurdity* in my judgement. But we can keep them at home and lye them in open shades, and send them to you in the spring if you don't wish to have them this summer or autumn. I am very glad that Will has brought the ewes so well up. Good driving is everything, and all the art is in taking time, a thing we *all know* and preach up but forget to *practice*. I offered our wool at 1s per pound but can't reach it yet, although it is given for some I am told. *Saturday 9 July* is a mild fresh morning. As I am going to Holy Island with my wife a few days I will send this to my brother to carry on. Matty also goes to Hexham sessions this week, they are on Thursday and a troublesome job we have with the sheep and skin stealer. There was one large lot of wether lambs belonging to Mr Walker of Wooler, sold for 27s per head. I am with every good wish Yours

Geo Culley

PS. Should you happen to want anything in particular our Matty will be at Mr Thomas Bates of Brunton on Wednesday night next, the 13th of inst. Mr Thomas Peacock and Mr Parker from Auckland I

---

63 Thin withered stalks of grass, left after the hay harvest.

think have come here in search of Mrs Peacock and go directly, so I send this letter.

194. *George Culley and Matthew Culley jr to John Welch*

Eastfield 12th July 1803

John

I sent a half letter to you by Mr Peacock and a Mr Parker the other morning, as they were returning so quickly home again, and did not know their errand at that time as they said nothing to me what they were come about. But Mr Peacock will inform you. But my reason for writing to you at present is that my brother has received a letter today from a Mr John Hog of Newcastle[64] dated 7th inst wherein he says he has received a draft value 30£ indorsed by my brother at Surtees's Bank Newcastle at the time my brother was paying the money to Mr Bates and which draft was one my brother got from Mr Peacock which you had paid to him (Mr P.). The man Mr Hog has not sent the draft to my brother but the memorandum my brother has is as follows: 'Geo Hog to Messrs Harrison Prickett & Co, London, Bankers. Manchester 27th April 1803'. However the following is the copy of Mr Hog's letter. 'Mr Mw. Culley, Sir. This day received a draft value 30£ from Messrs William Caswell & Co.of Paisley which is indorsed by you. It was paid to Messrs Surtees Burdon & Co. It was presented for payment to Messrs Harrison Prickett & Co on whom it was drawn (who refused to pay on account of "no order to accept". The expenses of protesting &c are £1. 8. 0, which with the bill please order payment off at your early convenience. I remain Sir your most obedient servant John Hogg. Newcastle July 7 1803'.

My brother has wrote to Mr Hogg to send the draft to you, and you must get Mr Peacock, if returned home, to advise you how to proceed in this business. I hope Mr Hogg will send the draft to you directly, and if the man that drew it is a good man there will be no loss, and if it is lost, we ought to be thankfull that it is so small a sum. However it shows what dangerous things bills or drafts are! and how little one should have to do with them. Yorkshire is a sad place for bills! and how to make a better of it I know not. You must know that my brother paid the bill with many others of yours into Surtees's Bank and got the notes for them after indorsing them, and it will be well if no more are returned to us. Yours in haste

Geo Culley

---

64 Possibly John Hogg, linen draper, Dean St, Newcastle upon Tyne: *Directory for the Year 1801.*

PS. Please tell Mr Peacock that I received his letter from Newcastle of the 10th inst., and my brother and I approve of his giving Mr N. Parker time to pay the money. Mr Peacock will understand the above. I am so busy at present that I cannot write to him, being just returned from Holy Island, and Matty going for Hexham sessions tomorrow morning by 8 o'clock, and we are very busy I assure you. What a plague these Surtees's people have been at this moment to this country. Few people here can get their servants &c paid for want of other notes, being all in possession only of Berwick notes G.C. Turnips grow charmingly, but our pastures are done. John I had forgot to say that my brother says you sow far too little rape seed. The poorer the land the more seed should be sown to a certainty, because to a certain age the plants assist one another, and then should be thinned by *harrowing, hoeing* &c &c or by any means in your power. But too little seed of either turnips or rape is very bad husbandry indeed.

[**M.C. jr**] Please to send us an account as soon a possible after you know the result of this awkward business.

195.   *George Culley and Matthew Culley jr to John Welch*

Eastfield 20th July 1803

Well John

Having been a good deal at Holy Island lately with my wife I don't get you so much plagued with letters. I don't know what to say about sending cattle to you. If you can supply yourself with cattle to tie up, we can do with all we have very well in the winter, and at any rate I cannot subscribe to you having our nice spayed queys to tie up which we can so well lie in folds or open sheds. You have given no answer to what I wrote to you about tying your cattle in open shades and folds. I am certain they will do the best that way and will make as much dung, although I confess that it may not be quite so good dung. But by turning it in the spring, it will do very well. This letter will go by William Wright, who takes 4 tups to Denton, one for Lord Darlington marked thus – – behind the shoulders, the lad also knows which is him, which you may send to Raby on Monday or not later than Tuesday at any rate. The other 3 you must be so kind as send to Mr William Smith, sign of Black Bull, Northallerton, by 12 o'clock on Monday and not later as Mr Marshall's servant will meet them at the Black Bull and your own servant can return home the same day. And it sets our man homewards on Monday morning again, and we want him home again very much. If you have not a cart ready prepared for tups, you ought to have one as you may frequently have tups to carry.

But you can easily get your carpenter to set up a few pieces of wood, and run a rope through a few holes bored in the upsets for this time, and a piece of wood can be put across the cart to keep 2 forward and one behind. However I beg that you will not spare time or expence in having the cart properly done, and sent by a carefull hand, so as no harm may attend them as they are valuable sheep and let at very good prices. We intended to have sent a tup for ourself and had one ready, but as Mr Marshall and another gentleman came the other day and [with][65] these 3, and Lord Darlington's one, it fills the cart, so you must be let alone until another time. We have had a great deal of thunder and lightning, and a fine shower this morning which will do a vast of good to turnips and corn. Yours in haste

Geo Culley

PS. If you *are* to have any cattle from hence, I think it would be the wisest way for you to make a stop over and chuse what cattle will best suit you, as we may send you improper ones.

[*M.C. jr*] What have you done about the 30£ bill which was returned to Mr Hogg of Newcastle and which my uncle wrote to you about lately? By a letter from Mr Sayle the Wakefield market was very good, beef from 8 to 9*s* very nice, mutton 8*d*, and Mr Deveril says wool is dearer but mutton only 7 ½ both dead and alive. I hope you have got your cattle sold, at least if you have not you have sit your market, and we wish you always to sell when you are fairly bid. The sooner you sell either them or any other fat stock you may have ready the better, as markets are sure to be worse and not better.

196.   *George and Matthew Culley to John Welch*

Eastfield 26th July 1803

Well John

I have this moment received yours of 21st and one from Mr Peacock of 23rd at the same time. I don't understand how your letters often happen to be later of coming than common. Well I will say no more about that in these *eventfull times*, but commend you for your caution and attention about Hog's bill, only you kept us longer in *suspence* than is pleasant. However by your clear well explained account I think there will be no loss by it, and I hope you have had a tollerable market at Skipton. I can't look for a good one this warm (nay hot) weather in the severest drought I ever knew I do think except the year 62, when people are under the necessity of going to

---

65 A word seems to have been omitted.

market for want of meat. I do assure you I can hardly tell you how our stock lives. But we had a blessed thunder shower last Wednesday which has done our turnips much good, they are doing surprisingly, and corn is much benefitted by it. But a vast of oats will be very bad here. Matty was at Alnwick fair on Monday last, a heavy market but not worse than expected 7 to 8s per stone sink. We should have shown a few cattle, but Morpeth was so very bad the Wednesday preceeding that we were affraid, and I do assure you we could have spared the cattle well, and have used the money. I do perceive that we shall be much pinched for money all along this year. However if you and Mr Peacock can make up the 700£ we may still do. But neither he nor you take any notice of it. I can excuse him because he is so mortifyed with this Glanton and Parker &c,[66] and I will excuse you too if you will only tell me by and bye what you can do. This Surtees's Bank is a sad business. We have a 300£ bill returned upon us which the Berwick bank drew, which Matty had sent to Mr Buckley for his tup. We shall now want about 700£ of them. I am very glad your rape and turnips are doing well. I am glad you are for keeping your own tup. Matty had nearly sent you one by William Wright. O how I wish you to sell your corn as fast as can be. You may be assured that it will be lower without a wet harvest which the Lord avert! I will leave my brother to answer about the house &c &c. At the Shildean a gable was drove in, some furniture damaged, with a bed where 2 children were, but thank God they no worse although the dog was killed by the fire side. A thrashing machine at the Troughs belonging Mr Hogarth Fireburn was set on fire, a house in Scotland near Norham much damaged and a dog killed also, while the people received no harm. How wonderfull! This all by thunder last Wednesday morning.[67]

[*M.C.*] Well John. Wark Wednesday 27 July. I am glad your masons are doing so well, the weather suits you over well, and our pastures are exceeding bad, but we must do as well as we can. You do not say that you have sold your oatmeal, which you should do as I think, do not try to sell over dear, be content with good prices. The money will do us good, and sell we must to carry on business as we get slowly forward here with selling corn, as we can sell little or none at Berwick. Do sell away, better me sell than me keep. I fear you have sowed the moor too thin with rape, the poorer the land the more seed is necessary. Our watered land has brought us a decent crop of hay which is

---

66 It has not been possible to trace this affair.
67 The signature is lacking.

not worth the expence. Turnips promising, have put 40 ewes on our rape, which is very good except ⅓ of the field. The bog here keep all our work oxen, some horses and 12 queys. The outfield has grass on it, and a field we laid in to cut for our horses but are obliged to put some cattle on it to keep their flesh on them. Have sent all our forwardest cattle to Thornington haugh to an excellent pasture which was watered a little while since. But the fear is that as the drought appears to have every sign of continuing here, that we may run fast before rain come. Wishing you well I am with respect yours

Matthw Culley

PS. Matty will likely set out on Monday for school, but he will write word from there when you may send for it, the mule to Houghton.

197. *Matthew Culley to John Welch*

[Wark c.31 July 1803][68]

John Welch

So Mrs Middleton and Joseph Birkbeck[69] have left this stage of approbation, for such I believe it is. We must follow in a little time, and then let us employ our time the more we may expect. Virtue alone is happiness below and will meet its reward. Mr Peacock and you will settle things, not in a mercenary way but according to justice. The widow I fear is a vain weak woman. If she could get a cheap cottage house, say at Houghton, at less money than you can afford her for the house and stable (for Mrs Robinson), say 10s a year or 20, it may assist her in bringing up her children. Locky will keep his field till the term ends May 1784 [sic].[70] Did you get the drift up to the wall at High House? Am glad the watered land was a tollerable crop of hay. It's a pity you have not got the rafters before you are ready for them. I wish for ways to saw out good deals for you from such wood as young Starforth approves off. Be geting forward a little. Did you pay Wass any money for wood as he expected? I am glad you got money for Hogg's bill. How will you succeed for the bill sent to John Hodgson? It was not well managed by J. Hodgson. You must submit to sell the oxen and queys, the 5 I mean, 2 of them may gain weight but you should sell 3 of them, the black ox and 2 spayed queys. We

---

68 The address and date are lacking; but the hand and the signature are Matthew Culley's, and the content indicates a date c.31 July 1803: see No. 196.

69 Joseph Birkbeck of Denton died on 26 July 1803 aged 85: DUL, BT Denton.

70 '1804' has been pencilled in above by a later arranger.

are pinched for grass, but our sheep and cattle are not much amiss and we should and will put them in the watered fields some time, where they may do well. Corn is likely to be a good (valuable) crop, droutht generally suits the crops of grain in our cold climate, although some light soils have suffered much. But what are they compared with the cold lands. Wheat seems to get a full head, turnips are likely to be a good crop. Ours are now prospering and we have more than 100 acres extra this year on our farms here and with you. Have put 40 ewes on our rape, as we are generally too late of eating it in part. We have some cattle here of a small size, may suit you to winter part of them from our casten cattle, but then we must buy when oportunity offers here or with you. If we could get some good or well bred steers about Raby then we might keep Buston's, and we ought if you can buy them which I think you should not miss, although dear are of a valuable sort, and where they are to be kept forward should be bought of a good kind. We have 1 or 2 working steers and I mean to cast them. It's a pity to work them any longer. We must try to put some in their stead by buying more good steers or oxen in their room. You wish us to send for the mule, but nephew Matthew will write to you from Darlington as he goes to Yorkshire. I mean to try if he and Mr Jobson (William) will come your way all night in going, and stay some days on their return home. We have no tares for cutting, and no second field for setting of clover for horses, and we have set the low haugh with cattle. It had no chance to be fitt to cut unless we get rain. I hope you had some when the wind blew from you to the north 2 or 4 days ago, it went off with drought. Corn seems to fill nicely, will probably be much superior to that in hand. Sell away. Oats at 21s is a good price here I believe. Angus may be found the best, a strong grain and good straw and in quantity. Our beans are very short, peas good and likely to load well, but our Dry Tweed was not plowed over before sowing so often as was necessary for good tillage. It gives me pleasure to see your church fields in such nice and clean grass and corn. You'll put cattle on your watered land directly, and if heavy rains fall run on the water if you can. John Colling, Punton and Sage drank tea here one afternoon on their return from a tour in Scotland. Sunday afternoon. A fine shower about midday, glass falling. I hope you have showers. Have got led into pike and hope to stack some nice hay tomorrow and mix with oat straw. Our rape much mended within this few days, but our ewes are not in high condition. Your obliged master

Matthw Culley

PS. We had some fine showers yesterday, I think litening from the sea in the morning but heard no thunder then. Pastures looking greener, turnips do well, mercury dropped a good deal, it was very high. Our rape where the land is dry on Jock's Ridge flourishing beyond common, ordered 100 ewes to be put on today, the ends of the field are watered, want a little dung to start it. It's a most valuable thing and would with a little dung bring about your moors and poor soils sooner than anything I know off. Eat on would insure a crop of wheat almost to a certainty. Nearly the best crop of wheat we had on Jock's Ridge the high or hill part of rape (the lowest part turnip and a deep drain between them runs much water, should have been a hedge and good ditch on the under side as moor land requires hedges to drain and warm it such as yours). But it's a charming crop of wheat indeed. Holburn is doing wonderfully well since it was drained, hedged and well wrought. As it now is only in winter it wants warmth and dryness and does little in wet weather for stock or corn. But our calves did very ill with turnips and straw the last winter, only a cold shade to ly in. Old Mrs Compton died yesterday morning at Roddam at Mr Nisbet's. Most of her sons had gone to see her, asked Mr Anthony's pardon for using him so ill. I was told so by William Smith who Mr Compton told it too. Let us know what stock of large or English cattle you may want to winter, and we will try to get you some. We have 10 or 12 queys and small oxen of our own, besides the 12 queys 3 years old of our own breed kept from bull, and 2 good cows all feeding if they look well, and the cows with sore udder, will be very good when your pastures suit to take them, and things are bought cheaper here and many of them better than with you. Young Ned Smith sold 6 nice ones at Ninian fair last year. Barber and John Bell had some very good steers there also. Mr Smith of Berrington showed some good cattle, some 4 years old from Mr Charles Colling's bulls. For my own part I like a good blood country-bred cattle, especially from the white rather than grey bull. Messrs Compton breed nice cattle and Smiths from 2 bulls from Ketton. Ralph Compton's not near so good although larger. Ralph had this 4 years a great bull bred by Mr Taynes of Stamford.

198.   *George and Matthew Culley to John Welch*

Eastfield 1st August 1803

Well John

All things considered, I don't think you met with a bad market at Skipton. Hot weather always injures flesh markets. If turnips prove a fair crop, I should not wonder if fat markets quicken a little when the weather becomes cooler, which will take place when the days shorten more after Lammas. Meat not only keeps very badly at this season, but people are fonder of vegetables than flesh meat. You are a clever as well as a lucky man to get your bill business so nicely settled. Bank of England notes will do, if anything do, if not forged! We had many forged ones hereabouts in the winter, but I have heard little about them lately, indeed we seldom see any so far north. But these very unpleasant failures or stops in the banks may occasion more to come north. There was bad accounts here several times about Durham bank, but I was always in hopes that it would not prove true or you would have named it. Be sure you write if anything befall it or Darlington. And if one go the other must follow. Whether you lodge your money in the bank or not, I think you will be right in consulting Mr Peacock in all money matters. He is sadly down about these Glantons and Parkers. I wish we don't come in for a share of loss if that Auckland Parker turns out a thief. Mr Peacock will tell you when that 700£ must be raised to remit to London, and you should talk to him about it. Skipton is the best market I think, take it upon an average, and you have done wonderfully well at Skipton. So Birkbeck is gone at last. He has been a strange creature. Nothing like fair straightforward dealings, all *shifty tricks* fall short at last. Besides, what a *precious balm* is a self *satisfied* conscience. A *steady virtuous prospect* enables a man to resist all misfortunes, *especially* if he has a thorough trust in the *Almighty Being*, which he must have if he is *virtuous*. A *virtuous* person can be cheerfull under even adversity, while a *wounded spirit* who can bear. You may be assured that we will not only have no objections but approve of any bargain you may make with the trustees of Joseph Birkbeck's family, because I am convinced that you will only do it with caution and propriety. Do sell away your corn, let me beseech you. You may be assured that it will drop in price any day. The wheat crop is good over every part of England, but dry years are always favorable to wheat and I never knew so much in the yards in this country, and from the information I receive from different parts it is the case all over. Ralph Brown sold wheat on Saturday for no more than 38s, and much sold below that. It was decent red wheat. It is certainly right to get your roof on as soon as possible, but the season is rather too favorable for building. We had a slight shower yesterday at one o'clock, but it is all over again. However these slight showers help turnips. It is really surprising how well turnips are still doing. Matty and William Jobson are going to Driffield show of tups, bulls &c and will likely call on you as they return, and will put this into the post somewhere. They go tomorrow. Your mistress here is at Holy Island, and not so well as I could wish.

The bathing does not agree with her this time at all. I was there yesterday. Yours truly

Geo Culley

Ralph sells oats well at 20*s*, at least I think it a good price. Pease are also well sold, but this is famous weather for black corn. Without a wet harvest, which Heaven avert, corn of all kinds will be down. Oats I believe will be the best sold, but one grain always brings down another. By a letter received this moment from Mr Peacock I am glad to find that he says you and he will be able easily to raise the 700£ which is to be funded on Nelly Culley's account. However I would wish you to see Mr Peacock and have a conversation with him upon it. It is always right to be prepared for all matters of consequence. Mr P. will inform you when it will be wanted, and I would advise you by all means to turn such of your stock into money as is *fat*, or that you think ready to *sell* for many reasons. The money will be wanted, part by us and part for you to buy wintering stock with. It will ease your pastures, and prevent an enemy from getting them, or you the trouble of driving that part of your stock you don't know whither if an enemy come. And we know not where the French will land, but that they will land somewhere and in great numbers as they possibly can there can be no doubt about! Therefore the less you have to lose the better. I don't say this to frighten you, but to put you upon your guard. If we are united, as I believe we are, I am not affraid of them conquering us, but they may and will make ugly work wherever they do land. I am glad to hear by Mr P. also that you had a shower of [rain][71] on Saturday, we had none. He also says he has offered you to give up to you immediately on a valuation all the land Birkbeck holds under us, and I am glad you have accepted it on these terms. It is wise and right so to do. Mr P. speaks of the knavery of Mr Parker, says either he or son will be over here in less than 3 weeks to concert proper measures to be taken with these people.

John I forgot to say in my last that a parcel went to White Hart Chester by William Wright for George Thompson, 4 new shirts and some clues[72] of yarn for stockings, which your mother can get knitt for him. William Wright said the parcel was not gone when he returned by Chester, but the man said he would have an opportunity of sending it in a day or two. Matty will enquire as he passes Chester I hope, and will put this letter into some of the post offices on the road

---

71 A word seems to have been omitted.
72 Balls of yarn or thread.

for you. We are sorry that George Thompson has been so ill, poor little fellow, but I have no doubt but every care was and will be taken of him. G.C.

[**M.C.**] John, be so good as to send for the mule to your father's at Penshaw as soon as this comes to your hand, and a letter I wrote to you and sent by my son, which the lad must call for at Houghton school, and Matty will give it to him for you.

M. Culley

PS. I was in hopes Matty and Mr William Jobson junior would have come your way, but now find they go by Sedgefield, likely they see you on their return.

199.   *George Culley to John Welch*

Eastfield 4th August 1803

Well John Welch

We had a very heavy shower yesterday between 5 and 6 o'clock afternoon and it rained a good deal in the night. I heard it last more moderately, and it is probable that the rains are coming now as we had some slight showers before. William Glass set of with the 10 ewes for Mr Marshall on Tuesday. He is to give them great time. One of them was a little lame at starting. Marshall is to send a person to meet them at your place, and Matty Culley will see him as he passes his place today. He and William Jobson set of on Tuesday for Driffield show, which is to be on the 9th inst. I expect that they will give you a call on their return, but they talk of being absent a fortnight at least. We had a rent day before Matty went, and our Akeld and Humbleton tenants both fell short a good deal. So much for your bank work. Please make my compliments to Mr Peacock and tell him I received his kind letter of the 30th ultimo and will be glad to see either he or his son whenever convenient. I am sorry to find by Mr P.'s letter that N. Parker is going to prove himself h – ish [*sic*], but I am rejoiced to hear that he and you are very equal to the paying the 700£ to London, which is a great relief to me in these times of tribulation, and I am also glad that Mr Peacock and the other trustees are so kind as to offer you the land &c which Birkbeck held of my brother. If Matty call, you may tell him that Mr Thomas Bates and Mr Charlton of Sandoe called here on Tuesday and breakfasted. They had been at Haugh Head all night. I went up and dined with them at Wark, and yesterday they came and dined here. My brother was obliged to attend old Mrs Compton's funeral, consequently could not come along with them. Mr Charlton wanted a tup, but the day they called our shepherd was gone in search of 2 ewes which we were affraid were stolen. However he got one at Alex Hunter Blink Bonny under Flodden Hill and wonderfull to relate the other was found lying aukward in a clover field (Middle Crane Lee) by the man that cut the horses' grass, and alive, and is now well after lying the greatest part of 3 days! But how they came to be so far separated is not so easy to account for. So this prevented them seeing the tups, as I [had][73] nobody to show them until Tom Cairns came home. He had been looking for the lost sheep every day until we found her, and then 'we rejoiced with our neighbors'.[74] Well I shall now come at it. Tell Matthew that Mr Charlton took Mr Nisbet's Blewker at 40 guineas and 20 ewes tupped to us before he go. He never made a word, only hesitated whether to take him or his son, a shearing that I think may be better than the sire, and would wish Matty to use him all or in part. Also let Matty know that I have received a very polite letter from Mr William Young, the gentleman who sent John Gregor (a young Scotchman we have)[75] inclosing a bill for 5£, wants 500 or 1000 bolls of oatmeal from Berwick. I shall show Clunie and Home his letter on Saturday. I am just going to Holy Island this morning, to St James's fair tomorrow, and Berwick on Saturday if all be well. Mr Young also wishes for information respecting the long-wooled sheep. Has had his tups from George Middleton hitherto, talks of coming here in a year or 2. Tell Matty that between 8 and 9 o'clock last night Mr Askew sent me a lad, requesting 'his account *if possible* this evening, as I am for Newcastle tomorrow, will thank you for 40 or 50£. I spoke to your son about some of your best ewes. I should like 6 or 10 of your best, will look at them when I return. G. Askew'. I sent Ralph with 50£ and the account, and wrote that we would be glad to show him the ewes on my son's return in about a fortnight. He never thanked Ralph, looked over the 10 notes and said 'I hope there are no Surtees among them'. He saw that all were safe, I was not going to do to him as he did to A. Cummins. But ill doers are ill dunners. I pray God to help me and mine from doing evil in turn. *Friday 5th August.* St James's fair, nobody to buy, but people asked great prices. The day was so completely wet that my brother and I never alighted but came home to dinner at Wark and got ourselves shifted. Mr Bates still at Wark, Mr Charlton gone. Mr Bates goes tomorrow or Sunday. Tell Matty that I

---

73 A word seems to have been omitted.
74 Luke 15.9: 'And when she hath found it, she calleth her friends and her neigh-
    bours together, saying, Rejoice with me; for I have found the piece which I had
    lost.'
75 As a farm pupil. See No. 62, n. 72.

have received a letter from Mr William Thompson London saying 'that he called at Sir Richard Glyn's &c house who informed him that £1115 consols at 3 per cent stand in the name of Matthew Culley Wark, for which they have received the dividend since it was invested and have placed the whole except the last to the account of Surtees & Co so that you will have two years dividend to claim of them'. Let Matty know that I can't well do anything in this business until his return, when I will show him Mr Thompson's letter. I have also opened a letter today directed to my wife and signed *Martha Welch*, which I suppose is your mother's name, and says the parcel [came][76] with 4 shirts and 2 clues of yarn, says George is recovering fast but looked badly when Matty saw him, had a bad cough which has quite left him which I am glad of poor fellow. Several pieces of oats cut in this neighborhood and on the road to Kelso, and much oats and barley coming fast on. Turnips look well all the way to Kelso, and the rain must do them much service. William Atkinson, John Miller and another Aitcheson came here last Tuesday when I was at Wark, dined here and called at Wark. Jane is better. Yours

Geo Culley

PS. Ralph came home late, and half drowned, but received a good piece *siller*.[77] Sold 35 bushells of wheat at 2s 7d and oats at 1s 1d. The tents had a bad day as Ralph says the gentlemen all went to the town to dinner. Thinks very little was done amongst the cattle.

200. *George Culley to John Welch*

Eastfield 6 August 1803

John Welch

I wrote to you and sent the letter to Berwick by Ralph Brown to put into the post, that you might receive it before William Glass arrives with the 10 ewes for Mr Marshall, and that Matty might know how we are going on in case he come your road. I think he and Mr Jobson will have a rainy journey of it, for we have constant showers one after another, although the wind has got into the north east. But when it is inclined to rain it comes from any quarter or any cloud, and when it is disposed to be fair it will not rain from any cloud or any quarter. I do think that turnips *now* have a particular good chance to be a crop, and if that should be the case all kinds of stock will keep up and you prob-ably have seen the lowest fat markets, because in time of war our soldiers and sailors make great consumption and they ought to be

---

76 A word seems to have been omitted.
77 Silver, money (Scots).

well supported to repel the hostile attacks of these *deluded followers* of the *infamous* Bonaparte. This wet weather will advance corn again if it continues a while. *2 o'clock afternoon.* I have this moment received Matty's letter dated 4th inst., covering Collinson's license for killing game. As a lad is to be here tonight for some meat for Wark I will send up the license. So your Darlington markets continue bad. I approve of your going to Skipton by all means if you are determined to sell. We had no rain of consequence until yesterday and shall not find fault for our pastures and fogs if it rain for a week. The wind Matty describes on Wednesday evening was very heavy and strong here, but we had very little rain. The wind took some of the early oats out. I am glad that some crops are promising although late, and I am rejoiced that you continue selling grain, and I beg you will keep going on. Had I gone to Berwick as I once proposed, but the wet morning prevented me, I should have met with Matty's letter and answered it in the one sent by Ralph today. I am glad also to hear by Matthew that your lambs are so good and healthfull and full of wool. Denton will breed and rear a much better sheep than any place here except Wark perhaps. I am also happy that you have got so nice a bargain with Birkbeck's trustees, and you have also acted a wise part by selling your oats at 3/9d to Miss Thompson. Not but oats are the most likely to *stand their price*, and these rains will advance all corn for the present. *Sunday 7th August.* Glass rising, and a north wind. Ralph sold more wheat yesterday at Berwick at 39s. I doubt Matty will not come back your way, but if he does tell him that Mr Brown Berwick has paid us 100£ lately which was for the 50 bolls wheat Ralph sold him a good while since. I am glad to see by yesterday's Newcastle papers that Surtees Burdon &c have now laid their accounts before some very respectable gentlemen, so that I hope their affairs will be brought to a good conclusion before it be very long. John your mistress here desires you will buy her 2 broad hooped milk skeels[78] and have them ready to come by the first cart that is coming from Denton here, to hold 22 quarts each. Your mother wrote to your mistress here that she had received the 4 shirts and yarn &c, and that George Thompson was much recovered. Poor fellow I did not know that he had been so very poorly. *Monday 8.* A fine mild morning, wind still north east and glass is steady. It will likely be fair so we are going to make a beginning to our harvest, one piece of Poland oats are all

---

78 Cylindrical milk pails, with a handle formed by one stave being slightly longer than the others.

about ripe, some are over ripe and should have been cut last week. I dare say that most of the Polands in this bottom will be cut if the week hold fair. If your sheep are not gone to Skipton before this reach you, *perhaps* it would be right to *pick out* all the last year's *gimmers* which have been once only at the tup, or young ewes to keep on *fogs*, rape and turnips until Christmas or later, as I am persuaded that they will swell and feed better than any you can buy. But I leave that to your own consideration. John I once said to you before that you had better come over to chuse what cattle you like amongst ours, and if you could come before your harvest it would be better. And you never gave me any answer about lying queys and light cattle in open folds and shades &c &c This day has proved rather damp and soft but not much wet. Everyone in this bottom are shearing Poland oats or barley. Crops in general rather light about Wooler and the Plainside very light and burnt indeed. My brother has stooked his rape upon Jock's Ridge, and Thornington rape is to be stooked tomorrow, William Brown is hardly ready. *Tuesday 9th* is a charming morning and has the appearance of settled weather again, wind westerly. I hope they will have a fine day for their exhibition at Driffield. *Wednesday 10th*. Fresh and growing, nice turnip weather, but not enough still for pastures, a slight shower last night in the evening and one this morning. *Thursday 11th*. A good harvest day all through, and the 4th day we have shorn. Many people mowing their oats, they are so short that they will not shear.[79] It may be right in you to grow a decent quantity of hay and always have some to spare, because you can sell with impunity if a dear time comes. Otherwise I consider hay the most expensive thing a farmer grows, injures the land more than anything except turnip seed or rape seed &c and makes the least return where people are not allowed to sell it. *Friday 12 August* is a very fine harvest morning. The poor grouse upon the moors may tremble for fear but not of our Matthew. I hope we shall get a letter from him today or tomorrow surely. Turnips have every appearance of a great crop. I have not often seen them grow faster. We had a nice little turnip shower last night. We have begun with potatoe oats today, very good and ripe, and it looks as though we should have constant harvest.

Berwick 13. Wheat market rather better, sold at 40*s* per. No letter from Matty which disappoints me. Yours in haste

Geo Culley

---

79 Mow, cut with a scythe; shear, reap with a sickle.

201. *George Culley to John Welch*

Eastfield 15th August 1803

Well John

   I sent you a letter from Berwick on Saturday which I hope Matty will see as well as the one I sent the week before, because I find by a letter received from him dated the 10th at Driffield that he intends returning your way and to be at Denton by the 18th or 19th inst., which will be Thursday or Friday in this week. Now as I have an opportunity by Mr Charles Colling, who is to be here this morning (I dined with him at Wark yesterday) I will hazard this letter by him as he talks of being at home on Wednesday evening. But if not it is not very material, as it will cost no postage. Mr C. Colling has been seeing Mr Compton's wonderfull cow bred by Mr C. Colling. It is very unfortunate that she has got something cancerous upon one of her eyes, which I am affraid will bring her to slaughter sooner than she should be, because she has every appearance of growing a great deal still. But she is now beyond anything I ever remember seeing except Charles's ox,[80] which he tells me is now at York. And if so Mr Jobson and Matthew would see her surely as they were to be at York last Thursday. This is charming harvest weather, and we are likely now to have pretty constant shearing in all this quarter. You may let Matthew know that his mother is much better thank God since she came home from Holy Island. He will be sorry to hear that our old Great A tup died the other day of a schirrous liver. His liver was as hard as a gizzard. I am not certain (as I very near had forgot to send you the letter on Saturday) that I told you Ralph sold 55 bolls of wheat at 40*s* to Mr Clay on Saturday. He sold only 20 bolls at Kelso, and no oats or pease. The oats are flat owing to the drop in oatmeal at Dalkeith and Edinburgh. Ralph sold 10 bolls oats also to Alex Hunter at 19*s* on Saturday. It is a good delivery, likewise 5 bolls pease to some miller at 36*s* per. May tell Matty that I have not seen Mr Askew since he returned from the assizes, therefore know not anything how the troop goes on. Mr Thompson (Revd) of Ford[81] attended the assizes about Lindsay's tryal with the parish about Dover, but Linsay did not bring on the tryal and will now most likely desist. George Rutherford

---

80 The Durham ox, bred by Charles Colling in 1796. He sold it for exhibition in
    1801, when it was thought to weigh 168 stone, to a Mr Bullimer of Hawmby
    near Bedale. The ox was then sold to John Day at Rotherham, who travelled
    with it round England and Scotland until February 1807, when it dislocated a
    hip and had to be killed: Bailey, *General View, Durham*, pp. 230–2.
81 Rev. Joseph Thompson, curate of Ford 1798–1806: NRO, Ford register.

the constable of Crookham was at Newcastle also and got pressed, and Mr Thompson had a piece of sad work to get him liberated again! 'Geordy will gang nae more tae Newcastle in these troublous times'. Turnips continue to grow apace except the little young ones, which did not come for the drought and have made their appearance since the showers. They are all killed or nearly by a *louse* which I have known in other years take place in warm droughty weather. Collinson only got 2 brace of moor game in two days, was obliged to shoot Mr Bates's dog after all, for although muzzled he preferred pursuing the sheep to grouse! We are going to pike the Westfield bog today, only a light crop. Tell Matthew that we shall be glad to see him home again and nobody more than his truly affectionate father

Geo Culley

PS. Miss Bessy Greathead is to go in the coach today. Perhaps I had better send this by her, as she will reach Darlington tomorrow noon.

PS. I think we had sold the 20 segs to Berwick before Matty left us at 57*s* per. Charles Colling came here from Newcastle fair, says cattle were from 7 to 8*s* per stone sink, few shown and few buyers. Good horses dear. Ralph Compton has bought *Honeycomb* of Mr Bailey, so you will have a blood stallion near enough. I fancy James Darling is to have a half. But Ralph gave too much for him, 200 guineas. I fancy he might have been bought much lower. Ralph was left 500£ by his old mammy, his youngest sister 200, Tom Nisbet's son Ralph 300 or 250, and the residue to Mr Compton supposed about 1500£, nothing to the Prest, Tom or Fenwick. The money was all lying by her in a *bureau*! Did not trust it to interest. She sent for Mr C. before she died and said she hoped he would *forgive her*. God knows when death stares us in the face we view things in a very different medium! May we all so behave as to appear at the last tribunal with a conscience void of offence toward God and man.

202.   *George Culley to John Welch and Matthew Culley jr*

Eastfield 15th August 1803

Well John

Just after I had sealed and sent a letter by Miss Bessy Greathead today who is gone per coach to Darlington, I received yours per Glass, who I am glad to hear got the 10 ewes safe and well up. Instead of dropping weather we have the finest hot sunny harvest weather ever known, and shall shear right forward now. We are cutting a very excellent crop of potatoe oats at Westfield today very especially ripe indeed. I am very glad you are selling away corn and at such famous prices too. John as we have plenty of apples at Wark and even here this year, shall have no need of yours, so you may do with them what seemeth good to you. My brother desired me to write this to you. *Tuesday 16th* is a charming harvest day and we are shearing with all the force we can raise. Indeed the corn seems to come faster on than one could expect. We have this moment received a long and kind, intelligent letter from Matty dated Snaith the 14, consequently has come very readily indeed. We received one from him only yesterday from Driffield dated the 10th which I think must have been detained somewhere. As he proposes being at Denton on the 18th or 19th I will finish this and send it to the post, as it may possibly reach you before he leaves Denton, so that he may get a chance to get a sight of it. I am sorry to tell him that Adam Sibbit, son of Mrs Sibbet of Berwick died yesterday morning.[82] Poor fellow he has gone off very fast indeed! Matty will be very sorry, and it is a most severe tryal for his poor mother. We were thinking that Matty would be invited, but we have just received a card inviting me on Thursday 18th 2 o'clock afternoon. He died at the Greensea, and we are affraid that his poor mother did not see him since he left Ancroft.

I was highly entertained and all of us with your agreeable letter, Matthew, but this bad news of poor Adam has damped it greatly, but such is the world. My dear lay your accounts to meet with disappointments daily, and indeed if we can only bring ourselves to think so, they are good for us, otherwise we shall become perhaps too fond of this uncertain world. But everything is ordered for the best, and God grant that you and all of us may be prepared to meet with that awfull moment when we are to give an account of the deeds done in the body. I think you were so far lucky in meeting with Mr Buckley to be a judge along with you, you certainly acted properly in not accepting any gratuity in the way and manner it was offered to you. Adam Sibbet had a blood vessel broke, the man says that brought the invitation, which no doubt of it would hasten his dissolution. I am glad that you received of Dickinson &c, perhaps you had better get your bills cashed at the Tyne or old bank at Newcastle as you return, or Durham. And pay the printers for publishing tups &c. We had Mr Askew yesterday for 50£ more, I thought him in tollerable good humor, and broke off about sowing wheat, and I do think we should have come to some agreement, but unfortunately your uncle and he got wrong. However he shook hands with uncle at parting and gave

---

82 Adam Sibbit of Greensea House, Ancroft, died on 15 August 1803, aged 20. His
    father Edward Sibbit had died in 1798. DUL, BT Ancroft.

me a finger. But more of this at meeting, which I hope will now be soon and soon as better, not that I would wish to hasten you, but your company at Eastfield is always acceptable to this family and particularly to your truly affectionate father

Geo Culley

PS. This melancholy event afflicts your mother much, and will do her no good.

203. *George and Matthew Culley to John Welch*

Eastfield 26th August 1803

Well John

These are critical and eventfull times both in the political and the agricultural or farming and grazing lines. We received yours this day with the account of Skipton market. We think you sold your sheep very well, and bad as the beast market was, it is more than probable that they would still be worse sold, and I would wish to caution you against buying much this back end except for straw, and even that should be well considered about before you do it. We thought that Whittingham was a bad fair on Wednesday last, but the few fat cattle were all or nighly sold at 7 to 7/6 or even better per stone, but not one man was buying lean or draft cattle. Old Mason bought nothing. Mr Thomas Armstrong bought none and said none he would buy except he could buy them at Dunse at his own prices. Dunse was yesterday. My brother and Mr Nisbet were both there and came to Grindon to dinner where we were taking our corn tithes of Mr James Bell, and we did take them very reasonably. Well they told us that Mr Armstrong had got letters from some great fair in the south (I have forgot the place) where 4000 head of cattle were shown and only 400 sold. Consequently the remainder unsold would all go to Woolpit. This decided Mr Armstrong, he did not buy one. A great show at Dunse and scarce any sold except what were bought by the neighboring butchers, and also by butchers from near Edinburgh. No wintering or lean cattle bought at all by the Scotch folks. For not only is *hay* a bad crop, straw *short and thin*, but everybody are now terrified respecting *turnips*, which in this very uncommon dry time are not only stopping and turning *yellow* in their under *leaves* and *brown* at top, and clapping to the ground with their leaves, but a nasty *green louse* is again attacking them. So that things have a very dreary appearance I do assure you. Jemmy Thompson is for going to Woolpit himself with 80 cattle. Now Thomas Armstrong always used to buy Jemmy's cattle. Mr A. says there is neither meat nor water on the road for them. Indeed the springs are all going dry here almost, and the Tweed and other rivers never were so low. However it suits harvest well, which will soon be done here, and plenty of room left in the stackyards for the bairns to play at football. We have led all our pease, and many of our early oats, cut a good deal of wheat and will be done everything at an earlier day than I ever knew. Crops are so light, and cut so fast away, and the days long and all working ones. And I may add that the severity of the drought fines the straw so that it really goes in less compass every day and may be led almost after the shearers. Since the 62 I never saw such a drought. I would not wonder John if you sell turnips in your country at 10£ per acre, and that's better than feeding them yourself. However buy nothing I beg of you at present, nor until you well consider it over. There can be no doubt of winter provisions being scarcer than we have known since the 62, and we don't know but turnips may all or mostly go still. They are gone, Mr Armstrong says, in Norfolk. That will make oats, beans and barley sell, because people will feed cattle with these grains. Our beans are all turning black and a bad crop owing to drought. They had not power to fill the pods which appeared, but pease are well loaded, and corn is advancing in the London markets, and Ralph sold more wheat than common yesterday at Kelso. I am glad that you and Mr Peacock will be able to manage the 700£ for London. We met him at Alnwick on Wednesday and got our business done, but I could not write to you as Mr Peacock was not to be home for some days. I was glad to see him in pretty good spirits again, says the Denton brewery goes well on. As Ralph waits I must conclude with cautioning you against buying stock.

Yours in haste Geo Culley

[**M.C.**] We think you have taken your tithes very well, as your crop is in general very good. You got a bad price for your maslin at B[arnard] Castle, only 12.3 is worse than our price of 14*s* at Kelso. There is very little damage done to the wheat in this country by the mildew. William Brown has a few acres. His wheat and barley crop is very good, his oats very bad, no straw. You did well in selling Miss Thompson your oats. Ralph sold new ones yesterday at 18*s* that is 3*s* per, Ralph is not quite content with the price at this time, but I consider it as good as 3/6*d* in summer. And we want money by so much being locked up in the Berwick bank, and we don't know when these affairs will be wound up. You may rest satisfied that lean or straw cattle will now drop in price considerably, even if rain was to come tomorrow, because though fogs and tails might come on pasture, there can be no more hay, and straw and turnips in a most critical situation. And how much does this Island depend on turnips now. In Scotland they are very much worse for straw than we are, and further north the worse, so you may expect to buy Kyloes much lower than in former years. It is said that in the shire of Murray, a fine flat coast of nice land fit for the turnip husbandry, and where they now grow a good deal of corn, the barley and oats will not shear. East Lothian is very bad, and indeed all dry countries, and no turnips in Scotland.

204.   *Matthew Culley to John Welch*

[Wark 28 August 1803][83]

John Welch

Well I was at Dunse fair and such a one I suppose as has not been seen since 1762. The crop of grain has failed by droughth throughout the east of Scotland and the turnips also. I am told in Lowdon,[84] many fields of turnips that were promising for good are now plowed down, and what shows it, not a head of cattle was sold north when I left town at 11 o'clock for wintering, and they used to buy many, and not one sold to go south of Newcastle. Armstrong and brother with Thomas Scott bought none, James Thompson junior intends going to Woolpit with his 80, Mr Scott said, what it would do for the sellers to take 3 or 4£ less than they could afford, and then it would not serve Mr A. and him, as most probable they could neither get meat nor water on the road. Not a buyer from the south neither to buy the winterers. Mr Low had a letter from Norfolk telling they would have a bad fair, Mr Armstrong has a driver and letter from Melton Mowbray in Leicestershire that 4000 cattle were shewn there, only 400 of them sold. What is to be the consequence who knows for want of meat for stock. The turnips have failed in Norfolk, the first sown are gone and I suppose the 2nd cannot do well without rain. Beef fallen to 7*s* in Smithfield they say. The Lowdon turnips are infected by a green louse which destroys the leaf, then the bottom must fail. Cowans, a butcher from Musselburgh who does a strange deal of business, bought 20 small cattle (pick of the fair) at 18£ per from Mr Low. A butcher of Dunbar bought 10 capital oxen big and full fat of Mr Fullarton of Hatchet Know, our neighbour called 75 stone or nearly so, at 27£ per. Edmund Cook bought 10 at 29£ per. Some Berwick men, and I believe Marshal, joined in and bought 20 Kyloes,

---

83 The address and date are lacking, but the date is presumably 28 August: see No. 205.

84 Lothian.

Bill Spours had bought 13, was I believe all the whole that was done. Milk cattle also fell greatly. Tomy Hope bought 6 queys of George Logan for fog, I had forgot, at 18£ per, John Mason had bought none, seemed afraid, but bid for some the cows 26£. Dunse is moist so the turnips and grass look as well as anywhere. Turnips in general not good, ours not to be complained off, but if no rain comes soon will give way. The small ones in the drills grow none, the headlands &c at a stand, our later broadcast lye from the sun and are recoverable if rain comes soon, but the Almighty knows best what He pleases to do with us. Cattle and sheep must drop greatly, likely a very great drop, and the sooner fat is sold for what you can get the better is my opinion. It has the appearance of a serious time, this is my opinion that things will come down of necessity. With respects

Matthw Culley

We can send you in a future day when you can take them 10 or 12 small oxen and spayed queys bred from ours, and servants' cows such as may please you, also we have some 10 or 12 queys kept from the bull which are now on Grindon clover and likely to feed well. They have (as I suppose) as many as you want to winter. If we or you want more there are likely to be plenty to be had when wanted in all probability (you will likely buy Mr Buston's steers) or any cattle you may approve off if wanted. We shall want several steers but are at present very scarce of money (I think Mr Buston's generally do well, they are well bred as any we can buy). We are pretty well off for straw (oat straw especially) at Wark as we have a stack or 2 good old, also our crop of oats are generally good, wheat as good as we expected but much of it red. Today Sunday the winds rather high, and I fear may hurt a little. Tuesday. A high gale of wind yesterday ended with some fine showers and some little shake on the red wheat, a little fresh of 2 or 3 inches in Tweed, will do good to the turnips. The louse has got on a patch of our garden, and seems to destroy it altogether. It is a spot of hot soil but is a pattern of what is done by them in Lowdon turnips. I hear Edmund Cook bought some cattle at Dunse said to be for Armstrong. This a bright fine harvest morning, going to lead wheat (7 o'clock) stack on the stadles,[85] hope the wet will not injure as it is full warm and dry, so as not to take harm. Wednesday. Dry, the mercury very high and inclined to dry, some harm on red wheat and early oats from the wind. Whenever you can take 10 or 12 queys spayed and small oxen for your winter feeding you should, as we can

---

85 Stands.

barely keep them up to keep the flesh on them, only by watering Thornington haugh has enabled us to keep them up, 2 lotts of feeders, and generally the keeping cattle also, 3 lotts on the haugh. The haugh for our cattle and the rape for our feeding sheep has enabled us to keep our stock better than could be expected this very dry season. Our lambs are good seemingly and our ewes went poor on to the rape are now greatly mended, 10 ewes an acre at least I think. Jock's Ridge 1st field cannot be more than 20 acres in rape, 205 ewes on it this 6 weeks, but we begun with it very early and the good part the hill has done much indeed. The field west of it has a good full crop of wheat on it, I suppose 30 stooks an acre, after rape, the under shotts decent enough. The fine shower on Monday (after a high wind) will help our turnips much. I hope we shall have a good crop, we have on Wark, Eastfield and Grindon more turnips than usual by 100 acres I believe, a great blessing this particular year of droughth. I believe the Lowdon people, contrary to Mr Bailey's rule, follow their land too quick with white crops in succession and turnip, land delights as well as nature, in variety. This I will explain to you if I live. If turnips use the land too often, the land surfeits of them. Mr Thompson &c of Lilburn grew turnips 2 years successively in the same fields, the 2nd year's turnips lost their tops before Michaelmas and went to nothing from disease. We fancy many things, may come to a belief, but time shews we are fallacious beings, they were deceived *in opinion*. Messrs Taylor grew turnips successively on a fine piece of ground near the sea, and good, but it was sea wared[86] every year, dung would not have done it, dung brings vermin, insects, grubs, lice &c &c. We have turnips on a thin gravelly soil in the garden compleatly destroyed by the louse, you would suppose ducks or some animals sat on them or were roled, but the louse has destroyed the top effectually. I hold it right to sell fatt cattle and fatt sheep as the prices must lower, stock most probably will fall in value, where is the meat to keep them forward, no fogs, and turnips failed. The sooner you submit to sell the better I am certain, as far as human probability goes. Have you sold the oatmeal? Do sell, be not too slow in selling. Mr Robson of Ellerton made it a rule to sell whenever he went to market with with [*sic*] stock, every one selling good, and they should sell in a bad market, as they, if they sell cheap, have an oportunity to buy as cheap again. He was allowed by all to be a clever man. Mrs Culley and Jane are going to Spittle for a little time to bathe, Nelly returns who has been at Tweedmouth 14

---

86 Spread with sea weed.

days or more, bathed at Spittle with Nelly from the Roads her cousin. We had a letter from Matty who is not content to stay at Houghton. I spoke to Mr Peacock to enquire about Mr Bow of Scorton. He should learn to write and account, also some matthematicks, before he come home. Rawes will not alter his plan, although I wrote to him to let him learn figures &c and less Latin. He shall leave Houghton at the term I think, although it is not pleasant to change his master so often. My brother advised to sell the apples, but use as many as you choose, only I wish you to save some of those which will keep, especially the 10 shilling ones and such as you choose besides. I hope you will farm the widow Birkbeck's stable, and Mrs Robinson will keep her house. Mr Peacock said the Mrs B. would leave town, it will be right I believe. I hope the house is now covered in, Starforth should get some flooring ready and should get the mill lofts finished properly, or Jack Dickinson should, which you approve most to do them, the mill only by Jack. X[87] Among friends I fear we shall want a carpenter. Robin Dickinson mends very slowly, his son is a fine lad but should go from home. Quietly look or enquire for one, honest master of his trade, can make wheels and carts &c, such a one as Jack Dickinson is but I fear such a one may be hard to get. Tomy should perhaps go to his cousin to be completed, his father put him forward, and he for a lad is knowing, but should be some time years with a good master so as to know himself. Lads should be from home, to know the difference between a master and a parent. Your watered land will be very dry I suppose, it has not such a moist bottom as ours, but ours keeps much stock at this dry season, we would have done badly without our boggy or moist land, and our pastures are not heavy stocked. The stubling will be very poor, last year did a great deal especially the clover, this year scarcely appears on poor ground. I wish you to take up the clows in the Fitts before the house and do them effectually, but I suppose they will be undone till I come over. When that will be I know not, as Harry is hardly acquainted with our farm and folks as yet. But we get some better on than we did, he is slow but sure. We have great difficulty in getting a man to manage our threshing since we lost Peter Fogo, the dust is so bad for them at times, especially the old wheat when it is threshed. Is Mr and Miss Harison at Walworth? Mr Peacock says Mr Midleton had a capital crop of hay on his watered land at first done part, do enquire. People are not so fond of pottatoe oats, ours are pretty good in general. My brother will or has

---

87 An 'X' here indicates the passage added at the end of the letter.

wrote for Angus oats, they will be scarce to get as they failed. They have a bad crop of grain in general we hear.

X[88] That is John Dickinson may or may not floor the mill loft, Starforth the main house, should have the windows done but this when we come over. You should bring up the windows Hunter made for us I apprehend. Did you give Wass any money, were you able to do it? It is never to wrangle with G. Culley or his son about housing your cattle, you and I will settle that when I come over to you. You must breed all the dung you can, for turnips &c. I hope you will be well off for winter meat. I hope you have taken your tythes well of corn, you should have deals sawn to dry for your clows, but we must tongue them with oak laths and make them stronger than Harry made them. The stones must also be taken up in the Fitts main carrier, so as to stop the swallows. M.C

205.   *George and Matthew Culley to John Welch*

Eastfield 2nd September 1803

When[89] I was at Wark the last Sunday my brother had a letter half finished to you, which he said he would bring down with him on that Monday but did not, and since he tells me that he finished and filled the whole sheet and sent it to you. This day week (Monday) we had a violent wind, but fortunately it did not continue long, consequently did not do so much harm although damage was done. It ceased with a hefty shower which was of use to turnips but is now gone, and they really look very sickly. I know not what we shall do if turnips go of. They can't now be a good crop, although warmer moist weather may improve them much where not totally destroyed by lice &c. How very fortunate that we nor you have bought no cattle. The few of them we may want, and which Kyloes you will have occasion for, are likely to be bought much lower than once was expected. I have not often wished for rain in harvest, but I have fervently wished for it this harvest, because rain would do much more good I think than it could possibly do harm. We have not above 7 acres of white corn to cut at Westfield, and I fancy it will be mostly cut this afternoon, and beans handy to reap, a bad crop. The pease, a capital crop, in long ago and covered. I have known many harvests begin earlier, but never knew one done so early as this will and must be, days long and not one broken day. I am certain that without a sudden and very wet change of weather indeed, everybody will be done or nearly done in this

---

88 The following passage relates to the text above indicated by n. 87.
89 The greeting is lacking.

bottom next week. Remember this is only the 2nd day of September, and so much got in too. But corn both cut and leads fast God knows 3 weeks. Shearers cut more than one acre per day this year. Did I write you word that a great deal of wheat is mildewed in the southern counties, and by a letter I received yesterday from Ireland they have suffered very much there by the same fatal disease. This will be better known soon and must have an effect by and bye on the markets. But so much old corn is in hand in all parts of this Island I understand prevents an advance at present, and Government will encourage all the importation they can at this time. But the crop is so defective in Scotland and this country for oats in particular, and the hay so bad a crop, will make more consumption in horse corn, and though wheat will I think be very good in quality here, yet it goes in small room. All corn *packs* into a little compass. By a letter from Mr Barf soon after we received your last, Wakefield had been worse than Skipton for sheep in particular, many as low as 7*d* and none above 7 ½ he says. Cattle 7/6 to 8*s*. I think that sheep must come down much this autumn. How wise or lucky then in you to sell your ewes at Skipton. It was a good market at Morpeth on Wednesday, but I suppose that was entirely owing to above 500 sail of ships coming into Shields. I fancy half the people in Berwick will fail by and bye, so fatal a stroke has the failure of Surtees's been. Adam Achison and Mr Brown have both stopped this week I was told at Wooler yesterday. We want 60£ of Brown but wanted much more a while since. But being officious kept drawing from him, and tomorrow he was to have paid the balance. Bill Embleton the stationer also stopped some weeks back, and I have my own fears of John the banker. But all these Embletons and Adam Achison and Brown &c &c, and whoever like them had their sole dependence upon that bank must go, because they had not 1*s* to start with and lived like petty princes, drank wine like water. How mad to live to the heights without ever looking before them or thinking that a fatal time might come. I would not wish to boast, but nothing ever gave me more pleasure than our mind not in that respect. I don't believe that we drank in Fenton in our first coming for the first 7 years one bottle of spirits, all ale, and never had a bottle of wine until we could well afford it. But is not that much happier than to have failed, as many thought we would because we enclosed, and hired tups &c at an expence which alarmed our neighbors but turned out a mine of wealth. May we be truly thankfull to that God that directed us aright, and may He always do so to us and any is the sincere and gratefull prayer of

Geo Culley

8 o'clock. We have been out shearing a piece of nice spring wheat in Planting's Close, and a shower has put us off, but I don't think it will be much as the wind is rather north west and it is breaking a little to the south west. I once thought we should not be able to raise the Chillingham rent which will be the first Monday in next month, but I think we will do as Matty received better than 100£ in Yorkshire and we have got a remittance from Mr William Thompson London of near 100£, which with what we have will make 800£ and we have more owed to us for corn than the remainder, and I hope we will sell as much corn old and new as will pay harvest wages &c which will come quick upon us this year. Mr Askew could not get his troop accepted by Government, so he and our Matthew have entered into Lord Osulston's troop. Most of the young farmers in this country have entered into Lord O.'s troop. Matty was out shooting yesterday, which was the first day for shooting partridges, and he shot 10 brace himself at Akeld where they are very plentifull this season. *Saturday 3 September.* A strong wind from the north west. What wonderfull harvest weather, but the *neeps,*[90] the *neeps* are in danger. After this day I think we shall not have above 4 days shearing with the few we have, beans excepted. How very lucky, after all we have a very decent crop considering the exceedingly dry year. We have not much demand for corn this year on these farms. Ralph sold 30 bushells spare wheat yesterday at Kelso at 42*s* and oats (new) at 18*s* per. 70£. I think we will now get the Chillingham great rent paid, and after that other rents &c I hope. *Sunday 4 September.* Nothing doing at Berwick. That fatal bank stopping has stopped everything I think. A very droughty morning, glass uncommonly high.

[*M.C.*] Sunday 4th September Eastfield. Well John I forgot in my long letter to desire you to buy 3 bushells of rape seed, we had hardly enough last year and William Air charged 9*d* per lb. to Mr Howey and the same perhaps to us. The wind changed to the north yesterday but as droughthy as ever. Cheviot clear as clear, no clould [*sic*] on it. We have led and shorn much but only begin to cover tomorrow, will be done shearing soon if this dry weather continues, except beans which are ripe and a weak crop. In haste Matthw Culley. My brother will give you an account of Mr Sayle's letter from Norfolk.

[*G.C.*] I had forgot to say that in a letter received yesterday from Mr Sayle from his Norfolk farm he gives a most melancholly account of the country respecting the mildew in wheat, but take it in his own

---

90 Turnips.

words: 'Here I am, and here I intend to stay until the harvest over, all the wheat is nearly got, and sorry I am to say very little good, all the way more or less mildewed, at Went[91] and neighborhood bad, some pieces good for nothing. About Bawtry better, and all the way to Newark. But from Newark all the way over Lincoln Heath to Sleaford very bad indeed. The same all through the fens and marshes, and the whole of Norfolk very bad indeed, I fear particularly in what is called the Woodlands where many pieces are mown and led into the fold yards for litter! I am also told that Huntingtonshire, Cambridgeshire &c are still worse. We have only one piece here (Titchwell) about 12 acres near the sea, to call tollerable out of 100 acres, yet no advance in old wheat, which convinces me that there must be a great quantity of old wheat in hand'. And I believe the same. Nevertheless I can't help thinking that wheat will advance by and bye. The effect of this immense mildew may not be felt immediately owing to the quantity of old wheat in hand, but I am much mistaken if it is not felt by and bye. Mr Sayle adds 'that there is scarce anything green except trees and a few turnip tops, and turnips are a bad crop through all the best turnip country. *Fat cattle* scarce, and the butchers are selling harvest beef, full as bad as your ship beef at *9/6d* per stone even weighs, mutton 8*d* and 8 ½ *d* per lb.'. One would suppose from his account of prices of harvest beef as he calls it that fat cattle are scarce, and I really believe that fat is nowhere very plentifull, only people are and have been under the necessity of sending them to market for real want of keeping. And how winter is to be got over if turnips fail I can't foresee. Which makes me recur to what I said to you in my last about turnips. I have no doubt but good turnips will be sold in many countrys at 10£ per acre and upwards. Here is an evening as frosty as can be, and every appearance of continued drought. I am glad to tell you that the report about Brown was false, at least he was going about on Saturday, Ralph says, and in better spirits than heretofore. Did not pay Ralph though, said he had had a sore run upon him for money from some evil reports, but hoped to pay on Saturday next. Adam Atchison is done, and no doubt but poor Brown is very precarious. However it gives us more squeak for our money. No sheep buyers out yet, but David Green is coming soon John Batters told me yesterday at church. What a strange price for beef and mutton in Norfolk by Mr Sayle's account. Mutton sold higher dead than alive here. If rain still come fat would advance. David Green is to buy ewes

---

91 Wentbridge.

this year. If it answers we must get you one made and sent by and bye. But I don't know whether our smiths can make the teeth right or not, for the excellence seems to consist entirely in the teeth which are long and tapered and anchored and steeled I believe. *Berwick* 2 o'clock. Very little business doing here today and I think without a new creditable bank this place will become good for nothing. I have bought 2 tons of oil cake at 11£ per ton, so much affraid am I of winter.

206.　*George and Matthew Culley to John Welch*

Eastfield 12th September 1803

Well John

I sent you a long letter from Berwick on Saturday about 12 days upon the stocks, and I forgot to say after all that we have had so many different applications about our wool within these 10 days that there certainly must be some particular reason for which has not transpired yet. First, a Wilkinson from Leeds, then Nott last Thursday appointed this day and tomorrow to look at it, presently afterwards Mr Vardy received a letter from the man that got it last year. But we were already engaged to Nott, and this moment a Mr Young from Alnmouth is gone to look at it, and I am of opinion he will buy it if Nott leaves it. Mr Young buys for a Mr Pollard of Leeds or there-abouts. Yesterday the wind turned north and is the same today, and is very cold and frosty. The country (now most of the corn is in the yards) has a most desolate appearance! I assure you the winter is often in my mind and will be in many people's thoughts by and bye, you may depend upon it. I would advise you to think and consider *well* before you buy any stock *once*. We are going to finish all our odds and ends this morning, the shearing which will not employ us all the forenoon, and then start with the beans. Mr Brown paid Ralph every shilling he owed us on Saturday and I hope he will go on poor man. However we don't think it quite prudent to deal with him yet a while again. It was very wickedly done whoever raised the report of his having failed. No bank established yet at Berwick and it is a poor shabby dead place I assure you. Nobody to buy but Alexander Hogg and Mr Clay, and they only bid Ralph 38s for 60 bolls of very good old red wheat. In short nothing is doing of any consequence at Berwick, and not much at Kelso. However we have now got nearly as much scraped togther as will pay the Chillingham great rent. G.C.

[*M.C.*] Well John I am just from Spittle where my wife is for a fort-night and Jane also, Bessy at Mrs Coulters at Tweedmouth, goes to Mrs Bellamy at Berwick to play on the piano. Harvest far on, we I believe got done on Saturday and now are begun our beans, which are very short, the peas in them will be the better ½ of the crop, which together will be ½ a crop, are too long of being shorn or pulled up rather, they are so very short. This is Durham fair, where I hope you are, and from whence we expect your account of it in full. We will want some steers for Wark to work, and where we have a very good stock of straw, both old and new, but suspect lean stock will come low in value by and by. Do let us know how Durham fair has ended. We suppose few to buy lean stock, as the drought now continues, it grows very alarming indeed. The straw is scarce on dry soil, and the turnips now are begun to fail on all hot dry soils. We must begin to give our fattest cattle turnips directly, as they cannot do longer with no benefit if we suffer them to loose condition. I fear some of them have lost more than they ought. Wark turnips nor Grindon have suffered any great loss or damage yet, yet I fear our Middle Close, or field west side of Dry Tweed, is at a stand as it is old land and hard. The louse has seized on our earlyest sown sweeds in the Park Close. I sent Jemmy Dippie and some small lasses to crush the vermin on the leaves, or rather where mildewed or spoiling. The leaves infected fall as flatt to the ground as if trod upon by people's feet, and I think so farr as I can see it is a means of checking their progress, but I cannot say yet what the killing beforehand may do. Turnips may very likely go high in price, 10£ per acre will be offered by some people I apprehend. This depends on the weather, if rain falls not soon who knows what injury turnips may get soon. Grass is all gone and the fields arid and dry, the windle straw is hay, yet sheep do well of it but should have some turnips, I mean fat ones to keep them up. Rape is very valuable. We bought 2 tons of oil cake at Berwick on Saturday 11£ per, but I wish we had more tons at the price.

M. Culley

PS. Accounts in the newspapers &c say fatt cattle stand up to their prices, yet I think it right to be selling fatt stock, however no stock can be made fatt without artificial meat at this dry time

[*G.C.*] *Tuesday 13.* A fine mild day, pulling our beans pretty well loaded, but the pease much the best and a weak crop on the whole, but much better than theirs at Wark where we went yesterday to meet Mr Handaside and sold him all our wool at 24s. Perhaps we have done wrong, but I think it too good a price to speculate upon. I am always willing to take a market price, indeed I have heard of none sold so high yet, and many of the buyers are out, and consequently something moves in the Borders than we are aware off. And I hope you will sell yours for more. We have put all our dinmonds to turnips, several cattle and the tups have had turnips for some time, also part ewes that are not upon rape. We think that we had better sacrifice turnips than the condition of our feeding stock, and they are evidently sinking. The stack yard at Wark is full of capital grain, perhaps the best we ever had, and several stacks made across the Goat. To be sure there are several old stacks in the yard, but will be more new ones out of it. In one field of barley I think there might be 60 bushells per acre of sprat barley,[92] and Harry says the common is better. Their wheat has also cut up well. But I think Mr Hogarth has the biggest crop in his haugh I ever saw, the stooks are uncommonly thick upon each other ridge, and 14 sheaves in every stook. The turnips eat half on half off, which I believe much the best way, although we have not been able to get to practice it, yet we are deter-mined to do it this winter if we injure our stock and be obliged to sell earlier. *Wednesday 14th.* Dry and rather windy, but the glass dropping a little and still very high. We are busy covering, which we will soon finish now, but I never saw so many stacks uncovered after the corn was all in. Nevertheless I think it wise to get covered as the rain may come sooner than we are aware of, and heavy too. I am happy to find the Wark turnips little injured and many of them a fair size, therefore we take the most stock here. Our first sown, which were confessedly the best and most promising in the whole country, are entirely done. But weak soils can't stand like deep loam. By a letter received this afternoon from Mr Deverell who tells me that wool is to be very dear, perhaps 14 or 15*d* per pound, so we are taken in. It is a pity that we did not receive this account sooner. But I am content, however I will send you this tomorrow that you may be upon your guard. Wheat is also upon the advance, Mr Deverell says, owing to the mildew being so prevalent in the midland and southern counties. He says wool has advanced most rapidly from 28 to 31/6 and expected much higher, and wheat is advanced 6*d* per bushell. But he says that they have much old corn in hand there, and I believe this to be the case all over England. Matty has sold Mr Askew 10 ewes today at 5 guineas each. And 10 before to Mr Marshall you know that came to you, and we have a chance to sell 10 or 12 more soon to go into Ireland. It is a good trade, and I have no objection to sell as many more every year at these prices. Mr Deverell says they have great crops and full yards, but the wheat very lean. Do let us hear from you again pray, which will oblige yours truly

Geo Culley.

---

92 A species of barley with short broad ears and long awms.

PS. *15th September. Severe, continued, unremitted, uninterrupted drought!* New corn will be as dry as old, and surely a blessed thing to have good corn, few things so important as good bread for mankind.

207.   *George and Matthew Culley to John Welch*

Eastfield 15th September 1803

Well John

Matthew took a letter for you to Wooler post today and I am then begun another. Matthew has joined Lord Ossulston and many of the leading young men in this district. I have been at Longknow, George Jemmison has a full yard of good corn. He has a good crop. We plowed all that good hollow round the herd's house, and it has produced a capital crop of oats. The sheep look well considering the excessive dry time. Andrew the herd says there never was the like of this in the world before. I told him that I had seen a much severer drought. We have had the misfortune to lose our best shearing tup this morning, and our best old tup died when Matty was in your country. *Friday 16 September 1803.* Dry and windy, a few drops about 10 o'clock glass dropping and has dropped the last 3 days, and we are again flattered that being so near the equinox I think we shall get it soon after. It is wonderfull to want and wish for rain in the middle of September. Well John the rain came on this day about one o'clock. We are all in but beans, and perhaps a little stack of wheat, and nearly all covered. The poor turnips seem to rejoice most of anything. I am affraid it is too late to benefit the pastures and fogs much. However they will turn green, and with a *tenth* it is to be hoped, and the land being so dry and consequently warm may perhaps assist vegetation more than we imagine. *Saturday 17th.* Tolerably fair, but like showers from north east, where the wind got to soon after the rain began. The great rain ceased about ten o'clock I am told. Ralph did none little business at Kelso yesterday, what wheat he sold was 2£ 2s and pease 30s, oats are much wanted, but people being unwilling to thrash them on account of fodder, few come to market. But they are certain to be high sold as they are the only grain that keep their price at London and are sure upon the advance. Col. Graham's shepherd came last night from Balgowan to chuse a sheep for the colonel. The price is left to ourselves. Well John, we have received yours of 15th this moment. Our beans have also hulled out a good deal, although we hand pull them always, and make them take low hold. Ours are a very thin crop, but much better than Wark haugh. It is as you say the *ripeness* and *dryness* of the beans which makes them shell out so much. I hope you will be right in selling what stock you have fat, that will suit

Skipton at this time, and I think a part will be better sold by and bye, as all our accounts tend to prove that the mildew has been exceedingly prevalent in all the seaside counties from Yorkshire to Kent, and in all the midland counties that I have yet heard from. But it is said that the west of England had escaped. I have wrote to Mr Boswell this day to know, and when I have his answer I will acquaint you with it. As to barley, I think people will give over growing it, as it cannot be turned into money scarcely at any price here. But oats which pay better, and much certainer, and the straw more valuable, are and will be very dear. This year in particular they are such a shocking bad crop in this country, and in all Scotland. I sent you a letter the day you had wrote yours, which will explain to you about wool, indeed we sent it of sooner than we intended on purpose that you might be upon your guard. Not but I would advise you to sell at a good price, say your last year's price, and I have little doubt but you will reach that. I am glad that you think that you can accomplish the London 700£ soon, but you are right to consult with and talk to Mr Peacock about it, and all money matters if you please, and he is very clever at these things, indeed it is his fort. I wrote to him the other day, and requested that he and you would consider, and contrive to get that 700£ sent of to London. O dear, I would wish you to go to Broughill fair by all means, and even to buy anything that you approve of. You have been very lucky at that fair, and cattle you must have to consume your fodder and make it into dung. And I am sure that by this time you are a better judge than I am what cattle suits you the best. Kyloes seem to answer best for Skipton, and you have succeeded wonderfully well at that place, and I am sure you have more sense and prudence than buy over dear at Broughill. It is only buying dear at the last. Perhaps the drovers will not be willing to submit at Broughill, but when they hear that turnips have failed so much in the south, they will likely agree to, before the close of the market. Upon my word my friend Charles will look big now, to deal with the King's farmer. But I believe that the King is not fond of laying out much money. All the royal family are said to be narrow and fond of money. You are a good lad for providing rape seed. It has done us an abundance of good this autumn, indeed I know not what we should have done without it. You have had rather too good a season for your house building and covering. I am glad poor Matty is well, your mistress thanks you for thinking about skeels for her but I begin to be affraid no more tups will go your way now this season. But they can come along with Lord Darlington's tups &c at home again. We shall have all covered that is in this evening I think. This has been a fine sharp hard harvest day after the rain, and the rain will do an abundance of good I trust, even suppose the wind is north and very frosty, but the land is warm with so long a dry time. Do be so kind as write again, as soon as you return from Skipton. If there are any very nice open quey Kyloes at Broughill I would wish you to buy some, as we have some thought of trying the breed in a small way.

Sunday 18th. Frosty and seems as though it would keep fair. Ralph can do nothing these several weeks at Berwick, 36 for new wheat and much the same for old unless very good. Jerry and David Green both out, David was at Falkirk, Kyloes not so low it is said as was expected. I have not heard what Jere &c have done. It is said ewes are to be 8*d* to 8 ½ *d* per lb., less than last autumn, and they will bear it pretty well. They were so exceedingly high the last season. Dinmonds it is supposed will not be so much wanted. *Monday 19.* A calm day, but not unlike a little rain. We are leading the last of our wheat, I think the turnips are looking better already. Col. Graham's man fixed upon a very nice shearing. As I wish you to have this in time I will consider and get it sent to post by some means, and am yours truly

Geo Culley

PS. I did not get this letter sent today. We have had nobody at the tup shew so we could not do any business. It is very well that Col. Graham's man was here and took one. It has been a soft daggy[93] day and I think will most likely be wet weather. We have led the wheat, and I wish we had the beans in. However the turnips look better and may be better than we expect. G.C.

[*M.C.*] John you say Matty is to be with you. Do caution him to carry his gun properly with caution and not to bring it home charged and the cock and hammer down and the loader on top of the pan. Many mischiefs are done by guns. I much wish Nelly Culley's money funded, it's high time, it's trust money. Am glad the house is so far forward, you must meet us at Luke fair and we will return with you to Denton. You take no notice of my saying we can let you have 10 or 12 queys of our own breed in good condition, nearly fatt some of them. May make them 20 by adding 6 or 8 open ones unbulled, have also 2 or 4 good fed cows in hand. This rain has eased Wark turnips, and we will want a stock of strong steers, some for work and some to keep on, if can be bought low. This is my opinion, but wish my brother to give his opinion on these matters. I prefer our country cattle if good sorts to Kyloes as a general rule, only such as you got at

---

93 Drizzly.

Broughill last year can not be spoke against, and Kyloes are small and proper for your moor land. No room to say more.

Matthw Culley

208.   *George Culley to John Welch*

Eastfield 20th September 1803

Well John

The game is up! Nothing but rain now. Indeed it did not require very great prophetick power to foretell the change now at the *equinox* when great rain and tempests and so on will prevail in general all over the globe. However we luckily have all our white corn in, except 21 stooks which were cut late in some holes in the Purdies holes, and our beans which I am affraid will waste and suffer much. However we have much reason to be thankfull, all covered too except a stack or 2, and much the same in all places except a piece of oats uncut and more unled, as well as a piece of wheat at Longknow. And no wonder when we consider the earliness of the season and Longknow is amongst mountains. We sent you a letter yesterday and I hope we shall hear from you tomorrow, or soon after your return from Skipton, and I wish it may be a good account. *Wednesday 21st.* Yesterday more *rain* with a very cold violent wind from the north east, winterish in the extreme. I dare say neither our ships nor the French can live upon the sea at present. This will surely kill these *vile lice* that have preyed so long and sore upon the turnips and every-thing green. *Thursday 22nd.* Very clear and frosty, wind west or north west. Yesterday proved a very drying day although slight showers now and then from north east. We are in hopes of leading some pease today (beans) but there are some ripe cuttings in the east which I don't like. By a letter from Mr Vipan this morning he confirms the account of the wheat being all or mostly affected by the mildew in the eastern and midland counties of Britain. Potatoe oats have succeeded with him and friends beyond compare. Beef at St Ives 8 to 9*s* per stone sink, sheep 7 ½ to 8*d* per lb. sink, wheat 7/6 to 7/9 per bushell. This proved a nice day and we have got in a good few beans. *Friday 23.* A very fine frosty morning, I hope more beans will come in. Ralph Brown bought Andrew Elliot's turnips the other day at 6£ per acre, a fine healthy thriving turnip, but I consider them a great pennyworth. They are for John Nisbet, but I would have taken them at all events provided Ralph would not, as I consider turnips will and must be very dear, although they are looking wonderfully better since the rain. *Saturday 24.* A very sweet morning and as timely for dry weather as ever. It is really remarkable what a propensity the weather has to be dry. We have part beans still to lead. Ralph could sell no wheat except 4 bolls of old wheat to sow yesterday. I hear that several ewes are bought here at from 8 to 10s lower than the last year's prices. I hope you will find time to take a trip over to us soon, as your harvest must be all over soon and no lime to receive for at Yarm, and then you could fix on what stock you would like best to be sent to you. Oats advance everywhere, and are expected to be uncommonly dear. The reason given is that so many are consumed by the dragoon and artillery horses. I understand that Morpeth markets are very bad now. We have sold 10 shot gimmers today for 50 guineas to go to Ireland.

Sunday 25th. Fresh and windy. Yesterday we received yours of 22nd, am sorry you met with so bad a market at Skipton, but it is the fate of war, and this is a season in which bad markets must be expected. Your cattle set up will improve, and I am disposed to think that real fat stock will advance a little now, as rain and fresh weather is comed and turnips &c are already improved, and it is possible that lean stock may be better sold than we are aware of as there is so much coarse fodder or straw in all parts of the south, and their hay a fine crop. However it would be very wrong in your taking lean stock to oppress you at present, except as you see any country steers fall in your way. I think John you had better give both your feeding sheep and Kyloes part turnips than suffer them to sink. A very little turnips will at this season do them much good, and I have no doubt but fat will advance before Christmas. I well remember that was the case in the year 1762, that stock, cattle in particular, were not plentifull in the country, only people hurry to market for want of meat and fear of winter, and high price of turnips which must be. I am of opinion that the Kyloes you have, if of a good sort, have a chance to pay you as much and more than buying fresh ones. But I only hint this as I am sure that you are better able to judge of that than I am. Late as it is, the grass and turnips will both mend more than people are aware of, provided the weather keep tolerably fresh. The land is so warmed with the long drought. I know it was the same in 1762. Don't fear but corn will mend, oats are to be uncommonly dear, and one thing always helps another. Ralph sold wheat better yesterday and will be sold better, only we have nobody scarcely to buy at Berwick. *Monday 26.* Cold and windy with slight showers, like the season. By a letter today from Mr Thacker[94] of Wilm Mills near Derby as follows: 'Our

---

94 In this letter of 19 September (ZCU 25) Thomas Thacker informed George
   Culley of the death of his father. In addition to the report on crops quoted

wheat crop in this district is much the same, a great crop but a large proportion of it very lean and indifferent indeed, owing to a blight in mild weather happened nobody knows when. The small part which escaped is very good and fine as I ever knew, it weighs 66 stone per part bushell 34 quarters and yields well from the straw. The other yields very ill and weighs from 60 down to 50 and less to the bushell, and I think is light. From the vast quantity of old in hand, must advance and at any rate must advance some time before next harvest. Oats a bad crop and advancing, barley a remarkable good crop, yet advanced from 26 to 31s per quarter lately owing to the great desire to feed pigs, a greater number of which are in hand than ever was known, selling from 6 to 7s per stone but dull sale'. There is a very good and valuable product of pork at Berwick, 6s per stone. How is it at Yarm? Be sure to let us know in your next. *Tuesday 27.* Fair but blusterous. By a letter from Balgowan Mr Thompson says the crops in the Carse of Gowry and Perthshire are like us in hay, light, but grain of good quality. He is selling fat at 10s per stone sink but in weight equal to 8s ours, stirks fat and plentifull. In a letter from my brother he says 'John never says whether he has sold his oatmeal. However if you have not sold, if you believe this country it is dearer by much than with us. Is the drain done up to High House before weather breaks? No mention of barley money received and the 700£ sent to London to be funded'. I could wish that was done, but Mr Peacock will not I hope neglect it in due time, and you must borrow of banks until you can repay it, for buying cattle &c &c. This is a rather fine day for the fair.

Wednesday 28. A dry frosty good harvest morning as ever. A very heavy fair yesterday for sheep, no shearings at all sold hardly, and ewes full 10s a head lower than the last year and also demand, a great number of sheep showed, not but we have seen as many, a very shabby show of steers and other cattle, fat or any way nearly fat were very readily sold indeed, never got leave to reach their stands. Billy Beany and Dickinson &c from North Shields, but I think Adam Atkinson of Ryal picked up most of the fat cattle, and it is now confessed on all hands that fat cattle are scarce all over the kingdom. A great demand at present to the coal trade, and it is said by some that the price of beef will be raised, and when we consider the great

---

above, Mr Thacker wrote that the market at Manchester was full of American flour, and more coming from Yorkshire by the new Rochdale canal. Grain could now come by water to within 10 or 12 miles of Manchester.

consumption that must necessarily take place in our navy, as well as what the soldiers consume, you may be assured that fat will advance considerably before Christmas. The coal trade and great slaughtering begins between this and Martinmas, and that will help sheep as one thing generally assists another. A Mr Caross, a jobber from below Yarm, said that very clever 2 year old steers rising 3 were bought at 12£ per at Guisborough, and he seemed to think that they would or might advance before and at Yarm. Now it may be wise and prudent in you to try to pick up some steers as soon as you can anywhere in your neighborhood, Barnard Castle or Langleedale &c &c. If a part were 3 years old gone they would suit us better for drawing, as they would yoke directly. We don't mind condition so much provided they are healthfull and tolerably fresh. If you are short of money borrow of the bank. I am sorry it is not in our power to assist you with cash at present, we are back with payments and likely so to be for some months yet, and the stupid Nott is now unwilling to pay us the 380£ that we were to have next Monday for the wool in part, although that was one of my principal motives for selling it. Mr Nisbet bought one very clever field of turnips at Crookham for 6£ per acre, but has been forced to give 8£ per acre for another little field just by it, and I will be bound for it that good turnips will be sold at 9£ and 10£ per acre by and bye except they mend more than they are likely to do this frosty weather. I think Adam Achison might give from 7/6 to 7/9 per stone for the cattle he bought, and if he did not expect an advance he would not have been so keen. Billy Beany said he bought them far too dear, but I think Adam as aware of what he is doing as 'croaking Billy'. John Jobson sold his wool the other day for 24/6 to the man he and us dealt with the last year, and he would have had ours but unfortunately gave Nott a promise before we received the other letters. Mr Adam Atkinson is for Yarm fair. Ratcliff bought 20 oxen and Beany 30 last week of Harry Taylor. I understand that the coal trade is very brisk indeed. The tup trade appears slack this year, but I believe we have done more than many of our competitors. It is said that Robert Thompson has done very little, Mr Bailey says so, and he was here all last night. Matty let a tup to give to his ewes yesterday at 40 guineas. Mr Thompson says lean stock (cattle) have also dropt 3£ in 10£ in Scotland, yet fat advance. I understand that Jemmy Thompson did no good by going to Woolpit. He would have done as well at home. The cattle drove go in so little meat and water on the road. *Thursday 29 September*. A very fine morning, but the glass is coming down, I will send this from Wooler. *Wooler*. All I can learn about Morpeth is that there never was so bad a market. None sold above 5s sink! However on coming along today I bought James Grey's ewes 80 and so more, we are to cast at 31/6 per head, 40 are from Dr Atchison of Clementwell near Musselburgh. We think of sending you these 40 to breed a fat lamb and then feed the ewes, and I believe they will pay you well. However this is only on condition that you and my brother approve of it, and my brother does not know of it. It only came into Ralph's head and mine as we came along, and finding we could buy them lower by 1s at least per head we ventured to do it, and I think they may pay you well as you can keep them on moor there. They are a pretty little ewe and sound as loaches.[95] We will keep them for you to keep as long as necessary. Yours in haste

Geo Culley

Oats are by today's paper 32 per quarter at London, that is 4s per bushell, and they will be 4s per here soon, and that will help barley and other things as folk will eat barley broke or boiled and sell oats certainly. Oats are to be very dear to a certainty.

209. *George and Matthew Culley to John Welch*

Eastfield 30 September 1803

I believe John that I neglected in my letter sent to you yesterday, to desire you to say *particularly* whether you would wish the 40 ewes bought of James Grey to be tupped for you and sent *afterwards* to Denton as inlamb ewes. Do be so kind as say in your next whether you would like to try them in that way or not. They are a few sweet ewes I do assure you. 'Either good roast or boiled.' Because if you are not for them we will sell them again, and for profit I have no doubt. A man might begin upon them to build as pretty stock as any ones. You know I think that he has a tup or tups from us every year, and though hardly kept they make sweet little sheep. I am disposed to think that they have a chance of making very handsome pay in your midling and even worst land by lying them *very thin* on upon your moor. The lambs will bring 20s per head, at least they have been worth more as keepers at Yetholm these two years, let alone fat, and if you have them I would recomend by all means to cut or *bib* the tup lambs. The glass still comes down and is now nearly at changeable, but does not show much yet for rain. However I think it must come soon. This is a mild pleasant fresh morning. I dare say both grass and turnips are growing. It is said that a great number of ewes were sold at the close of the fair, and only few shearings. James Grey sold 6 steers 2 years

---

95 A small river fish akin to carp.

old gone at 14£ 14s per head at home the day after the fair. We were to have looked at them but he was tempted with price, and I hope you will be able to buy such cheaper. We only got the last of our beans in yesterday afternoon, but the weather has been wonderfull. I go for Chillingham rent on Monday, Alnwick fair on Tuesday and the sessions on Thursday which I am obliged to attend as a juryman. Consequently I shall break the back of that week. They ought to cease me being upon the grand jury now at my time of day. It is very odd that they never troubled my brother at all but always me! However I will take this and give you an account of Alnwick fair. Jere bought 800 sheep of the Messrs James's and 300 of Mr Weddell of Mousen, I think he did not do much at the fair. He and David Green *both* bought George Purvis's ewes at 2£ 2s very big ones, and had a difference about them, but David had got them away of the ground. I know not how Tomy Bryce had blundered the business. Be so kind as say, in answer to this, what is the price of butter firkins and what is your opinion of them, rising or falling. We understand that they are worth 2s or 2s 6d per. They should be very dear after the uncommon dry summer, as well as cheese. *Saturday 1st October*. Clear and fine, we had a fine shower for about an hour yesterday but was much more at Kelso. Ralph only sold 5 bolls old wheat yesterday at 2£ 2. I know not what is to be done with our old wheat, indeed although good and wonderfully dry does not sell well 35 to 38s perhaps, except for seed. Several prices shown for seed from London, Ralph says *very indifferent samples*, which is to me a confirmation that wheat is bad in the south. If not they would certainly send none good for seed. Oats alone advances, and barley keep pace with them, but it will be very extraordinary if other grain don't advance by and bye, and there is so much old corn (wheat) in hand. Oats are now got to a *guinea* and very keen of them at the money.

[**M.C.**] John I would have you not to forget to have some acorns gathered at the proper season which is fast approaching.

[**G.C.**] *Sunday 2nd*. Overcast and like rain. Ralph sold 40 more bolls of wheat to Mr Clay at same price as last Saturday 39s per, some new 48s and upwards for seed. By a letter from Mr Peech Sheffield wheat is not mildewed in that quarter, oats have advanced rapidly 4 or 5s per quarter, trade dull yet fat cattle are 8 to 8/6, mutton 8 to 9d both small sheep at the great fairs in Derbyshire of Tidwell and Newhaven, have dropt 7s to 8s per head since the last year. This is not so much as our drop, and this is entirely attributed to the badness of turnips

Alnwick 4th September [*sic*: October] 1803

Well John

Here has been as expected a poor shew of everything, fat in particular, which was sold at not quite although very near 8*s*. But there was none to call quite fat. Perhaps the scarcity of fat cattle will mend the mutton market by and bye. I understand it is very bad at Morpeth at present. Yours in haste

Geo Culley

PS Let us hear from you soon. Bad pen.

210.   *Matthew and George Culley to John Welch*

Wark 5th October 1803

John Welch

Yours of the 2nd inst. is before me, in which you give an account of the fair at Broughill. It now has the appearance that lean cattle will be wanted, to consume the straw. The want of straw is in the north, no want of it in the south. We have plenty at Wark, Thornington, and abundance at Longknow where Jamieson had a good crop, a full stackyard, this year suited his land, also his turnips are good, so are ours at Wark. We shall in my weak opinion be short of working cattle as well as working horses unless we lessen our tillage land. Perhaps grass land pays as much as tillage now when labour of all kinds is so very dear. Draft horses and cattle are both exceeding dear, and so are labourers of all descriptions. Cattle were well sold at Ninian fair except for milk, sheep on the contrary very dull. Younghusband of Ellick refused (I was told) 40*s* per ewe at home by Green, and sold at N. fair for 36*s*/4*d* per ewe, lost by being too hard a salesman. Jere Clayton has bought some ewes &c since the fair. Money (or paper rather) is very scarce since the fair, at least many people have their money locked up in Burdon's bank. It has given the other banks more caution, are not at this time so fond of extending their notes. James Darling of Oak Hall has I am told sold his turnips at 8 guineas per acre, not a great crop. How are sheep to pay at such prices? Ours now grow apace, but on hot soils or thin land do not grow much, were so dreadfully mildewed or destroyed by insects, and our rape fields here &c much grown when not eat too close. Have got plenty of rain, our land lately plowed, too wet to plow this day, totally put off at Grindon on their fallows which are unsown. We have all our fatt cattle and sheep on turnips, and were too late of giving to them as they had shrunk much for meat, least so on Thornington haugh which kept fresh as did our watered land, not equally with Thornington haugh which lies low and is soily. I wish you a good market at Skipton with your Kyloes, but the fall in sheep will drop and injure fatt cattle. In our great concerns it is hardly prudent to speculate much except when things are low. Do not neglect to buy Mr Buston's steers, at least I think you should not miss them as they are of a good kind. We can let you have 12, 15 or 20 feeding cattle but can keep them on as long as you choose, when perhaps you should come over and choose them before they are sent. We should keep them at the door, so as they may be able to bear the weather if it prove bad when they are on the road for you. We are selling barley at 21s. It is said to be the greatest crop of any grain this season. I can suppose much of it will be broke, or boiled to feed horses on. Some say many cattle in the south will be fed on it this season, there is more food in barley before husk than on oats, but should be broke before giving it to stock of any sort. A good substitute for oats it is allowed, but all grain should be crushed or milled for all cattle in general, especially aged horses which do not ruminate or chew the cud. What is not broke before it gets to the stomach affords no nourishment to the animal. I hope Matthew got safe to you, but what right has he to shoot without a licence, he should take out one as it is wrong and shamefull and he may be informed off. I will pay for it. Give our and his sister's love to him and charge him to be very cautious with the gun. Your obliged master

Matthw Culley

Matthew should write to his sister from Denton, she intends writing to him soon. Our turnips now grow very fast, as do our cattle and sheep on them, but not so well as those on rape by farr. Our watered land keeps much stock. I fear you do not set Bob to get all ready before the floods come, the first are the best by much, are busy with ours, and the wheat, and some at the highways. I think you should not object to the 6d per £96 for the road from Darlington to Staindrop, although what you say of the overseer may be true, and I know not how or when it may be got put right, although it ought to be done, but the 6d I leave to my brother who may object to it. I am not satisfied that Mr Peacock does not send the money to London, you ought to have your money for the barley, too long credit is bad for seller and perhaps buyer also. Yours &c

Matthw Culley

[**G.C.**] Sunday 7th October 1803. So far my brother, now I begin John. I am just returned from Alnwick Sessions where I was obliged

---

96 A rate levied on occupiers in a parish for the upkeep of local roads. The overseer was appointed by the justices.

to attend as a juror. I sent you a letter from Alnwick fair on Tuesday, and if I had read and seen this of yours of 2nd of October I could have [. . .][97] 2 years old steers at Alnwick fair reasonable enough, only they were not so good [. . .] as your steers about Yarm. Alnwick was just the reverse of what you write [. . .] well and readily sold at Alnwick from 7/6 to near 8s. But a very shabby show [. . .] to all ripe or near ripe. Tommy Blacket was there, and one Swan [. . .] him. You may easily think that there was not so many as the butchers [. . .] or Ratcliff would not have bought *above* 50 since. He bought 40 of [. . .] the day after the fair, and 13 of Mr Ralph Fenwick. He bought 20 first of [. . .] as he went to Detchant by appointment to look at some of Mr Prideaux Selby's, which he did not buy as they wanted beef (a general want here I assure you), and on his return he bought 20 more, in all 40, of Mr Taylor. This I had from his own mouth, for he had seen 9 of Mr Ralph Fenwick's at Embleton on his road going or coming, and 6 more at Shortridge, 4 of which he bought along with the 9, 13 in all, at 28£ 10s per head 3 years old gone. I was present at the bargain at Warkworth, and helped to put them together, and after he had bought them he said he was determined to buy them if he could to top the oxen he has bought of Harry Taylor. I think Mr Fenwick's were hardly 8s per stone but very near it, strong *bumpy fruitfull* steers of his own breed, about 74 stone each by what I could learn. I did but see 4 of them and I thought them 74 good. Mr Ratcliff says fat cattle are very scarce and they have a very considerable consumption. He makes no secret of it at all, and everybody think so in this country and expect that they will be scarce and dear all winter. Indeed what can make them other ways? Cattle are laid on turnips very thin in general, and turnips are not of so feeding a quality as we have known them. And if the demand continue, and the slaughter must be great *surely* to the army and navy, what can prevent them being scarce and dear? Only 2 things. If Bonaparte come, which pray God avert! Or the vast quantity of pigs in every part of England. I believe there are an immense number of pigs in every country, and if fat cattle fall scarce it must help mutton sheep. I am of opinion that your Kyloes would have paid you plenty of money to have kept at turnips until Christmas. They never feed so well (turnips) as this season of the year, and you may depend upon it that they will be much wanted thereabouts. However I am far from blaming you, as I am sure you have done it for the best, and you must best know how you are

---

97 There is a large blot on the page here, affecting 12 lines of text.

circumstanced for meat. We can keep the cattle which you are to take from us a long while (I hope) through the winter, but I assure you that they are much shorter of beef than I would wish. I don't mean *all* the winter but a good part of it until you can conveniently take them. By mere chance I bought 2 stots at the latter end of the fair in the afternoon and no money in my pocket, because I had expected that you could buy cheaper and better steers than what are generally shown here. They are to be no more than 16£ the 2, one a healthfull clever smart steer, the other little and has been a late calf, but healthfull I think. I bought 3 more this morning of Ella Reed of Prendwick out of 5 at 12£ per, not over cheap I assure you, although I cast 2 I could have cast another 2 of them, are smart canny stots, but not such as I like by far. By a letter from Mr Boswell he says their wheat is very good in the western counties, but turnips bad, and sheep dropped 10 and 12s per head. So you find that turnips are an indifferent crop all over the Island? How are cattle to be got fat then so as to become plentifull. I own that I can't find it out. However turnips are and will mend wonderfully where they are not mildewed or turned white, but can't be a good crop at any rate. We will send this tomorrow, and I must beg of you to say whether you will have James Grey's and the shepherds' ewes, in all 52, because I have no doubt but we can sell them for profit, and I also am persuaded that they will pay handsomely upon your moor and out ground laid [. . .][98] I would not on any account have them to interfere with your feeding cattle, more than [. . .] and outfield seeds to make them feed their lambs in the spring. I know that they mean [. . .] and laid them on even *rotting land* will not easily take harm [. . .]

8 October. I did not go to Berwick John and I entirely forgot to [. . .] Ralph Brown, so must send it tomorrow. We have very [. . .] day now, and very likely to continue so I think, and all the [. . .] that were not mildewed or eat with the *lice* are really improving very [. . .], and one does not know how much they may improve if the severe frosts keep of a good while. Mr Edward Pease, with a Mr Stickney from Holderness and a Mr Maud from Bradford, all Quakers, have been here 2 nights but went away this morning. The other 2 had taken a ride to see the country along with Mr Pease. Mr Stickney is a very eminent farmer in Holderness below Hull, Mr Maud is a doctor. One thing which prevented me sending this today was that we fully expected a letter from you today or even yesterday after your Skipton

---

98 There is a large blot on the page here, affecting 9 lines of text.

market, which I thought would require some answering. But not receiving a letter from you made me quite forget to send this. But I am affraid you have had a very bad market. However murder will out, therefore you might as well have told us now as afterwards. However wishing you better markets I am yours sincerely

Geo Culley

PS. Ralph Brown could sell no wheat yesterday at all, but sold a few oats at 1£ 1s and a piece barley at same price, and pease at 30s. What can be the reason that there is no demand for wheat at all? William Brown's land is got far too wett now to sow wheat on, however he has luckily sown a good piece, but still has a good deal of fallow and rape land to sow.

211. *George Culley to John Welch*

Eastfield 9th October 1803

Well John

We have sent you a letter this morning which I neglected to send by Berwick yesterday. This is a fair day so far 2 o'clock, but very cold and not unlike rain. David Green has been in the country ever since Ninian fair but is now gone I believe. He has been buying dinmonds lately, bid James Darling 47s per for his but James would not take less than 50s. They are particularly good, large and fat, I have seen them at different times. Mr Barf bought George Archbold's of Dudlaw at 39s at the fair, very large and he says fat, but I think them coarse and slow feeders. However it was a slow price. *10th October*. Thick as shroud, a saving day. *11th October.* A fine dry morning, I hope people will get to sow wheat again upon damp soil. We had a very decent day yesterday, let above 100£ worth of tups. But what is the matter that you have never wrote us an account of your Skipton market? We have just let 8 tups today at no more than 50 guineas, but they are only midling outside ones. We sold a midling shearing tup also to go to Ireland at 20 guineas. Mr Walker had bought one of Gibson of Grindon Ridge and he died the 2nd day's journey. I believe he was over drove and very fat, not fit to drive at all. Ours is young and not over fat. I received a letter from Mr Peacock today, that he did not wish to find the 700£ until November as stocks are expected to be very low then. So I wrote to him directly to do what seemed best to him. He says he can raise the money pretty easy. *12 October*. A very fine calm morning. The weather seems as though it would settle again and set people to sow wheat again. I consider it blameable in you not to give an answer about James Grey's ewes, whether you incline to take them or not, because it keeps us in suspence, and if you are not for them we could sell them at Wooler fair. *Thursday 13.* A wonderfull fine morning, quite a Michaelmas summer, but no papers nor no letters this morning, when we are in such a critical and alarming state! But Staward the Kelso post generally uses us so in these most *eventfull times,* and I fully expected a letter from you too John to say whether you were inclined to try James Grey of Milfield's few ewes for fat lamb or not. The Wooler fair is on Monday too, when we could have shown them, but are in real suspence. I have to lament your not writing on business now that you can't be so exceedingly busy I should think, but you might spare half an hour or even 20 minutes to have explained that matter. We let John Burton of Darlington 4 tups yesterday. We have come on pretty well latterly although we shall not make so good a season as for a few years last past. But the tup trade is and must be a very fluctuating trade, and the sheep dropping so much in price has a wonderfull effect, whilst rising markets have a contrary effect. G.C. As I wish to hear from you soon I will conclude and send this per post today, and am yours in haste

Geo Culley

212. *Matthew and George Culley to John Welch*

Eastfield 16th October 1803

John Welch

My brother received your letter of the 14th this morning (it now blows a hurricane of wind and now rains). Turnips are much mended wherever the vermin had not much injured them, are selling for 6 to 8£ per acre. Ours at Wark tollerably good and we can keep the cattle you are to have from here for feeding as long as it may be thought prudent before they are sent to you, say Martinmas or so on, sooner or later if weather permit. Pork at Berwick 6s, supposed to get some higher, beef at Shields 56s per quarter. By a letter from Ireland sheep fell 10s per head below last year's prices at Ballinasloe fair. Cattle, fatt ones, were well sold, lean cattle also. Mr Ralph Forster says Kyloes were higher at Falkirk than last year, though bought best the first day. At Berwick wheat sells no worse. Ralph sold some small ill fed red new wheat at 37s, the worst we have I hope. 50 bolls old red at 38s per boll. Good new, of which good nice samples are shewn, are worth more I believe although I heard of no higher prices. I was in hope you would have gone near Raby and tryed to buy some young cattle for straw. I hope you have or will buy Messrs Buston's steers, as probably cattle will be very high, at least appearances are such. Fatt cattle seem scarce, and how can they be otherwise, where was the grass to feed them on? Hope you did well taking fog for them at Skipton. Hope Matthew may find time to write to his friends now when he gets to school, at least we expect he will after being indulged so long at Denton. We are obliged to Mr Harrison for taking so much notice of him. Remember me to him with our thanks for his great civility to our son. Likely we shall proceed to Denton from Newcastle fair, you must meet us at Newcastle to assist us to buy some work horses. How many we may want I know not so far.

Matthew Culley.

[*G.C.*] John you don't say a word about the mule, but I hope you have found him again You certainly acted wisely in taking fogs, because I have little doubt but cattle (fat) will be well sold by and bye. They are well sold in this country at present, and in Ireland, consequently must take place in this country by and bye, and by the account from Falkirk lean Kyloes were very dear. Therefore it is better to keep your fat ones than sell them at lean prices. Turnips as you say will and do go fast away, consequently it would be wrong to buy any more fat, or feeding stock for turnips, because it would be most unwise to put too much stock to turnips at these prices, and when winter meat is so scarce and likely so to be. Upon this account I would caution you against buying too much stock. I only speak in regard to ourselves. For your own self you can best judge what you may want. However should Yarm be a slow fair you can't be wrong in buying a good few. I would advise not to be in a hurry to sell our firkins as I am sure they will be high sold. We are bid 55s here. You are very right, that turnips never feed when coarse necked and woody. The yellows have been fatal on red clover to many sheep this year. Ewes never or seldom take it. We will sell James Grey's few ewes tomorrow if we can. The wheat in Ireland is much mildewed.

G.C.

[*M.C.*] I believe all or most of the wool is now sold in this country, 23s 6 to 24s 6 per stone of 24 lb. M.C. The weather seems settled again. The Yorkshire folks say their turnips are much mended.

Eastfield Tuesday 18th

[*G.C.*] In my hurry at the fair yesterday I quite forgot to send away this letter. We had an advance in the price of dinmonds, and I believe few of them were left unsold. Ewes were not so well sold as Ninian fair. However as you did not approve of James Grey's going to Denton, we sold them at 1s per head profit. We bought 6 steers, rising 3, very suitable for us to draw &c &c, price 14s/10. I saw Messrs Armstrong and Scott and many others who had been at Falkirk, and all agree that Kyloes were as dear as last year, and by a letter from

Ireland I find that sheep dropped 15 to 20*s* per head, wedders, ewes 10s but cattle had a most brisk sale. I had forgot to say that we shewed 70 Longknow ewes, asked 36*s* but were never bid although we had your friend Rawling Sherborn often. But as they are handsome mutton, having been on rape long, I am not affraid but they will sell very well and soon as markets mend about Christmas or so. Yours in haste

Geo Culley

213.   *George Culley and Matthew Culley jr to John Welch*

Eastfield 21st October 1803

Well John

I neglected sending you the letter from Wooler fair, and I am affraid it would not reach you before you went to Yarm. Indeed there was no post of Wooler that day, but I might have sent it by Mr Bugg &c to Belford. However I sent it the next day. I hope you had a fine day at Yarm fair yesterday as we had here. Indeed the weather is now remarkably fine, fresh and warm showers, which must and does assist the turnips much, and as the Yorkshiremen bought all or nearly all of the dinmonds that were shewn at Wooler fair we think we shall not have more sheep left in Northumberland than usual, consequently that our sheep markets will probably mend about Christmas, as they generally have done other years. We sold 10 gimmer and twinter[99] ewes to Mr Witham to go after tupping to Cliff, and I was drawing them yesterday when I was very glad to find the ewes and gimmers very handsome mutton indeed. To be sure they have been going, and are part of them still upon rape, which is wonderfull feeding food for ewes *perhaps* in particular. We are to meet Mr Witham's people with the 10 sheep when all tupped at Minsteracres. I can't help repeating what wonderfull weather it is. Such a harvest and seed time have seldom happened, particularly after so very dry a summer. I hope you will find leisure to come out here after a while, to fix upon such cattle as are most suitable to send to you afterwards. I wish they had been forwarder in condition, but the season was such that it is my amazement how they kept any condition on them at all. One thing must help them, they have now been 4 or 5 weeks on turnips, and although these were mildewed yet I can perceive them much mended. Let me beg John, that you will not neglect to send an account in time before the 22nd of next month of your *stock* and *cash*,

---

99 One and two years old.

and remember that we consider *turnips, potatoes* and *rape* the same. And pray tell me the price of skim milk, cheese and butter firkins at Yarm fair. Tommy Frank bid me 58*s*. I would have taken 3£, and he confessed the best dairies were selling in Yorkshire at 3£ 3*s*.per. Mr Nisbet has been ill, but thank God is so much better as to be able to begin on Monday with Messrs Bailey and Bates to value the estates in this county, and then they go to Newcastle fair and from there to Denton to value it, when you must pay them every attention and give them every information in your power to such questions as they may have occasion to put to you in regard to the advantages and disadvantages attending the Denton property.[100]

John I am so much affraid that you will neglect to inform us what cattle you have bought at Yarm or before Yarm fair, if any, that I will send this per post from Berwick tomorrow and must request that you will write directly, if you have not already wrote saying what steers you have bought and at what ages, to come into this country. Because upon William Brown's looking over the steers at Longknow &c today, we find that we shall still want about 8 steers 3 years old for present yoking. Matty bids me say that he will be at Newcastle on this day week (Tuesday) in the morning, and he wishes you to be there on the same day as he will want your assistance to buy a horse or 2. Ralph sold nothing today but 20 bolls barley from Wark at 20*s* per. We sowed most part of the East Flat bean stubble today with wheat, and being scuffled[101] before plowing and the weeds led off after harrowing, it worked like the nicest fallow. I never saw anything work prettyer. I hope we will soon be done sowing our bean stubble if this uncommon nice weather continue a few days, I am yours truly

Geo Culley

PS. Don't forget to send your annual account of stock and crop &c. I would gladly hope you will find time by and bye to come over and chuse such of our fat cattle as you wish to have.

[*M.C. jr*] If you have not already bought the wanted number of steers gone 3 years old it would be right for you to bring what money you can spare to Newcastle in case we need it, as we are very poor, at any rate it is good company and easily carried home again if not wanted. Be so good as give us a full account of Yarm fair.

---

100 The Culleys were beginning to prepare for the dissolution of their partner-
     ship: see p. xxxi.
101 Scarified with a scuffler: see No. 159, n. 102.

[*G.C.*] *Saturday 22nd October1803* is as sweet morning as ever was *born* at this season. Pray God send us fine moderate weather, and then our fleets will be able to stay before the French ports. Was Patey ever found yet? If you can do without him, he would be very usefull to us here. Berwick Saturday 11 o'clock. As there is no letter from you John I will seal this and send it away, particularly as the market is not likely to furnish any information worth notice. Pork is got a fall of 6*d* per stone. We have been too late of bringing in our pigs. Do write immediately on receipt of this if you have not done so already. G.C.

214.   *George Culley to John Welch*

Eastfield 24th October 1803

Well John

I am much obliged to you for your full intelligence at last respecting Yarm fair. We received yours of 22nd this morning. We are very happy that you have bought Mr Buston's and Mr Best's steers, particularly as we think them well worth the money. Respecting steers, I have now only [to][102] recomend to you to buy no more now *upon our account here*, except 4 more steers gone 3 years old, and I would wish you to perfectly comprehend me that 3 years old steers are what we want at present. We have as many 2 years old as we can find meat and room for. Indeed 8 steers either *gone 3 years* or the *strongest* and *fittest* to go yoke at present, are all we want. The rest you may keep if you chuse, to tread down your own straw, or if they be too many for you we will still take a part. But that my brother and Matty and you can settle to determine upon when you meet at Newcastle, where you will receive this as I intend to send it by our Matthew. I think from what you say that if it had not been for our high-mettled North country boys, steers would have been better to buy at Yarm. I should not wonder if part of the fat steers bought by Anderson and Bolton be for Adam Atchison at Ryal. I think Mr Best was very kind in taking so much less of you than Bolton bid him. You were perfectly right in buying those queys I will be bound for it. Nothing like things well bred. I am glad that pork is 6/6 ½ *d* with you at Yarm, and butter firkins so dear. I thought you were wrong informed about pork when you spoke of 5*s* per stone, because pork is generally from 6[*d*] to 1*s* per stone higher at Yarm than Berwick. But when I say that we want no more than 8 3 years old steers for ourselves here, I would not by any means to prevent or hinder you from buying 2 years old steers or

---

102 A word seems to have been omitted.

even *stirks* for your own use, and if cheap we will even struggle hard to winter a few more here, because I am confident that cattle cannot nor will not be cheaper during this war, nor afterwards if this blessed Island retains its liberty and constitution which I pray God grant that it long may! Notwithstanding the bad markets at Skipton at present, I have no doubt of the markets being good by and bye. Mr Armstrong bought 80 Scots of Harry Morton, 40 of them fat, and they were very keen of buying I understand. Matty and brother and you will fix about steers and the cattle for your meeting, but it is much better for them to come and go all the way, and we can send one of the Glasses, but had better defer it untill you see whether you can buy a few more 3 years old steers, as they will come best together. I am glad that you had occasion for no more than 100£ from the bank. You may be sure that Newcastle will be dear for both Scotch and English cattle. I am glad that the mule cast up again. I am rejoiced that my friend Charles Colling's ox still keeps well, I hope he will come to Berwick. Yours in haste

Geo Culley

Tuesday 25. Still surprising weather, I am as warm with walking up to the Westfield as though it was middle of summer. Matty's horse being lame he has my horse to the drills.

215.   *George Culley to John Welch and Matthew Culley*

Eastfield 29 October 1803

Well John

It is not very material whether I address my brother or you, as I am sure you will both see it. I was told by Ralph Brown that he met my brother and sister yesterday as going south as he went to Kelso, and what would be extraordinary of any other person, my brother said he had wrote a letter to me which Ralph would get on his return at Wark. But the letter could nowhere be found at Wark by his amiable daughters, and one thing I have to lament, a Mr Trench wrote to me from Ireland requesting that we would one of us take a young man, an Irishman who has been at Kitt Mason's some time, to work for his meat, and my brother said something about taking him, and I believe his determination is contained in the said letter. I would not mind, if Mr Trench was not kept in suspence, because Mr Jobsons or many others might take *Irish Jack*. However there it rests, and I must wait until either this Wark letter pops out his head somewhere, or my brother write from Denton. As far as I see we must keep our old wheat as no one has offered to buy any at Kelso of some weeks, and I am affraid it will be difficult now to sell it at Berwick as Mr Clay, the only person who has done any business of some time at Berwick, has had the misfortune to have his mill burnt. And what adds to the misfortune is that they had been grinding all they could lately as they will be obliged to break the mill stones if the French should land near to this country, and a great deal of wheat flour were in the mills. I speak in the plural number because Oswald, a baker in Berwick, had a mill adjoining, which has suffered the same fate. Either the one or the other is supposed to have taken fire for want of cooling when stopped, and being so near each other, had set the other on fire. This unfortunate event took place on Tuesday night last we are told. Ralph sold 24 bolls barley at 20s at Kelso to millers, and Mr Clunie offered him the same price. Now I conceive that Mr Clunie must be buying it for Scotland, because 20s for our boll and 26s for the quarter in London is only a ½ [d] per bushell difference, and 26s is the best London price for barley, 6 times 3/4d is 20s, and 8 times 3/4½d is 26s. Therefore it can't pay to London. I only name this to prove the propriety of selling barley at present at 3/4d or 20s per boll, because it is clear to me that barley will be lower and not higher, as it is confessedly the best crop of any grain in this Island, whilst oats are the worst crop, and nothing that I foresee can advance barley except the scarcity and probable advance in oats. Still I have said before and I remain in the same opinion, that either oats will advance barley or that barley will injure oats in price, for it is so obvious that few will eat oats while their price is higher than barley. Barley and oats are now as near to the same price here as may be, with you it may be different. Ralph only preferred selling the barley to the millers because the delivery was nearer than Berwick. The barley is from Wark. Kelso fair is on Wednesday next, and if Ralph had adverted to that he would not have gone, for he neither received much nor sold any quantity. Our bean lands and pease also work as clean as any fallow, cleaner than many, this wonderfull weather. By scuffling, some of it only once and some twice, the weeds and few quickings come out so charmingly that I never saw wheat put into the ground in a nicer state, the seed so good, so well matured as I have seldom seen it before, yet we know not how the crop may turn out. We got the hill or Longknow dinmonds from Wark the other day in good health and spirits I dare say, but I think in worse condition than they were in the spring when I wished them to go to you. This may be owing to the want of meat at Wark this uncommon dry summer, but I can't conceive why they could have more the appearance of Cheviot sheep than they have had for some years back. This is the case or either Ralph or I are mistaken. Ralph goes so far as to say that Mr Morton's or Mr Somebody's tups must have got to the ewes. Certain it is, that they were confessedly the best hogs this spring we ever had bred upon Longknow, but they are not *now* the best sheep we have had. Excuse me for reminding you again, about sending me an account of your stock, crops, turnips &c &c, because you have generally been too late of sending me the account and that business devolves upon me always. I mean the putting it upon paper, and I must also beg of my brother to inform me what I am to say to Mr Trench about Irish Jack. This is a very rindy morning indeed. Cornhill will be white with it and yet it will probably be a warm day, at least I sweat yesterday on horseback without stirring. I have determined to send this letter per post today in hopes my brother will inform me about the Irish man. It is so unbusinesslike not to answer letters that I can't subscribe to it, but will subscribe myself in haste your well wisher

Geo Culley

216. *George Culley to Matthew Culley*

Eastfield 30th October 1803

Dear Brother

I sent a letter to Denton yesterday by Ralph to put into the post office at Berwick, and I hope to have regular answers now when you are at Denton as you carry the pen of a ready writer. Ralph sold a little stack of wheat from Wark (red) at 37s very soft he says. I was in hopes it should have been better than the last. Clunie and Home bought it. The fire began in Oswald mill at the stamping machine which they had set to work and gone to bed, and by 12 o'clock all was in a blaze. By good [luck][103] Mr Clay's people saved about 50 or 60 sacks of flour. It had taken fire at the bolling machine for want of oil or grease it is supposed. Poor Alex Davison's estate of Whiterigg was seized upon by the British Linen Company one day in the last week, it is said for 14,000£, what a hard story poor man! They say Tom Hall had a near escape, or he would have had his body taken, but he got of on horseback. Mr Davison's tenant John Thompson is said to have underwrote several of T. Hall's drafts, and that is given as a cause for his bad health. How cautious people should be, to avoid becoming bound! Solomon's proverbs are worth reading on many accounts, and that amongst the rest. Some people say that all T. Hall's effects and Whiterigg &c will not liquidate the immense debt he is in. If so another Tommy will be brought in, who lives near the collieries as we

---

103 A word seems to have been omitted.

go to Berwick by Grindon, and 2000£ will be a serious loss to his large family. *Monday morning 31st October* is a very fine moonlight morning, not cold although the wind is north east, with large frosty, snowy-like detached clouds driving south west. Until this morning we have neither seen sun nor moon for some time, it has been so hazy and thick but no rain, and still a wonderfull propensity for drought, and were it long days the drought would be as severe as ever. The ponds are drying up again, yet our pump has as much water as ever. Matthew got home about 3 o'clock yesterday. He and John have got a very good looking black mare 26/[£]3s well worth the money if she prove an honest drawer. I fancy it was a dear fair for most of things. Matty says they got a good deal of wet as they came along yesterday so as Tom Smith was wet through. We had nothing here but a haze with a very peevish east wind. I never knew a better time for goods at turnips, they should do much good only many turnips have not such richness and feeding in them as usual. I observe that when the sheep scoop them they fill full of water or *sap* as they do when *making top* in the spring, which is the case with our early turnips at this moment. The old withered and mildewed top is dead off and a fresh top is come, which I presume has occasioned that *flux* of *sap*. However the bottoms are swelling *confessedly*, as well as the top growing, and I never observed the wether or male sheep eat the tops so clean even where they eat them as they grow. Ewes always eat the tops better than any other sheep, or any other animal I know of, but the reason why I do not understand. *Tuesday 1 November.* A fine day although snow has laid upon Cheviot all day and will not go away now 4 o'clock pm. I went up and breakfasted this morning with your sweet daughters. Nelly had found your letter and gave me, so I shall now write directly to Mr Trench. Everything appears to be doing very well at Wark. Indeed stock have such fine weather that they must do well on midling meat. I recomended to them to begin ruta baga with the tup lambs. They can never learn in finer weather, and learn they must. They are eating Leemingside turnips, getting some wethers and a lot of cattle. I shall certainly go to Wark once or twice a week at least, and offer my advice if needed. I shall likely send this per post from Berwick High Market tomorrow. Our respects to the gentlemen and with every good wish to my sister, yours

Geo Culley

Tell John that there are 13 spayed heifers and cows &c going below the Gallow hill which will in general suit him well, but I am persuaded that he or few other people have such turnips to eat. I never saw any better, few so good any year. The queys are not in such forward condition as other years, but that was owing to the summer being so dry, and they have a chance of being as good and forward as other people's. But if John tye them up they will not do. Let them either go *out and in*, or lie loose in a fold yard, and then they will keep their hair and turn out in the spring without suffering from cold or wet &c &c. That is my opinion, but remember I only give it as my *opinion*, not as a *command*. It is what I never wish to do, hardly to Ralph Brown or my own servants that are immediately under me. There are also several nice steers besides for John to pick from, our Blue Cap for one. He is growing like a trout, and deserves oil cake or any cake. But all the cattle and sheep are thriving well at Wark, this fair season and excellent turnips while ours are poor withered rotten or thin, Blue Bell excepted and a few at Westfield.

Berwick Wednesday 2 November. Here is an exhibition of every sort but one. Indeed I have priced one steer, rather a good one of George Davison. He asked 19£ 19s only 2 years old, He is comed to 15£ and if I can get him to 14£ I may happen buy him. I also bid 6£ for a *bull* for John Welch. This is a fine mild day, but the snow increases upon Cheviot. I hope we shall hear from you soon. G.C.

217. *George Culley to Matthew Culley and John Welch*

Eastfield Thursday 3 October [sic: November] 1803

Dear Brother

I sent a letter per post yesterday from Berwick High Market, where I did nothing, and Ralph did as little at Kelso fair same day. I know not what is to be done with the old wheat, but Ralph can sell none. He sold nothing but a stack of oats from Wark at 20s, and received something above 70£. But William Mabel wants 100£ for him and Longknow, which with other payments keeps us very low indeed. Your friend Ralph Foster must wait a little longer until the cake be baked. Mrs Bell 27£ 10s and some more interest money is due on the 7th inst. However if we can only keep the wheelboard in the nick[104] I am content, but such a heap of men making water meadow at Thornington &c at the present high wages employs a good deal of money, and corn does not raise money so fast as we have known it. Indeed this old wheat lying in hand is very inconvenient, and no stock sold yet. However we sold 40 ewes today to John Batters junior at a little above 50s per. He is taking away only 10 a week is the worst of it. I really think that John Welch should keep all the steers he has

---

104 In the groove, i.e. keep going straight.

bought to tread down the straw, except the 3 years old ones for immediate yoking, as we find on examination that we have as many cattle, young and old, to the full as we can find wintering for. I hope John will also by and bye take a lot of our feeding cattle, as I see plainly that with the very great stock of sheep and cattle that we have feeding, notwithstanding great economy and wonderfull weather, that the turnips go fast away. However I would gladly hope that sheep markets will mend about Christmas, and perhaps we may be lucky enough to sell a lot of oxen at an earlier time than usual, as the demand I hear continues pretty brisk. I really think that we should be content with 8s per stone sink if we can get it, and 8d per lb. sink for sheep. The 40 we have sold today are thereabouts. The ewes are excellent mutton. The rape has set our ewes a-going, and I do highly approve of giving the sheep the first fruits of the turnips, and letting the cattle follow. Observe this John Welch. I am persuaded that the cattle don't do *much* or *any worse* for the ewes &c taking the turnips tops off first and breaking the turnips. The cattle seem to eat them with as much pleasure, do exceedingly well, and to a certainty the sheep do better and the turnips go much further as the sheep, *ewes* in particular, eat all the tops clean and compleat, which would be nearly all wasted if the cattle go first to them. I hope John this will meet with your approbation. I never saw turnip stock grow so fast as at present, although turnips are not so good as we have had them by a deal, here in particular, but the weather is such as never was. Yet snow lies close upon Cheviot these several days. *Friday 4th* is a most astonishing fine morning. Mr William Spours came home from the drill with Matthew, says Morpeth was a little better on Wednesday, is going to Low Lin to try to buy old Gregson's ewes. The wethers are upon the Hackney Close turnips at Cornhill, 300 good looking sheep. Bill Spours thinks markets will turn for the better between [now][105] and Christmas, says more sheep than common went south this autumn, is much dissatisfied with Lord Ossulston's troop going to Newcastle to stay a fortnight, not only expensive but extremely inconvenient to people who have much business, and so it is to be sure. But these young noblemen wish it and carried their wish. Besides, it often leads to drinking and various dissipations. Let people be ever so cautious and guarded, when a set of young men get together the heart warms, and one bottle produces another and sometimes leads to unpleasant consequences, and at any rate does not suit either the time or pockets

---

105 A word seems to have been omitted.

of many of the troop. William Pattison's bill amounts to upwards of 500£, your friend Brown's to several hundreds, besides taken Hatton &c &c. These *contractors* all lie on, and it is an ill wind that blows no one good. Even 25£ per man besides expences at Newcastle &c &c will be hard upon your Harry Potts &c. However if they can only keep away the Corsican I shall be content. Privations of various kinds must and ought to be borne without grumbling in these perrilous and eventfull times. Only every economick plan should be used that can possibly be hit upon, so said Sheridan,[106] and it is highly agreeable to the sentiments and long experience of considerate people as well as

Geo Culley

Grindon Saturday morning 5th November. Met with your and John's letter at the bridge end as we came here this morning, so shall answer it and send this by Ralph to post. I am sorry that Mr Nisbet was so ill as to return home before they had done, hope to hear how he is by Ralph tonight. Mr Reed's steers once paid are not to pay again. I dare say they are dear indeed. I bought them against my own *conviction* which is wrong in general. I have told William Brown and Ralph to see about your wine, spirits &c soon at Berwick, but the ships are very slow of coming from Newcastle. We wish John Welch to send no steers here except such as are fit to yoke 3 years old without he can't keep them, but they may pay for treading down his straw as well as Kyloes which are very dear I am told. However John is the best judge what suits him, but I am clear we shall be sadly whipped for winter meat in most of our farms. If John's moor rape be so bad, is it not better to cut it down and sow wheat if it can be done, as wheat will pay better than barley upon that land. But remember I only suggest that, John and you must be better able to judge of it. I am very glad John that you can sell your old corn at any price at Barnard Castle. Ours can't be turned into money. Matty thought you would buy the horse. Pray John buy no more steers for us except you can get 4 3 years old ones. We have as many or more than we can winter, and I would have you come and fix on what queys and oxen are to go to you before they go, as it must [be][107] best known to you what suits you. After you fix we will keep them as long as we possibly can for you, but certainly John you should come and chuse for yourself. I do think there are several that will please you. With respect to all I am, in real haste as Ralph waits, your affectionate brother

Geo Culley

---

106 It has not been possible to identify this reference.
107 A word seems to have been omitted.

218.   *George Culley and Matthew Culley jr to Matthew Culley*

Eastfield 6th November 1803

Dear Brother

I am happy to say that Mr Nisbet is quite well, Ralph saw him at Berwick yesterday. Ralph did no business yesterday, indeed very little business is done now at Berwick. This is a very fine morning, and we have had no rain except some very slight showers indeed, although the mercury has dropped more in 3 days than I have often known it. I answered the letter wrote by you and John yesterday in one I had wrote before in part and sent it by Ralph to Berwick. As your daughters dine here today we will give them your letter. 4 o'clock. Your daughters are just gone, and were so kind as take our Nelly with them to Cornhill. *Monday 7th November* is a wet morning and has rained it is said most of the night but not heavily. However it is more than possible that this *long wonderfully dry time* is at an end, as the mercury has comed down for several days and fallen from very high to very low and is still dropping. We once knew a more severe drought, but never one that continued so long even into a *winter* month. Our ground is really in want of moisture, in many places very hard to plow, and the pond is lower than has been remembered. Mr Askew's pond nearly dry, and most of soft marshy and even boggy places so dry as to bear *riding* over, which in many seasons cannot be walked over with safety.

Baker Wonderfull News

Robert Tate son of Robert Tate of Wark in the county of Northumberland blacksmith, has commenced *master baker* and has given or is bound to give 400£ for the benefit of the business to his mistress, who it is said ought to have had nearly as much more if she had chose. That this mistress has made an independent fortune and has one only child, a daughter, which is not unlikely may some day, perhaps not far distant, change her maiden name of *Brown* to *Tate*, and that more unlikely things have happened in that large city than Bobby Tate becoming *Robert Tate Esqr* and 20 years after this drive his charriot! Well well, but Bob was always an extraordinary Bobby. Ralph Brown was told this by some people who were at Berwick on Saturday and know Mrs Brown and *Mr Tate* well. The latter part is certainly only conjectural, but if Miss Brown has as favorable an opinion of Mr T. as her mother, nobody knows what may be the result. This has turned out a very charming day, and very little wet except what had fallen in the night.

Matty if you have bought no leather as you went through Newcastle, you are requested to bring on your return what you think may be needed, as William Brown, who was here today, says he will want a good piece and intends to have the sadles soon. However it may be sent by sea with your spirits, iron and wine &c, because I will be bound for it the ship will not be sailed to Berwick when you return to Newcastle, If you have not an opportunity, would Mr Thomas Bates do it for you any time when he is at Newcastle on account of this said leather. I will send this letter to the post tomorrow that it may catch you before you leave Denton. William Mabel got 100£ today to pay at Thornington, and William Brown will want nearly as much soon. However labor must be paid, and I hope we will get other things paid by and bye. Bill Spours bought Mr Gregson's 200 ewes on Saturday last, but I don't know the price, and Ralph Brown met Tommy Scott of Coxswold tonight, enquiring for sheep, and I think I wrote you word that Tommy Meek was at Morpeth last Wednesday buying all the best ewes &c. This looks as though turnips were mending and sheep wanted in Yorkshire. May that long continue to be the case says your affectionate brother

Geo Culley

Tuesday 8 November 5 o'clock. A fine fresh morning. As I am going around by Thornington to Wark and have to shave, I have started rather early, but I do tire of lying in bed most sadly these long nights and Matty takes this to Wooler to the post as he is going to the drill. I am much dissatisfied with their going (the troop) to Newcastle for 10 days or a fortnight. To be sure it is a time of year that farmers can be best spared, but I don't like it for many reasons, and at this moment newspapers say that a most dangerous typhus fever is very fatal in that town. But it seems the great ones of Chillingham are bent upon it. But surely if that fever story be true they will have some regard to their own health, but perhaps they will lodge at Fenham and exercise in the Town Moor, consequently will not affect them. John Welch has never once said to my recollection what fat cattle he means or wishes to take from us, and if this soft weather continue any time we shall think of tying up our larger cattle. Could John not find leisure soon after Martinmas to take a ride over? Nobody can so well chuse the cattle he would wish for as himself surely. Let him turn this in his mind and either say he will come or not. If not he must then give us such instructions as we can chuse for him. But surely he may spare all or a part of one week to come north. It will be well worth his while I think, as many things can be said in conversation that cannot be conveyed upon paper so well. G.C.

[*M.C. jr*] You may bring 6 hides of limber leather well dressed within, and 6 or 8 pair of thin light bellys, the lighter the better, also a good few bazels if you can get them good, and not more than 2*s* 2 or 2*s* per. William Brown and us will take half of the above, and you can order what more or less you like for you and Thornington.

219. *George and Matthew Culley to John Welch*

Eastfield 10th November 1803

Well John

 I fancy I may as well address you now again both outside and in, as my brother will have left you before you receive this, perhaps before the last reached you which was sent on Tuesday last, and we fully expect to hear from you now what you did at Midlam Moor. Tom Tulip was at Edinburgh Hallow fair, says there were a prodigious show of Kyloes, but nevertheless dear. However they were lower by 20 to 30*s* per head than last year's fair he says. But then they were out of all bounds dear this last year. John now that we have the fat cattle to turnip we are quite at a loss. You must either come over at all events to fix which cattle you will have, or we must run all hazards. And it would be *ridiculous* for us to tie up such as you would wish to have, and then to turn them out again would be doubly *ridiculous*. Therefore I beg that you will come as soon as possible, and write us word when, because we will still put of tying up any except the very heavy cattle until you either fix the time of coming or say that you will not or cannot come at all. Now if Martinmas were over I cannot see anything to hinder you to come. Surely plenty of time between one fortnight day and another, indeed half the time will suffice to come in. To be sure the cattle by rights should have been fixed upon long ago, and been sent when the weather was so good, and plenty to get on the road for them. However by time taking they may do very well, and it would be very absurd to meet. Never meet. Let the same driver take them all the way, both going with the cattle and coming with the steers, and pray send no steers but 3 years old gone fit to yoke, because we already are well enough stocked and the young steers will make your straw into dung as well as Kyloes or other cattle, and either feed the summer following or may be sold. *Friday 11th* is as fine a fresh day as ever was, never been much rain yet, and the roads are drying fast. Yet the glass is lower by much than I ever saw it. Met with Mr Thomas Scott who had bought 400 sheep for turnips, mostly ewes as people had put their dinmonds to turnips, those that have any, and were unwilling to sell them. He very candidly said that turnips were so much improved in Yorkshire as nobody would believe who had not seen them. Therefore I hope we may get pretty good sheep markets still after a while. Ralph could sell no wheat yesterday at Kelso, oats and barley much in demand but he had no samples of those grains. We met your master and mistress this evening going home. I had been at Wark where everything is doing well thank God, only you must come over as soon as possible, otherwise we are quite handfast until you have fixed upon what cattle you wish to take south.

Saturday 12. Rather frosty, and Cheviot white with snow, but a fine-like day and no appearance of rain although the glass is still lower than was ever remembered by me. It is a phenomenon! The turnips are wonderfully mended and good at *Wark, Thornington* and *Grindon* and our Blue Bell is very good, which shows the superiority of deep or cool soily land over gravels or sand this dry year. And I really think notwithstanding the very *great stock* of *turnip* goods which we have this year more than common, and the *early starting* with turnips, that we shall (except a rot in turnips take place) be able to keep *most* of our stock a long time yet. I name this to show you that we can keep the cattle you may take from us a good while still provided it will be any advantage to you in making your turnips go further &c &c. And as cattle can travel even if snow was upon the ground I name this also, that if the keeping the fat cattle here a while longer be any advantage to you. You may either keep the lean cattle until the fat go to you, or send the lean ones sooner as is most agreeable to you. But certainly it will be the least expence to send all the way and take the fat ones back, but wish you to judge for yourself. But don't think that by my saying this on the other side that we can excuse you coming over. No, by no means. We can do nothing until you come, and cannot tie up or *assort* or *arrange* our cattle as we ought to do, until yours are *fixed* upon. Therefore let me entreat you to arrange your matters so as to come over for a few days, be it ever so short a time. We cannot get forward at all until your cattle are fixed upon. We don't desire you to *take them* away, or have them *sent* to you when fixed upon, but only to come and chuse them, and then we can go forward. Otherwise we are quite at a standstill, and it will be so awkward to have the cattle out that ought to be tied up, if the weather become very coarse, and it cannot stand for ever. I would not wish you to come by Newcastle by any means, as there is so shocking a *fever* there at present and so dreadfully *infectious.* Your master and mistress luckily did not come by Newcastle.

John we have for breaking a very nice and valuable horse out of my old mare and got by Archer. Tommy of Grindon had tryed to break him last year, but being a very high-mettled horse he quite got the better of Tommy and threw him several times. The man we now have breaking him thought he had quite got the better of him, but the other day he unfortunately threw him in spite of all he could do, and got from him. However the man has been on him since, and is in no way affraid of him, and we have every reason to believe that the man has managed and behaved to him very properly. But so bad are *misteaches* or bad *habits* to conquer that I name this to you to know if you have any clever *genius* in your country who could *conquer* this bad habit in a fine animal. And I do assure you he is perhaps the likeliest horse we ever bred to sell for a deal of money. Matty had promised himself great pleasure in riding and hunting him, but I can't think of letting him try, and he seems now to be satisfied to have him only so managed as to be shown so in a fair or to be sold to advantage. Indeed the man we have seems not to be affraid to do that, as the horse seems to have no real *vice*, but it is a pity not to have him well managed if it is possible to be so done.

Sunday 13 November is a showery morning, the glass rising again, but John my brother was sure there would be a letter from you today at any rate giving an account of *Midlam Moor* and Darlington *Neat fair day*, and we were to send it up to Wark directly. But John you are a sad man for neglecting to write so often as you ought. My good lad would you only write ever so little it would be satisfactory, but you too frequently *put it of* and *put it of* for days, and I had almost said *weeks*! Is there no cure for you John? However we shall have a personal account from you soon, because if you do not come over in a very little time after receiving this letter the cattle *will* and *must* be very much injured in this cold climate by lying out so long. Now the weather is turned *cold, raw* and winter-like. Do my good lad come as soon as Martinmas day is over, I am sure you have time enough to return again before the great Monday after Martinmas. I was in real hope from what my brother and sister said, when we met them on Friday evening, that there would have been a letter from you this morning, and nothing prevents my sending this away but the hopes of having one from you every morning, and I am determined to keep teazing you with letters until you do come, because come you must if well, otherwise we may as well give over altogether. Nothing so unpleasant as to be kept in suspence, and we are entirely at a loss what cattle to tie up unless you fix upon such of them as suits you, Ralph sold some wheat yesterday at 39*s* per boll, something we must sell to keep the wheelboard in the *nick*. We are still very back with our *payments*, and when we shall get up with them I really can't see. Not that I am uneasy, because thank God we have plenty of both corn and cattle to turn into cash, but corn is so bad to sell and the time of selling cattle and sheep is not yet arrived. However if Bonaparte only keep of us I am in no fear but that we shall have pretty decent markets, and I am determined to begin as soon as ever the markets stir, and I would gladly hope that it will not be long, as the Yorkshiremen are still taking away sheep from this fruitfull breeding country. It is really surprising what corn, cattle and sheep this district affords. *Monday 14th.* A very hard frost indeed and some slight showers of snow fall up and down G.C.

[*M.C.*] John my brother told you we got home on Friday, and hope we found all well. The horse breaker must be fixed about when you come over, which time you must fix yourself. Bob Dippie speaks of coming to Wark at Christmas, you cannot both be from home together, therefore come before that time, as we can and do think of keeping the fat cattle until the days grow longer, and the steers will eat few or no turnips from you compared with what fat cattle will take from you. You can tye up or keep the steers in courtains as you chuse, as we must tye up the 3 years old ones when they come here, those that are to work this season at least. Hope the miller now has water enough, as we have had great plenty of rain. I wished you to take out the ground below the Hall bridge sufficiently before you stop the beck or swallow. Steers were very dear sold at Hexham Tyne Green fair,[108] much higher that what you had bought. A man came to Wark for a bull calf bought of Mr Buston or some [. . .][109] Mr Buston's neighborhood, but we could give him no account of a calf [. . .] I hope you got the race cut out, or cleaned over the [. . .] Your obliged

Matthw Culley

[*G.C.*] John I know not whether you will quite understand my brother, but the matter is, as I said before, that we wish you to send no steers here but those that are 3 years old and fit to yoke, as we have 2 years old plenty, as many as we can well winter, and I should suppose that steers although only 2 years old can tread down straw as well as Kyloes. And we will keep the fat cattle or feeding cattle intended for you until Christmas or even longer if you wish it, as our turnips at every place but this are very good now. And I still think you will make a mistake if you tye up the nice queys which we intend for you, because if tied up you are obliged to sell before turned out, but if kept in folds or courtains you can sell them whenever the markets are best,

---

108 8 November.
109 There is a blot on the page here, affecting three lines of text.

and I think that cattle and sheep markets will be good in the spring, perhaps best in May or beginning of June.

Geo Culley

220.   *George Culley to John Welch*

Eastfield 21st November 1803

Well John

I wish you a better fair at Skipton than you expect, and I hope it may prove so, as I by letters from both Messrs *Sayle*[110] and Barf I find that Leeds and Wakefield fairs were better than had been a good deal, 7/6 to 9s per stone sink, and the last market at Wakefield 8*d* per lb.sink and better for sheep. The only thing against you John is that in the neighborhood of Skipton this *particular year* there are more fat Kyloes &c &c in proportion than anywhere else, owing to plenty of grass in that district when there was none anywhere else. And if we had received your first letter in time I would have wrote directly and advised you to *try Wakefield*, as I believe it will be found the best market for fat cattle, for the reason given above. But that letter as you say must have met with some delay somewhere or other. Indeed I do suspect that your letters frequently *are delayed* somewhere or we would receive them sooner than we do. Now if you send them by Piercebridge they may be delayed there, but hardly at Darlington, However it is worth your while to *investigate* this matter a little, as *letters* or their *contents* are most commonly of high importance. Now you say that letter was wrote on the *Friday*, and I do assure you we did not receive it until the *Thursday after*! Which was 7 days, and it ought to come from Darlington to Berwick always in one day and we get it the next morning. Now you know best, but if you send it to Darlington, or even to Piercebridge, on the Friday before the Barnard Castle post returns, it would be at Darlington still on the *Friday evening* and leave Darlington next day (what time I know not) but in certain time to reach Berwick the same evening. And we almost always have someone or other past Tilmouth every day. However that week we had one every day to a *certainty*.

As a proof of the other being *delayed* or *stopped* somewhere, the letter we have received from you this morning is most *notorious*,

---

110 In this letter of 18 November (ZCU 25) Benjamin Sayle reported that sheep had been plentiful but were now fetching better prices. Cattle had been rather plentiful at Leeds and Wakefield, selling from 7*s* to 8*s*. Corn was looking up except for wheat. Cattle were still going south to Norfolk and Suffolk.

because it is dated the 18th *Friday* and we have received it on the 21st *Monday*, which is exactly as it ought to be. The other letter I sent to Wark by Ralph last Friday morning, so I cannot answer its contents as my brother did not return it. But I will answer this directly. First I thank you for your very accurate account of stock, crops and rest. We will keep you the fat cattle until Christmas or later, which at any rate can be settled when you come over, and which I still hope you will contrive to do before Christmas as we want certainly very much to see you. The steers to draw we can do without until then very well I dare say. The main thing is that you have never yet said which, nor how *many cattle* you can take from us. However the longer they are a-going, the more I hope you will be enabled to take. Or at any rate you can be better able to calculate upon your turnips &c &c, and perhaps a more unlikely time than this to drive cattle may not take place, because we have every reason to expect very changeable wet weather at this time, and this very wet morning, with many changeable days lately, is a proof of it. Besides the days will be as long in the beginning of January as now. The only thing I say is that we may have tied up some cattle that would have suited you. You certainly are the best able to judge of the value of your own time, and we must leave you to come when most suitable to you. But remember one thing, as your master said lately, that you and Bob Dippie cannot both be absent together well. I think that Joe Coarl will not be very keen of going to Skipton again very soon any more than Jemmy Thompson to Woolpit fair. I have no doubt now but both cattle and sheep markets will mend now, provided Bonaparte don't prevent it. He has put of his visit so long that people begin to have more confidence, and even to think that he will not invade us at all. I own I am not so sanguine. He will come if he can, but the longer he is of coming the better we shall be prepared. By a letter from Mr Barf of 17th inst. he says 'that at Leeds and Wakefield great fairs they had a very small show of fat oxen and those very indifferent, what were good were sold at 9*s* sink but more at 8*s*. However these are very good prices for this time of the year. Sheep at Wakefield full 8*d* sink if not better'. A letter from Mr Sayle corroborates the above, particularly in regard to sheep markets. Mr Sayle does not speak so high of the fat cattle at Leeds &c, but both agree that few are shown and few of those are good. I am glad you have sold so much corn to Monkshouse. I think your oatmeal will sell well enough *if good* by and bye, because to a certainty oats are and will be *scarce* and *dear*. Archer horse we will consider about.

Now for your first letter. I am glad you bought steers, I am disposed to think they will pay as much as dear Kyloes even upon your moor laid thin on. Kyloes were sure to be sold dearer at Midlam Moor than anywhere owing to the plenty of fogs and rough pasture in the district of Skipton and westward. I hope you have bought the steers and stots &c worth your money, and beseech you to keep them all through winter except those fit for yoking. The stirks or stots will run nicely in your moors next summer too if not too thick laid on, and you will be able to judge whether they or Kyloes pay better and then hold to that which is best. The 5 steers 3 years old and quey will do for us, at least we will make them do. Will your Kyloes stand quiet and do well enough tied up? I wish they may, and then you can keep our queys &c in open folds or sheds &c &c, because I still think they should not be tied up on any account. I am glad that you only had occasion to borrow another 50£ of the bank. I hope both you and we will get straight by and bye. We are still short of cash, but I hope we will do after a while. We are much obliged by your attention and anxiety about my son. I believe they are not to go to Newcastle at any rate now, and I would not have consented to his going to Newcastle willingly. John don't *neglect* to bring the account from your mother for George Thompson when you come over. Be sure don't forget it. I have a particular charge from your mistress here for you to do so. You know she has a great wish to pay when and where due. Yours in haste

Geo Culley

PS. I hope you will write immediately according to your promise, as soon as the Martinmas day fair is over. This is a thorough wet day, I doubt you will feel it at Skipton more ways than one.

221. *George and Matthew Culley and Matthew Culley jr to John Welch*

Eastfield 28th November 1803

Well John

We have this moment received yours of the 26th which is just as early as it can possibly come to us. It is not at all impossible but the letters may lie at *Stawards* Berwick some times, as we are often disappointed of our London paper by Staward. However I will examine Staward the first time I am at Berwick, which will be on Saturday first I hope. Bad as the market was, I am glad you have got through your cattle, and you may depend upon it that when a person profits *once* by setting up *stock* he loses 3 times. At least I always found it so, and I have had a good deal of experience and far of markets like Skipton are still worse. To bring home from Darlington is nothing. You acted wisely and properly in selling your 4 large cattle to Peacock, and it shows that I was right in suggesting that at this time Wakefield was much better than Skipton for fat cattle, particularly for large cattle. But that letter of yours lying so long prevented me from giving you that advice in time. There cannot be a doubt of fat cattle as well as sheep advancing everywhere in this Island by and bye, but in the west of England the last, for this obvious reason that they had grass when we had none in the north, and all the east and even south coast. By all means begin selling your fatt ewes John now that prices are got to 8*d* sink and better in Yorkshire. Not that I am very fond of your taking them to Skipton at this time of the year, if you can sell them to either home butchers or Wakefield or Skipton jobbers, and the sooner you can begin the better. A reason also occurs to me why you should sell a good many of your ewes together. It is a well-known fact that all sheep fed upon rape die badly in tallow. We have the fullest proof of it this very year from a few I sold in this neighborhood which prove exceeding ill indeed. I don't wonder at all at wool advancing, nor have I any objection to your selling your piece. Only wool eats no *turnips* but ewes do, and I would advise you to sell your ewes directly. Should they (sheep) keep advancing, sell your wethers also and we will send you more, either ewes or wethers, as sell something we must, or turnips will fall short with us all. I want very much, I do assure you, to begin to sell either sheep or cattle, but sheep are lower by much here than with you. Consequently I repeat, sell as soon as you can ewes or wethers *or both*, and in quantity if you can. *Offals* are certainly a vast encouragement to butchers, they *are* and *will* I think be *high*. If you cannot sell your ewes at home at 8*d* sink I am agreeable for you to go to either Skipton or even Wakefield, which last place seems to be the best market at present. We got on Saturday 5/4*d* for pork, and advancing. Certainly Mr Peacock has sent the 700£ to London before this? John I think this and December the very worst months to sow wheat in. *January, February best*, and even March to the 20th better than December or end of November. I think oats will keep their price unless barley hurt them, and barley must fall in price or I am mistaken. This is a sweet day, the snow wasting on Cheviot for the first time this autumn. Yours in haste

Geo Culley

PS. Before Escomb is let to Brown or any person you must act in conjunction, or rather Mr Peacock must do the business as he acts as our agent in letting our lands in your neighborhood. John the Property Tax[111] is to be paid in every place where due, and as you

---

111 Pitt introduced the first income and property tax in 1799. On the renewal of

may not quite understand it we would have you apply to Mr Peacock to lend you his assistance, as he is more conversant in these things than you or us. You will have to pay 1s for us as proprietors and 9d as tenants *per pound*, distinguishing between my brother's own property and our joint property. I shall write to Mr Peacock in a day or 2 on this account also. He or Mr Graves will pay the tax for the houses in Darlington, and the tenants at Sandforth Moors and Escomb will pay their respective taxes of 9d per pound as well as Joseph Hodgson at Denton. Yours again

Geo Culley

PS. Pray write again soon.

[*M.C. jr*] The tenants pay 1s 9d in the pound and deduct the 1s from us when they pay their rents.

[*M.C.*] Well John. You should have a knife as the cloggers have fixed in a clog to cut your Sweedish turnips in 3 or 4 pieces for your horses. We think you should take a horse to Matty to bring him at Christmas and come north with him, it is a vacant time and suits his coming home and he will like it. This you must fix with Matty. Bob Dippie may come at some time else. May bring the plans with you. You should get those pieces in Sir Ralph's farm surveyed by Cumming, which you suspect to be wrong surveyed. Matty and you must call on Mr Grieveson's chamber maid. I left my whip at Mr Day's where we slept all night. She is a widow and lives in a street turns from the foot of Westgate.

Matthw Culley

222. *George and Matthew Culley to John Welch*

Easttfield 11th December 1803

John

I was very happy to receive yours of the 9th inst. this morning, which is as early as it could possibly come. We sent you a letter from Berwick by post desiring you to send Bob Dippie with Matty if you could not come then yourself, and many other instructions copied from a letter sent me by my brother, which I hope you will receive in

---

war Addington brought in a new act in 1803 (43 Geo III, c. 122), taxing income from land including houses, farming, professions, businesses, public office and the funds. The rate was 1s in the pound for incomes of £150 and above, abatements for incomes between £60 and £150, and exemption for incomes below £60. Occupiers of property paid 9d in the pound on the annual value for themselves in addition to the owners' 1s. The rates were raised in 1805 and 1806. Dowell, *History of Taxation*, vol. 3, pp. 92–102.

due time. But if the mule go for Matty he will do well enough. I am rejoiced to find that you showed and sold your half dozen Kyloes so well, I only wish you had taken more, not but they will be well enough sold I hope now. But you have sold your pigs very well I think, at least much better then we could. We only got 5/3*d* yesterday. And I hope you will sell your ewes by and bye at 8*d* sink. Indeed I would not advise you to sell your ewes below 8*d* per lb. sink. It is not a bad price for your wool, but if Bonaparte keep away you will sell higher by and bye. By all means show and sell the remainder of your Kyloes at 8*s*. I am very well satisfied with the price, and we must be selling something as both you and we will likely fall short of turnips. What a blessed strength to have such a nice cover of snow to save our turnips. In the neighborhood of Edinburgh there is no snow but a most bitter frost. John you are quite mistaken about barley, you may depend upon it that barley will will [*sic*] be much lower, and if you have any to sell I would recommend to you in the strongest terms to sell at 30*s* let alone 32*s*. We sold a good deal from Wark at 20*s* and 21*s* per boll, although we shall need to buy afterwards I dare say, but I never was more convinced that barley will be low sold than this year from its being the most plentifull crop of any other grain. I have all along said that barley would hurt oats, or oats would assist barley. And it is evident that barley has hurt oats in this country, that is our farmers mostly eat barley with their horses and sell oats, which has made more oats come to market. Barley got up a little while here, but it is now down again several shillings per boll, and depend upon it will be the same with you or I will be more mistaken than I ever was in my life. Indeed every grain is as flatt as a flame here. The highest price for wheat on Saturday was 35*s*, barley 15*s*, very good if it reached 17*s*, oats that are good are still worth 17*s* or 18*s* or even more. But I have as little fear of oats dropping much as of barley advancing, for that very obvious reason, barley is the greatest crop that perhaps ever was known, and oats the worst. Besides there were no old oats left in hand scarcely, but an abundance of barley if I am well informed. I shall leave you entirely at liberty to do as seemeth best to you about your wool. 19*s* is certainly far from a dispisable price, especially to people in want of cash, but we shall get through our payments fast I trust, and as you are selling cattle and may sell sheep soon, I hope you will also be able to pay your bankers soon, and the wool will not be lower I think but generally advance in your neighborhood about Candlemas or towards spring.

G.C.

[*M.C.*] Well John if you can spare Bob Dippie a horse of any kind he may come with it. This I leave to you. We had a letter from Matty desiring to know what way we would send for him, Nelly had wrote to him, which he has got ere now. You must write to him and send the mule and spare Bob a horse if you can. This I leave to you to do as well as you can, as we shall not write to Matty again. I hope Mr Peacock and you have settled the barley account and that he has sent the money to London. When Bob comes away someone must attend the water without fail as this is the main season for watering the land. There should be a new dam made above the conduit in the tenement that goes into the glebe, then the spare water can be run over that high part on the west side of the tenement. Also I hope you have stubbed up the hedges and bushes on the east side, then it may be watered with the spare water &c. The carrier or hedge gutter will be to lower &c. Also the planting in Fitts Howl should be filled up with trees, and the pasture howl &c and the old and new clowers should be put right, also the water should be dammed and run over the Stubs's Carr as well as you can, and wherever you can by stopping the gutter on the west side which Tomy Hutchinson was cuting away the thorns, also the runner in the east side of Stubs's Carr to the best of your judgement. Bob should have leisure to shew that man before he comes away, indeed his time should be adapted to watering the land. From your well wisher

M. Culley

223. *Matthew Culley jr to John Welch*

Eastfield 21 December 1803

John Welch

I have received a letter this moment from Mr Marshall of Newton who has two tups from us, and one his neighbor has, and he says that his cart will be at Northallerton next Thursday the 27th instant, when you must send to meet him, and after we hear from you that they are arrived, and that Lord Darlington's tup and them are ready we will send for them. Perhaps Mr Marshall will only send 2 sheep as he talks of keeping one of them another year. If that is the case there will be only two to come from him. I hope you will get this before you go to Skipton, but for fear you might not I have directed it to Bob Dippie also. I am in haste Yours &c

Matthw Culley

I hope in your answer you will say when you are to be north.

224.   *Matthew Culley and Eleanor Culley to John Welch*

Wark December 26th 1803

John Welch

Well John it is at last agreed that the fine colt by T. Gregson's horse is to come to you to be broke in by Frank, which should have come ere now. He rears very much and is full of spirits. I am truly sorry for your sister's death.[112] Matty got well home on Wednesday, but was very wett and daubed so as his great coat and saddle were both to wash. Bolam would not send a chaise over Rimside Moor the night they got there, with the 2 Hogarths, young Air and 2 Baileys, so Matt and them stayed all night and found Rimside sore drifted. With daylight he out-travelled the chaise on his mule. I am afraid you forgot to send the plum tree wood to Newcastle for Mr Bates when you sent for the barley mill stones. I now think you should put in a sluice or dam in the well springs immediately below the ford. Perhaps it should have 3 or 4 boards to lift up, and be set on broad good limestones, then well walled with limestone side stones or check stones from Belwood from Houghton bank, I wish it well done, but Richardson will best judge whether it can be done water tight with limestones or sandstone best. If not done this season I fear the field will grow little or no hay. If you think it too late in the season must refer it till next year. But I wish you to set Bob to attend to the watering as his sole business all winter, if not it disapoints all my reasons for sending him to Denton and it's not certain that he may continue with us another year, and we may have difficulty to get one who knows so well as he does how to lay out the land for that purpose. The well cores, pasture howl, and tenement should all be done, as well as the dam in the tenement above the conduit that runs over to the glebe, which will enable Bob to water the west side of the tenement which I have given Bob some directions about. I have also told him how to turn the water on Stubs's Carr when there is more than the mill wants, that is the superfluous water. I fear you have not sold the oat meal. Oats here are fallen from 21 to 16 some lower I am told, and barley has fallen much. Ralph Brown sold 100 bolls of wheat at 35*s* per to Mr James Forster. He did not sell our wheat last year, and now it will not sell, why no one will buy it. I believe there is more lost by speculation than gained, and we have been whipped for money and hardly yet can pay our way. I am a hard seller for want of good

---

112 Martha Welch died 15 December 1803 aged 23. She had married a neighbouring Penshaw farmer, William Pearson, the previous year. DUL, BT Penshaw.

judgement, but better we sell than we keep. Wishing you many happy returns of the season I remain

Matthw Culley

[*E.C.*] My father desires me to say that he wishes you to be selling your stock when you have an opportunity, as we are afraid of running short of turnips here.

E. Culley

225.   *George Culley and Matthew Culley jr to John Welch*

Eastfield 27th December 1803

John Welch

This will come with William Wright with a cart for the tups, and I am to desire you to send *the skeels* for my mistress if they are ready. William also and Bob Dippie have the mischievous horse, as he is called, along with them, which you are to send to the horse breaker *Frank*, or any other as appears to you properest, to break him in so as to be quiet. Although he is called *mischievous* I really and truly believe that he has no real *vice* about him. He unfortunately had been very badly managed at the first by Thomas Thompson of Grindon, who he *threw* off once at least if not oftener, and then I suspect Tommy had used him harshly. However I have no proof of this but only suspicion. But what makes me suspect it is that the brute now wishes to get quit off every one who mounts him by throwing them off. Nevertheless he does no more than jump up and down, setting up his back and rearing sometimes, but what he wishes most of all is to make off as soon as he throws his rider, or at all times whenever he has the least *opportunity*. If the stable door is open while they are putting the bridle on him, he will bolt out if possible, and when he is off he does not run, he flies and *bounds* like a *deer*. He is wonderfully mettled, and a horse of seemingly uncommon powers, and I have not a doubt but he will make one of the first hunters in the north of England if he could only be made *tractable*. But I must observe that he threw the man we got to break him, and got from him at different times, and we have reason to think that the breaker became affraid of him. However he foolishly put the stable lad upon him just before he, the breaker, was going away, and he threw the boy in no time. Then young William Brown mounted him unknown to me, and has now rode him 3 weeks or a month without ever being thrown by him although he tried many a time and oft, and although William B. thinks he would have got the compleat master of him in time, we thought it right to send him over to your country at this time, having so good an opportunity by William Wright and Bob Dippie. Indeed yesterday while William

Brown was doing something about his head and not suspecting, he bounced *back* and *out of doors*, and then there is no turning him. For I believe he would run over anything, *hedges, gates* or *rails* he goes over like a swallow. But by turning out my grey horse he was got into the stable again, and this determined us to send him at all events to you. But I think it right to give this history of him, which I believe a *true bill*, however I beg you may ask William Wright about him although I do assure you I have told you all that I know or have heard about him. Not but Willy Wright may know some things of him that I do not. It is evident that the horse, by throwing everyone who had mounted him until he came into William Brown's hands, began to think that he could master everyone. But I sincerely believe that he never did throw Willy, but made many a serious attempt. But Willy declares that he was become much quietter lately, and had he not got out of the stable yesterday I have reason to think that Willy would have completely cured him. But there was such a *hubbub* with all the women of the town being out, and Peggy Brown, William's wife amongst the rest and she big with child poor thing, decided us in sending him away. However Willy Brown got upon him yesterday after he was got in and demmed[113] again, and although he reared and played all his pranks he could not throw him. On the contrary he rode him until he was quite *quiet*, before he got of him. But I promised his wife and Ralph Brown that Willy should mount him no more. He has been leading him with another horse since that, both yesterday and this day, but I hope there is no fear but he will lead well enough, even with quiet Bob Dippie.

John we want much for you to come over. Matty has all the annual accounts done now but yours, but can't get finished for want of yours. You don't write. Why did you not write by Bob Dippie? Suppose Darlington was a bad day, there was the more reason for you to write about it. It will be a bad job if prices of fat stock don't mend soon, as we must begin to sell soon. I had forgot to say that the horse is the quietest animal in the stable that possibly can be. You may rub him as you will, how you will, and when you will, you may in short creep through between his legs or under his belly. No horse can or need to be quieter. And if he had not been spoiled at the first would have been as easy broke as any horse whatever. Wishing you many happy returns of this season I remain yours truly

Geo Culley

---

113 Shut in.

John after all I have said, I must beg that you will not attempt to ride the horse yourself upon any account, as you may depend upon it that he is a very dangerous horse, especially to strangers. Nor should anyone offer to ride him until the horse breaker has got him made so quiet and handy that he can be rode with safety. And I think it would be wrong for *Frank* to be hurried in breaking him, and we wish him to be well paid provided he succeeds in making him tractable.

The weather has been very wet of late but fresh, and much of the snow is come of the hills, and the rivers have been much swelled, and a bad time for outlaid stock of all discriptions. However turnips have kept quite fresh although they are said to eat fast, which makes us very desirous to be quitting some fat stock whenever an opportunity offers. It had been very lucky if you had sold all or a part of your ewes when the market was 8*d* per lb. sink. I don't say this to blame you, but only to *impress* you with a thorough conviction of the necessity of quitting the very first opportunity that offers. We never had so many fat sheep in our lives, and we are pretty well for cattle. But we never had anything like the number of sheep, and as your markets are very likely to take the first rise I would not have you on any account ommit any opportunity of selling. We will take care to supply you, and in truth you must be convinced of the *urgent* necessity of selling whenever you can. I see no probabillity of our markets getting up here, therefore there is the greater *necessity* for you to sell whenever a *chance arises*. Never mind what it is, sell when you can get a decent price, ewes, wethers, or cattle. And if they should get even much dearer afterwards, we have plenty to follow with. Better sell a part *midling now* than worse *afterwards*, and if severe weather should come further on, which is a ten to one it will, then people will hurry to market. At least that must be the case in this country in my opinion, where the turnips are going now far over fast already.

[**M.C. jr**] Well John, my father has said so much about the disposition and temper of this said horse that I can say little more than inforce the necessity of neither you nor any other person mounting him until the breaker has got entirely the master of him and he is become perfectly quiet. In regard to the other points I have to add that there never was a quieter horse in a stable, any person may rub him or take care of him, and all that ails him is that he has got such a fright by some means that he is determined to be quit of every person that attempts to mount him, either by foul or fair means. And that I think is his only fault, which he certainly is very wicked in. He should certainly not be well kept, as he cannot bear high feeding. He has a small bump on his knee which he got when he threw off the breaker and fell by leaping over a hedge. We have used means to reduce it, and it is now quite gone, and all you have to do with it is to rub some sheep suet on it down back to bring the hair on again. However if it should seem to grow larger you can reduce it by rubbing oil orignum[114] on it but sew nothing else. The hair was clipt off his knee. He also is chafed between his forelegs, which may easily be taken away by being rubbed night and morning with the white of an egg or with fuller's earth made moist with water.

Wishing that you may succeed in getting this devil made quiet, I am yours &c

Matthw Culley

PS. I hope you have got some acorns to send by the cart. You may send Matthew's grey hound with the cart.

1804

226. *George Culley to John Welch*

Eastfield 5th January 1804

Well John

William Wright arrived here last night all well, and I was very much rejoiced once more to receive a letter from you. It is such a favour to receive one. It matters not, summer or winter, Sunday or Monday, you never can spare time to write hardly. Would you only be advised and write letters when you have 5 minutes to spare. But you are like all those [. . .]¹ used to have much of their own way. Had you been as often *whipped in* as many I could name you, you would have been more open to and easier to be advised. Well I am glad that you have sold your Kyloes, but as mortifyed that you have not sold your 1000 bushells of barley at a much better price than [. . .]² or any place I know off. Notwithstanding [. . .] selling your barley if you had any to

---

114 Oil of origanum, essential oil of marjoram, an ingredient in a blister ointment: James White, *A Compendium of the Veterinary Arts*, London 1802, p. 214; Thomas Boardman, *A Dictionary of the Veterinary Arts*, London 1805.

1 There is a blot on the page here.
2 There is a very large blot on the page here, substantially affecting 14 lines of text.

sell [. . .]. What a [. . .] unpleasant [. . .] you forget the old proverb [. . .] However I am not without hopes that [. . .] because you are possessed of a good [. . .] and the experience you have had in not selling your [. . .] and must impress you with the impropriety of not taking [. . .] when offered. Now when I think of it John [. . .] Eales that lives at Coldstream told me the [. . .] fat market is very much [. . .] season. It ought to tell [. . .] able to judge of it better every day [. . .] if you had kept your tup, as you would have got [. . .] of your neighboring butchers. Your mistress here desires me to thank you for your attention to her, and is very highly pleased with the two skeels, and my daughter Mrs Darling, who is now sitting by me, would take it as a particular favor if you will procure her two, and send by the first conveyance when any carts are passing. I sincerely wish that John Frank may make the horse quiet, and I hope he will. But I would have you not to be too forward in mounting him after that. Remember your life is of serious importance to your family! Frank should be with you if you are to ride him, and he must have full exercise and not *over high feeding*, as you may depend upon it that he is a horse of *wonderfull constitution* and great *comportation*, and if properly cared will make a most valuable horse. I am glad your ewes are good. Certainly nothing equal to rape for feeding a few ewes, especially of the Dishley blood, and I am glad that you are going to try Skipton. You are best acquainted with Skipton, and it is much nearer to you than Wakefield, consequently less expence and time taken up. But if Wakefield be found the best, I dare say you will have no objection to try it also, because if our markets do not get better we shall be under the necessity of sending you a good many sheep, as both you and us to run or fall short of turnips will not do. I assure you that our ewes are also good, and if George Nicholson would come to try us again we will use him well, he may depend upon it. It is likely that he would buy a few of you, and pray John why not sell or try a few shearings to Skipton along with a few ewes? I always find that we can sell a part of both at Morpeth better than all of one kind, and if we could once hear of mutton being got at 8*d* per lb. sink at Morpeth I will certainly begin with both ewes and wethers. Ralph sold 8 ewes the other day well enough to a Wooler butcher who pays and takes at 56*s* per or near 8*d* per lb. sink, but bless you these are wonderfully good, and what is 8? Only small fish are better than none, and we are glad to be quitting if it is only 8 at once. We have also sold 3 cattle to William Marshall and 5 to Mr Lilly, some at 7*s* some at 8*s* sink. Most of these only midling beasts, but we are glad to mix some good ones and get the bad ones off. But whenever we shall have the pleasure of your

company we have a great deal to say to you on many accounts particularly [. . .]³ fat sheep selling, as we have so many this year that I [. . .]
with a part. But freeing our cattle from [. . .] that we will be [. . .] the
markets are [. . .] and the large cattle at [. . .] improving fast. Notwithstanding [. . .] with the turnips, but I do [. . .] that [. . .] season never
was better, so very juicy and fresh as though it were [. . .] This I
attribute to the rains setting them a-growing so [. . .] then such a nice
cover of snow during the [. . .] and now although we have 3 or 4 days
[. . .] turnips are preserved with so much top as they have this year
[. . .] snow has fallen, and now is increasing [. . .] the turnips will not
suffer from frosts this season [. . .] there is no saying what may
happen. Only when turnips stay fresh and growing they don't easily
take harm. Our ewes eat the tops through and through as if it was
Michaelmas time. We always allow our sheep to have the *first fruits*,
or the first of the turnips, before the outlaid beasts go to follow them,
and I not only think it very great *economy*, but I am of opinion that the
cattle do as well as they would to go first to them, while the sheep do
half as well again. You see the sheep eat only the tops, and break a
part of the turnips, when the cattle are allowed to come to them,
while the sheep again are put to fresh turnips a little distance off, a
boy attending to keep them separate, which is both well worth the
expence and highly necessary except you have the benefit of two
fields, but what I suppose you and people will not do. I always make
our girls pull the turnips with their hands, and fill them same way
after knocking of the dirt or mold by *dadding* or knocking them one
against the other, which keeps them quite clean.

I had a letter from Mr Peacock telling me that the money was sent
to London, and I will thank you to tell Mr Peacock that I received
both his letters, as it will save me writing and him postage, and I have
nothing of consequence to write to him about. I have no doubt but
you will pay of the bank as soon as you possibly can, and I am happy
to tell you that we are now well up with our payments and will soon
be clear until rents put in again, a thing you are clear of. Nor have I
any doubt but you will do right enough with your wool. That article
is not sinking but advancing, if my information from Mr Deverell be
just, and he is a steady clever man and never once deceived me, and I
think I told you that he sold his lately as nearly as could be 14*d* per
pound, and being a tup man his wool is not so good as yours. Indeed
I consider yours as a particular nice piece, no tups and few ewe

---

3   There is a very large blot on the page here, substantially affecting 16 lines of
    text.

fleeces. And you may say that neither it nor barley beat either turnips or hay, but barley is in every part of this Island sinking in value and will waste partly by keeping in the loft, while wool will [. . .]⁴ little or none. Not that I wish you to keep it, No No! I am always for selling John at fair prices. I have seen much loss by keeping stock and corn, but seldom any loss by selling at fair prices. Many a time my brother has told me that I had a *soft spot in me* and was a cheap seller. But he has as often honestly confessed that he believed things would never have been got through or *sold* if it had not been for me. By the bye, it is a very foolish way for people to sell wool or anything by another man's prices. I have known it done by our prices here, but always decryed it down. It is a very foolish way and totally wrong. Grain is certainly very dull everywhere and very likely so to be, because Government are determined to keep down prices by every means that they legally can, and for wise reasons. We are selling now at 34s *wheat* our boll, and gets lower every day with us. It is certainly wise to sell yours in flour when you have it in your power, which is the case in the winter I suppose, and I have no doubt but you will overcome all your enemies in that way very soon provided your flour is as good as others. It is a vast affair, that you can by that means save your oats. We must consider about the quicks when you come over. As quicks are so very high here, I fancy they would pay for a cart coming on purpose. But as I say that can be considered off. We certainly wish much to see you on many accounts, but you must take the best opportunity for yourself certainly. One of the things is, Matty can't get the annual accounts settled for want of your books. And would you but write, but you promise so to do, but forget to perform! If George Nicholson can sell his mutton at or near 7*d* per lb. he can well afford 8*d*, because the offals never were better sold. They are well worth 2s per lb. overheads. Wishing you and yours many happy and peaceable new years I remain yours truly

Geo Culley

PS. Is anything said about how Crozier Surtees⁵ has left his affairs or what state they are in? Is his wife coming back to her deserted

---

4   There is a small blot on the page here.

5   Crosier Surtees of Merryshields 1740–1803, a first cousin of the Culleys, married his (and their) first cousin Jane, daughter and co-heiress of Robert Surtees of Redworth. They became estranged, and she left him in 1800. He was found dead on 31 December 1803 at a ford, having apparently fallen off his horse on the way home from a dinner at Raby. Surtees, *Records of the Family of Surtees*, p. 98.

family now he is dead? One of the cattle sold to Mr Lilly is a bull, another a bad ox, and a bad cow, 3 out of 5 bad. But we have some spayed queys that I hope will please you still.

227. *George Culley to John Welch*

Eastfield 14th January 1804

Well John

Yours of the 8th inst. came in due course, and only shows when you are disposed to write how soon communications can be made. And I really wish that you may have had a decent market at Skipton. A good one I could [?not][6] expect in these critical times. Not but I think the fat stock will be all ways yet before June. Perhaps pigs and cheap grain may have an effect upon fat at present, but if beef be wanted and scarce, it will have an effect upon mutton by and bye. However I am very glad that you sold George Nicholson a few, and at plenty of money, more than you would get per pound at Skipton I am affraid. But I beg that if George Nicholson have cause of complaint that you may make him amends.

John I rejoice that you have so much barley, but then I lament that you will not be prevailed upon to sell it at a higher price. More is given in many other parts of this Island. I shall [say][7] no more than that you may depend upon it you are wrong. Otherwise I will be glad to be mistaken. What does it amount to, that barley is scarce, and consequently dear with you? It can be had at a great deal less either from the *north* or *south*. And as to giving them a year's credit, if the money only be safe we can afford to do it and the price of 30*s* will require it. I am disposed to believe that you will be obliged to take some shillings per quarter less, or keep it, and the remedy is worse than the disease. But it matters not John, you are determined to do your own way if you suffer for it as well as your master. Not but I am convinced that you intend right, because you persuade yourself that you are right. At same time one would think you would like to do as your masters advise, because in that case you steer clear of blame. In the other way, if you lose money you both mortify yourself and them. However I am glad that you are determined to sell your sheep. They certainly cost more keeping than corn. You should not be angry with Eales, because I dare say he had no other meaning than for any information, and knew not that I would acquaint you. If Bonaparte keep

---

6   A word seems to have been omitted.
7   A word seems to have been omitted.

away the wool must advance. Not but I think it would be unwise to keep wool at these times. I am sorry that you did not keep the tup for many reasons. As the Archer colt is now with you, you can do to him as seemeth best to you. Only allow me to advise you once more to be cautious for your family's sake, as I am perswaded that he will not be easily got the better of. I sincerely thank you for your rebuke for not writing single upon the last letter, because I am to blame.

John I have this moment received yours by post, and will answer it and send away tomorrow per post again, so that you may get it I hope at Darlington on Monday. I did suppose you would have a terrible journey to Skipton, but a bad market was still making worse. However we have had very many good ones, therefore must expect bad ones sometimes and be thankfull. These things are good for us! You were certainly right to sell, let prices be as they would. It would seem as though sheep are to be ill sold at present. However my opinion and wish would be to sell cattle and keep sheep, for two reasons, 1st because cattle are now decently sold, and 2d because sheep consume far less turnips in proportion. Cattle butchers never had such a time in my remembrance. It is as good as the farmers had in the dear years of corn. Morpeth was an exceedingly bad market on Wednesday, and many goods of both kinds set up, *7d* sink for sheep and 7/6 for cattle, I am sorry that you still postpone your journey. However we must wait with patience until you come. If hay sell at 10£ per ton! You will never come at the price you have been bid for barley again.

I thank you for information respecting Crosier Surtees's death and family and circumstances. If you hear any more particulars I will thank you for them. Are you correct about the widow cohabiting with another man? The snow went away very nicely here, and we now have very fine weather and turnips growing nicely, and not now at all likely to rot. Our corn markets are now so bad that we have quite given over selling anything, Barley in particular is not saleable hardly. If you wish for a cargo or 2 sent to Stockton, we can buy you plenty very decent at 20*s* per quarter. John take advice for once and sell if you can have 29 or even 28*s* now. But I am affraid you are now too late. That you may amend your ways is the wish of both your masters, but particularly your real friend

Geo Culley

You have really behaved well in writing so often lately, and I hope you will be as good as your word in writing as soon as possible again after Darlington.

228.　*George Culley to John Welch*

Eastfield 21st January 1804

John I thank [you][8] for writing, we received yours of 19th this moment, and will lose no time in answering it. I advert to what you say about sheep being so badly sold at Darlington. They are still worse sold at Morpeth! There has been 2 shocking bad days, I fancy many good sheep were sold the last day at 8*d* and little more per lb. sink. Yet I don't despair but still hope for a good time if Bonaparte keep away, because the greater consumption now the better, and I am clear that we shall have no rot in turnips. On the contrary I think they will make top very fast in this country, which will make them go far, and the plenty of spring meat will then prevent people from hurrying to market. Turnips are already becoming plentifull, and people who are sellers are beginning to be affraid already, and many turnips are now offered to sell. We have a capital lot of oxen at Wark, which if we could sell in a month, to be then clear I mean, I do think that we could then keep all our sheep if necessary as long as turnips will be usefull to them. And I am not affraid of selling them because cattle that are tollerably good are readily sold at 8*s* or nearly at Morpeth, 8*s* per sink. I am told that Ratcliff is rather in want of good cattle, and I wish Ratcliff to have a refusal of ours as he bought all ours last year. By a letter from Mr Barf the last week he says 'For 2 market days before this sheep could hardly be turned into money, but were much better sold last Wednesday and a good demand. Ewes from 7*d* to 7 ½ and wethers good to sell if fair mutton at 8*d* sink, beasts 7*s* to 8*s* very nice 8/6*d* sink. Corn worse every day, wheat 7*s*, oats 2/6 to 2/8'. He does not name barley, nor will I say more to you than that I still think you wrong. I am glad that your mill answers so well, and to be sure you are a man of spirit and exertion. No doubt but your mill in the winter will do an immensity of business, perhaps consume your whole crop or nearly. If so what a saving in oats by eating the dust of barley and beans and pollard of wheat &c. Go on and prosper. I am glad that you sold your wool, and I hope you will very soon be out of debt. I rather think that our weather has not been quite so wet as yours. I never knew finer weather than at the present for the season. We could sow wheat, but I would rather do it in February for many obvious reasons. Come you must John, but come certainly when it is most convenient to you. We must talk about Angus oats when you are over. I will make enquiry of our men about them. I have wrote for some from Angus

---

8　A word seems to have been omitted.

but it is very uncertain when they may come. A few of those pease might be right enough to try here, but this is not a country for selling boiling pease. I am happy to hear that Frank thinks so well of the Archer colt. I don't doubt that he might be too harshly used, and I am glad that he uses him different. My brother will be rejoiced to hear that Bob manages the watering well. I assure you he, my brother, has nice *sport* at Thornington haugh. Nothing like plenty of water. Now my good lad do continue to write, and you will much oblige your real friend

Geo Culley

PS. The late wet time has hurt turnip outlaid stock much, but these fine days will mend them again. Tom Nisbet of Rodden is marketing some famous ewes 20 to 24 lb. per quarter, and sold at 8*d* per lb., 44 or 45*s* the most. But everyone thinks, and the markets confirm it, that fat cattle are not plentifull. I am told that Ratcliff bought the last Mr Taylor had very bad, and bad they must have been at no more than 12£ per. Our ewes are remarkably good, and full of tallow. The butchers allow that some skins are likely to be loose. I can hardly think so myself. We have not heard of more than 6*s* here and of our own price too. What a money butchers must make now! They have as good a time as farmers had when corn was so dear. I fancy they often make more per head than the offals, and the offals in good sheep are worth from 16*s* to 20.

229.   *George and Matthew Culley to John Welch*

Eastfield 24th January 1804

John

The letter of 19th from you we answered immediately that day we received it, which was 21st. But I forgot in my hurry to remark that it appears to us that you sell your wheat flour badly. You say '20 stones at 37*s*'. Now Ralph Brown says that the Berwick bakers sell the same weight at this time at 44*s*. How this is, perhaps you can explain but we can not. Ralph says the bakers and millers here always calculate upon 5*s* per boll of 6m over the prime cost of the wheat. *Monday 30th.* As I have not heard any more from you, and not having much to say, I have not wrote anything of some days. Mr Nisbet told Ralph on Saturday that the market on Wednesday at Morpeth was very good. He got 7 ½ per lb. and asked no more or he might as well have had 8*d*. Cattle were sold at full 8*s* sink or better, but not 8*s* 9*d* as newspapers said. There is no relying on newspaper intelligence on these matters, but on Mr Nisbet who was there you might. We have had an abundance of wet weather, very good as you say for millers and loaders, and I suspect that we are not quite done with it yet. Indeed such a remarkable dry summer was to be followed by a contrary set of weather in the winter or autumn, it is perfectly consistent and natural. I am disposed to think that markets will not be so bad for stock as was once expected, although they may not be quite so high as some late springs. If the pork had not been so very plentifull I believe they would have been a good way quit, and now pork time is wearing away, and the vast demands amongst our seamen and soldiers may have a wonderfull effect. Remember one thing, that our soldiers are mostly at home and in Ireland, and our fleets are victualled almost entirely in England and Ireland. There is a much greater consumption about Berwick than usual, which can only be attributed to a regiment of soldiers lying there, and more shipping than common. The Lyles of Berwick have now bought 200£ worth of cattle of us, 5 I named to you before and 4 cows on Saturday, and we expect the company today as they all called, 4 men who join, and buy and pay well, to handle a few ewes which we shall ask 7 ½ for.

We have this moment sold the Berwick Co. of butchers 60 ewes at 52/6 per , to take 20 per week, just 7 ½ per lb. according to our calculation. They take 20 away this very week, also a quey clyered at 29£. She weighs 50 stone.

[*M.C.*] Well John we have an uncommon wet weather, but we make use of the water. We never had our watered land so well saturated. We had a good crop of hay last harvest and may expect a better next season, as it's completely watered in most places, and hope to do the rest well, and are now swiming the turnips down the carryers to the end of the south one, to the road near Lambsknow. M.C.

[*G.C.*] John I would not have you *now* run away from any stock, except your turnips should run short, because I think I may venture to say that we now have turnips equal to serve all the stock we have without sending you any fat stock. Nevertheless I do think that we have a few cattle which will tempt you when you come over. Not but I think they will be well enough sold even in this country. I am now determined to send you this letter today by William Brown to the post office that you may know what we have done. Corn is not worth naming, but I would fain hope it has seen the worst except *barley*, which is lower every day. Yours

Geo Culley

PS. It is just as rainy-like as ever, but perfectly fresh and warm, and our dry land grows grass surprisingly, and I should think your dry and watered land will have a chance of being very forward in the spring.

PS. I hope I shall hear in your next that you have sold your barley even if you have 2s per quarter. If you could say to within 2 or 3 days when you think you may be here, it would be better as we could continue to be in the way better.

230. *George Culley to John Welch*

Eastfield 1st February 1804

John

I sent you a letter the other day Monday that we had sold 60 ewes at 52/6, also a quey. But Ralph Brown the same day went to Wark to meet William Marshall of Kelso, and sold him 3 more cattle, 2 steers at 20£ 10s per and a very bad quey at 14£, 55£ the whole. The steers were no great things. This is a charming drying day and I would gladly hope that we shall get some settled weather soon to enable us to get some wheat sown. *Thursday 2nd.* I have just seen George Thompson on his return from Morpeth, says the beast market was not so good as last week but the sheep were very well sold. *February 3rd.* A great deal more rain fell yesterday evening, and the ground is now pretty well *saturated* as brother Matthew says. From different people yesterday at Wooler I heard that sheep were well sold, full as well as the week before, 8d per lb. sink for good sheep, but cattle not so well, hardly 8s per. We had such an alarm here yesterday as set all this corner in a *hubbub*. William Selby of Biddlestone, a very worthy gentleman who is captain of one of the companys belonging to the Cheviot Legion, had requested *Waterman*, the landlord at Whittingham, to send him *without fail* the very earliest intelligence of an invasion. Waterman on Wednesday afternoon asked the guard of the Charlotte coach what news. The fellow was drunk and answered *Oh!* says he, *40,000* French are landed. Poor Waterman without further enquiry sends an express to Mr Selby, Mr Selby to Mr Collingwood of Lilburn another captain. Mr Collingwood himself enters Wooler a little after 2 o'clock yesterday morning, and with a most vociferous horn awaked all Wooler. The drums beat *to arms*, the trumpets *sounded*, the old women *fainted*, the young women *wept and howled* for the loss of their *sweethearts, husbands* &c &c &c, the dogs *barked*, and in short poor Wooler was never in such a state before! Mr Collingwood despatched an old stupid fellow to Cornhill, Crookham, Brankstone &c &c &c that they must come away, the French were just at hand! In short carts were yoked and the soldiers in them in no time and away they went. The first of Lord Ossulston's troop that arrived at Chillingham rather flustered *Philosopher Bailey*. However he (although he had heard nothing before) soon began to suspect that the intelligence seemed doubtful, and he and Captain Reed (Mr Bennet is in London the commander) despatched an express to the Duke of Northumberland[9] like wise men, and stopped Lord O.'s people until the messenger returned. His Grace's answer was 'That he was highly delighted and gratified with the *zeal, promptitude* and *alacrity* of the different volunteers, but begged they might return quietly home again as no invasion had yet taken place'. And I assure you that nothing could show more readiness than the soldiers. What is very remarkable, the south of Scotland had been equally alarmed only the day before, and the volunteers were in arms and marched to Dunbar most of them. This was occasioned by the beacons on the different hills being all lighted about one o'clock on Wednesday morning, but it has not yet been made out who fired the first *beacon* or what occasioned it. But it was naturally supposed the enemy were upon the coast, and Sir John Riddle of Riddle sent an express to Mr Bailey to know if the beacon at Ross Castle (above Chillingham) was lighted or where the first fire began. But luckily the beacon on Ross castle had not been lighted at all. Where the first lighting took place may perhaps *now* not easily be made out.[10] Be so kind as show Mr Peacock this letter which will save me the trouble of writing to him and he postage, and tell Mr Peacock that I received both his kind intelligent letters. However we all slept very sound last night and had no unpleasant dreams. But I do assure you, it is very pleasant to reflect with what *readiness* and *good spirits* the soldiers turned out, both rich and poor, although it was a false alarm and did not come in the regular notice that it should have done. Nobody was to blame but the drunken rascal of a *guard*! It was laudable and praiseworthy in both Mr Selby and Mr Collingwood, only they were too precipitate perhaps, as they should have had fuller information and in a more regular manner, and the poor animal Mr C. sent from Wooler had no letter and could scarcely *tell* why he *was sent*. I saw him myself on his

---

9   The Duke was Lord Lieutenant of Northumberland, and therefore responsible for home defence.

10 According to the press the alarm began on 31 January in Berwickshire, triggered by the sighting of a fire, possibly the burning of whins, being supposed to be a signal beacon. Beacons were then lit all along the coast, and troops, both militia and volunteers, were called out to Berwick and other centres. The report, and consequent calling out of troops, reached as far south as Durham. Of the Newcastle papers only the *Tyne Mercury* gave an account of events in Glendale. All the papers commended the alertness of the brave defenders of the country. *Edinburgh Courant*, 2 Feb., *Newcastle Chronicle*, 4 Feb., *Newcastle Chronicle*, 4 Feb., *Tyne Mercury*, 7 Feb. 1804.

return. Old Col. St Paul is confined to his bed. I saw him yesterday and he thought with me that it was all nonsense before we knew the mistake. His son[11] who commands the Legion was in Edinburgh, so we wanted both the Glendale commanders.

Saturday 4th February. We had a slight grinning of snow fell last night in the evening. I heard yesterday that Jemmison and partner of Newcastle, who bought Will Wilson's wethers a while ago and should have bought George Nicholson's, went along with John Barber of Doddington to Lesburry where Barber's sheep are turnipping with Bob Gibson, and bought 300 wethers of him at 900£! I have no doubt of the number nor of their having bought them, but I am affraid that the price was not so much although I sincerely wish it may be so. However it is nearly 400 sheep that cannot now hardly appear in Morpeth market, and I think it will do good to prices in particular, as prices at Morpeth have been better ever since. I will send this by Ralph to Berwick to the post, and hope to see yourself or have a letter from you, either of which will be very agreeable to yours sincerely

Geo Culley

PS. It is rather a smart frost and wind northerly, and I should [not][12] be at all surprised at a fortnight or 3 weeks storm. Be sure and let Mr Peacock see this.

231. *George Culley and Matthew Culley jr to John Welch*

Eastfield 5th January [February][13] 1804

Well John

I sent you a letter yesterday morning to Berwick, and last night received yours of 2nd inst. It is a little extraordinary that we should have sheep at a higher price at Morpeth than Darlington, and cattle rather lower. Your turnips have certainly eat very fast, while ours have lasted better than we could well have expected. I think if this hard weather don't stand over long that turnips will be plentiful enough in this quarter. I wish with all my heart that you may meet with a good market at Skipton, you have luck for bad weather however, if the weather is as bad with you as here. And I think you wrong in sending the small weathers, they will certainly sell in warm

---

11 Henry Heneage St Paul, 1777–1820, second son of Col. St Paul, second in command to his elderly father, but actually commanding the Royal Cheviot Legion: *History of Northumberland*, vol. 14, pp. 188, 189.

12 A word seems to have been omitted.

13 From the days of the week given, the month should be February.

weather when the large ones will not be so acceptable. But you have your own reasons, and no doubt have done it for the best. Therefore let them take their chance. I doubt also the last market being so very good at Skipton, may set too many to market this one. Your barley must be particularly good, it weighs so well. I am glad the horse goes so well on, but I must beg you not to be too rash in riding him yourself. We will be glad to hear from you once you [. . .][14] after Skipton, but you will scarcely be able now to come until after Candlemas fair at Allerton. The worst thing is, we cannot get our annual accounts settled. *Wednesday 8th.* This day appears inclined to fresh weather again after 2 or 3 days of extreme frost. Turnips have not been so hard frozen this winter I think. Mr Scott of Scremerston was here yesterday and was at Morpeth last Wednesday, and told us that good sheep were fully 7 ½ sink, and many sheep butchers, not so many beef butchers, yet that oxen for ships were best sold. Too many small midling cows and queys. If we hear of this day continuing as good for sheep as the last, I think of going with ewes the week following. *Thursday 9th.* A very nice fresh day indeed, snow all gone again and the frost much out of the ground. We have been drawing 50 ewes for next Wednesday in case we have another pretty good account of Morpeth by Matthew this evening. He is gone to Wooler. We hear that Ratcliff came as far as Mr Thomas Howey's last week and bought his cattle it is said at 8. He was also at Mr Jobson's but we have not heard whether he bought of him or not. If we once begin with our fat sheep we shall keep going on, so you may perhaps contrive so as to lend your assistance some of the Wednesdays coming. We have such a number of fat sheep that we must make a beginning soon. Not but we are still pretty strong of turnips, and I have no fear of their rotting at all. These 2 or 3 very severe frosty days humbled their tops a good deal, but I see they are rising today again. A fine settled morning *Friday 10th* but no letter John, so I will bett you a penny that you have had a heavy market at Skipton, because if a good one you would have put a letter into the post on Wednesday and it would have been here this morning. *Saturday 11* proved so dark and misty a day that we could not see much of the *eclipse*.[15] Now rain, and we had much rain last night.

One o'clock. Have just received yours of 8th for which I thank you. You have had a good market, but you don't say what sheep were sold

---

14 A word seems to be missing here.
15 No eclipse is recorded in the press on 11 January. There was an eclipse of the moon on 26 January.

at per pound sink in general. It would appear that yours if only 18 lb. per quarter is rather more than 8 ½ per lb. Are you not wrong in the weight think you? If not it is a wonderful price, certainly a good price at any rate. It is a wonderful turn in regard to sheep, while cattle seem standing still. However if sheep really fall scarce, cattle must advance by and bye. To a certainty cattle are not plentifull in this country, at least everyone are of that opinion, especially on the Scotch side, and sheep are now beginning to be thought not over plentifull in this country. However a few weeks will show how that matter is. We had sent 50 ewes to Morpeth before we got your letter this day, and perhaps it may be as well to sell our ewes if markets keep tolerable, because ewes do not make flesh good at this season it is generally thought, and we have a good few gimmers which with the wethers we can keep longer until we see whether markets mend or not. And the selling our few ewes will relieve us a little, and if the weather keep fresh we will have turnips plenty I believe. Besides, if your Darlington market get up for sheep which it now surely must, it will send the Wearside butchers to Morpeth. You should take care to be here and back in time for your first Monday in March, which will be on the 5th of March. I thought you would attend Northallerton fair for cattle which is on Monday I think, and that made me think you would not come north until after that fair. I am glad to hear you say that you will be able to clear the bank off on Monday. If you come to Morpeth the Tuesday evening come a week you will very likely fall in with and assist Matty there and get his company home. But you are so kind as say that you will write again before you come. Well John I am very happy that the country is so united loyal and keen of volunteering. If it had not been so we would have had Bonaparte or least his adherents here before this. If they are now so fool hardy as make the attempt, they will not only have many hazzards by sea but will meet with a more severe reception by land than they have been accustomed to, and with the blessing of God will be completely discomfitted. That this little Island may long retain its constitution unimpaired is the serious prayer and ardent wish of yours truly

Geo Culkley

Wet and more wet suits water meadows but retards spring wheat sowing, yet there's plenty of time. Your watered land should keep a good few wethers after turnips. You need not be afraid of cutting them. It is now as sower an ill-natured night as can be, wind a good deal east south east with a driving rain. G.C.

[*M.C. jr*] John we will thank you to say positively in your next if you will be at Morpeth on Wednesday week's market, because if you will come there it will save me the trouble and expence of going, and if you say in your next that you will be there as well, take care to send an account of the weight and value of the sheep by the lad who drives them. Ralph is just returned from Berwick and has sold 400 bolls of old red wheat at 29/6, a very bad price, it's bad corn.[16]

232. *George Culley to John Welch*

Eastfield 15th February 1804

Well John

Yesterday we sold Mr Ratcliff 20 housed oxen at Wark at 31£ per, to go away together on Monday come a week, some way very near 8*s* per stone but perhaps not quite 8*s* although very near I do think. However we are all very well pleased with the price. He also bid us 48*s* for 80 ewes, we offered them at 50*s* and parted. He had bought 146 dinmonds of Mr Smith, they say 8*d* per lb. sink, to go away all on Monday. He also bought 6 oxen of Baily Hare, and 4 at the market on this day week, and he bought a large lot of Mr Taylor lately, and was going to buy of Ralph Fenwick and Mr Smith of Togstone. But he does a vast I fancy. They say he kills 80 sheep per week, besides many pigs and more oxen than any 2 in Shields. We have been weighing our Thornington wool, very light, Grindon and Wark next week. People now begin to think fat sheep and cattle will both fall scarce. However I can tell you more about it when the two Matthews come from Morpeth tonight. If this prove another decent market, I think we may conclude that sheep in particular are falling scarce. I understand that fat got an advance last week at Dalkeith and Edinburgh, and by a letter from Mr Sayle (sheep) were sold pretty well at Rotherham. He says nothing about Wakefield. But says turnips are plentifull and that will help sheep. John your dinmonds weighed badly if they only weighed 18 lb. per quarter! You had no hill hogs. We have all the Longknow dinmonds and will be more than 18 lb. per quarter I have no doubt although sore pinched in summer. *Thursday 16 February.* Well John the two Matty Culleys got home at 8 o'clock last night after a good market. They sold for full 8*d* per lb. sink 50 ewes. We shall send 60 next week, and if you can meet them at Morpeth it will save both trouble and expence. But of this you must not neglect to write us directly, that no mistake may take place. But I hope you have wrote already, or will before this reach you, as we mentioned the matter to you in our last letter on Sunday morning last, which I trust you have

---

16 There is no signature.

received before this time. If not pray write as soon as you receive this, because we should receive it before or on Tuesday morning at latest. This is a decent morning, although the wind is northerly, yet we have no frost and are going to sow wheat for the first time this season. Matty says many sheep in Morpeth but very few good ones. I fancy sheep have not fed well in general this wet unsettled winter. Mr Sayle says so and Mr Barf both. Few cattle shown in Morpeth, and those not good. Some nice queys of Mr Ralph Fenwick sold at 8s 4d per stone sink. Yours in haste

Geo Culley

233. *George Culley to John Welch*

Eastfield 3rd March 1804

Well John

We now have more severe winter weather than we have expected at any time this winter. It not only retards the spring wheat sowing but is particularly hard and bad for our young lambs &c. I received your letter from Morpeth by Will Glass at Wooler, which was the fullest town of servants ever seen. A small advance was given in the bondage wages. Matty set down your accounts and of the money paid to Mr Bennet, which his letter is a receipt for. People agree that it was not so good a market as had been the preceding weeks, but still I think your sheep were well sold upon the whole. I fancy you took the morning prices, which are the best 6 times out of seven. By a letter from my brother this day since writing the above, he says 'We have sent the 10 bolls of oats this morning to Berwick for John Welch to the care of Mr Moffatt Newcastle. Mr Moffatt and John Welch should be wrote to' (8 sacks containing 8 bolls marked G.C. or 48 bushells and 3 packs containing 2 bolls marked M.C.). Now I suppose these *sacks* and *poaks* have been thus marked before this. Probably the sacks have gone from this place some time and have been marked with the initials of my name. My brother does not explain them, nor does he say that they are directed, but I have no doubt but they are properly directed, as my brother was requested to do so. Ralph Brown will see about a ship today at Berwick, and I hope they will go safe and soon to Newcastle. However I will not send this letter until I see Ralph Brown, and then give you all the information in my power. But as to writing to Tommy Moffatt I think it quite unnecessary, because from what I said to you, you would not fail to see Mr Moffatt, and I wrote to my brother so today. However if you neglected to call on Mr Moffatt, be so kind as either write to him or let me know that I may write to him. But I am certain you would not forget. *.Sunday 4 March.*

It was very unlucky that the oats for you did not go 2 days sooner, because a ship had sailed the very morning (yesterday) that the oats went, and perhaps they may have to wait a good while for another. There is only in time and out of time. However we will make enquiry every week, and you may depend upon it that I will give you the earliest intelligence in my power by letter. And Ralph desired the Berwick wharfinger to be sure to send them by the first ship. The frost is extremely severe and winterish. *Tuesday 6th.* Most unpleasant weather indeed! Snow, sleet or rain every day. However I do think that fresh is coming, as the wind has got south east and seems as though it was working about south. It neither suits stock nor seed time. We have got the troublesome job of the double servants hiring, except for joiner which we will get somewhere before Whitsuntide. As there is no hurry about this letter, and we shall get one from you on Thursday morning I hope, I will not send this until we receive yours. I am affraid that I have been the cause of the butter firkins being kept a bad time. I forgot to bid you make enquiry, but perhaps you would recollect to do so. *Wednesday 7th March*, thank God is a very drying day although still exceedingly cold. I think it was lucky that Matthew Culley of Wark did not go with you, as his uncle Bates came to Wark the last Wednesday. But as there was so much snow he went away home again.

Thursday 8th. We have this moment received your letter of 6th inst. and it appears that your markets are worse for fat sheep and no better for cattle. By all means sell your wethers, there is nothing so unwise as running fast forward of meat at this season. I am also inclined to keep selling our fat stock here, as our markets may become worse and not better. The sheep butchers will go to Darlington if they can buy things cheaper than Morpeth. However if John Forster's account be true, the butchers must have great proffits at 3£ per cwt. I am happy to hear that George Thompson is better. I think the doctor right in giving him bark and port wine. He should not begin sea bathing too early in the season I think, as he is so thin and delicate. Let the weather be warm before he go. Your mistress will consider about the great coat and night shirts. We thank you for the accounts of tithes and mill. I will take this to Wooler, and send you an account of Morpeth market if I can before I seal and send it away. We are starting to sow wheat today again in Blue Bell hill. I will send this letter to Wark, that my brother may see what you say about Mr Peacock going to Scorton, and am with every good wish yours truly

Geo Culley

Wooler one o'clock. A full market for sheep at Morpeth and not quite so well sold as weeks before I am told. Cattle much the same. G.C.

234. *George Culley to John Welch*

Eastfield 18th March 1804

Well John

I am much obliged by your attention in writing so often and readily. I received yours of 16th this moment. Your Skipton and Wakefield accounts are bad, and still worse than Morpeth which dropped to about 7 ½ for mutton sink, and prevented us from going this week. Indeed we are very strong of meat, but I am really unable to form any judgement whether markets for fat stock will mend or not. However they can't well be lower I think at this season, and they are certain to gain weight, dinmond sheep in particular, and cattle, and we are very strong of turnips. You are certainly right in struling with your sheep rather than submit to 7*d* per lb. sink. I should think your water meadows would assist now along with your seeds. To put off is a great affair at certain periods, let alone improvement. I have known things take a sudden turn for the better, when hardly expected. But the Yorkshire fat markets being below ours is *phenom-enon* never before known, and entirely damps my hopes. There is only one thing, and that is the pork, which must be over as soon as warm weather take place. Perhaps the advance in grain which I do look for and is now advancing slowly, will operate a little in favor of fat selling. We have got a wonderfull deal of wheat sown, while many good farmers have sown very little. Whether we are right or wrong time will determine, but when things are done with a good intention it is always right. At least it succeeds 9 times out of ten. We have quite a cover of snow this morning, and have had a few exceeding wet days. We have sown no beans yet, and the season advances. However His will be done that gives the weather, and let us do the best we can and be thankfull, and God will bless the increase. Ralph Brown found a ship yesterday at Berwick, loading with potatoes for Newcastle, and prevailed on them to take the oats for you. As soon as I hear of their sailing and who consigned to I will write to Mr Moffatt who will not fail to acquaint you when arrived at Newcastle. Ralph sold red wheat yesterday at 34*s* from Grindon. You are really a persevering man John to get 30*s* for your barley after all, while it is only worth 14*s* our boll here! You are fit for *a general*. I commend you also for selling your oats to Miss Thorn. It is a fair price and nice delivery. But as I fancy there can be no fear of Miss T. I think you should not hurry her in paying, as we can't want money here until about the term. We can now only get 50s per for our firkins, therefore if you can get a better price, allowing and calculating the terrible expence of carriage to your country, pray be doing. I will I think meddle no more with butter. It is an article I don't understand, and consequently had better not meddle with it. The shoemaker should not go beyond his last. I should suppose you have acted wisely about letting your lime kilns, particularly as you reserve one for your own use. I am rejoiced to hear of the good behavior of Archer, may it continue. But let me advise you to be cautious, remember you have a family. I will enquire about the carriage of the books and name it. *20th March*. Your books were not paid carriage for. As we have got the *letter* or power of attorney executed and it may be wanted by the agents to the Darlington bank, the Everetts in London, I will inclose it in this letter, and must beg you to give it immediately to Mr Peacock, to whom I must request you to make my compliments and say that I will write to him very soon, and am in the meantime yours sincerely

Geo Culley

PS. I sent your letter and this answer up to Wark in order that my brother might add anything in it. But he returned it to me without any addition. We have most serious weather John, and will or may be of serious consequence to the next crop, especially wet lands.

235. *Matthew Culley to John Welch*

Wark 18th March 1804

John Welch

Well John, you have said nothing about the firkins of butter, perhaps forgot to make enquiry how they sell. Now I have orders (not without reason) from Mrs Culley for you to sell ours as well as you can to a good man, if you have it in your power. We have and it's more than time that they were sold for what they will bring. We cannot get so much now at Berwick as we once could have had. The weather has been too wett for wheat sowing in general, but yet we have sown all that we proposed except some small pieces which we intend to sow tomorrow or Tuesday. The land was too wet yesterday by much, yet we sowed all the afternoon, when it rained. The burns are runing much today, and the hills covered with snow, not a good time for lambs. We are I think pretty strong of turnips, and fodder of all sorts, shall eat all our hay. Store stock, especially work oxen, improved in condition, they work well on Sweedish turnips. Oats pretty well sold both for seed and millers, wheat a little better to sell 34 or 35s per boll, half old half new sold at Kelso, but the old is a draw back on the new. Have sown very few blendings, and this day's snow and sleet with an east blow retards much. Grass now is very forward, and so it is on the watered land, which we are now giving a 2nd turn, which will ensure a hay crop most probably. I fear yours is not so good on your watered land, ours was never so good. Morpeth was very slow for sheep on Wednesday last I hear, in so much my brother George sent none of our wethers to market. I think we will be able to clip the hill wethers on Thornington haugh, as we cannot safely feed the hogs on the watered land, as it would likely taint or rot them. My brother and son say our best bred sheep will be too fatt if we were able to keep them forward to clip. Our 8 few outlaid oxen and queys have not done much here, the weather not favoring them although the queys get oil cake once a day when they come home in the evening. Theirs at Thornington are doing well seemingly. 19th a cover of snow, and we have some small spotts to sow with wheat, and although it is late in the season we may be blessed with a valuable crop of grain if it please the Almighty to keep away war, the scourge of mankind. I am with real regard your esteemed friend

Matthw Culley

The bolls of oats were sent to Berwick exactly as you requested, but the ship was full loaded, and they said they could not take them in that vessel. Ralph Brown was there, but whether shipped now I cannot say. The winds have been unfavorable and the sea stormy. Write about the butter, sold or not, to Wark.

236.   *George Culley and Matthew Culley jr to John Welch*
Eastfield 12th April 1804

Well John

We yesterday received your long-expected letter of the 8th inst. I have not seen Robert Thompson of many weeks. Indeed I have been longer confined and more distressed with a violent rheumatick pain in my right ear and all that side of my head, than I ever was confined in my life. I have been better and worse several times, and now for 2 days, after a large blister between my shoulders, I have been pretty free of pain. I have been 4 times blistered, upon the head and within my right ear. Poor George Thompson, I think he must be in a bad way indeed! I am sure your people will not let him want any comfort in their power, and it appears that the doctor treats him very properly.

You may depend upon it the seed time has been worse, *wetter, colder* and *later* everywhere that my information leads, than I have remembered, and the further north all the way the worse. We have got no beans sown at all, and shall now sow early pease, with a few beans by way of sticking them. Our oats are nearly done, and our spring wheat, of which we sowed much, is coming above ground very prettily. A worse lambing time could not be, and few twins it is said. However we have lost very few lambs. We beat you for sheep markets. Matthew got within ½ a farthing per lb. of 8*d* sink, and cattle full 8*s*. The 2 weeks before were 7 ½ only and cattle 7 to 7/6*d*. We are determined to continue marketting our sheep, and the few cattle we have we will try to keep as long as we can. Ratcliff was out and bid us 7/6, he had bought 27 of Tom Nisbet of Roding, and was easy about more I believe then. I am not affraid of cattle dropping, but sheep we think more plentifull. You certainly sold your steers well. I don't know the young man from Newcastle. Are you correct in Lincoln buying *100*? I am glad you are convinced of the value of ruta baga. We know not its merits sufficiently, and your land suits it. Our land is too weak and dry for it. They can grow it well at Wark, and William Brown. We will send you 3 or 4 stone of seed, it is pure and good. We have hitherto sold it at 2/6 per lb., but I think of taking 2*s* this year. But don't take less than 2*s*. We will attend to your instructions about the blue pease, and you must not forget the rape seed a 2nd time. We now sell wheat at 36*s* our boll, and oats for 20*s*, can sell for more at 19*s*, I believe they will be 20 soon. The demand for both wheat and oats is considerable into Scotland, yet I see London is lower for all grain again last Monday. Nevertheless I am inclined to think that corn will be dearer. I could not make out who Thomas Eales was at the first. I should not wonder if his father never told him. If occasion, I can not only make an affidavit of the debt, but produce our books. But don't use legal means without he absolutely refuse payment, nor need you to hurry him. I have had no opportunity to consult either my brother or our Matty about Archer, but you are like to get him done and William Rain I believe understands it. The weather is now dry but exceedingly cold. I am with every good wish yours truly

Geo Culley

PS. John your mistress my wife begs that you will write to your mother to know if G. Thompson wants an upper coat, or any other necessity, to get it for him, as it is quite inconvenient for us to send. And request your mother to write to us once a fortnight how the boy is. You can give her directions. I hope he is not confined to school or anything, but quite at liberty.

John Welch. As things are now *circumstanced* the whole of Denton must come to either my brother or me, and that will now be known in a few days.[17] I have therefore only one simple question to put to you,

---

17 In their award of April 1804 Messrs Bailey, Bates and Nesbit divided the leases

viz. If the whole of Denton fell to me, have you any objection to serve me in the same capacity that you have served my brother and me jointly? And to this I must beg your immediate answer without you taking any notice of this matter to anyone else. If Denton fall to my brother I certainly have no right to interfere nor will I. I hope you understand that I expect an answer to the above as soon as may be, without speaking upon any other business whatever. G.C.

[*M.C. jr*] John I am sorry the Archer horse is likely to be spavined. If he is better when you have to come north soon, we hope you will ride him and I will keep him to ride all summer if he is quiet. I have a high opinion of William Rain's judgement about horses, but I differ from him in regard to pinching. I believe it may be sometimes of use, but it often proves otherwise, and always leaves a blemish. I would advise not to punch him at first but by blistering, and then if that fails try pinching or any other means that will make him sound. As to the curb, I think nothing of it. I am sure that nothing is so simple and effectual for taking off a curb as oil orignum will leave no blemish, while blistering (unless very nicely done and well attended to) is very apt to do. However I leave you and William Rain to do as you like. I am yours &c

Matthw Culley

PS. Is he lame of the spavin, because if he is not, oil orignum will cure him of both complaints. I hope you ride him and he is quiet.

The skeels are come. Say the price in your next.

237. *George Culley to John Welch*

Eastfield 8 June 1804[18]

Well John

My brother and I met at Wooler yesterday, for the first time since my return, when he was very ill-natured indeed. I had some difficulty to keep my temper. We have spoken to ---,[19] who is very willing

---

between the Culley brothers, Wark and Thornington going to Matthew, Eastfield, Westfield, Longknowe and Grindon to George. The properties were also divided: Akeld went to Matthew, Easington Grange and Denton to George, but as most of Denton was Matthew's property, George was to pay him a nominal rent. See above, pp. xxxi–xxxii.

18 This letter is preserved only as a typewritten transcript in ZCU 43 (see p. xxxiii). There is no reason to doubt the authenticity of the letter; but it is particularly unfortunate that the original has been lost, since the transcriber was apparently unable to read the name of the man who was to go to Denton. It may have been Thomas Thompson: see Nos. 238, 239, 240.

19 Thus in the transcript.

to go to Denton, and I have no doubt but you will give him all the information and instruction in your power. I really think you should see the harvest done or nearly before you leave; but if we should suffer so much less [*sic*: ?loss], I am determined you should come north. And I am certain your own good sense will shew you the very great hardship of leaving us at this time, in such a manner, and from a place at such a distance. But I will depend upon you making a stipulation with my brother, 'that you are to go now and then to Denton to see how ---[20] is going on, and assist him in arranging matters'. You must take him to market, and make him acquainted with your customers &c., and try him about the weights of cattle and the value of lean stock, and particularly the *handling*, and those *shapes* or *forms* which are most likely to make fat. Certainly, that is a matter to be acquired, only by practice, but instructions will do good, and the sooner a person begins, the sooner he will learn. Be so kind also as to let your servants know that he is to succeed you and that they are to pay him the same attention and obedience they have shown to you or me. Yours

Geo Culley

238.   *George Culley to John Welch*

Eastfield 18th June 1804

Well John

I keep pestering you with letters, but I by no means desire you to write as many in return. It would be very unfair and unreasonable. I have much more leisure than you have, and I am sure you will not be offended at my suggesting hints and offering advice to you, so long as I have the benefit of you at Denton now that I consider it all my own, and am consequently the more particularly interested about it. I wrote one letter to you yesterday and the day before, and now begin another. My reason for writing thus soon again is to remind you of what a vast of mowing you will have this season, therefore I would recomend to you not to be over long in starting, either in those parts of your water meadows which are likely to *lodge* or get down (which will be the case by the time you receive this) or in some of your seeds, particularly where you have much rye grass and little clover, except where you intend the rye grass for seed. There it must be allowed a sufficient degree of ripeness. Otherwise where it is for *hay only* without any regard to the *seed* you can hardly cut rye grass too young,

---

20 Thus in the transcript.

because it makes very indifferent hay when allowed to be too ripe before cutting. If you have not mowers to spare of your own, try to get them from the neighboring villages. Remember I will not *begrudge* expence when necessity *demands despatch*. I have frequently known more lost by *delay* than by *zeal and exertion*. *19 June*. I was glad to hear by Harry Rutherford yesterday that a letter had comed to Wark from you. This is a dark gloomy day, like wind and rain in the west. John you need not send or remit my brother's money over here, which you have to receive for his 3 beasts &c. But write us an exact account of it and we will give him as much of our own here, or any other money which may become due to my brother for half the corn sold or to sell. We will give him equal to it here so long as we can, and when we fall short it is only for you to remit it then. I was on horseback at 4 this morning, to look at a few clipped hogs of John Batters's, but he asked me 38s, I bid nothing, but came home again. You had much better spare some of your pastures and make fogs of them for your cattle, than buy things which may be worth no more in the autumn than when bought. Not one of Mr Batters's hogs were nicked in the least. *22 June*. Yesterday was our high market at Wooler, but a very stupid one indeed. Only 3 wool buyers from Yorkshire, and are unwilling to give 24s per stone or 1s per lb. We are not likely to meet with *any hogs*, and the drought *coming* on us again (we have had some very blowing *fining* weather lately), might it be right to end you 5 or 6 feeding cattle, which will all get fat before winter? 3 of them are pretty fair meat now, and it would ease us greatly. Do give me an answer to this directly. We heard that Darlington was very bad for both cattle and sheep. Morpeth was good for sheep, 8d per lb. or nearly, but cattle 8s to 8/6 and heavily sold. The glass is got very high, and it has much the appearance of drought in the extreme, not but this is a calm morning. How fortunate you were at Skipton compared with Wakefield, which it is said was not so high for sheep by 4 or 5s per head. I own I really think you wrong in not putting a few sheep into each of your feeding pastures, say 20 in Hall field and 10 in the Fitts. More might be put in without much injury to the cattle, but an immense benefit to the sheep. I only name this for your consideration, and I am certain you will see the propriety of it. It is what we have done all our lives long, and what Mr Wastell always did who was the best grazier I ever knew. The Taylors never did, I know, but Mr Wastell blamed them very much for not doing it. A mixed stock always eat a pasture truer and with more economy than *all cattle* or *all sheep*. In Lincolnshire they used to stock almost altogether with sheep, whilst in other counties they stock all with cattle. But Mr

Bakewell, Mr Wastell and all the best, cleverest and most sensible men I ever knew in that line, were always for mixing sheep and cattle, and a single horse or 2 in some pastures. Your master and I differed in that I think one horse in 20 or 2 in 40 acres are never known, they eat where cattle do not, &c &c. In short it is so ordered and *contrived* by nature that one animal is fond of one plant and other animals of other plants. How wonderfull are thy works O Lord! I am sure you will be glad to hear that Mrs Darling is doing wonderfully well, and nurses her child and has plenty of milk, and to all appearance in high health thank God.[21] My brother told me he wished you to sell his 3 beasts, and when you can sell ours I shall have no objection as I am pretty certain that things will drop now. Tell us if you have heard how Borroughbridge fair has proved. We are for trying to buy you a few Border hogs on Wednesday at Yetholm. I have known some kindish little hogs from Lauderdale shown there sometimes, which if they will but make fat are good to sell, and nice eating, got by our tups or of that sort.

John I have opened this letter again to say that we could meet you with our cattle if you approve of it at Stagshaw Bank fair on Wednesday come a week July 4th. I only wish to name this for your consideration, as you will likely be there and at any rate have plenty of time to give me an answer. If we send them they will be at Mr Gibson's, James Laidler's master. John we named to you about John Dickinson making a kitchen table and a few common chairs, but I don't recollect whether we named two bedsteads for the *servant men* to sleep upon. If we did not, be so kind as get them made. *23rd June.* As Tomy Thompson can't get away before Monday and be at Denton some time on Tuesday, I will send this letter to Berwick by Ralph today, because I would wish you to write directly on receiving it whether we may send you the half dozen feeding cattle or not. The weather has every appearance of being droughty. Nevertheless I would not wish to send you them if you can't conveniently take them. Ralph sold 45 bolls of wheat at Kelso, half red half white, at 43*s* per, and a few oats at 1£ 1*s* per. I think corn will keep. The mercury is exceedingly high, and the atmosphere has every appearance of being droughty. Consequently if you have anything you can mow and any men to get or spare, I would advise you to be doing a little. It is most likely that Matty will be going to Denton soon after Monday first, which is our show day for tups. Robert Thompson's was yesterday,

---

21 Eleanor and James Darling's daughter Jane was born on 18 June: NRO, Cornhill Register.

and he has not comed home last night. They have a drill today upon Whitsun Bank, so our lad is to meet him at Wooler with his regimentals. I am your sincere friend

Geo Culley

PS. Tell me what is said, and what you think about wool.

PS. Give me your answer to this letter directly. If I have to write to Lord Darlington or his steward, how do I direct to his steward? Tell me.

239. *George Culley to John Welch*

Eastfield 24th June 1804

Well John

I expect that you will receive this by Thomas Thompson. As I sent you a letter yesterday I have very little to say, only that we are so feared of falling short of grass that we have determined to send away the cattle on Tuesday morning, and as the Grindon herdsman who drives them is an utter stranger on the road, we would wish you to send a person of on *Friday* to meet him. I think if all be well he will be that night at Mr Gibson's Stagshaw Close House where James Laidler is, and there can be no difficulty in your man reaching that place on Friday evening. We propose his going to Haugh Head the Tuesday night, Moor House Wednesday night, Cambo Thursday night, and Stagshaw Close House on Friday night where I hope your man will find him. Should he not be there, he cannot be far of except something has happened wrong with him. Should your man be at Stagshaw Close House pretty early on the afternoon and before our man reach there, he may as well go on until he meet our man and cattle, and that will enable our man to get back as far as Cambo and so home the next day. Ralph Brown sold 50 bolls of red wheat from Grindon yesterday to Messrs Clunie and Home at 40s per, and 50 bolls oats at 50£. We have had thunder today, but very little rain. But it is warm and fresh, and likely of rain I hope. We need it much. I am yours truly

Geo Culley

25th A very warm mild morning and has been uncommonly warm all night, and I should suspect more thunder were it not that the mercury has rather risen if anything. The rain it seems fell in torrents upon the river Teviot about Crailing &c yesterday. This is our tup show day and we expect a good deal of company, and our tups are not so forward in condition as we have often showed them.

Tommy is just comed, and I cannot help repeating that you must not leave Denton until the harvest be in or nearly in. I must make a point of that, indeed my brother has said nothing to the contrary to me, whatever he may have said to you. But let him say what he will, I am determined that you shall stay until the harvest is in at all events, as that is an important matter. And I trust my brother will allow you to go over again about Martinmas or when the principal fairs are, to assist in valuing and selling anything that may then be ready, and assist or advise about tying up the cattle to turnips &c &c &c.

240.   *Matthew Culley to John Welch*

Wark 28th June 1804

John Welch

I have wrote several times to Denton but have only received 1 answer to mine. I want not long letters, only wish you to answer the material parts. James Dickinson has married a woman, which we fear may injure the family if they live altogether. I wish you to look for a decent young carpenter, if one can be got who may assist our work here &c. I have a Scotch lad, works well and is we think very attentive to the work, but probably a better may be got for Akeld. At last I have a young man there from Humbleton, who does the work pretty well but will hardly work under 2s 2d per day, which from necessity am necessitated to give him as I know not how to help it. This with many other things should be considered about, and I have occasion for assistance. Was obliged to go to Yetholm fair yesterday to buy our highland or Cheviot bred wethers for the out land, bought 35 at 14s per head, 38 at 16s 9d per better bred hogs, also of Jemmy Scott 40 wethers or 2 shear ones, at 18s per. Am going to send our ewe lambs to Akeld lower hills to wean them on the middle pastures. The highland sheep are for the Skaws or Cold Hill furthest of Tains Kirknewton and High Commonburn farm. I said we have 3 or 4 oxen should meet you at Alnwick fair the last Monday in July. My brother says he is not willing you should come there. (Pray did you sell my 3 head of cattle, heard it was only midling market, hope you sold them, good or bad market.) I've no notion he should continue my master. We have about 60 barren ewes here 40 of them good, if your brother-in-law Nicholson would buy them. We still have our firkins, wish they could be sold for a bad price, as it's a shame to have them on hand. I wish we had 40s a piece or even less for them. I owe Ralph Forster for 50 ewe hogs at 30s per, 75£, but have not that sum by me, and am going part of the way to Brunton, shall I hope see Stagshaw fair, should be glad to see you there. I have [no]²² objection for your

---

22 The wafer has obscured one or two words here.

showing T. Thompson Denton and putting him in the way of managing farms &c and collecting the debts, but yet I have occasion for you here, and you say you come when I please yet G.C. says not till after harvest. Ralph Brown was to assist me, he was not at Yetholm. Matty I did not see until I had bought the sheep. This is assisting me in buying stock with a witness, and yet G.C. says they will assist me whenever I give them notice. We are mowing an excellent crop of hay on the watered land, weather too good to turn it in. Corn crops appear now to be thin, none will lye except some barley. Excellent weather to work for turnips. Mr and Mrs B. also return tomorrow, and we go with them. Pray when does Peacock come here? But why ask you, who seldom writes to me. Your only letter wrote to me of a long time gave me great pleasure. But perhaps it is wrong to ask you to write soon. I do not find Mr Bailey will allow me what I have laid out on the house. In such case you are to sell all the bricks and wood which was provided and not used at May last. It appears to me somewhat hard on me, who gave up my estate to accommodate G.C. without asking a premium which Mr B. said I should have had. Surely I have acted imprudently in doing anything to the house, although the laths were failed, but it's now too late to say anything about it. Yours truly

Matthw Culley

Am for Berwick today to forward the award. You must if you can meet me at Stagshaw Bank with such writings as are necessary belonging to the purchase of Denton bought of Sir Ralph Milbank and Messrs Bowser and Middleton &c. Mr Peacock will give, or tell you where to get the necessary deeds, as I want to have things settled as soon as convenient. I now hear by Mr Peacock's letter that you intend to come to Alnwick fair. Now from Wednesday at Morpeth to Monday Alnwick fair is an intervening time of 4 days. If you find this an inconvenience it may be better to come 2 or 3 days later to Alnwick only, and so to Akeld or Wark. I say this but to leave it for your consideration. Mr Compton's shearings seem big and good mutton, and if we can make twice 50 or 60, or 40 each time you will be best judge how many can be sold in one day, but we can say more at the Bank where I hope we will meet. The deeds are to go to Mr Liddle, attorney at Hexham, and you will be better put up at the Black Bull where Mr Bates lights when at Corbridge. M.C. We have but little done at Akeld, only some repairs and necessaries as it's a busy season. As soon as turnips are sown must get forward with the east end of the house. Mr P. sayth they have sold Escomb to Mr Dobson at 180£. It is well, I wish the brewery &c sold, so as to bring things into less room. I shall have enough to do here with farming &c for me and you.

241. *George Culley to John Welch*

Eastfield 29th June 1804

John

Perhaps we should have bought you a few hogs at Yetholm fair, where they were very low. But we are so discouraged with drought in this district that we are affraid you may fall short of grass also. We are going to have a bad crop of all kinds of grain, and pastures are gone on our dry lands. But at Grindon they are still very good. The wool buyers seem in better spirits than they were, a good deal sold at 1*s* per lb. I would like very well to know your opinion, and what prices are given with you. Wheat is selling everywhere better than Berwick. Without a bank we shall go to nothing. At Alnwick I understand it has been sold at 45*s* per or 15*s* new boll some weeks. *Sunday 1st July 1804.* Well John we received yours of 27th on Friday, but in consequence of a letter on business from Mr Peacock the same day, I was obliged to go directly to Berwick, where Mr Bailey and the other commissioners were going to finish and execute the award on my brother's business and mine. So we have had no opportunity to answer yours until this morning. However I wrote to Mr Pease from Berwick yesterday to thank him for his kind offer for our wool, and to say that we could not take the price. Not that I think it is to be sold for much more, but it will be convenient for me this year to have ready money. I fancy it is full as well that we have bought you no hogs. The weather is much milder, but not much appearance of any rain. Our wool buyers are now very keen, and a great deal sold at 1*s* per lb., and none scarce at a lower price, some at 25*s* our stone of 24 lb.

I am glad you are mowing your watered land. It is right for rye grass to be tollerably ripe, only you have a very large mowing this year. I am glad there are no bugs in the buds. I am also glad that Purdy Starforth is getting forward with something. You are very kind in saying we may have the oven and furnace pots. We can settle these matters when my wife and I come over. I fancy Matty will give you this himself, or at any rate he will be at Denton soon. I will write to Mr Scareth, and thank you for his address. Mrs Darling and her child thank God are so fat doing wonderfully well. Our markets are flattening again much for corn. The merchants say a great importation is expected. Yours in haste

Geo Culley

Monday 2 July. I have this moment parted with Mr and Mrs Culley of Wark and Mrs Thomas Bates at Wooler in their road to Brunton, Mr Bates went on Saturday. We had a fine shower yesterday afternoon for 3 hours, which will help us much. It had reached no further to the south than Flodden. We have done nothing in the wool way yet, but several have sold at 25s, and never so much sold so early. John you have never sent me an account of the carts, plows and implements of husbandry as per value. You are to get Mr Wright and the 2 Mr Collings to value your crop of corn in proper time, and if they or any of them are unwilling to act, chuse any other respectable men. It is too far for the 3 gentlemen to go again to value a few fields of corn.

242.   *George Culley and Matthew Culley jr to John Welch*

Eastfield 19th July 1804

John

By a letter dated the *11th* and only received this moment from your mother, I find that once fine-looking boy *as ever I saw* George Thompson died on the 7th. How her letter had been so long of coming I know not, nor is it at all material as I find your mother has buried him decently and according to the custom of the place.[23] I have only to request that you will *not fail* the very first opportunity, to go to your mother and discharge everything that we owe on the above account. Afterwards when we see you we must consider and make your mother some reasonable compensation for the tedious toil and trouble she necessaryly must have had. I understand you are to be north before Alnwick fair. Could you not come that way, or return by your mother and settle the above account? Be sure you keep that account separate from any other, that no mistake may take place. Has George Nicholson ever paid you for the sheep he bought of you, as that money would assist you in paying the above account. Have you perceived any caterpillars on your turnips? They have made their appearance upon a part of ours and on many people's in this district, and this very hot weather will engender them very quickly. Turnips grow apace here in general. I see by the London papers of this morning that grain of every kind keeps advancing in the London markets. Nevertheless I would recomend to you in the strongest terms to be selling your grain as fast as convenient now when prices are so decent. My brother has sold his wool at 25/6 which is a famous price I think. Indeed it was lucky he was out of the way at the

---

23 George Thompson was buried at Penshaw on 10 July: DUL, BT Penshaw.

commencement of the market, because all the way the last sold wool has been the best sold, and I don't know of a clip unsold.[24]

[*M.C. jr*] John we want some iron axeltrees and thurrow metal bushes for carts. You got us the last bushes of somebody at or near Darlington, will thank you to order bushes for 6 pair of single horse cart wheels, and bushes for one two-horse cart. The single horse cart bushes to be 11 inches long, 2 ¼ in. by ⅞ of an inch within, the 2–horse cart 12 in. long and the size within you most approve of, and I desire the maker not to chamber them, and also to make them as smooth as possible in the inside, as our blacksmiths had much labor and trouble to get the last smoothed. You must also order for us as you come through Gateshead of Messrs Hawks & Co.[25] 4 pair of single horse cart axeltrees to fit the above bushes, and also one two-horse cart axeltree to fit the above 2-horse cart bushes. The reason we only want 4 axeltrees for 6 pairs of bushes is that when you sent the last bushes there were only 4 pair come instead of 6, and I had ordered 6 pair of axeltrees. They must all be ready by the beginning of September next, when our cart will be over with the furniture and will bring them home, and can call at Hawks for the axeltrees. You can have the bushes brought up to Denton to be ready. I hope you get on well with the buildings and also the pulling down. I hope to have a good account of it when you come north. I am in haste yours sincerely

Matthw Culley

---

24 The signature is lacking.
25 William Hawks & Co, ironmongers and edge tool manufacturers, Gateshead: *Directory for the Year 1801.*

BIBLIOGRAPHY

Unpublished sources

Berwick upon Tweed
Northumberland Record Office:
    Simpson Papers
    ZSI 46, 50,51,57 George Hughes accounts and wages books
    ZSI 344–56, 388–9 Royal Cheviot Legion

Durham
County Record Office:
    Parish registers
    Strathmore Archives
Durham Cathedral Muniments:
    Registers of Leases
    Renewals Books
Durham University Library, Archives and Special Collections:
    Church Commissioners, Bishopric Estates: Lease Enrolment Books
    Durham Diocesan Records:
        Bishops' Transcripts
        Tithe Awards, Denton
    Durham Probate Records
    Hudleston Papers: Clergy Index

London
British Library:
    Arthur Young Papers, correspondence: Add. MSS 35,126–35,133
The National Archives:
    IR 23 Land Tax Redemption Office: Assessments 1798

Morpeth
Northumberland Record Office:
    Parish Registers

Newcastle upon Tyne
Northumberland Record Office:
    Culley Papers
        ZCU 2 Letters and papers from Sir John Sinclair

ZCU 3 Letters and papers from Arthur Young
ZCU 4 Correspondence with R. Honeyborn
ZCU 5 *Observations on Live Stock*: drafts and papers
ZCU 6 Letters to John Welch, Denton
ZCU 8 Letters to George Culley, 1783–4
ZCU 10 Letters from Lord Darlington, 1785
ZCU 11 Letters from William Walker, 1786
ZCU 12–30 Letters to George Culley, 1787–1812, with some replies
ZCU 31 Letter book 1788–1811
ZCU 32 Miscellaneous copies of letters
ZCU 33 Farming accounts
ZCU 43 Transcripts prepared by Dr A. Macdonald, 1951
ZCU 44 Miscellaneous transcripts
ZCU 45 Family deeds and probates
ZCU 47 Leases
Tankerville MSS, Box 4, Bailey Letters 1784–1809
Tithe Awards, Crookham, Wark upon Tweed

Staindrop, Raby Castle
Raby Archives, account books

Published works cited

The Agrarian History of England and Wales. General Editor Joan Thirsk.
Vol. 5, *Agrarian Change 1640–1750*, ed. Joan Thirsk. 2 parts. Cambridge 1984, 1985.
Vol. 6, *1750–1850*, ed. G.E. Mingay. Cambridge 1989.
Vol. 7, *1850–1914*, ed. E.J.T. Collins. Pt 1. Cambridge 2000.
Bailey, John. *General View of the Agriculture of the County of Durham.* London 1810.
——, and George Culley. *General View of the Agriculture of the County of Cumberland.* London 1794.
—— ——. *General View of the Agriculture of the County of Northumberland.* London 1794.
—— ——. *General View of the Agriculture of Northumberland, Cumberland and Westmorland.* 3rd edn London 1805; reprinted, ed. D.J. Rowe, Newcastle upon Tyne 1972.
Banks, John. *A Treatise on Mills, in Four Parts.* London and Kendal 1795.
Barnes, Donald Grove. *A History of the English Corn Laws from 1660–1846.* London 1930.
Bates, Cadwallader. *Thomas Bates and the Kirklevington Shorthorns.* Newcastle upon Tyne 1897.
Bayley, Thomas B. 'On a cheap and expeditious method of draining land.' In Alexander Hunter, *Georgical Essays*, vol. 4. London 1772.
Belcher, R. *General View of the Agriculture of the County of Stirling.* Edinburgh 1796.

Bell, Thomas. *The History of Improved Short-horn or Durham Cattle, and of the Kirklevington Herd, from the notes of the late Thomas Bates*. Newcastle upon Tyne 1871.

Boardman, Thomas. *A Dictionary of the Veterinary Arts*. London 1805.

Boswell, George. *A Treatise on Watering Meadows*. London 1779.

Brayley, E.W., and J. Britton. *The Beauties of England. An Original Delineation, Topographical, Historical and Descriptive, of each County*. London 1803.

Buchanan, Robertson. *Practical Essays on Mill-work and other Machinery*. Glasgow 1814.

Buchan-Hepburn, George. *General View of the Agriculture and Rural Economy of East Lothian*. Edinburgh 1794.

Burke's Landed Gentry. 15th edn. London 1937.

Burke's Peerage. 105th edn. London 1975.

Butler, George Grey. *Colonel St Paul of Ewart, Soldier and Diplomat*. 2 vols. London 1911.

Butler, Josephine E. *Memoir of John Grey of Dilston*. Edinburgh 1869.

The Commercial Directory of Ireland, Scotland and the four most Northern Counties of England for 1820–1 and 1822. Manchester 1820.

The Complete Baronetage, ed. G.E. Cokayne. 6 vols. London 1900–09.

The Complete Peerage, ed. Vicary Gibbs et al. 14 vols. London 1910–98.

Culley, George. *Observations on Live Stock, containing Hints for choosing and improving the Best breeds of the most useful kinds of Domestic Animals*. London 1786. 2nd edn 1794; 3rd edn 1801; 4th edn 1807.

————— *See also* Bailey, John and George Culley.

Culley, Matthew and George. *Travel Journals and Letters 1765–1798*, ed. Anne Orde. London 2002.

The Directory for the Year 1801, of the Town and County of Newcastle upon Tyne, Gateshead, and Places adjacent. Newcastle upon Tyne 1801.

Douglas, Robert. *General View of the Agriculture in the Counties of Roxburgh and Selkirk*. Edinburgh 1798.

Dowell, Stephen. *A History of Taxation and Taxes in England*. 2nd edn. 4 vols. London 1888.

Evans, Eric. *The Contentious Tithe. The tithe problem and English agriculture 1750–1850*. London 1976.

Fenwick, Thomas. *Four Essays on Practical Mechanics*. Newcastle upon Tyne 1801.

Fussell, G.E. *The Farmer's Tools: AD 1500–1800*. London 1952.

Gall, John. 'The search for the Vardy horse', in *The Ark*, Feb. 1993.

Gilly, W.S. *The Peasantry of the Border. An appeal on their behalf*. Berwick upon Tweed 1841.

Granger, Joseph. *General View of the Agriculture of the County of Durham*. London 1794.

Gray, Andrew. *The Experienced Millwright, or, A Treatise on the Construction of Some of the Most Useful Machines, with the Latest Improvements*. Edinburgh 1806.

Grey, John. 'A View of the past and present situation of agriculture in Northumberland', in *Journal of the Royal Agricultural Society of England*, 2(1841).

Hay, Douglas, and Francis Snyder, eds. *Policing and Prosecution in Britain 1750–1850*. Oxford 1989.

Henderson, John. *General View of the Agriculture of the County of Caithness*. London 1812.

History of Northumberland. *See* Northumberland County History Committee.

Hunter, Alexander. *Georgical Essays*. 4 vols. London 1770–72.

Johnston, John. *An Account of the most approved modes of Draining land; according to the System practised by Mr Joseph Elkington, late of Princethorpe, in the County of Warwick*. Edinburgh 1797.

Johnston, Thomas. *General View of the Agriculture of the County of Selkirk*. London 1794.

————. *General View of the Agriculture of the County of Tweedale*. London 1794.

Kain, Roger J.P., and Hugh C. Prince. *The Tithe Surveys of England and Wales*. Cambridge 1985.

Kirby, M.W. *Men of Business and Politics. The Rise and Fall of the Quaker Pease Dynasty of North-East England 1700–1943*. London 1984.

Longstaffe, W. Hylton. *The History and Antiquities of the Parish of Darlington in the Bishoprick*. Darlington and London 1854.

Loudon, J.C. *An Encyclopaedia of Agriculture*. London 1825.

Lowe, Alexander. *General View of the Agriculture of the County of Berwick*. London 1794.

Macdonald, Stuart. 'The role of George Culley of Fenton in the development of Northumberland agriculture', in *Archaeologia Aeliana*, 5th ser. 3(1975).

Mackenzie and Dent's Triennial Directory for Newcastle upon Tyne, Gateshead, etc. Newcastle upon Tyne 1811.

Marsden, Angela, ed. *A Raine Miscellany*. Surtees Society vol. 206. Newcastle upon Tyne 1991.

Marshall, William. *General View of the Agriculture of the Central Highlands*. London 1794.

————. *A Review of the Reports to the Board of Agriculture from the Northern Department of England*. 2 vols. York 1808.

Mathias, Peter, and John A. Davis, eds. *Agriculture and Industrialization*. Oxford 1996.

Mitchell, B.R. *British Historical Statistics*. Cambridge 1988.

Naismith, John. *Observations on the Different Breeds of Sheep, and the State of Sheep Farming, in the Southern Districts of Scotland, being the result of a tour through those Parts, made under the direction of the Society for Improvement of British Wool*. Edinburgh 1795.

The Newcastle and Gateshead Directory for 1795. Newcastle upon Tyne 1795.

Northumberland County History Committee. *A History of Northumberland*. 15 vols. Newcastle upon Tyne 1893–1926.

Orde, Anne. *Religion, Business and Society in North East England. The Pease family of Darlington in the Nineteenth Century*. Stamford 2000.

Overton, Mark. *Agricultural Revolution in England*. Cambridge 1996.

———. 'Land and labour productivity in English agriculture, 1650–1850', in *Agriculture and Industrialization*, ed. Peter Mathias and John A. Davis. Oxford 1996.

———. 'Re-establishing the agricultural revolution', in *Agricultural History Review*, 44(1996).

Oxford Dictionary of National Biography. 60 vols. Oxford 2004.

Pawson, H. Cecil. *Robert Bakewell, Pioneer Livestock Breeder*. London 1957.

Perkins, W.F. *British and Irish Writers on Agriculture*. 3rd edn. Lymington 1939.

Pevsner, Nikolaus. *The Buildings of England. County Durham*. Revised edn. London 1983.

———. *The Buildings of England. Northumberland*. 2nd edn. London 1992.

Phillips, Maberley. *A History of Banks, Bankers and Banking in Northumberland, Durham and North Yorkshire*. London 1894.

Raine, William. *The History and Antiquities of North Durham*. London 1852.

Redhead, William. *Observations on the Different Breeds of Sheep, and the state of Sheep Farming, in some of the principal Counties of England*. Edinburgh 1792.

Ringsted, Josiah. *The Cattle-Keeper's Assistant*. London n.d., 9th edn 1776.

Rowe, D.J. 'The Culleys, Northumberland farmers', in *Agricultural History Review*, 19(1971).

———. *See also* Bailey, John, and George Culley.

Russell, Nicholas. *Like Engend'ring Like. Heredity and Animal Breeding in early modern England*. Cambridge 1986.

Sinclair, James, ed. *History of Shorthorn Cattle*. London 1907.

Sinclair, Sir John. *Hints regarding certain Measures to improve an Extensive Property, more especially applicable to an Estate in the Northern Parts of Scotland*. London 1802.

———. *A Sketch of the Improvements now carrying on by Sir John Sinclair, Bart, M.P., in the County of Caithness, North Britain*. London 1803.

Smeaton, John. *An Experimental Enquiry concerning the natural powers of Water and Wind to turn Mills and other Machines depending on a Circular Motion*. London 1760.

Stanby, Pat. *Robert Bakewell and the Longhorn Breed of Cattle*. Ipswich 1995.

Surtees, H. Conyers. *Records of the Family of Surtees, its Descents and Allies*. Newcastle upon Tyne 1925.

Surtees, Robert. *History and Antiquities of the County Palatine of Durham*. 4 vols. London 1816–40.

Trow-Smith, Robert. *A History of British Livestock Husbandry 1700–1900*. London 1959.

Turner, Nicholas. *An Essay on Draining and Improving Peat Bogs*. London 1784.

Ure, David. *General View of the Agriculture of the County of Roxburgh*. London 1794.

Ward, W.R. *The English Land Tax in the Eighteenth Century*. London 1953.

White, James. *A Compendium of the Veterinary Arts*. London 1802.

Wight, Andrew. *The Present State of Husbandry in Scotland*. 4 vols. Edinburgh 1778–84.

Young, Arthur. *The Farmer's Tour through the East of England. Being the Register of a Journey through various Counties of the Kingdom, to enquire into the State of Agriculture. etc.* 4 vols. London 1771.

———. *On the Husbandry of three celebrated British farmers, Messrs Bakewell, Arbuthnot and Ducket*. London 1811.

———. *A Six Months Tour through the North of England*. 6 vols. 2nd edn London 1771.

Newspapers and periodicals
Annals of Agriculture. 46 vols. London 1784–1815.
The Farmer's Magazine. 26 vols. Edinburgh 1800–25.
The Newcastle Chronicle. Newcastle upon Tyne.
The Newcastle Courant. Newcastle upon Tyne.
The Tyne Mercury. Newcastle upon Tyne.
The York Herald. York.

INDEX

The index is not completely comprehensive of place names

Achison, Adam, 523, 525, 527, 537, 549
Achison, George, 408
Adams, Mr, 431
Adderstone, xxin, 241
Addington, Henry, 481, 567n
Air, William, 370, 426, 428, 431, 432, 524
Air, Mrs, 317
Akeld, xxi, xxxi, 97, 224n, 480, 508, 595n,
    600, 601
Alder, Counsellor, 298, 300, 307, 388
Alder, Tom, 294, 350
Alder, William, 79
Allan, Mr, 30
Allen, Stein, 125
Allison, Tom, 279
Alnwick, 202, 323, 438, 541
    fair, 75, 76, 86, 88, 97, 100, 147, 193,
        245, 250, 254, 255, 258, 266, 268, 270,
        274, 275, 279, 280, 281, 283, 310, 319,
        356, 357, 448, 451, 452, 454, 455, 502,
        539, 540, 542, 600, 601, 603
Ancroft, 368, 374
Anderson, Ralph, 55, 386
Anderson, Mr, 298
Archbold, George, 379, 544
Armstrong, James (Eastfield), 402
Armstrong, James (Yorkshire), 428, 455,
    516, 518, 550
Armstrong, Thomas, 102, 135, 228, 333,
    348, 352, 364, 428, 444, 516, 518, 546
Armstrong, Mr (Cambo), 141, 387
Ascough (Ayscough), George, 192, 313,
    435, 439, 452
Askew, George Adam, xx, 23–4, 81n, 185,
    188, 289, 346, 465, 466, 487, 509, 513,
    515, 524, 530, 557
Atchison, Dr, 538
Atkinson, Adam, 536, 537, 549
Atkinson (Atcheson), William, 46, 53, 57,
    85, 139, 142, 510
Atkinson (Atcheson), Mrs, 326
Auckland, Bishop, 130, 135, 345, 356, 427,
    428
    St Helen, 2, 131, 135, 155

West, 15, 37
Aynsley, Tom, 88, 465

Bailey, John, xxi, xxvii, xxx, xxxi, 27, 32n,
    67, 110, 131, 135, 153, 177, 180, 236,
    241, 248, 327, 328, 402, 419, 438, 463,
    465, 466, 468, 469, 470, 481, 485, 488,
    491, 520, 537, 548, 583–4, 594n, 601, 602
Baker, Mr, 50
Bakewell, Robert, xv, xix, xxvi, xxxi, 39,
    97, 236n, 598
Bamburghshire, xxv, 299, 360, 364, 401
Barber, John, 334, 356–7, 367, 495, 497,
    505, 585
Barber, Tommy, 334
Barf, Mr, 193, 298, 349, 350, 351, 388, 411,
    443, 445, 451, 455, 523, 544, 563, 580,
    589
barley,
    condition, 24, 28, 84, 108, 180, 189, 198,
        289, 327, 329, 335, 360, 378, 454, 530
    cultivation, 135
    prices, 42, 44, 55, 59, 64, 117, 120, 121,
        124, 125, 130, 135, 143, 177, 195, 199,
        203, 205, 212, 218, 219, 220, 236, 245,
        251, 254, 260, 269, 283, 291, 298, 306,
        361, 363, 376, 389, 393, 398, 405, 536,
        539, 568, 570, 578
    sales, 130, 164, 174, 214, 229, 236, 240,
        328, 335, 378, 379, 405, 421, 536, 544,
        548, 551, 574, 577, 579, 583, 591
Barnard Castle, 59, 302, 563
    market, 184, 191, 334, 517, 537, 556
Barisford, Mrs, 432
Barugh, Henry, 101, 112, 183, 211, 212
Bates, George, 50, 54
Bates, Thomas (Aydon), 39, 50, 54
Bates, Thomas (Brunton), 200, 315, 558,
    570
    assistance with horses and stock, 8–9,
        22, 24, 117, 125, 154, 155, 206, 244,
        247, 255, 258, 261
    division of Culley leases, xxxi, 548,
        594n